Applied Theoretical Organic Chemistry

Applied Theoretical Organic Chemistry

Editor

Dean J Tantillo

University of California, Davis, USA

World Scientific

NEW JERSEY • LONDON • SINGAPORE • BEIJING • SHANGHAI • HONG KONG • TAIPEI • CHENNAI • TOKYO

Published by

World Scientific Publishing Europe Ltd.
57 Shelton Street, Covent Garden, London WC2H 9HE
Head office: 5 Toh Tuck Link, Singapore 596224
USA office: 27 Warren Street, Suite 401-402, Hackensack, NJ 07601

Library of Congress Cataloging-in-Publication Data
Names: Tantillo, Dean J., editor.
Title: Applied theoretical organic chemistry / edited by Dean J. Tantillo
(University of California, Davis, USA).
Description: New Jersey : World Scientific, [2018] | Includes bibliographical references.
Identifiers: LCCN 2017024260 | ISBN 9781786344083 (hc : alk. paper)
Subjects: LCSH: Chemistry, Organic.
Classification: LCC QD251.3 .A67 2018 | DDC 547/.12--dc23
LC record available at https://lccn.loc.gov/2017024260

British Library Cataloguing-in-Publication Data
A catalogue record for this book is available from the British Library.

First published 2018 (hardcover)
Reprinted 2021 (in paperback edition)
ISBN 978-1-80061-064-4 (pbk)

Desk Editors: Anthony Alexander/Koe Shi Ying

Typeset by Stallion Press
Email: enquiries@stallionpress.com

Preface

When I think of Applied Theoretical Organic Chemistry, I immediately think of Ken Houk, Roald Hoffmann and Paul Schleyer. I am lucky enough to have personal connections to each: Ken was my Ph.D. advisor, Roald was my postdoc advisor (and the first person I remember hearing use the term "applied theoretical chemistry") and Paul and I collaborated briefly before he passed away. I learned many important lessons from working directly with all three and from reading their many (there are very very many) published works. Two of the most important lessons are relevant here:

(1) Results obtained by employing a "high level of theory" do not equate to insight. Corollary: Deep insight can be derived from "low level" results + chemical intuition. Ken, Roald and Paul all derived models of chemical structure and reactivity that are still useful today despite originally arising from results using theoretical methods that are no longer considered "sufficient" or even "valid." I tell anyone who will listen about these successes to drive home that computational chemistry is not automated, it's not done by computers — it's done by chemists! Computers just compute.
(2) Experimentalists are the perfect audience. Most of Ken's, Roald's and Paul's papers were not written for physical/theoretical/computational chemists. Instead, they were written for experimentalists: synthetic organic chemists, bioorganic chemists and

enzymologists, materials scientists, spectroscopists, etc. This book has been constructed with the same audience in mind. It is a book that will hopefully find use by theoreticians wanting to help and inspire experimentalists and experimentalists whose research may be enhanced by theoretical models.

This book is dedicated to Ken, Roald and Paul. I am grateful to all of this book's contributors, many of whom are "descendants" of Ken, Roald and Paul (some of more than one), to my students and collaborators for their encouragement and inspiration, and to my wife and son for support and distraction!

Dean Tantillo
Davis, CA, USA
2017

About the Editor

Dean Tantillo applies the tools of theoretical chemistry to problems in mechanistic bioorganic chemistry, chemical biology, organic and organometallic reactions of synthetic relevance, and natural products biosynthesis and structure elucidation. He is also working to help make Applied Theoretical Chemistry research accessible to blind and visually impaired students.

Dean received an A.B. degree in Chemistry in 1995 from Harvard University and a Ph.D. in 2000 from UCLA. After receiving his Ph.D., he moved to Cornell University, where he carried out post-doctoral research. He joined the faculty at UC Davis in 2003, where he is now a Professor of Chemistry.

Contents

Preface v

About the Editor vii

Chapter 1. **Modeling Organic Reactions — General Approaches, Caveats, and Concerns** 1
Stephanie R. Hare, Brandi M. Hudson and Dean J. Tantillo

Chapter 2. **Overview of Computational Methods for Organic Chemists** 31
Edyta M. Greer and Kitae Kwon

Chapter 3. **Brief History of Applied Theoretical Organic Chemistry** 69
Steven M. Bachrach

Chapter 4. **Solvation** 97
Carlos Silva López and Olalla Nieto Faza

Chapter 5. **Conformational Searching for Complex, Flexible Molecules** 147

Alexander C. Brueckner, O. Maduka Ogba, Kevin M. Snyder, H. Camille Richardson and Paul Ha-Yeon Cheong

Chapter 6. **NMR Prediction** 165

Kelvin E. Jackson and Robert S. Paton

Chapter 7. **Energy Decomposition Analysis and Related Methods** 191

Israel Fernández

Chapter 8. **Systems with Extensive Delocalization** 227

L. Zoppi and K. K. Baldridge

Chapter 9. **Modern Treatments of Aromaticity** 273

Judy I-Chia Wu

Chapter 10. **Weak Intermolecular Interactions** 289

Rajat Maji and Steven E. Wheeler

Chapter 11. **Predicting Reaction Pathways from Reactants** 321

Romain Ramozzi, W. M. C. Sameera and Keiji Morokuma

Chapter 12. **Unusual Potential Energy Surfaces and Nonstatistical Dynamic Effects** 351

Charles Doubleday

Chapter 13. **The Distortion/Interaction Model for Analysis of Activation Energies of Organic Reactions** 371

K. N. Houk, Fang Liu, Yun-Fang Yang and Xin Hong

Chapter 14. **Spreadsheet-Based Computational Predictions of Isotope Effects** 403

O. Maduka Ogba, John D. Thoburn and Daniel J. O'Leary

Chapter 15. **Stereoelectronic Effects: Analysis by Computational and Theoretical Methods** 451

Gabriel dos Passos Gomes and Igor Alabugin

Chapter 16. **pK_a Prediction** 503

Yijie Niu and Jeehiun K. Lee

Chapter 17. **Issues Particular to Organometallic Reactions** 519

Gang Lu, Huiling Shao, Humair Omer and Peng Liu

Chapter 18. **Computationally Modeling Nonadiabatic Dynamics and Surface Crossings in Organic Photoreactions** 541

Arthur Winter

Chapter 19. **Challenges in Predicting Stereoselectivity** 583

Elizabeth H. Krenske

Index 605

Chapter 1

Modeling Organic Reactions — General Approaches, Caveats, and Concerns

Stephanie R. Hare, Brandi M. Hudson and Dean J. Tantillo

Department of Chemistry, University of California — Davis, Davis, CA, USA

1.1 Introduction

Computational chemistry is unique in that the researcher who uses it as a tool is not bound by the confines of reality. This freedom has both benefits and drawbacks because it not only allows the researcher to be boundlessly creative in determining how to answer chemical questions but also requires an understanding of all the assumptions being made in the analysis (which is often nontrivial). This chapter aims to describe the basics of applying quantum chemistry to organic structures and reactions and to highlight "everything that can go wrong" in quantum chemical calculations — that is, when the assumptions break down or are incorrectly applied.

1.1.1 *What can the Schrödinger equation do for YOU?*

The primary goal of any quantum chemical computation is to associate an energy value with a particular distribution of electrons around nuclei in particular positions. This goal is achieved using the Schrödinger equation, shown below in its least threatening form (physically oriented chemists, please forgive):

$$H\Psi = E\Psi \qquad (1.1)$$

In this equation, a distribution of electrons/electron density, expressed as a wave function (Ψ) is associated with an energy (E) by a Hamiltonian operator (H; an operator is a function that acts on another function). The Hamiltonian (for short), also the "functional" in density functional theory (DFT), is a function that contains terms associating energy values with the relative positions of electrons and nuclei (see Chapter 2 for a survey of different Hamiltonians). The Hamiltonian is also referred to as the "model chemistry".

1.1.2 *Putting the Schrödinger equation to work*

How is the Schrödinger equation used to predict a distribution of electrons/electron density? Generally, a prediction is arrived at by using the following procedure. (1) A *basis set* of atomic orbitals (usually Gaussian functions) is used by the quantum chemistry software of choice to construct guesses at the shapes (mathematical forms) of *molecular orbitals* (MOs). This step corresponds to the famous *linear combination of atomic orbitals* (LCAO) technique. In this step, electrons localized on individual atoms are transformed into electrons *delocalized* over molecules (or portions thereof). Note that basis sets are not restricted to one 1s, one 2s, three 2p orbitals, etc.; for example, many *p*-orbitals of differing "shapes" (e.g., short and squat, tall and diffuse) are generally used. By doing so, electrons are allowed to more freely explore larger regions of the space around the nuclei, thereby increasing their chances of finding ideal positions/distributions. While this all sounds rosy, the user is limited by the computational time required to perform such sampling, which is proportional (generally not linearly) to the number of basis functions used. (2) Once one has a guess at the MOs, the Hamiltonian then operates on the corresponding wave function that describes this orbital arrangement to calculate its energy. (3) Steps 1 and 2 are then repeated, changing the coefficients in the LCAO each time to produce different MOs and associated wave function, and correspondingly, a different energy. This iterative procedure (called the *self-consistent field* (SCF) procedure) continues until the energy is minimized (given particular convergence criteria built into the quantum chemistry software or set by the user). At the end, one has the "best" (lowest energy) distribution of electrons/electron density (wave function) for the given set of atomic/nuclear coordinates. From the wave function, myriad useful properties can be derived, e.g., spin densities, charge distributions, and NMR chemical shifts. As a whole, this calculation is called a *single point* calculation, since only one molecular geometry was considered.

1.1.3 *But what is the shape of the molecule?*

If the user would also like to optimize relative atomic positions for the structure in question, i.e., derive ideal bond lengths, angles, and dihedral angles, then *geometry optimization* is carried out. This procedure involves a series of single point calculations. After carrying out the single point calculation on the initial geometry, the relative atomic (i.e., nuclear) coordinates are varied slightly, a new single point calculation is carried out and the energies associated with the old and new structures are compared. This goes on and on until a geometry is found with the lowest possible energy (Fig. 1.1). Thus, the optimization of electron positions/distributions is decoupled from the optimization of nuclear positions — which is allowed by the Born–Oppenheimer approximation, which notes that electrons move/adjust their positions much more quickly than do nuclei, because the former have much smaller masses than do the latter.

What control does the user possess over this process? First, the user must specify the Hamiltonian, e.g., HF, MP2, B3LYP, etc. (see Chapter 2). The Hamiltonian to be used in a particular calculation is generally determined on the basis of: (1) past experience (of the user or others, as reported in the literature) in applying particular Hamiltonians to molecules with particular functional groups and particular types of reactions and/or (2) test calculations in which the performance of particular Hamiltonians is assessed against experimental data or data from "high level" calculations (on small model

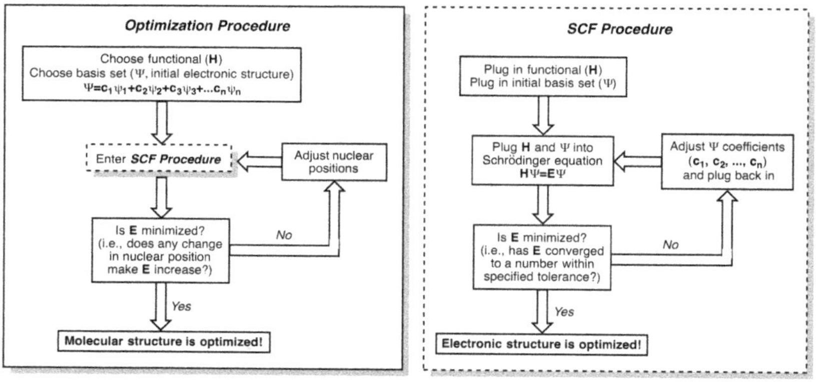

Figure 1.1. A flowchart showing the process by which quantum chemical software calculates the electronic structure and optimizes the geometry of a molecule. The SCF procedure occurs at each iteration of a geometry optimization to calculate the energy of the electronic structure of each nuclear geometry.

systems, since such "high levels" are usually prohibitively expensive for organic molecules of sizes relevant to organic synthesis, bioorganic chemistry, etc.). Second, the user must specify the basis set. Again, this choice is generally made on the basis of precedent or model calculations. Some functional/basis set combinations are better than others for calculating the geometry of a molecule, while some are better than others for calculating the energy of a molecule. Third, the user must specify the initial atomic (nuclear) coordinates of the system to be studied. These coordinates are generally obtained from X-ray structures (on the off-chance that one is available) or result from the user's chemical intuition piped through (nowadays) a graphical user interface attached to the quantum chemistry software of choice. If one is performing a geometry optimization, then the closer the initial geometry is to the optimized geometry the faster the calculation will be, i.e., *there is a direct pay-off for having well-developed chemical intuition.*

1.2 Energy

Let us talk about the energy that is calculated in the procedure described above. Arguments regarding the reactivity of organic molecules almost always boil down to *energy*. There are various forms of energy that can be calculated using computational methods, with each step of a calculation bearing its own assumptions. Here we will briefly discuss how each type of energy is calculated and what major assumptions an applied theoretical chemist should know about to make valid arguments about a molecule's reactivity. For a more detailed mathematical description of these calculations, specifically as they are implemented in the Gaussian program package, see the white papers by Frisch *et al.* and Ochterski *et al.*[1] on quantum chemistry and thermochemistry in Gaussian.

1.2.1 *Potential energy surfaces and stationary points*

We will focus first on geometry optimization. The *electronic* energy (E), also called the potential energy, is the only type of energy calculated in a simple geometry optimization. This is the energy of all nuclei and electrons combined into a molecule (or complex) compared to the energy of all nuclei and electrons at infinite separation. Optimized molecular geometries correspond to points on a calculated potential energy surface (PES; Fig. 1.2). A PES describes how the energy of a molecule changes with changes in its geometry (i.e., nuclear positions and accompanying electron distribution). A molecular structure can be optimized to either a minimum (corresponding to a reactant, intermediate, or product) or first-order saddle point (a minimum with respect to all geometric changes *except* those that occur along the intrinsic reaction coordinate [IRC, *vide infra*], corresponding to a transition state structure [TSS]). Both minima and TSSs are "stationary points" on the PES, because the slope of the tangent to the surface at these points is zero. There has been some debate over the years about the terms "transition state," "transition structure," and "transition state structure." We employ the convention that "transition structure" and "transition state structure" are synonyms, meaning a first-order saddle point on the PES, i.e., the high energy point along a particular optimized reaction coordinate. However, we reserve the term "transition state" for describing an ensemble of structures around a TSS, i.e., it is a

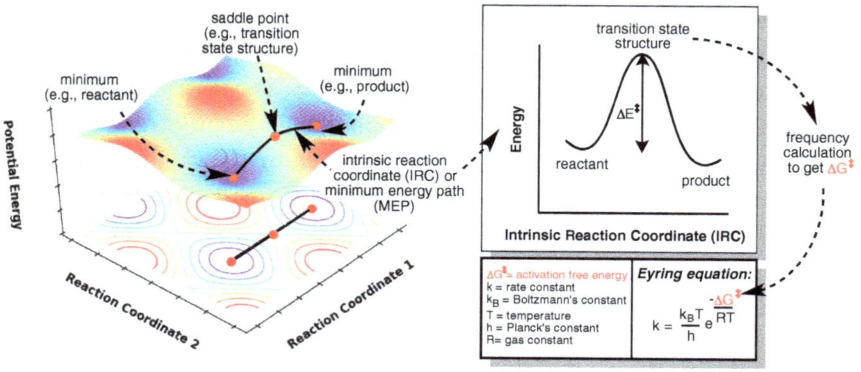

Figure 1.2. A theoretical 3D PES (left) showing two minima and the TSS connecting them. The minimum energy pathway (MEP) between two minima is called the IRC, which has the 2D form shown on the right. A frequency calculation at each stationary point (minima and TSS) leads to calculation of the free energy barrier, which can be used in the Eyring equation[2] to calculate the rate constant of the reaction.

state not a single structure. Transition states are particularly relevant in dynamics calculations (see Chapter 12).

No matter how many separate molecules are included in an input file for a quantum chemical calculation, the software will treat the entire system as a single electronic structure. This means that the basis functions for each individual molecule are "accessible" to all other molecules in the calculation, which artificially lowers the total energy of a complex vs. separated species by giving the electrons of each component molecule more space to sample. This phenomenon is called basis set superposition error (BSSE).[3] There are a couple of methods to mitigate this error, namely, the counterpoise (CP) method[4] and the chemical Hamiltonian approach (CHA).[5] This error is often not corrected for, because its effect is small (especially with large basis sets), but nonetheless should be considered in the case of modeling complexes or bimolecular reactions.

Let us talk about PESs in more detail. If we have a molecule constructed of N atoms, in 3D there will be three ways all N atoms are moving in the same direction (along the x-, y-, and z-axes) and three ways all N atoms are rotating in the same direction (around the x-, y-, and z-axes), leaving $3N - 6$ ways the nuclei can move with respect to one another to affect the *internal* potential energy of the molecule. While it is straightforward for a computer to do this calculation, in order to be able to visualize a PES, one would need to be able to see in $(3N - 6) + 1$-dimensions (where the additional dimension is the energy), which is impossible for even small organic molecules and exceptional humans. What can easily be visualized, however, is the variation in energy along the IRC.[6] The IRC is the MEP between a TSS and its flanking minima, calculated by determining the path of steepest descent along either phase of the reaction coordinate (i.e., toward reactants or products).

The foundational assumption of transition state theory (TST)[7] and Rice–Ramsperger–Kassel–Marcus (RRKM) theory[8] is that the activation barrier for a reaction is related directly to the rate constant. This corresponds to the assumption that a molecule remains *on* the PES throughout the course of a reaction. These conditions are sometimes referred to as "quasi-equilibrium" because there is no time-dependence of the rate and rate constant.

IRCs provide evidence of the connection between minima and TSSs along a reaction pathway and the assumptions of TST and RRKM theory are often reasonable enough to reproduce experimental data. Additionally, the shapes of IRCs can give valuable insight into the nature of a mechanism. For example, a "shoulder" along an IRC is often indicative of a concerted reaction with asynchronous events (see Sec. 1.4.1), which may possess a so-called "hidden intermediate".[9] However, unique features of certain PESs can preclude the quasi-equilibrium assumption. In fact, shoulders in IRCs are also sometimes indicative of post-transition state bifurcations (PTSBs, Fig. 1.3; see Chapter 12 and Refs. [10, 11]). A PTSB would not be visible in the 2D TST model. PTSBs occur in systems with two products that are directly connected to the reactant through the same TSS. Such a PES exhibits a "valley-ridge inflection (VRI) point," where the exit channels leading to the two products diverge. If the product distribution shows anything except equal

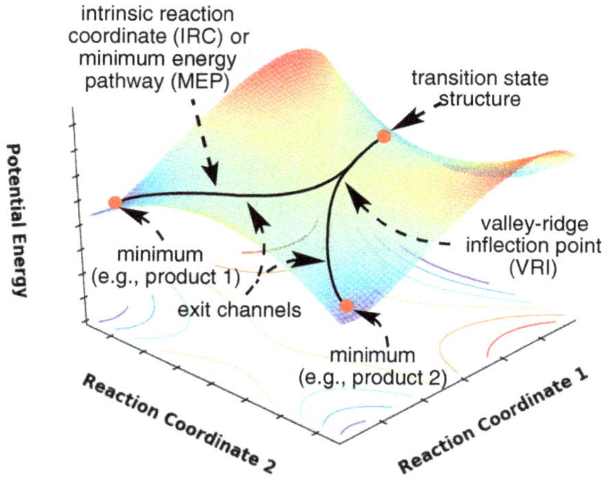

Figure 1.3. A theoretical 3D PES showing the features of a PTSB. A single TSS connects two possible products, though only one of these products is along the IRC path.

amounts of the two products following the PTSB, traditional TST will fail to accurately model the system.[11,12] Molecular dynamics (MD) simulations, which incorporate kinetic energy and thus do not assume quasi-equilibrium conditions, are helpful for these more complex systems (see Chapter 12). In MD simulations, instead of assuming that the energy of activation is the sole indicator of reaction outcome, many (often hundreds or thousands of) trajectories are run in order to get a statistical distribution of pathways to products.

Another important caveat is that IRCs lead to minima that are in conformations that are productive but are not necessarily the lowest energy conformations (Fig. 1.4). This issue is important because, for example, only the lowest energy conformations generally will be detectable by spectroscopic techniques (*vide infra*). Neglecting this fact can lead to a severe underestimation of the barrier of a reaction. Conformational searching is critical in determining the energy of a molecule of interest, and as such, Chapter 5 details many methods by which one can conduct a conformational search. In short, there are two major approaches: (1) calculating the energy of each possible

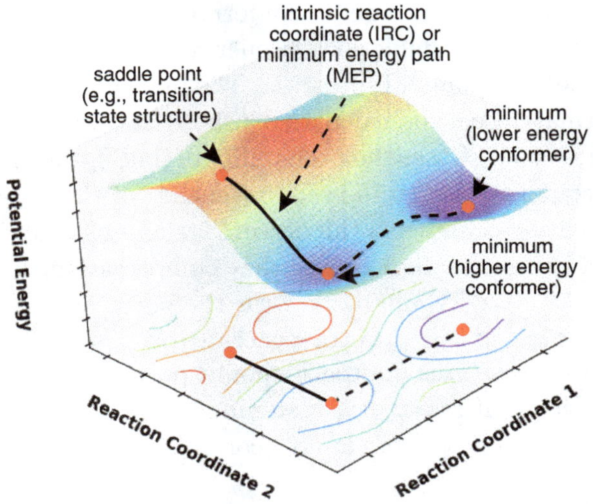

Figure 1.4. This 3D PES qualitatively depicts the consequences of an IRC leading to a *productive* conformation of the molecule, which is a minimum on the PES, but not the *lowest* energy conformation.

conformation, or a Monte Carlo sampling[13] of conformations if a systematic search is prohibitively large, using a molecular mechanics (MM) force field and (2) running MD simulations at high temperature. Once the relevant conformations are located, their energies can be calculated with a quantum chemical method and ranked. Most flexible organic molecules will require only a conformational search, but some molecules, particularly metal–ligand complexes, will require additional sampling of different *configurations* (i.e., the different ways the ligands can be coordinated around a central metal atom). These issues apply not only to minima but also to TSSs.[14]

Conformational searching also is particularly important in the calculation of spectral data, particularly nuclear magnetic resonance (NMR) spectra (see Chapter 6). If several of the lowest energy conformations are close in energy (typically within a few kcal/mol or so of one another, depending on reaction conditions), all of these conformations are likely present in some measurable quantity. This means the NMR spectrum obtained for this molecule will contain contributions from all low energy conformations, and calculating the spectrum of only the single lowest energy conformation would not reproduce the experimental spectrum. The proportion of molecules in each conformation can be calculated using the Maxwell–Boltzmann distribution, which relates the probability of finding the molecule in a particular conformation to the relative energy of that conformation. The probability calculated is the percentage of molecules in the total ensemble of molecules that are in that conformation. Then, multiplying the calculated probability with the corresponding spectrum and summing all of the spectra generates a "Boltzmann-weighted average" spectrum. This approach should provide the best estimate of the experimental spectrum, assuming that the relative energies used for each conformer are close to correct.

1.2.2 *Frequency calculations and fretting about free energy*

After a geometry optimization, the final energy calculated is the electronic energy (E), which is *only* the energy of the electronic structure for a single (although optimized) nuclear configuration. In order to calculate the zero-point energy, internal energy, enthalpy, and free energy of a molecule, an applied theoretical chemist will conduct a frequency calculation on the optimized structure. A frequency calculation will determine the frequencies of all $3N-6$ vibrational modes (again, these correspond to the degrees of freedom, or the ways the atoms can move with respect to one another, that affect the internal energy of the molecule). Doing so amounts to *computing the IR spectrum* of the molecule (or complex) in question, which can be useful in the identification of, for example, reactive species isolated in frozen matrices.[15] An approximation that quantum chemical programs generally make is that the frequencies calculated are for a quantum harmonic oscillator (QHO). This approximation is usually decent, because, unless significant changes are made to the bonding structure of a molecule during a reaction, the anharmonic corrections would mostly cancel out when looking at the difference in energy between, say, a TSS and a reactant. Anharmonic corrections are usually only computed in practice if an extremely high-level calculation of frequencies is necessary.[16]

The software uses these frequencies to determine thermodynamic properties of one mole of the molecule at a certain temperature (usually 298 K, unless otherwise specified). The most important approximation to remember in the calculation of these thermodynamic properties is that they are all usually calculated, at least initially, for an *ideal gas*. Modeling of a solution-phase reaction can be done in a variety of ways (*vide infra*), but additional corrections to the calculated energies need to be considered. A correction to the energy that is made for all types and phases of molecules is the zero-point energy correction. QHOs can never have exactly zero energy as a consequence of the Heisenberg uncertainty principle. The lowest possible energy of a QHO is called its zero-point energy, which is very straightforward to calculate for each QHO and can be summed over the molecule to get the total zero-point correction. The molecule's partition function, which determines how the energy of the system is partitioned between different translational, rotational, vibrational, and electronic states, can then be used to calculate thermodynamic quantities. The internal thermal energy (U) is the kinetic + potential energy associated with motion of the nuclei. The enthalpy (H), which is just $H = U+PV$ where P is pressure and V is volume, also includes, in a sense, the energy that was necessary to get the system into its current "state." The Gibbs free energy (G), the most commonly used quantity to compare computational with experimental results, takes the enthalpy and subtracts a factor of entropy multiplied by temperature ($G = H - TS$). It is important to remember that TSSs are found by determining the location of a saddle point on the PES, but such points are only necessarily saddle points with respect to the *electronic energy*. A subsequent frequency calculation determines the free energy barrier to get to that TSS. The activation free energy is then used to calculate the rate constant via the Eyring equation[2] (Fig. 1.2). In some cases, the TSS on a PES differs significantly in structure from the TSS on a free energy surface. In fact, it is not always the case that a barrier on the free energy surface would be detectable on the PES, thus there are methods of locating a TSS on the free energy surface if necessary.[17]

Frequency calculations are also used to confirm whether an optimized structure is a PES minimum, TSS, or neither. A fully optimized structure with zero imaginary frequencies is a PES minimum. A fully optimized structure with a single imaginary frequency is a TSS. The vibration associated with this imaginary frequency (which can be animated and viewed using most quantum chemistry software packages) will correspond to movement along the reaction coordinate. A structure with more than one imaginary frequency is generally not relevant to a reaction (although some such structures have been proposed as relevant from time to time[18]).

1.2.3 *Issues with entropy*

Because the entropy correction calculated for an individual molecule is only due to the relative motion of nuclei *within* the molecule, there are additional entropic factors that need to be considered when looking at *ensembles* of molecules. In order to obtain a more accurate entropy value to compare to experiment, there are several corrections that can be easily applied to the calculated entropy that are specific to particular experimental systems. First, if a calculation is performed for an ideal gas when the experimental system is in solution, the standard state of the molecule needs to be changed from a pressure of 1 atm to a concentration of 1 M. This correction is necessary only if the number of molecules changes along the reaction path (e.g., two molecules reacting to form one molecule or vice versa) in the solution phase. If the reaction is unimolecular, or the number of molecules never changes along the reaction path, this correction would cancel at each stationary point along the reaction path. If the reaction is in the gas phase, then a 1 atm standard state applies and there is no need to apply this correction. If one of the reagents is a solvent molecule that comes from neat solvent, the standard state correction for *only* that solvent molecule needs to include an additional correction for the concentration of the solvent in its neat form. This treats the standard state for that solvent as neat solvent, rather than 1 M. If the reaction pathway involves a mixture of structures at any stationary point, an entropy correction must also be included to account for the mixture. There are two components to this correction, and either just one or both could be applied depending on the system. The first is a correction for an intermediate or product that exists as a chiral species (assuming both stereoisomers are in solution). The second is for a molecule that could adopt multiple low-energy conformations in solution. A Boltzmann-weighted average is used to account for this mixture of conformers (*vide supra*). Lastly, if the symmetry number of a reactive species changes over the course of a reaction, there is an additional entropic factor. The symmetry number is the number of different, but indistinguishable, arrangements of a molecule. For example, water would have a symmetry number of 2 because its C_2 axis allows for two indistinguishable arrangements of the molecule.

These corrections are often neglected because they tend to amount to a free energy correction on the order of 1 kcal/mol, which corresponds to a higher degree of accuracy than is called for in many research studies. However, this is not always the case, and there are scenarios where accurate calculation of entropic factors is required. For example, in 2015, Kazemi and Åqvist[19] studied the deamination reaction of cytidine to uridine that occurs within the enzyme cytidine deaminase. This enzyme is purported to catalyze the reaction by significantly decreasing the entropy of activation by decreasing the entropy of the substrate on binding. This concept is referred to as Jencks' "Circe effect"[20]: if the substrate's configurational entropy is significantly reduced on binding, the entropy loss associated with reaching the TSS can be eliminated. Kazemi and Åqvist wanted to develop a way to model this effect on the cytidine deamination reaction by computing the activation entropy in solution, which is beyond the scope of the corrections outlined above. To do this, they made computational Arrhenius plots from MD and empirical valence bond (EVB)[21] simulations, which do extensive configurational sampling of arrangements of the substrate and a number of explicit solvent molecules. The Arrhenius plots showed the temperature dependence of the thermodynamic (specifically entropic) activation parameters for the reaction, which had very good agreement with experimental data.

An important logical fallacy that this study was able to avoid is the assumption that one can model an entire flask's worth of molecules by modeling a single molecule or multi-molecule system using quantum mechanics and then extrapolating the results. Just because something has a low barrier does not necessarily mean that process corresponds to what is happening in solution. More importantly, just because a pathway is energetically viable does not mean it is the *only* mechanistic pathway. The problem arises from incomplete sampling of the possible scenarios that have a traversable barrier at the specified temperature. Additionally, the more components that are in solution, the more difficult it becomes to accurately predict a mechanism. Plata and Singleton[22] conducted an important study that illustrates this point. It was found, by obtaining definitive mechanistic data experimentally for Morita Baylis–Hillman reactions[23] that, despite accounting for all of the entropic factors described above, no level of theory examined (there were many) could predict a mechanism that is consistent with all experimental data. The important take-away message from this study was that computational results should not be analyzed (or believed) without guidance from experimental data. Computational organic chemistry can reveal important mechanistic information not visible by experiment but *must* be benchmarked with experimental evidence to confirm its physical relevance.

1.3 Worrying about solvent even when entropy is not on your mind

Solvation models move a molecule from the gas phase to a condensed phase, which can make modeling an organic reaction mechanism more physically relevant, since many reactions of interest take place in solution.[24] Solvent can affect a reaction through short-range interactions such as hydrogen bonding or long-range interactions such as polarization. Solvent effects are also responsible for differences in bond dissociation enthalpies (BDEs), which affect the reactivity of molecules to varying degrees.[25] There exist many flavors of implicit models that treat solvent as a continuum and explicit solvation models that describe the solvent molecules atomistically.[24] This section will discuss general guidelines for using solvation models (see Chapter 4 for an expanded discussion).

When does including solvent in your calculation affect your results to a degree that cannot safely be ignored? Under certain conditions, solute molecules may change the structure of a solvent in the vicinity of the solute resulting in a change in free energy.[26] Solvent continuum models use a homogenous medium to describe the continuous electric field of the solvent. The effects of this field may be seen in the kinetics of a reaction if intermediates and TSSs undergo changes in charge or polarizability.[27] For example, selective stabilization (see Sec. 1.4.2 for a discussion of problems with this term) of a TSS for the rate-determining step of a reaction by a solvent can lead to different rates of reaction in different solvents (Fig. 1.5).[28] For this reason, it is important to include solvent, whether implicitly or explicitly, in calculations on solution phase reactions during which charge distributions are likely to change significantly.

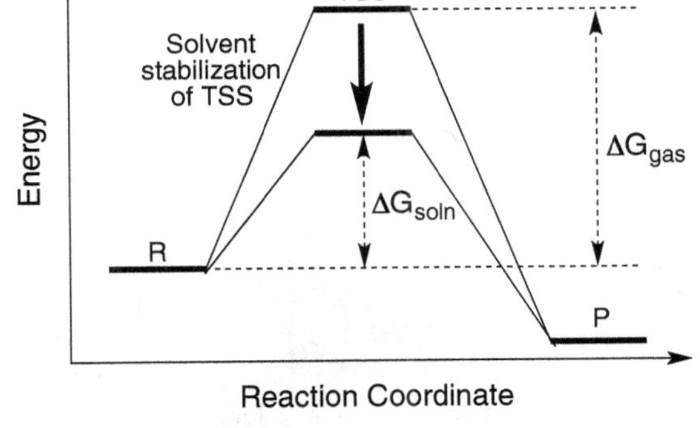

Figure 1.5. Example of solvent effect on a PES. If the solvent stabilizes, the TSS more than R, the relative energy of the TSS decreases, minimizing the free energy barrier, so that $\Delta G_{soln} < \Delta G_{gas}$.

1.3.1 *Solvation by implication*

An implicit solvent model is one in which the solvent is represented as a medium with a fixed dielectric constant with a cavity into which is placed a solute molecule. The cavity of vacuum is introduced into the solvent, the solute charge density is then placed in the solute cavity, the solute polarizes in response to the solvent polarization, and the solvent polarizes in response to the solute charge density (Fig. 1.6).[29] Implicit solvent models come in multiple flavors (see Chapter 4), but the most commonly used are the density-based solvation model (SMD) and the polarizable continuum model (PCM).

The SMD model uses the full solute electron density and separates the solvation free energy into an electrostatic component and interactions between solute and solvent in the first solvation shell. It uses the electron density to calculate the solvent accessible surface area (SASA) and determines dispersion-repulsion energies.[30] If one is interested in calculating solvation free energies or dispersion–repulsion energies, SMD is a good choice.

PCM[31] or dielectric PCM (DPCM) generates multiple overlapping van der Waals spheres for each atom of the molecule. The free energy of solvation calculated by this method is primarily the electrostatic energy. The conductor-like PCM model (CPCM) is a variation on the PCM model, using a group of nuclear-centered (rather than atom-centered) spheres to define the dielectric continuum.[26b] The solvent is treated as a conductor, which changes the polarization

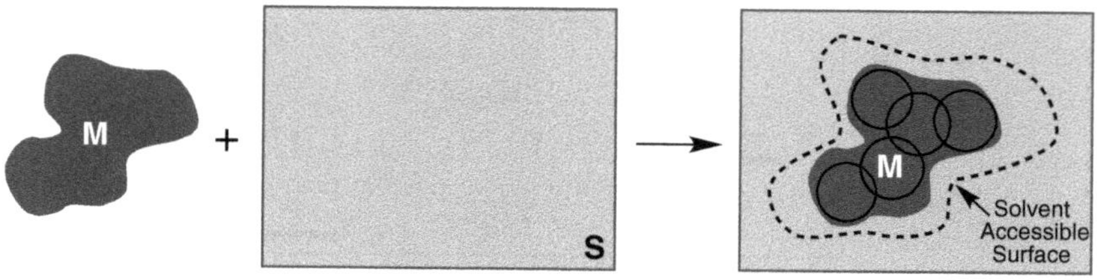

Figure 1.6. Pictorial representation of a solvent continuum model. A cavity for molecule, M, is placed into a solute charge density, S. A solvent excluded area surrounds the cavity. The SASA allows the solvent and solute molecules to polarize in response to the other.

charges at the SASA between solute and solvent. Solvents with higher permittivity behave more like ideal conductors, providing better results.[30c] CPCM provides improved results for solvents with high dielectric constants. Because the math is simplified in CPCM, this is a less computationally expensive method than is PCM. Neither PCM nor CPCM calculate dispersion-repulsion energies, however, which is a limitation of these methods.

In continuum models, a self-consistent reaction field (SCRF) is used. A solute cavity is created and dispersion (mainly van der Waals) and electrostatic interactions between solute and solvent are calculated.[27] Together, these terms provide the solvation free energy of the system.

Some issues to take caution of: If the reaction you are modeling uses a nonpolar solvent (low dielectric constant) with little to no dipole moment, such as hexane, cyclohexane, benzene, toluene, or chloroform, you are perhaps better excluding a solvent continuum model from your calculations because (1) CPCM models do not provide high accuracy for such solvents and (2) modeling the reaction in the gas phase will likely provide comparable results while saving on computational time.[32] Another concern is that implicit solvent models ignore short-range effects such as hydrogen bonding or hydrophobic effects, which, if relevant to the chemistry being explored, may demand the use of an explicit solvent model.

1.3.2 *Being explicit*

An explicit solvation model, one in which solvent is represented atomistically, can illustrate how noncovalent interactions contribute to the free energy change of solvation (Fig. 1.7). However, the addition of solvent molecules can increase computational time dramatically. Exactly how many solvent molecules should be added, and where they should be placed, are fundamental questions still up for debate.[33] A common practice is to begin with a single solvent molecule and optimize your reactant or TSS. Then, increase the number of solvent molecules one at a time until the solvent molecule does not remain coordinated to the solute. This is a crude approach,

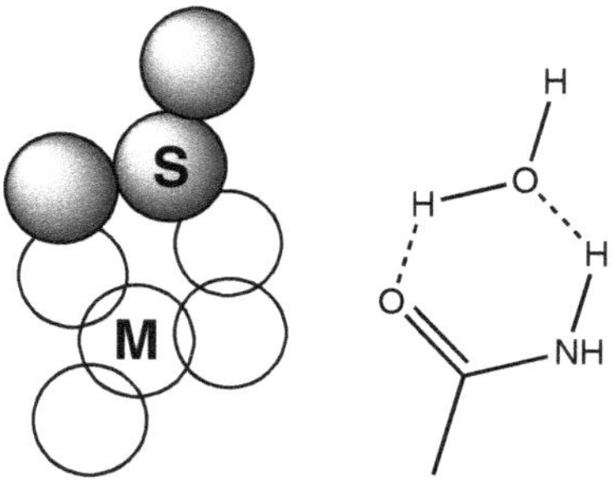

Figure 1.7. Pictorial representation of an explicit solvent model. Circles represent the molecule, M, and shaded circles represent the solvent, S. An example is shown on the right of an amide that is hydrogen bonded to a single water molecule.

but provides some confidence that you have considered the minimum number of solvent molecules that may significantly affect your system. MD calculations involving explicit solvent molecules that surround the solute (i.e., a box of solvent surrounding the solute) can also be used, but the size of such systems forces one to use levels of theory with relatively low accuracy (see Chapter 4).

1.3.3 *Beware proton transfers*

A common scenario where one may want to model solvent explicitly is in reactions where the solvent may be assisting in the progress of the reaction mechanism through short-range interactions or through coordination with solute molecule(s). A proton transfer or proton shuttle pathway is one such example.[34] However, Plata and Singleton[22] recently demonstrated the pitfalls of computing such pathways using quantum mechanics and warned strongly that one must be aware of the errors associated with carrying out such a calculation (*vide supra*).

1.4 What is a mechanism?

Before carrying out any quantum chemical study, one should ask oneself several questions. These include: (1) Is quantum chemistry required to answer the question of interest? The answer to this question may be "no" if tight error bars on energies are not required or if the system in question is well described by classical mechanics methods (i.e., force field/molecular mechanics methods) or even by the use of plastic models (effectively a tactile version of a force field with steep potentials). (2) If the answer to question (1) is "yes," and the issue is a structural one, then one should plunge ahead. If, however, the issue is mechanistic in nature, one should step back and ask what is meant by "mechanism," i.e., which mechanistic framework will be used to interpret the results?

A reasonably noncontroversial definition of a mechanism is an accounting of all structural changes that occur as a reactant is transformed to a product. This definition is, of course, ambiguous enough to lack utility. First, one need consider which sort of structural changes are relevant to the question at hand: geometric (nuclear positions), electronic (e.g., orbitals, spin densities, electrostatic potential), vibrational (dynamics), etc. Second, one should consider whether the question at hand can be addressed by modeling isolated molecules or ensembles (of both reactants and solvent).

One should also consider the language in which the results will be conveyed — from the theoretician to the experimentalist, from the experimentalist who posed the question to other experimentalists, from faculty to students, etc. For example, some prefer to converse in terms of curved/curly arrows, some in terms of stationary points on PESs, some in terms of IRCs, some in terms of MOs, some in terms of electrostatics, and some in terms of dynamics. Ideally, one is comfortable conversing in all of these flavors of chemical language.

Several comments are perhaps warranted on the dependence of organic chemists on "arrow-pushing mechanisms." First, the proclivity for arrow-pushing is, perhaps, a result of our approaches to undergraduate organic chemistry education, which results in many students proposing a particular arrow-pushing mechanism because "it gets me to the product".[35] Second, proficiency with arrow-pushing is not necessarily a bad thing. In fact, proposing an arrow-pushing mechanism is often the first step in a theoretical study on an organic reaction; it is a means of defining which intermediates (along with the TSSs that connect them) one will attempt to locate as PES stationary points. And whether or not such stationary points are found along a reaction coordinate at relative energies commensurate with experimental conditions is one means of assessing the aforementioned proficiency.

1.4.1 *Time, distance, and synchronicity*

There have been many debates during the history of mechanistic organic chemistry centered around whether one reaction or another is concerted or stepwise, and if the former, how synchronous the events occurring during the reaction are[36] Often, these debates were complicated by the use of loosely defined terminology. We advocate for the following definitions. *Concerted* is used to describe a reaction in which reactant and product are connected by a single TSS, i.e., no PES intermediates/minima are found along the reaction coordinate connecting reactants and products. *Stepwise* is used to describe a reaction in which one or more PES minima are found along the pathway from reactant to product. While others may prefer to define conceitedness in terms of lifetimes of intermediates and their associated "chemical relevance" (e.g., the concept of "dynamically concerted"[37]), we prefer definitions based on the PES, since these are unambiguous (assuming tight enough convergence criteria are used in optimizations). *Synchronous* and its opposite, *asynchronous*, are adjectives that should only be applied to compare different "events" occurring during a single chemical step,[38] not a whole reaction, and the relevant events should be made clear. Such events might include formation/breaking of a bond, changes in a particular dihedral angle, etc. In order to avoid ambiguity, it should also be made clear whether one is assessing synchronicity in terms of time or in terms of "distance" (however defined) along a reaction coordinate. Standard quantum chemistry only provides information on the latter, i.e., one calculates energy vs. structure, not structure vs. time. To obtain information on synchronicity in terms of time, one needs to carry out dynamics calculations (see Chapter 12). As an example, many carbocation cyclization/rearrangement reactions proposed to occur during the biosynthesis of terpenes (including steroids) are predicted by standard quantum chemistry to be concerted with multiple bond forming/breaking events occurring asynchronously along their reaction coordinates, i.e., formation/breaking of one bond is nearly complete in terms of structure (e.g., bond distance, bond order) before the next gets going.[38,39] The synchronicity of these events in terms of time has only been examined for a subset of these reactions.[10]

1.4.2 *Three particularly troublesome words: Stability, proof, and truth*

Here we discuss three words that cause endless misunderstandings (and arguments) — between theoreticians and experimentalists and between theoreticians and other theoreticians.

The first of these problematic words is *stability*. When someone asks a theoretician if a molecule is stable or if one molecule is more stable than another, we believe that the appropriate response is "what do you mean by stability?" This is not (merely) an attempt to be pithy. Without appropriate modifiers, the word is ambiguous. First, it should be made clear whether one is concerned with thermodynamic (i.e., which side of an equilibrium is favored?) or kinetic (i.e., how quickly will one molecule transform into others?) stability. Second, one should always remember that stability is defined *relative to something*, and not just anything convenient. For example, one cannot decide which of two molecules is more thermodynamically stable by simply computing the energy of each, if the two molecules do not have the same number and types of atoms. This situation can cause immense frustration; the question of thermodynamic stability seems simple, but sometimes one must resort to complicated isodesmic equations to arrive at a reasonable answer, and, given that the construction of such equations is a subjective process, the answer is likely to be controversial (see Chapter 15).[40] If one is concerned with kinetic stability, it is not sufficient to characterize a molecule as stable just because it is a PES minimum, an issue that was the subject of an essay by Hoffmann, Schleyer and Schaefer that demanded "more realism, please!".[41] Being a minimum means that there is a barrier for conversion to other structures, but the magnitudes of such barriers are important! A barrier of 0.5 kcal/mol has essentially no relevance, a barrier of 20 kcal/mol would lead one to predict that a reaction will be rapid above room temperature but slow at −78°C.

In short, there is no such thing as inherent stability; *stability is always relative.* Molecule A may be lower in energy than molecule B but higher in energy than molecule C. Molecule D might have a very high barrier for conversion to molecule E but a very low barrier for conversion to molecule F. Accepting that stability of any sort is relative can cause significant angst. How does one know that there isn't another isomer or conformer that was not considered but is lower in energy (e.g., the well-known problem of finding a global minimum)? How does one know that there will not be an unexpected fast reaction leading to an unwanted product, that was not dreamt up during the theoretical study, but which reveals itself when a prediction is put to the test in the flask?[41]

On to the second word, *proof.* It is dogma that a mechanism can be disproven but cannot be proven.[42] While we agree with this statement, some do not; a fascinating series of short commentaries on the subject appeared in 2009 in the *Journal of Chemical Education.*[43] Why do we argue that a chemist who states that his research group has proven a mechanism should not be trusted? The issue is similar to that just mentioned for kinetic stability: How does one know that there is not another unexpected mechanism at play that was not dreamt up during the mechanistic/theoretical study? You just never know.

An even more problematic word is *truth.* This word elicits different responses in different people, many of these responses tied to personal philosophy, spirituality, and emotion. To steal from Isaac Asimov, truth is "based on assumption and adhered to by faith".[44] While we are not opposed to seeking truth, we do not believe that doing so is something that a mechanistic study (theoretical or experimental) can achieve.

1.4.3 *A more useful word?*

So where does this leave us? At best, we arrive at a *model* that describes the conversion of reactant(s) to product(s) that is consistent with all available data on the reaction in question, be it derived by experiment or calculation. To be most useful, the assumptions underpinning such a model should be clear, ideally stated explicitly. Models are also strengthened by being "productive" and "portable" — terms applied by Roald Hoffmann to describe "the best theories".[45] *Productivity* here is equated with the ability to prompt new experiments. In other words, a model will ideally have predictive value. If one constructs a model that supposedly explains some experimentally observed phenomenon, then it ought to be useful in predicting related but different phenomena — and the theoreticians who constructed the model should welcome experimental tests of these predictions! This situation is actually win-win for the theoretician. If the predictions are borne out in the wet lab, hooray! If not, the theoretician stays in business, with a new mystery to solve. *Portability* implies that the model can be applied by those who did not construct it. Ideally, a model derived using quantum chemistry will be usable by experimentalists without the need for additional quantum chemical computations, i.e., although quantum chemistry may be needed to make quantitative predictions, the model should at least lead to qualitative predictions in the absence of calculations. This goal may seem lofty, but we believe that all theoretical studies should strive to achieve it. It may seem that doing so will put theoreticians out of business, but there are many additional systems with interesting problems to tackle, many other models of structure and reactivity to construct.

1.4.4 *Beware beauty and simplicity*

How does one assess the validity of a mechanistic model? It is not by proving it, but by acquiring evidence consistent with the model or evidence that invalidates it. On the experimental side, there are many approaches that may provide useful data. These include kinetics experiments, testing for linear free energy relationships (LFER) using "Hammett plots," kinetic or equilibrium isotope effects, and solvent effects on rates. A nice overview, with a pedagogical bent, is provided in Ref. [35d].

The results of such experiments can also be predicted using quantum chemistry, providing a link between the experimental data and specific chemical structures along particular reaction coordinates. In addition, quantum chemistry obviously can be used to predict barriers for reactions. As a result, a particular mechanistic model can be ruled out if the overall barrier for the associated transformation is predicted to be too high to correspond to the experimentally observed rate (often known only as "fast at room temperature," "required refluxing in benzene," etc.), assuming one did not encounter one of the computational pitfalls described above.

We end with a caution that is relevant to both experimental and theoretical studies: the simplest mechanism, even the simplest mechanism that is consistent with all available data obtained by theory and experiment, is not necessarily the mechanism followed in a reaction vessel. There is no valid reason to assume that nature will always follow the simplest path of which a human can conceive. One need not invoke complexity if simplicity will suffice, but one should not assume that simplicity implies correctness. Hoffmann,[42] Minkin,[45] and Carpenter[46] have expounded elegantly on the origins and implications of applying this fallacy, but in short, all models that are consistent with all available data are equally valid until refuted. And all models are beautiful, to someone.

References

1. (a) Frisch, M. J. T., Trucks, G. W., Schlegel, H. B., Scuseria, G. E., Robb, M. A., Cheeseman, J. R., Scalmani, G., Barone, V., Mennucci, B., Petersson, G. A., Nakatsuji, H., Caricato, M., Li, X., Hratchian, H. P., Izmaylov, A. F., Bloino, J., Zheng, G., Sonnenberg, J. L., Hada, M., Ehara, M., Toyota, K., Fukuda, R., Hasegawa, J., Ishida, M., Nakajima, T., Honda, Y., Kitao, O., Nakai, H., Vreven, T., Montgomery, J., J. A., Peralta, J. E., Ogliaro, F., Bearpark, M., Heyd, J. J., Brothers, E., Kudin, K. N., Staroverov, V. N., Kobayashi, R., Normand, J., Raghavachari, K., Rendell, A., Burant, J. C.,

Iyengar, S. S., Tomasi, J., Cossi, M., Rega, N., Millam, N. J., Klene, M., Knox, J. E., Cross, J. B., Bakken, V., Adamo, C., Jaramillo, J., Gomperts, R., Stratmann, R. E., Yazyev, O., Austin, A. J., Cammi, R., Pomelli, C., Ochterski, J. W., Martin, R. L., Morokuma, K., Zakrzewski, V. G., Voth, G. A., Salvador, P., Dannenberg, J. J., Dapprich, S., Daniels, A. D., Farkas, Ö., Foresman, J. B., Ortiz, J. V., Cioslowski, J. and Fox, D. J. (2009). *Gaussian 09* (Gaussian, Inc., Wallingford, CT); (b) Ochterski, J. W. (2000). *Thermochemistry in Gaussian (White Paper)* Gaussian, Inc., Wallingford, CT, USA); (c) Ochterski, J. W. (1999). *Vibrational Analysis in Gaussian (White Paper)* (Gaussian, Inc., Wallingford, CT, USA).

2. Eyring, H. (1935). The activated complex in chemical reactions, *J. Chem. Phys.*, 3, p. 107.
3. Kestner, N. R. and Combariza, J. E. (1999). Reviews in computational chemistry. In *Basis Set Superposition Errors: Theory and Practice*, eds. Lipkowitz, K. B. and Boyd, D. B., Volume 13, Chapter 2 (John Wiley & Sons, Inc., Hoboken, NJ, USA) pp. 99–132.
4. van Duijneveldt, F. B., van Duijneveldt-van de Rijdt, J. G. C. M. and van Lenthe, J. H. (1994). State of the art in counterpoise theory, *Chem. Rev.*, 94, pp. 1873–1885.
5. (a) Mayer, I. (1983). Towards a "chemical" hamiltonian, *Int. J. Quantum Chem.*, 23, pp. 341–363; (b) Mayer, I. and Vibok, A. (1988). SCF equations in the chemical Hamiltonian approach, *Chem. Phys. Lett.*, 148, pp. 68–72.
6. (a) Gonzalez, C. and Schlegel, H. B. (1990). Reaction path following in mass-weighted internal coordinates, *J. Phys. Chem.*, 94, pp. 5523–5527; (b) Fukui, K. (1981). The path of chemical reactions — The IRC approach, *Acc. Chem. Res.*, 14, pp. 363–368; (c) Maeda, S., Harabuchi, Y., Ono, Y., Taketsugu, T. and Morokuma, K. (2015). Intrinsic reaction coordinate: Calculation, bifurcation, and automated search, *Int. J. Quantum Chem.*, 115, pp. 258–269.
7. (a) Pechukas, P. (1981). Transition state theory, *Annu. Rev. Phys. Chem.*, 32, pp. 159–177; (b) Laidler, K. J. and King, M. C. (1983). The development of transition-state theory, *J. Phys. Chem.*, 87, pp. 2657–2664; (c) Truhlar, D. G., Garrett, B. C. and Klippenstein, S. J. (1996). Current status of transition-state theory, *J. Phys. Chem.*, 100, pp. 12771–12800.
8. (a) Marcus, R. A. (1952). Lifetimes of active molecules. I, *J. Chem. Phys.*, 20, pp. 352–354; (b) Marcus, R. A. (1952). Lifetimes of active molecules. II, *J. Chem. Phys.*, 20, pp. 355–359; (c) Rice, O. K. and Ramsperger, H. C. (1927). Theories of unimolecular gas reactions at low pressures, *J. Am. Chem. Soc.*, 49, pp. 1617–1629.
9. Rzepa, H. S. and Wentrup, C. (2013). Mechanistic diversity in thermal fragmentation reactions: A computational exploration of CO and $CO_{(2)}$ extrusions from five-membered rings, *J. Org. Chem.*, 78, pp. 7565–7574.
10. Hare, S. R. and Tantillo, D. J. (2016). Dynamic behavior of rearranging carbocations — implications for terpene biosynthesis, *Beilstein J. Org. Chem.*, 12, pp. 377–390.

11. Ess, D. H., Wheeler, S. E., Iafe, R. G., Xu, L., Çelebi-Ölçüm, N. and Houk, K. N. (2008). Bifurcations on potential energy surfaces of organic reactions, *Angew. Chem. Int. Ed.*, 47, pp. 7592–7601.
12. (a) Collins, P., Carpenter, B. K., Ezra, G. S. and Wiggins, S. (2013). Non-statistical dynamics on potentials exhibiting reaction path bifurcations and valley-ridge inflection points, *J. Chem. Phys.*, 139, p. 154108; (b) Hong, Y. J. and Tantillo, D. J. (2009). A potential energy surface bifurcation in terpene biosynthesis, *Nat. Chem.*, 1, pp. 384–389; (c) Rehbein, J. and Carpenter, B. K. (2011). Do we fully understand what controls chemical selectivity? *Phys. Chem. Chem. Phys.*, 13, pp. 20906–20922; (d) Oyola, Y. and Singleton, D. A. (2009). Dynamics and the failure of transition state theory in alkene hydroboration, *J. Am. Chem. Soc.*, 131, pp. 3130–3131.
13. Jorgensen, W. L. and Tirado-Rives, J. (1996). Monte Carlo vs molecular dynamics for conformational sampling, *J. Phys. Chem.*, 100, pp. 14508–14513.
14. Wheeler, S. E., Seguin, T. J., Guan, Y. and Doney, A. C. (2016). Noncovalent interactions in organocatalysis and the prospect of computational catalyst design, *Acc. Chem. Res.*, 49, pp. 1061–1069.
15. (a) Kunze, K. R., Hauge, R. H., Hamill, D. and Margrave, J. L. (1977). Studies of matrix isolated uranium tetrafluoride and its interactions with frozen gases, *J. Phys. Chem.*, 81, pp. 1664–1667; (b) Bernstein, M. P., Sandford, S. A., Allamandola, L. J. and Chang, S. (1994). Infrared spectrum of matrix-isolated hexamethylenetetramine in Ar and H_2O at cryogenic temperatures, *J. Phys. Chem.*, 98, pp. 12206–12210.
16. Meyer, M. P. and Klinman, J. P. (2005). Modeling temperature dependent kinetic isotope effects for hydrogen transfer in a series of soybean lipoxygenase mutants: The effect of anharmonicity upon transfer distance, *Chem. Phys.*, 319, pp. 283–296.
17. (a) Peters, B. (2016). Reaction coordinates and mechanistic hypothesis tests, *Annu. Rev. Phys. Chem.*, 67, pp. 669–690; (b) García Martínez, A., de la Moya Cerero, S., Osío Barcina, J., Moreno Jiménez, F. and Lora Maroto, B. (2013). The mechanism of hydrolysis of aryldiazonium ions revisited: Marcus theory vs. canonical variational transition state theory, *Eur. J. Org. Chem.*, 2013, pp. 6098–6107.
18. (a) Hrovat, D. A. and Borden, W. T. (1992). CASSCF calculations find that a D8h geometry in the transition state for double-bond shifting in cyclooctatetraene, *J. Am. Chem. Soc.*, 114, pp. 5879–5881; (b) Mauksch, M., Schleyer and P. v. R. (2001). Effective monkey saddle points and berry and level mechanisms in the topomerization of SF4 and related tetracoordinated AX4 species, *Inorg. Chem.*, 40, pp. 1756–1769.
19. Kazemi, M. and Åqvist, J. (2015). Chemical reaction mechanisms in solution from brute force computational Arrhenius plots, *Nat. Commun.*, 6, Article No. 7293.
20. Jencks, W. P. (1975). Binding energy, specificity, and enzyme catalysis: The Circe effect, *Adv. Enzymol. Relat. Areas Mol. Biol.*, 43, pp. 219–410.

21. (a) Warshel, A. (1991). *Computer Modeling of Chemical Reactions in Enzymes and Solutions* (John Wiley & Sons, Inc., New York, NY, USA); (b) Åqvist, J. and Warshel, A. (1993). Simulation of enzyme reactions using valence bond force fields and other hybrid quantum/classical approaches, *Chem. Rev.*, 93, pp. 2523–2544.
22. Plata, R. E. and Singleton, D. A. (2015). A case study of the mechanism of alcohol-mediated Morita Baylis–Hillman reactions. The importance of experimental observations, *J. Am. Chem. Soc.*, 137, pp. 3811–3826.
23. (a) Morita, K.-I., Suzuki, Z. and Hirose, H. (1968). A tertiary phosphine-catalyzed reaction of acrylic compounds with aldehydes, *Bull. Chem. Soc. Jpn.*, 41, p. 2815; (b) Baylis, A. B. and Hillman, M. E. D. Verfahren zur Herstellung von Acrylverbindungen. DE2155113A1, 1972.
24. Skyner, R. E., McDonagh, J. L., Groom, C. R., van Mourik, T. and Mitchell, J. B. (2015). A review of methods for the calculation of solution free energies and the modelling of systems in solution, *Phys. Chem. Chem. Phys.*, 17, pp. 6174–6191.
25. Borges dos Santos, R. M., Costa Cabral, B. J. and Martinho Simões, J. A. (2007). Bond-dissociation enthalpies in the gas phase and in organic solvents: Making ends meet, *Pure Appl. Chem.*, 79, pp. 1369–1382.
26. (a) Cramer, C. J. and Truhlar, D. G. (1996). Solvent effects and chemical reactivity. In *Continuum Solvation Models*, eds. Tapia, O. and Bertrán, J., Volume 17, (Kluwer, Dordrecht) p. 1; (b) Cossi, M., Rega, N., Scalmani, G. and Barone, V. (2002). Energies, structures, and electronic properties of molecules in solution with the C-PCM solvation model, *J. Comp. Chem.*, 24, pp. 669–681; (c) Truong, T. N. (1998). Solvent effects on structure and reaction mechanism: A theoretical study of [2+2] polar cycloaddition between ketene and imine, *J. Phys. Chem. B*, 102, pp. 7877–7881.
27. Tomasi, J. and Persico, M. (1994). Molecular interactions in solution: An overview of methods based on continuous distributions of the solvent, *Chem. Rev.*, 94, pp. 2027–2094.
28. (a) Kostal, J. and Jorgensen, W. L. (2010). Thorpe-Ingold acceleration of oxirane formation is mostly a solvent effect, *J. Am. Chem. Soc.*, 132, pp. 8766–8773; (b) Nguyen, M. T., Raspoet and G., Vanquickenborne, L. G. (1997). A new look at the classical Beckmann rearrangement: A strong case of active solvent effect, *J. Am. Chem. Soc.*, 119, pp. 2552–2562.
29. (a) Tomasi, J., Mennucci, B. and Cammi, R. (2005). Quantum mechanical continuum solvation models, *Chem. Rev.*, 105, pp. 2999–3094; (b) Takano, Y. and Houk, K. N. (2005). Benchmarking the conductor-like polarizable continuum model (CPCM) for aqueous solvation free energies of neutral and ionic organic molecules, *J. Chem. Theory Comput.*, 1, pp. 70–77; (c) Cramer, C. J. (2004). *Essentials of Computational Chemistry: Theories and Models*, 2nd edn. (Wiley, West Sussex, UK).
30. (a) Floris, F. and Tomasi, J. (1989). Evaluation of the dispersion contribution to the solvation energy. A simple computational model in the continuum approximation, *J. Comp. Chem.*, 10, pp. 616–627; (b) Floris, F. M.,

Tomasi, J. and Pascual Ahuir, J. L. (1991). Dispersion and repulsion contributions to the solvation energy: Refinements to a simple computational model in the continuum approximation, *J. Comp. Chem.*, 12, pp. 784–791; (c) Marenich, A. V., Cramer, C. J. and Truhlar, D. G. (2009). Universal solvation model based on solute electron density and on a continuum model of the solvent defined by the bulk dielectric constant and atomic surface tensions, *J. Phys. Chem. B*, 113, pp. 6378–6396.

31. Mennucci, B. (2012). Polarizable continuum model, *WIREs Comput. Mol. Sci.*, 2, pp. 386–404.
32. (a) Cammi, R., Mennucci, B. and Tomasi, J. (2003). Computational chemistry: Reviews of current trends. In *Computational Modeling of the Solvent Effects on Molecular Properties: An Overview of the Polarizable Continuum Model (PCM) Approach*, ed. Leszczynski, J., Volume 8, Chapter 1 (World Scientific Publishing Co. Pte. Ltd., Singapore); (b) Takano, Y. and Houk, K. N. (2005). Benchmarking the conductor-like polarizable continuum model (CPCM) for aqueous solvation free energies of neutral and ionic organic molecules, *J. Chem. Theory Comput.*, 1, pp. 70–77.
33. Bachrach, S. M. (2014). Challenges in computational organic chemistry, *WIREs Comput. Mol. Sci.*, 4, pp. 482–487.
34. Dub, P. A., Henson, N. J., Martin, R. L. and Gordon, J. C. (2014). Unravelling the mechanism of the asymmetric hydrogenation of acetophenone by [RuX2(diphosphine)(1,2-diamine)] catalysts, *J. Am. Chem. Soc.*, 136, pp. 3505–3521.
35. (a) Bhattacharyya, G. and Bodner, G. M. (2005). It gets me to the product: How students propose organic mechanisms, *J. Chem. Educ.*, 82, pp. 1402–1407; (b) Friesen, J. B. (2008). Saying what you mean: Teaching mechanisms in organic chemistry, *J. Chem. Educ.*, 85, pp. 1515–1518; (c) Grove, N. P., Cooper, M. M. and Cox, E. L. (2012). Does mechanistic thinking improve student success in organic chemistry? *J. Chem. Educ.*, 89, pp. 850–853; (d) Meek, S. J., Pitman, C. L. and Miller, A. J. M. (2016). Deducing reaction mechanism: A guide for students, researchers, and instructors, *J. Chem. Educ.*, 93, pp. 275–286.
36. (a) Jencks, W. P. (1981). How does a reaction choose its mechanism? *Chem. Soc. Rev.*, 10, pp. 345–375; (b) Beno, B. R., Houk, K. N. and Singleton, D. A. (1996). Synchronous or asynchronous? An "experimental" transition state from a direct comparison of experimental and theoretical kinetic isotope effects for a Diels-Alder reaction, *J. Am. Chem. Soc.*, 118, pp. 9984–9985; (c) Dewar, M. J. S. (1984). Multibond reactions cannot normally be synchronous, *J. Am. Chem. Soc.*, 106, pp. 209–219.
37. Jimenez-Oses, G., Liu, P., Matute, R. A. and Houk, K. N. (2014). Competition between concerted and stepwise dynamics in the triplet di-pi-methane rearrangement, *Angew. Chem. Int. Ed.*, 53, pp. 8664–8667.
38. Hong, Y. J., Ponec, R. and Tantillo, D. J. (2012). Changes in charge distribution, molecular volume, accessible surface area and electronic structure

along the reaction coordinate for a carbocationic triple shift rearrangement of relevance to diterpene biosynthesis, *J. Phys. Chem. A*, 116, pp. 8902–8909.
39. (a) Hess, B. A. (2004). Formation of the B ring in steroids and hopanoids from squalene, *Eur. J. Org. Chem.*, 2004, pp. 2239–2242; (b) Tantillo, D. J. (2008). Recent excursions to the borderlands between the realms of concerted and stepwise: carbocation cascades in natural products biosynthesis, *J. Phys. Org. Chem.*, 21, pp. 561–570; (c) Tantillo, D. J. (2010). The carbocation continuum in terpene biosynthesis–Where are the secondary cations? *Chem. Soc. Rev.*, 39, pp. 2847–2854.
40. Wheeler, S. E. (2009). A hierarchy of homodesmotic reactions for thermochemistry, *J. Am. Chem. Soc.*, 131, pp. 2547–2560.
41. Hoffmann, R., Schleyer, P. and Schaefer, H. F., 3rd. (2008). Predicting molecules–more realism, please!, *Angew. Chem. Int. Ed.*, 47, pp. 7164–7167.
42. Carpenter, B. K. (1984). Chapter 1. In *Determination of Organic Reaction Mechanisms* (Wiley, New York).
43. (a) Buskirk, A. and Baradaran, H. (2009). Can reaction mechanisms be proven? *J. Chem. Educ.*, 86, pp. 551–554; (b) Brown, T. L. (2009). A discussion of "Can reaction mechanisms be proven?" *J. Chem. Educ.*, 86, p. 552; (c) Lewis, D. E. (2009). A discussion of "can reaction mechanisms be proven?" *J. Chem. Educ.*, 86, p. 554; (d) Yoon, T. (2009). A discussion of "can reaction mechanisms be proven?" *J. Chem. Educ.*, 86, p. 556; (e) Wade, P. A. (2009). A discussion of "can reaction mechanisms be proven?" *J. Chem. Educ.*, 86, p. 558.
44. Asimov, I. (1950). *I, Robot* (Fawcett Publications, Greenwich, CT, USA).
45. Hoffmann, R. (2003). Why buy that theory? *Am. Sci.*, 91, pp. 9–11.
46. Hoffmann, R., Minkin, V. I. and Carpenter, B. K. (1997). Ockham's razor and chemistry, *HYLE Int. J. Phil. Chem.*, 3, pp. 3–28.

Chapter 2

Overview of Computational Methods for Organic Chemists

Edyta M. Greer and Kitae Kwon

Department of Natural Sciences, Baruch College of CUNY, 17 Lexington Avenue, New York, NY 10010, USA

2.1 Introduction

This chapter provides an overview of computational chemistry methods of potential interest to organic chemists. Methods we will discuss include *ab initio*, density functional theory (DFT), and semi-empirical calculations, as well as the ONIOM approach that defines two or three regions of a molecule to be treated with different levels of theory. Various basis sets will be discussed, including minimal basis sets, Pople basis sets, and Dunning's correlation-consistent basis sets. Our chapter is brief by design, but we recommend Refs. [1–8] be consulted for more detail.

2.2 *Ab initio* methods

Ab initio methods are used to predict molecular interactions and properties, with molecules ranging in size from small organics to large biologics and extended materials. The term *ab initio* is Latin for "from the beginning," implying that the calculation is performed using fundamental laws of nature without assumptions or empirical data. *Ab initio* methods can be used to calculate electronic energies, electron densities, and a number of thermodynamic properties.

Ab initio calculations can be computationally costly, but various methods have been developed to address this issue.

2.2.1 *Hartree–Fock (HF) theory*

The exact wave functions of the hydrogen atom and H_2^+ cation can be determined, but for practical purposes, determination of wave functions of larger systems requires approximations. Hartree[9] introduced a method that provided an approximation for solving the Schrödinger equation, called the Hartree method. Fock[10] and Slater[11] had used a mathematical approach of introducing the antisymmetric wave functions to fulfill the Pauli exclusion principle to the Hartree method, which gave rise to the Hartree–Fock method (HF).

HF calculations are carried out using the self-consistent field (SCF) approach to approximate solutions to the Schrödinger equation. A problem that comes with using the SCF method is the consistent calculation of closed- and open-shell species. To calculate closed-shell systems, restricted Hartree–Fock (RHF) theory, with all spins being paired, can be used. Later, unrestricted Hartree–Fock (UHF) method[12] was developed to treat open-shell systems.

Roothaan[13] and Hall[14] developed a way for computers to perform calculations based on the HF method. Boys[15] introduced contracted Gaussian-type basis functions, or contracted Gaussian-type orbitals, which Roothaan also employed, that reduced the computation time dramatically. Contracted Gaussian-type basis functions, commonly referred to as basis sets (*vide infra*), are sets of one-electron functions associated with atoms that are used to build delocalized molecular orbitals.

Since electrons are (electrostatically) repulsive toward each other, this repulsion should contribute to the electronic energy of a system, but this contribution is not treated appropriately in HF theory. The difference between the electronic energy from HF and that of the "true" electronic energy is called correlation energy, a term coined by Löwdin.[16] The correlation energy is also known as electronic correlation. Despite this issue with HF theory, HF was successfully applied to reproduce and rationalize experimentally observed high

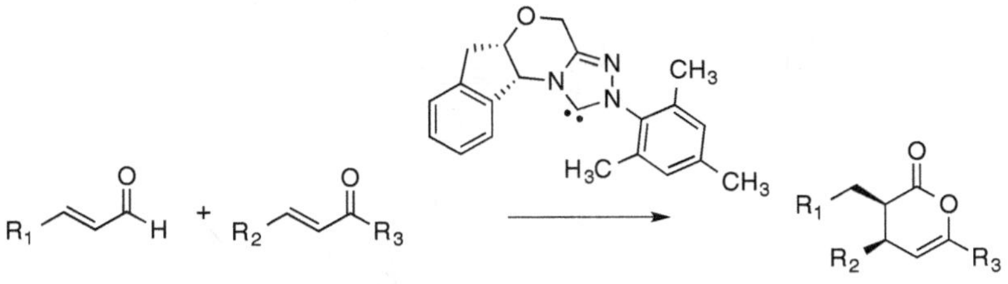

Figure 2.1. Reaction of acetaldehyde enolate with acrolein in presence of a *N*-heterocyclic carbene catalyst.[17]

enantio- and diastereo-selectivities in the reaction of acetaldehyde enolate with acrolein in presence of a *N*-heterocyclic carbene catalyst (Fig. 2.1).[17] The HF method has also been used to calculate kinetic isotope effects to examine mechanisms in Cope rearrangement.[18] Although studies using HF are now rare, such studies played an important role in the development of applied theoretical organic chemistry.

2.2.2 *Post-HF methods*

2.2.2.1 *Møller–Plesset methods*

Møller and Plesset[19] (MP) developed a perturbation theory by modifying the HF method to include electron correlation. As the order of MP theory increases, electron correlation is increasingly accounted for. Thus, a series of MP methods have been developed, although the most widely used is MP2.[20] Higher-order MP methods such as MP4[21] are also used but can be less practical due to their computational expense. Through the years (i.e., ∼1990–2010) MP methods have been used for predicting reactions involving "weak" interactions, such as hydrogen bonds[22–26] and deducing relative stabilities of isomers.[27,28] For example, MP2 has been used extensively in the past to study organic reactions, such as Diels–Alder (DA) reactions of isoxazoles (Fig. 2.2),[29] and heteroene reactions (Fig. 2.3),[30] although MP2 methods are used less frequently these days due to the rise of DFT (*vide infra*).

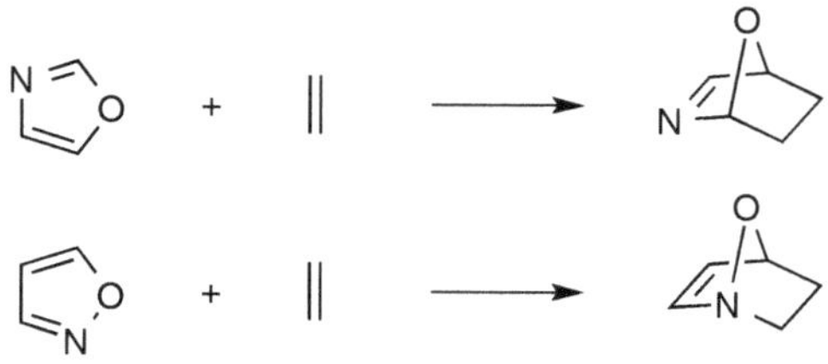

Figure 2.2. DA reaction of isoxazoles.[29]

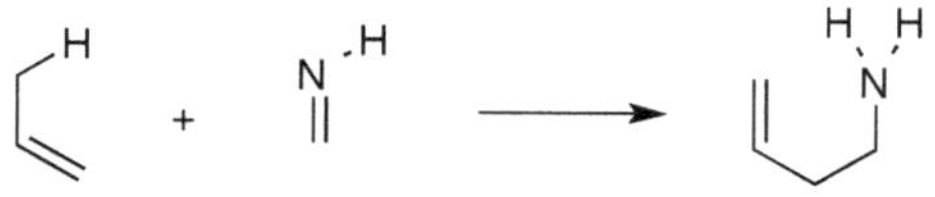

Figure 2.3. Ene reactions.[30]

2.2.2.2 *MCSCF and CASSCF*

Multi-configurational self-consistent field (MCSCF) theory allows orbitals to have nonintegral occupation numbers, which allows it to perform well for bond-forming and -breaking reactions. MCSCF theory utilizes multi-configurational wave functions constructed from *inactive* orbitals and *active* orbitals. Various electronic configurations are used for the active orbitals, i.e., using all possible configurations in these orbitals is referred to as carrying out a complete active space self-consistent field (CASSCF) calculation.[31] CASSCF frequently is used for studies dealing with organic reactions involving radicals and biradicals, and for excited states.

CASSCF was used extensively in the study of pericyclic reactions. For example, Bernardi *et al.*[32] computed concerted and stepwise pathways for simple DA reactions (Fig. 2.4(a)) and found that the former was energetically favored. Li and Houk[33] also used CASSCF to investigate concerted vs. stepwise paths of D-A reactions between 1,3-butadiene and ethylene (Fig. 2.4(a)), and the D-A dimerization of 1,3-butadiene (Fig. 2.4(b)). Hrovat and Borden[34] used CASSCF to study the mechanisms of the cleavage and ring inversion of bicyclo[2.2.0]hexane and the natures of the structures

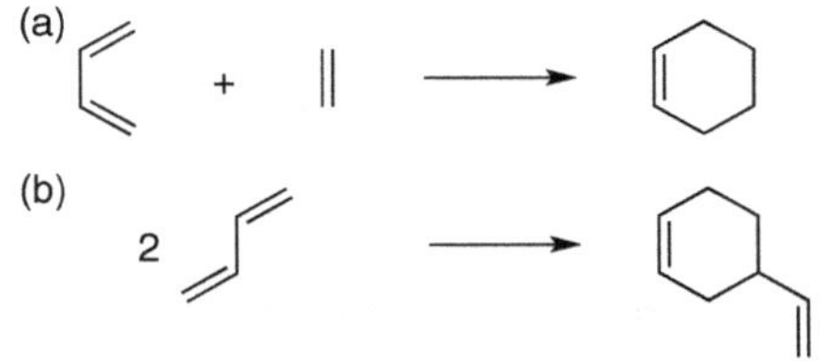

Figure 2.4. Simple DA reactions.[32,33]

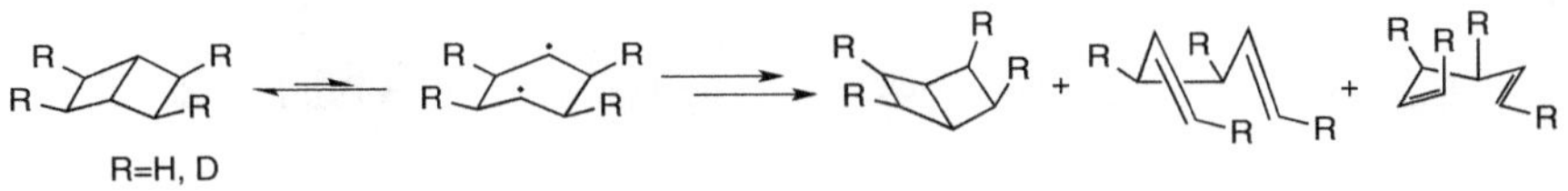

Figure 2.5. Ring inversion of bicyclo[2.2.0]hexane.[34]

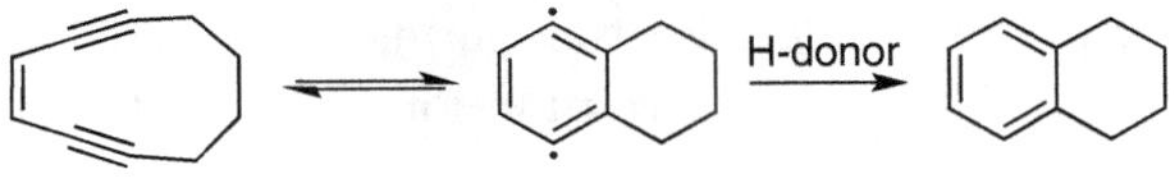

Figure 2.6. The Bergman cyclization of 10-membered ring enediyne.[35]

involved (Fig. 2.5). Greer *et al.*[35] used CASSCF to study vibrationally activated tunneling in the Bergman cyclization of ten-membered ring enediynes (Fig. 2.6). These are but a few examples where CASSCF calculations were useful in the study of rearrangement reactions where biradical character of reacting species was possible.

As mentioned earlier, CASSCF is a useful method for studying excited molecules. For example, Bearpark *et al.*[36] applied CASSCF to probe details of the decay of singlet-excited fulvene to the ground state. Also, Doubleday used CASSCF calculations for geometry optimizations and reaction paths of excited-state trimethylene (Fig. 2.7)[37] and tetramethylene[38] biradicals. Fantacci *et al.*[39] also computed the isomerization of an excited retinal chromophore model (Fig. 2.8). Methods such as CASSCF are still frequently used to treat excited state species.

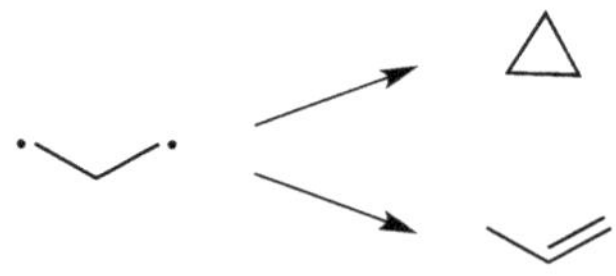

Figure 2.7. Studies of the lifetime of trimethylene.[37]

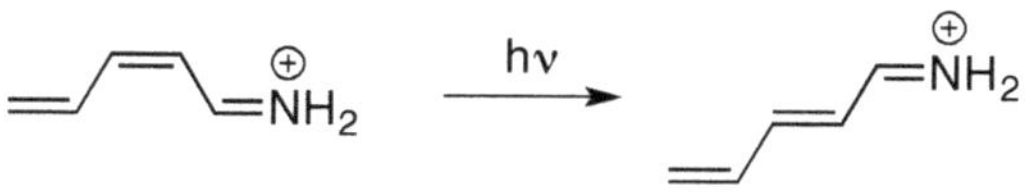

Figure 2.8. Isomerization of an excited short-chain retinal model.[39]

2.3 Density functional theory

"Electron density" with different numbers of electrons per unit of volume in different parts of a space can be described as "electron gas".[40] Hohenberg and Kohn[41] derived that an electron density determines the external potential and a wave function, forming the basis of what is known as DFT.

2.3.1 *Ground-state DFT*

Hohenberg and Kohn treated the many-body Schrödinger equation in terms of electron density, which simplified the solution of the equation. Kohn and Sham[42] utilized the theorem of Hohenberg and Kohn to show that the total energy can be divided into energies such as kinetic, potential, and Coulomb energies, and the exchange-correlation energy (calculation of which is difficult). The fundamental mantra of DFT calculations is that the energy of a molecule is a function of the electron density. Therefore, there exists an electron density where the molecule is at its ground state, or the global minimum of the energy function. Many efforts have been made to find exchange-correlation functionals that can provide comparable results to those of costly higher-order *ab initio* calculations, since DFT methods enjoy the advantages of being fast and often comparable in accuracy to post-HF calculations. There are many types of DFT functionals.[43] Commonly used functionals for organic chemistry are the B3LYP, ωB97X-D, PBE0, and M06 family of functionals, which will be described in the following subsections.[44]

2.3.1.1 *B3LYP*

Becke[45] introduced a hybrid approach combining HF and DFT. While some problems were encountered in calculating the total energies of molecules, ionization potentials, and proton affinities, the popular hybrid functional called B3LYP ultimately arose.[46,47] B3LYP combines Becke's treatment of exact-exchange energies, and Lee, Yang, and Parr's method of obtaining correlation energy. B3LYP has gained popularity amongst organic chemists because of its relatively low computational cost compared to *ab initio* methods while still providing accurate results.

2.3.1.2 *M06 family of functionals*

The M06 family of functionals (also called the Minnesota functionals) was developed by Zhao and Truhlar.[48] These functionals consist of either local or nonlocal functionals in which calculations depend on spin densities, spin density gradients, as well as spin kinetic energy densities.[48] There are four types of M06 functionals: M06-L,[49] M06-HF,[50] M06,[51] and M06-2X.[51]

M06-L was the first and only M06 functional with no HF exchange. M06-L was shown to be useful due to its ability to provide reasonable energetics for barrier heights of reactions involving transition metals, metal–metal, and metal–ligand bonds, and hydrocarbon-bond–dissociation energies. Moreover, this functional performed well for geometry prediction and vibrational frequencies. M06-L is able to perform calculations on moderate- to large-sized molecules in a relatively short time.

Figure 2.9 shows an example where M06-L was used in a study of a gas-phase synthesis of a gold–carbene complex, where results were compared against those from B3LYP.[52] Experimentally, the dissociation of triphenylphosphine leading to the gold–carbene complex required an energy of $E_0 = 51.7 \pm 3.3$ kcal/mol. The B3LYP calculated binding energy was 30.5 kcal/mol. When M06-L was used, the

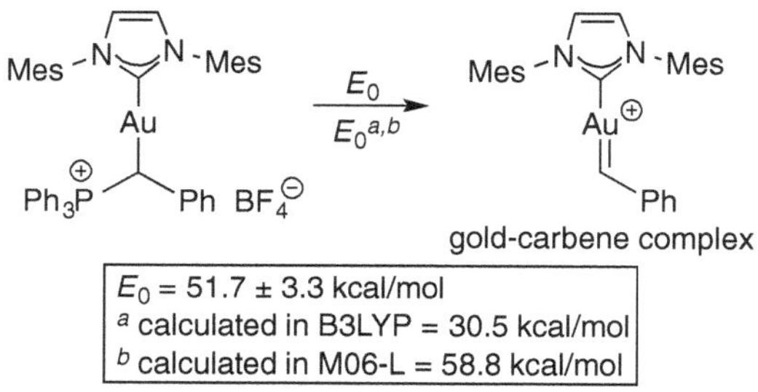

Figure 2.9. Synthesis of gold–carbene complex.[52]

obtained binding energy was 58.8 kcal/mol, which is in reasonable agreement with the value obtained from the experimental procedure. It was concluded that the discrepancy was due to M06-L performing better than B3LYP in terms of addressing noncovalent interactions, namely dispersion forces, between dissociated phenyl fragments (see Chapter 11).

The second M06-class functional is M06-HF. The "HF" indicates that the functional contains full HF exchange. What that does is eliminate long-range, self-exchange interaction error, which is important for calculating Rydberg and charge-transfer states in time-dependent DFT (TD-DFT; *vide infra*) calculations. In calculations for excitation energies, M06-HF did not perform in calculating the energetics of valence transitions compared to other functionals. However, M06-HF performed well for Rydberg and charge-transfer excitations.

Subsequently, Truhlar and Zhao published M06 and M06-2X. The "2X" in M06-2X is in reference to the functional's percentage of HF exchange (54%), which is two times more than the percentage of HF exchange of M06 (27%).[53] The two functionals are parameterized differently from each other, which makes each suitable for different purposes. Although both are parametrized for nonmetals, M06 is also parametrized for transition metals. For that reason, M06, and M06-L mentioned above, excelled in calculations that involved transition metals, such as bond energies and excitation energies. M06 is also particularly useful for treating noncovalent interactions, but not for charge-transfer interactions. Of the M06 family, M06-2X was the best for calculating thermochemistry, energy barriers, and noncovalent interactions involving nonmetal atoms.

Much like M06-L, M06 is used frequently in organometallic studies. For example, the M06 functional was employed in mechanistic studies of the $(4+2)$ and $(4+3)$ cycloadditions of allene-dienes catalyzed by a Au(I) complex that has sterically hindered ligand.[54] The M06 approach aided the authors in predicting the mechanisms of $(4+2)$ and $(4+3)$ cycloadditions.

The M06-2X functional is popular since it performs well for a wide range of organic structures and reactions.[55–58] The functional can also handle large organic systems — such as graphene and fullerenes.[59] For example, the M06-2X functional was employed in a study that involved elongating the C–C bond through steric effects and increasing its strength via noncovalent (dispersion) interactions.[60] The M06-2X method reproduced the X-ray–derived length of the central bond in diamantane–diamantine and gave a bond dissociation energy (BDE) of 71 kcal/mol, which agreed with the high stability of the compound observed experimentally. This stability was ascribed to attractive interactions arising from intramolecular H–H interactions between the bulky groups. In contrast, B3LYP calculations without a dispersion correction yielded a BDE with an error of around 30 kcal/mol.

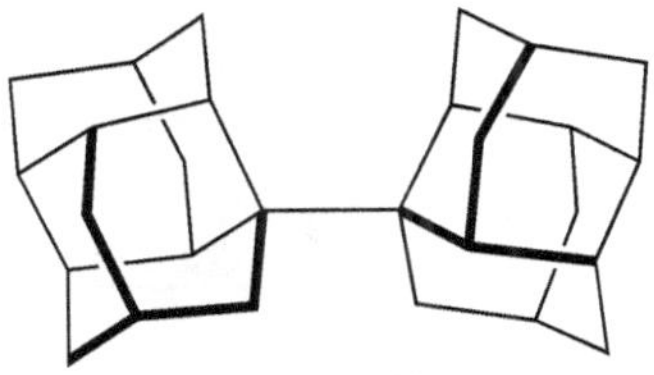

diamantane–diamantane

2.3.1.3 *ωB97X-D*

ωB97X-D is a functional developed by Chai and Head-Gordon.[61] This functional is a revision of two predecessors developed by Chai and Head-Gordon, ωB97 and ωB97X.[62] All three functionals are long-range corrected (LC), except ωB97X, which contains short-range exchange. There were some problems with ωB97 and ωB97X, one of which was the inability to treat dispersion interactions. The ωB97X-D functional adds an empirical dispersion correction to the ωB97X functional so as to include long-range dispersion interactions without much computational cost. ωB97X-D demonstrated improvement in calculating the energetics of noncovalent interactions, atomization energies, and reaction energies, as well as predicting molecular geometries, among other areas of improvement. The authors recommended that this functional is ideal for calculations "... where noncovalent interactions are expected to be significant".[61]

A representative study in which ωB97X-D was used effectively concerned the DA cycloaddition, catalyzed by squaramide-based aminocatalysts, of anthracene and nitrostyrene (Fig. 2.10).[63] The ωB97X-D functional was employed for single-point energy

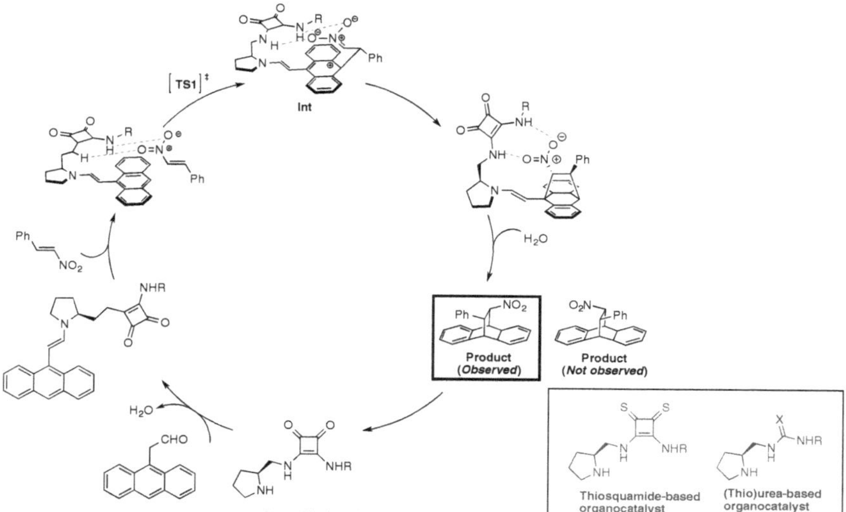

Figure 2.10. The DA cycloaddition catalyzed by squaramide-based aminocatalysts.[63]

calculations on species optimized with B97-D. Among the three organocatalysts considered, a (thio)squaramide catalyst was predicted to decrease the energy barrier of the DA cycloaddition the most because it had the strongest hydrogen-bonding interactions with nitrostyrene in the transition state. The authors considered the influence of solvent (dichloromethane) on the DA cycloaddition by using polarizable continuum model (PCM; see Chapter 4).[64,65]

2.3.1.4 *PBE0*

PBE0 is a hybrid functional developed by Adamo and Barone,[66] which is based on the Perdew–Ernzerhof–Burke (PBE) functional.[67,68] In the literature, the PBE0 functional is also referred to sometimes as PBE1PBE.[69] Unlike the functionals we have seen thus far, PBE0 contains no parameters (aside from those in the Local Spin Density treatment). Among the benchmark tests in the original paper, an area that PBE0 excelled in was the calculation of excitation energies *via* time-dependent (TD) approach (*vide infra*). PBE0 was able to reproduce experimental values of excitation energies quite well. There is a large body of literature on studies that have used PBE0 for calculations that require the use of the TD approach,[69–73] but PBE0 has been used for studying energetics of ground state organic reactions as well.[74,75]

2.3.2 *Time-dependent DFT*

Problems can arise when using DFT for computing excited systems (see Chapter 19) and TD processes. Thus, Runge and Gross[76] modified DFT and introduced time-dependent density functional theory (TD-DFT). Since the introduction of TD-DFT, chemists have found it useful in predicting absorption spectra and optical properties of organic compounds. The applications of TD-DFT include predicting the absorption wavelength of organic compounds (see Fig. 2.11 for representative cases)[77,78] investigations of π–π^* transitions in dyes and sensitizers,[79,80] and other aspects of photoactivated compounds.[81,82]

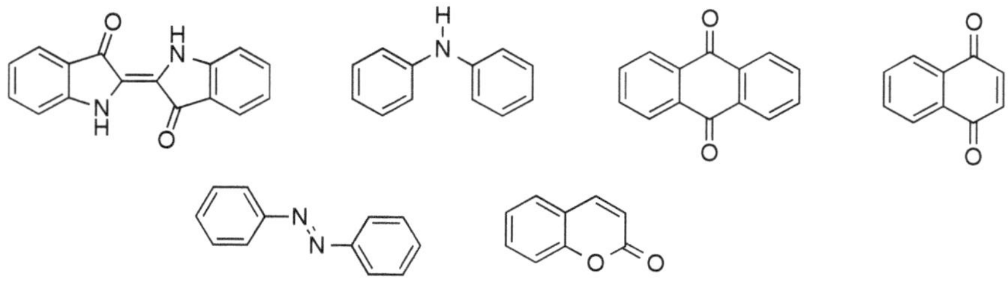

Figure 2.11. Representative compounds studied with TD-DFT.[77,78]

2.4 Semi-empirical methods

Semi-empirical methods are rooted in Molecular Orbital (MO) theory and use small basis sets to treat valence electrons. The absence of calculated integrals, such as two-electron integrals, is compensated with empirical parameters derived from experimental reference data. Therefore, semi-empirical calculations can be inaccurate when experimental reference data are absent or incomplete. However, semi-empirical methods do not involve many rigorous aspects of *ab initio* calculations and, as a result, they are much faster. For that reason, semi-empirical methods have found use in the study of large molecules (e.g., biological systems). The errors from semi-empirical calculations are smaller for most organic compounds than for compounds that involve heavy-atoms — an example that illustrates the general rule that semi-empirical calculations should only be used when one trusts that the parameterization associated with a particular method is appropriate for the chemical system to be investigated.

The original semi-empirical methods include the Hückel molecular orbital method (HMO) and the Pariser–Parr–Pople method (PPP).[83,84] They were developed solely for calculating conjugated hydrocarbons and to treat only π-electrons. These methods will not be discussed further. Instead, this section will cover semi-empirical methods that are designed to treat all valence electrons.

2.4.1 *Extended HMO method*

Hoffmann[85] found that the HMO method could be improved to treat systems that are neither planar nor conjugated and developed a modified version of HMO called extended HMO theory (EHT). EHT was effective in calculating charge distributions, finding equilibrium conformations, and predicting the geometry of molecules. The method was reasonably successful in finding the total energy of a system as a function of internuclear distance, although the calculations fell short for some diatomic and triatomic species. Hoffmann noted that the method was not successful in determining energies from steric interactions, for which the energies were overestimated, leading to wrong isomerization energies for some species. The most notable application of EHT in organic chemistry was to support the Woodward–Hoffmann rules for predicting of stereochemistry of electrocyclic reactions.[86]

2.4.2 *Neglect of differential overlap methods*

In the early days of semi-empirical calculations, Pople *et al.* developed methods that were called neglect of differential overlap methods, including complete neglect of differential overlap (CNDO)[87,88] and CNDO/2,[89] intermediate neglect of differential overlap (INDO),[90] and neglect of diatomic differential overlaps (NDDO).[88] Although these methods are no longer used much in computational chemistry, they provided the basis for the development of other semi-empirical methods discussed below.

2.4.2.1 *MNDO*

Introduced by Dewar and Thiel,[91] the modified neglect of diatomic overlap (MNDO) method was developed in order to address some of the effects caused by lone pair repulsions. MNDO also provided an improvement in calculating unsaturated molecular geometries. However, MNDO did not perform well in calculations on carbocations

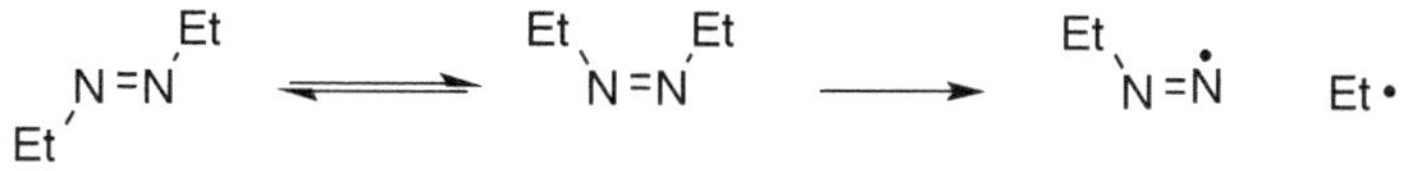

Figure 2.12. Application of MNDO to the mechanistic study of thermal decompositions of azoalkanes and 1,1-diazenes.[92]

and four-membered ring compounds. MNDO was not appropriate for molecules containing *d*- and *f*-orbitals as the method only uses *s*- and *p*-orbitals. Nonetheless, MNDO has been used successfully. For example, MNDO was used for a mechanistic study of thermal decomposition of azoalkanes and 1,1-diazenes (Fig. 2.12).[92] Better relative energies were found compared to other semi-empirical and some *ab initio* methods. For a time, MNDO was a method of choice for calculations involving azo compounds.

2.4.2.2 *AM1*

Dewar *et al.*[93] introduced a version of MNDO called Austin model 1 (AM1). AM1 was introduced and parameterized for the elements hydrogen, carbon, oxygen, and nitrogen. The AM1 method involved modifying the core repulsion function in MNDO by introducing additional attractive and repulsive Gaussian terms. Both types of Gaussians were present for parameterizing hydrogen, carbon, and nitrogen, but only repulsive Gaussians were used for oxygen. AM1 showed improvement in calculations of heats of formation compared to MNDO. For hydrocarbons, most of the results were satisfactory, although the heats of formation of cyclopentane and cyclohexane were too low.[93] Problems were also noted for diatomic molecules, where the heat of formation of singlet oxygen was found to be too low and that for carbon dioxide was found to be too high. It was noted that rotational barriers around saturated single bonds were underestimated, and activation energies were sometimes overestimated. However, in terms of optimizing geometries, AM1 gave good results.[93]

Nonetheless, AM1 is still sometimes used for studying organic molecules. An example of a study that demonstrated the superiority of AM1 over MNDO is the DA reaction between β-angelica lactone and cyclopentadiene.[94] Compared to MNDO, AM1 yielded a more reasonable activation energy value and the correct preference for an *endo* orientation in the transition state (Fig. 2.13).

A study of ozonolysis of ethylene (Fig. 2.14) and 2-butene was also carried out using AM1.[95] AM1 was found to outperform other semi-empirical methods including MNDO, which yielded an overestimated activation energy, presumably due to the high repulsion between heteroatoms.

Other studies that were performed using AM1 include the investigations on electron transfer reactions,[96] computation of structural and electronic properties of porphyrins,[97] and mechanistic studies of dihydrogen transfers (for representative cases, see Fig. 2.15).[98]

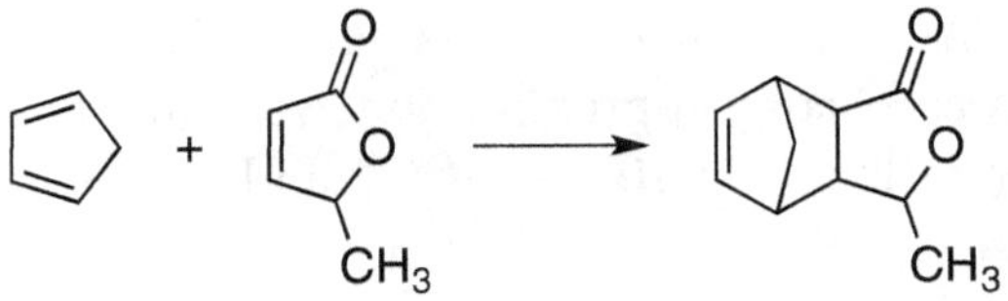

Figure 2.13. The AM1 study of DA cyclization of β-angelica lactone and cyclopentadiene.[94]

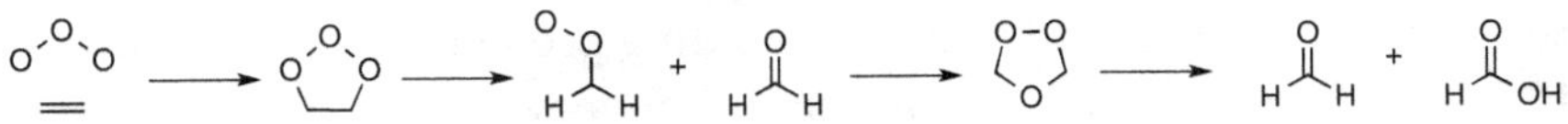

Figure 2.14. Ozonolysis of ethylene.[95]

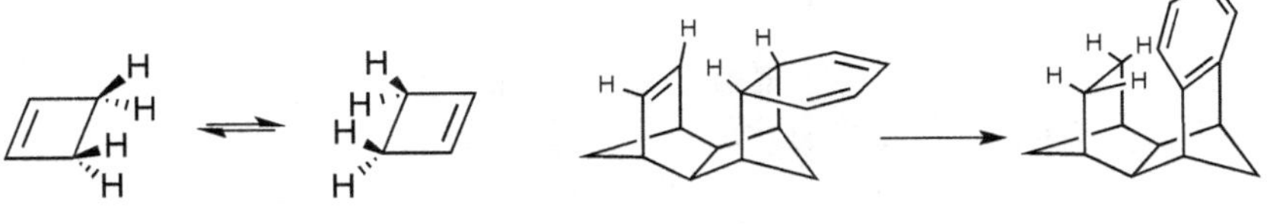

Figure 2.15. Dihydrogen transfer reactions.[98]

2.4.2.3 *PM3, PM6, PM7*

Stewart[99,100] introduced the parametric method 3 (PM3) that bears similarity to AM1. PM3 is better for calculating heats of formation of normal molecules but gave lower-than-experimental heats of formation for hypervalent molecules. Compared to AM1 and MNDO, PM3 provided overall good results for organic molecules, including those containing nitro and polynitro groups, and was able to accurately predict the orientation of the water dimer, with a near linear O–H...O hydrogen bond. In terms of geometry prediction, although bond length estimates were improved, some bond angles were inaccurate.

PM3 has been utilized in various studies on organic molecules. In particular, organic chemists have chosen PM3 for investigations of properties of nitrogenous heterocycles,[101–109] such as hydrogen bonding in DNA bases (Fig. 2.16).[101]

PM6[110] is the next-generation version of PM3, in which modifications to the core–core approximation were made. In PM6, there are parameterizations for 70 elements, and *d*-orbitals were added to the main-group elements for hypervalent species. Compared to AM1 and PM3, improvement was shown for all main group elements, including conventional organic elements. PM6 allows one to introduce additional diatomic parameters. Therefore, more types of interactions can be added without changing the fundamental features of the method. Moreover, hydrogen-bonding predictions were improved. PM6 itself cannot properly predict all noncovalent interactions, such as those between halogens, and dispersions, but for simple organic

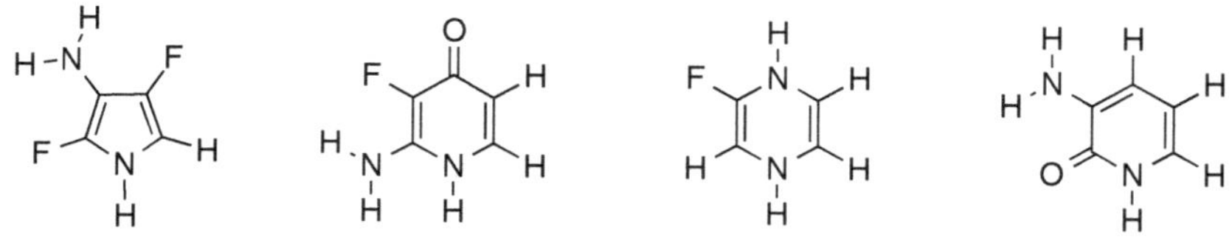

Figure 2.16. Representative nitrogen heterocycles used in predicting self-associated hydrogen-bonded pairs with PM3.[101]

compounds, PM6 has shown decreased average unsigned error (AUE) for their heats of formation.

The next generation of PM6 is PM7.[111] In PM7, the parameterization involved experimental data and data obtained from high-level *ab initio* calculations. Modifications were made to approximations in order to improve the treatment of noncovalent interactions and resolve the minor errors in the NDDO formalism. Compared to PM6, PM7 showed improvements in predicting geometries and heats of formation, with 10% lower errors for simple organic compounds. PM7 corrected the treatment of noncovalent interactions between halogens, although significant error was observed for Cl–Cl interactions. Both PM6 and PM7 were able to reproduce gas-phase hydrocarbon structures pretty well, and decreased error for solid-phase hydrocarbons was observed for PM7. PM7 was slightly inaccurate for elements and compounds for which there is scarce reference data, as expected for a semi-empirical method.

2.4.2.4 *RM1*

Stewart *et al.*[112] reported a method called Recife model 1 (RM1), which was parametrized for H, C, N, O, P, S, F, Cl, Br, and I. Available experimental data from 1,736 compounds for properties such as enthalpies of formation, dipole moments, ionization potentials, and various geometric parameters were used to build the RM1 method. Use of RM1 led to low errors for enthalpies of formation, dipole moment, ionization potential, and bond lengths for hydrogen-, carbon-, nitrogen-, and oxygen-containing molecules. For species with phosphorous and sulfur atoms, the errors in enthalpies of formation, dipole moments, ionization potentials, and bond lengths were also reduced compared to PM3, as well as the prediction of geometries. Calculations involving halogens also showed improvement, where reductions of errors were observed with RM1 compared to AM1 and PM3 in calculating the enthalpies of formation, dipole moments, ionization potentials, and bond lengths.

Because RM1 was parameterized only for 10 elements, it has been used less than AM1 and PM3, but some chemists have

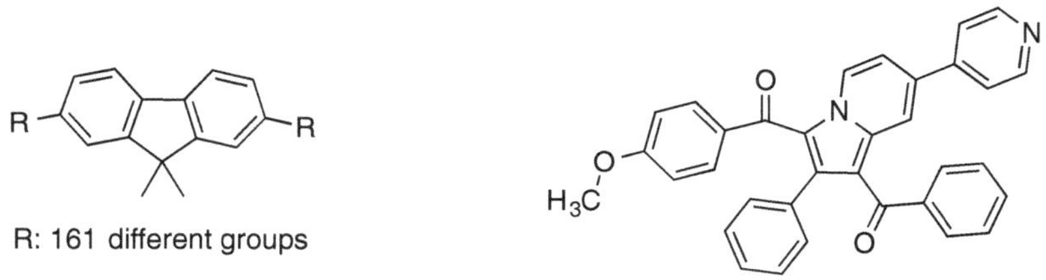

(a) (b)

Figure 2.17. (a) Representative collection of small molecules[113]; (b) representative large molecule[115] investigated with RM1.

favored it. RM1 is not very computationally demanding, thus it was used in studies of two-photon absorption of fluorene derivatives (Fig. 2.17(a)).[113] RM1 was also used to examine the Boyland–Sims oxidation of arylamines with peroxydisulfate[114] and to study the molecular structure and electronic properties of a derivative of pyridylindolizine (Fig. 2.17(b)),[115] among other molecules.[116,117]

2.5 The ONIOM approach

When dealing with large molecules, a dilemma that is faced is whether to use low-level theory at the expense of accuracy, or to use high-level theory that may exhaust computational resources. One solution to this problem was developed by Morokuma *et al.* using an approach called ONIOM, namely "Our own *N*-layered Integrated molecular Orbital and molecular Mechanics".[118] ONIOM comes in two major types, ONIOM2 or ONIOM3, where the number refers to the number of layers (i.e., theories) used to treat a system. ONIOM3 involves three layers, which involves applying three levels of theory, from high to low, to different portions of a molecular system. The nomenclature of the ONIOM3 method is as follows — High: Medium: Low. An example of such nomenclature is, ONIOM(CCSD(T):B3LYP:HF). The way ONIOM3-based calculations run is that the molecule is divided into three types of systems: "real systems," "intermediate model," and "small model." The real system refers to the whole molecule, which is approximated with the low level of theory. The intermediate model is the part of the molecule that can affect the area of interest where the heart and soul of the reaction takes place. This model is treated with the medium level of theory. The part of the molecule that is directly involved in bond making/breaking (and other important interactions) is the small model, which is treated with the high level of theory. ONION2 calculations are analogous, but use only two layers, high and low. A review[119] was published on the topic of ONIOM, and the use of ONIOM in organic chemistry was divided into three major categories: (1) locating the geometry with the lowest energy, (2) investigation of physical and chemical properties of a system, and (3) organic reactions.

A representative example: ONIOM(B3LYP/6-311+G(d,p):AM1) was employed for a geometric investigation of moxonidine species (Fig. 2.18).[120] Compounds were divided into two systems: a real system of moxonidine I (A), and a model system of 2-amino-2-imidazoline, for the amine tautomers (bottom of Fig. 2.18). For the imine tautomer, the real system was moxonidine II (C), and the model system was 2-imino-2-imidazolidine. The study revealed that neutral moxonidine (C) was the most stable of the three isomers.

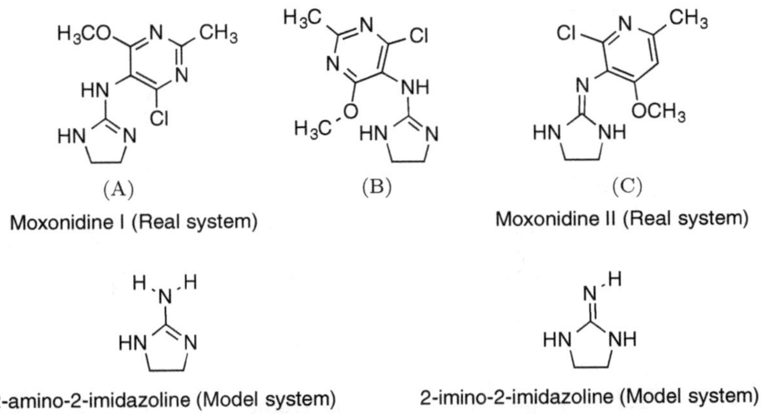

Figure 2.18. Moxonidine species investigated with ONIOM(B3LYP/6-311+G (d,p):AM1).[120]

2.6 Basis sets

The choice of basis set is an important factor in carrying out computations.[121] The choice of basis set depends on the nature of the molecules and reactions that are being investigated. A list of basis sets is available,[122] and a thorough review on the topic has been presented by Davidson and Feller.[123] Here, we review the major types of basis sets employed, including minimal, Pople-type, and Dunning's correlation-consistent basis sets.

2.6.1 *Minimal basis sets*

The STO-3G basis set requires minimal computational resources.[124,125] "STO" refers to Slater-type orbitals and "3G" refers to three Gaussian-type orbitals (GTOs) that are used. STO-3G is small and therefore leads to fast calculations, but its speed comes at the price of relatively low accuracy. Despite the inverse relationship between speed and accuracy, STO-3G has been found useful for geometry optimization strategies. Namely, STO-3G can be used to pre-optimize a molecular geometry and find a stable wave function, before optimizing with a higher, more reliable, level of theory (e.g., Fig. 2.19).[126,127] Thus, STO-3G has also been used in ONIOM calculations for studying the aspects of reactions and properties distal to a reaction site.[127–130]

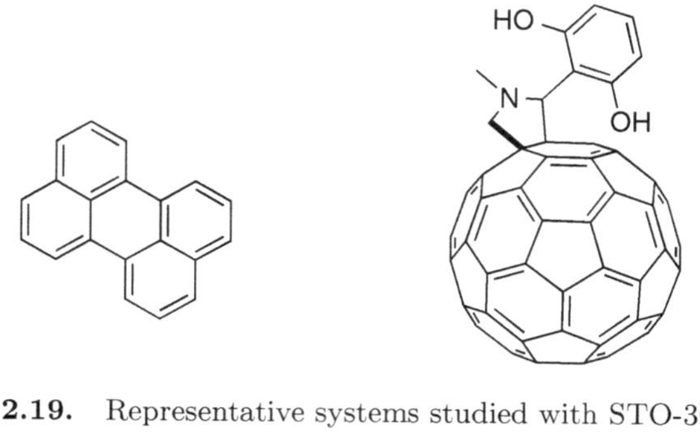

Figure 2.19. Representative systems studied with STO-3G.[126,127]

2.6.2 *Pople basis sets*

Pople introduced a type of basis set which is made up of core orbitals and valence basis functions that are split into those using contracted Gaussian-type orbitals (CGTO) and uncontracted Gaussian-type orbitals (GTO). Such basis sets are often called split-valence basis sets. The first split-valence basis sets published by Pople *et al.*[131] was 4-31G, where "4" refers to four CGTOs used for core orbitals, and "31" refers to three CGTOs and one GTO for valence basis functions. Pople has published 3-21G,[132] 5-31G, and 6-31G.[133] Also used in computational chemistry is the triple-split basis set,[134] such as 6-311G, which adds more GTO to the valence orbitals.

Polarization functions can be used with basis sets and these are denoted by *, also indicated by (d), or **, also indicated by (d,p), labels. Polarization means the introduction of higher-order orbitals to hydrogen and nonhydrogen atoms (excluding *f*-orbital containing transition metals) that will yield more accurate results.[135] Let us breakdown the common polarization basis set, 6-31G* or 6-31G(d). The * or (d) indicates that first-row atoms (Li to Ne) have *d*-type Gaussian orbitals. Another common polarization basis set is 6-31G** or 6-31G(d,p). The ** or (d,p) indicates that hydrogen atoms have *p*-type Gaussian orbitals as well as *d*-type Gaussian orbitals for nonhydrogen atoms. In general, larger basis sets lead to more accurate results but are more computationally demanding, i.e., calculations with larger basis sets require longer times and the time required is not linearly related to basis set size — it is much worse.

When computing anions, cations, radicals, and diradicals, diffuse functions are often added to the basis set,[136] denoted by + or ++. For diffuse (and polarized) basis sets such as 6-311++G(d), the ++ means that two d and one f diffuse functions are implemented on all atoms excluding hydrogen atoms. Moreover, the hydrogen atom has two p and one d orbital functions. Diffuse functions also yield more accurate results, but at the cost of computational time. Some researchers believe that diffuse functions should always be used with DFT calculations.

2.6.2.1 *3-21G basis set*

Much like STO-3G, 3-21G can be used as a starting basis set for obtaining "ballpark" molecular geometry predictions, which are then refined at higher levels of theory (e.g., Fig. 2.20).[137]

Furthermore, the 3-21G basis set has been shown to give reasonable molecular structures, including transition state structures of various hetero-DA cyclizations[138] and the Bergman cyclization leading to 2,5-didehydropyridine biradical (Fig. 2.21),[139] but it is problematic for calculating reasonable activation energies.[138,139] With the advent of more sophisticated computers and increased computational capacity, using 3-21G as a starting point has become less common.

2.6.2.2 *6-31G and related basis sets*

There is a large body of organic chemistry literature that has used the 6-31G(d) basis set. As an example, the Bergman cyclization and

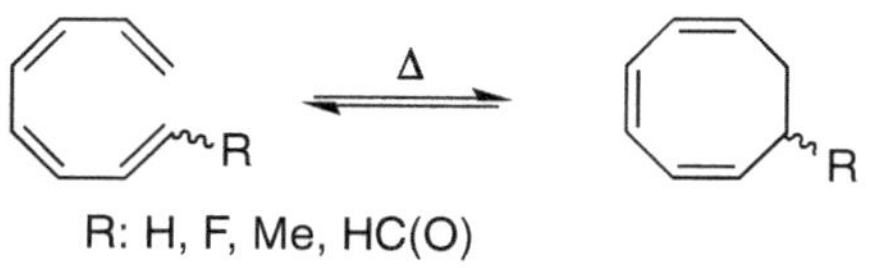

Figure 2.20. Electrocyclizations of 1-substituted 1,3,5,7 octatetraenes.[137]

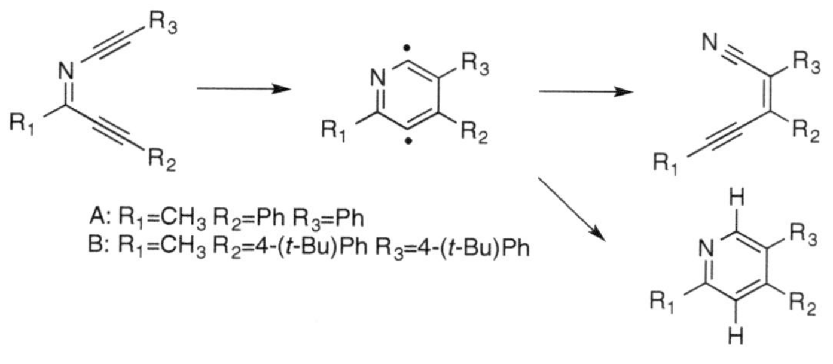

Figure 2.21. The Bergman cyclization leading to 2,5-didehydropyridine biradical.[139]

other representative pericyclic reactions treated with such basis sets are shown in Fig. 2.22.[140–142]

The 6-31G(d,p) basis set has been used to compute highly π-conjugated systems, including cation–π interactions between Na^+, Ca^{2+}, and benzene (and other arenes), or between amino acid residues, some with biological importance (examples are shown in Fig. 2.23).[142–145]

The 6-31G(d,p) basis set has also been used for studying the properties and mechanisms of diradical-forming cyclization (e.g., Figs. 2.24[147] and 2.25[149])[46,146–149] and hydrogen shifts (e.g., Figs. 2.25 and 2.26).[149–153]

Diffuse basis sets are relevant for studying cationic and anionic molecules, as well as free radicals and diradicals. Recently, the 6-31 + G(d,p) basis set was employed in a series of papers on the

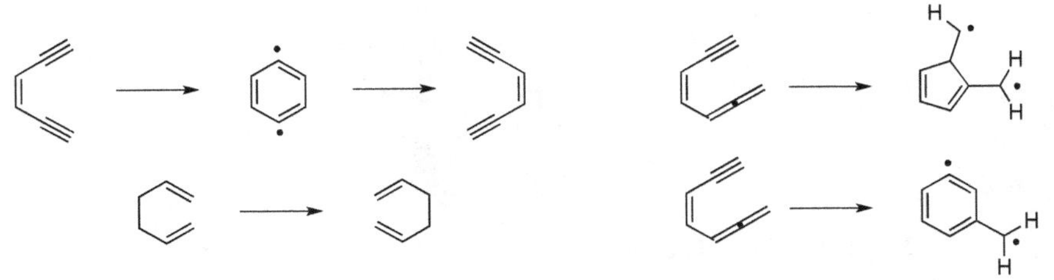

Figure 2.22. Representative reactions studied with 6-31G(d) basis set.[140–142]

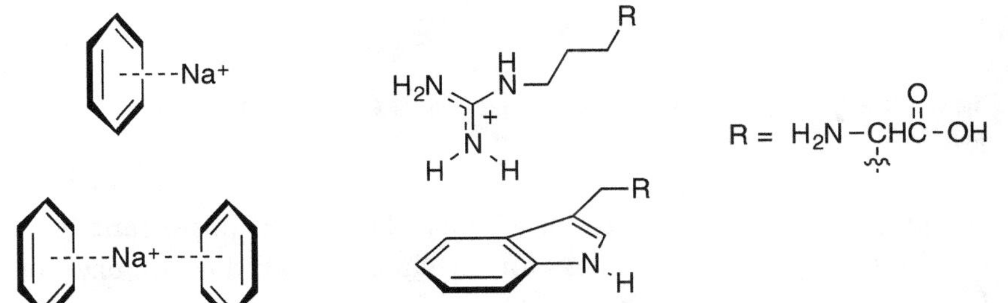

Figure 2.23. Representative cases of cation–π interaction studied with the 6-31G(d,p) basis set.[142–145]

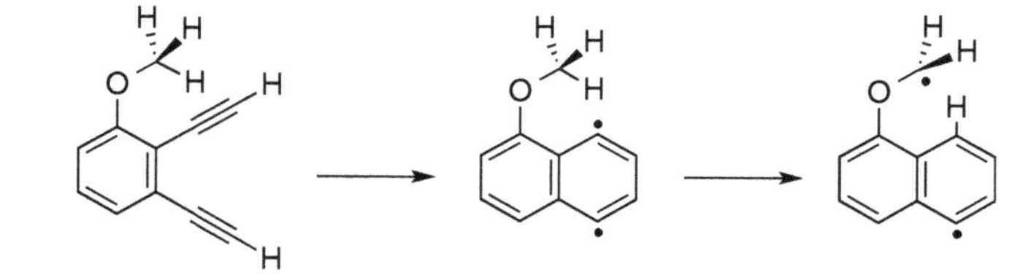

Figure 2.24. Study of the Bergman cyclization followed by intramolecular hydrogen abstraction intercepting *p*-benzyne.[147]

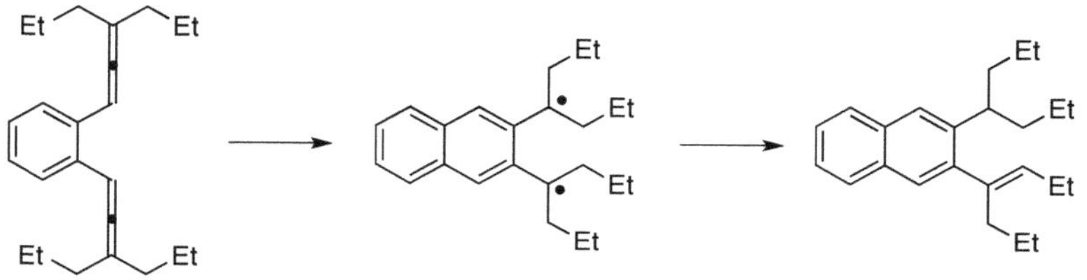

Figure 2.25. The Garratt–Braverman/[1,5]-H shift studied with 6-31G(d,p).[149]

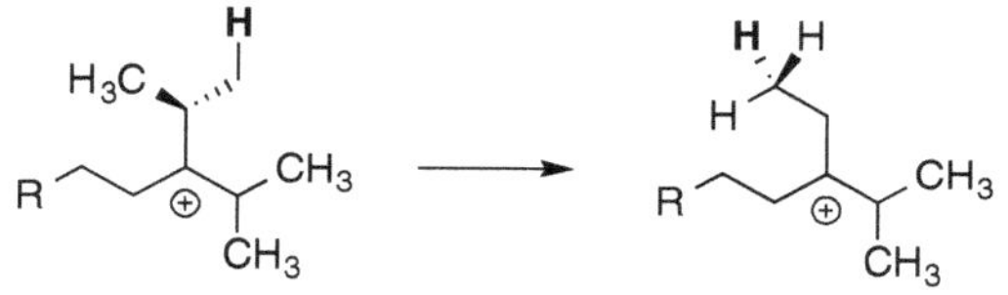

Figure 2.26. Carbocationic triple shift rearrangement.[154]

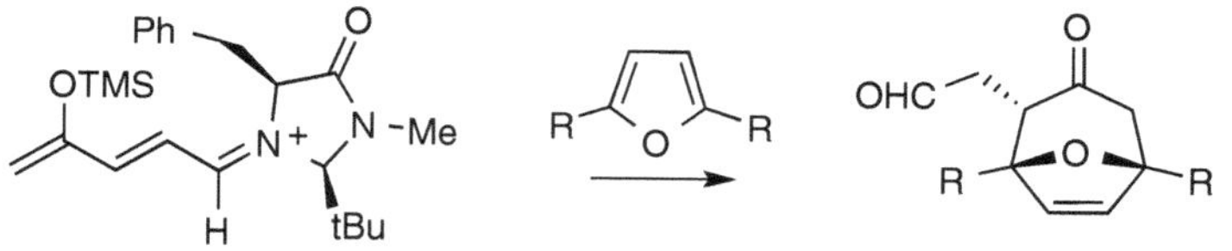

Figure 2.27. Imidazolidinone-catalyzed enantioselective (4 + 3)-cycloaddition reaction.[158]

rearrangement of carbocations, which holds an important position in bio-organic chemistry for mechanistic studies of enzymes (Fig. 2.26).[154–157]

The 6-311+G(d,p) and related basis sets have also been used to study other cations (Fig. 2.27)[158] and cation radicals (Fig. 2.28).[159]

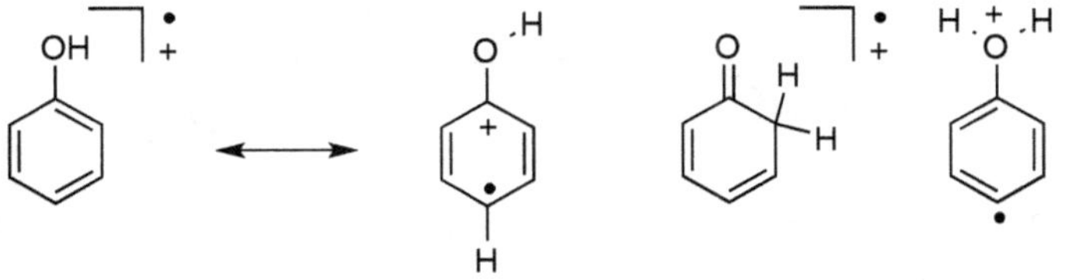

Figure 2.28. Ionized phenol and its isomers.[159]

One recent study[160] suggests that diffuse functions can provide reasonable results for the energies of systems as well as their correlation energies.

2.6.2.3 *Correlation-consistent basis sets*

Correlation-consistent basis sets were introduced by Dunning.[161] These are basis sets used with post-HF (or DFT) methods. Correlation-consistent basis sets have polarization (correlational) functions, for the purpose of correlating valence electrons. The nomenclature of correlation-consistent basis sets is as follows: cc-pVXZ, where "cc" stands for correlation-consistent, "pV" indicates polarized valence shells, "XZ" can be "double zeta (DZ)," "triple zeta (TZ)," "quadruple zeta (QZ)," and so on up to "sextuple zeta." To add diffusion functions to these basis sets, one needs to include "aug-" before the basis set of choice. The "aug-" indicates that the basis sets have been augmented "*with additional functions optimized for atomic anions*".[162] There are also correlation-consistent basis sets that include core and core-valence correlations with the nomenclature of (aug-)cc-PCVXZ.[163] A recent application of correlation-consistent basis sets was the study of the Stone–Wales rearrangement in polycyclic hydrocarbons (Fig. 2.29).[164,165]

For post-HF calculations, the higher the level of the basis set, the more accurate the results should be. The correlation consistent basis sets are meant to converge to the complete basis set limit (where increasing the size of the basis set further does not change the results). This is so because the more complete the basis set, more exact the total energy is.

Figure 2.29. An example of the Stone–Wales rearrangement calculated by applying the cc-pVDZ basis set.[164,165]

2.7 Practical considerations

Clearly one has many choices for methods — semi-empirical, HF, post-HF, DFT — and basis sets — minimal, Pople, Dunning, etc. The theoretical recipe to use for a given system depends on the system in question, the accuracy required, and the time available for a project. Sometimes others have validated particular methods for systems closely related to the one of interest, but if not, testing of multiple methods to look for convergence in results is strongly recommended.

Acknowledgments

Acknowledgment is made to the Donors of the American Chemical Society Petroleum Research Fund. We also thank the City University of New York PSC-CUNY Research Award Program, and the Eugene Lang Foundation for financial support.

References

1. Szablo, A. and Ostlund, N. S. (1996). *Modern Quantum Chemistry: Introduction to Advanced Electronic Structure Theory*, Revised Ed. (Dover Publications, USA).
2. Young, D. (2001). *Computational Chemistry: A Practical Guide for Applying Techniques to Real World Problems*, 1st Ed. (John Wiley & Sons, USA).
3. Levine, I. N. (2013). *Quantum Chemistry*, 7th Ed. (Prentice Hall, USA).
4. Koch, W. and Holthausen, M. C. (2001). *A Chemist's Guide to Density Functional Theory*, 2nd Ed. (Wiley-VCH Verlag GmbH, DEU).
5. Harget, A. J. and Murrell, J. N. (1971). *Semi-empirical Self-consistent-field Molecular Orbital Theory of Molecules*, 1st Ed. (John Wiley & Sons, USA).
6. Cramer, C. J. (2013). *Essentials of Computational Chemistry: Theories and Models*, 2nd Ed. (John Wiley & Sons, USA).

7. Lewars, E. G. (2011). *Computational Chemistry: Introduction to the Theory and Applications of Molecular and Quantum Mechanics*, 2nd Ed. (Springer, The Netherlands).
8. Jensen, F. (1999). *Introduction to Computational Chemistry*, 1st Ed. (John Wiley & Sons, USA).
9. Hartree, D. R. (1928). The wave mechanics of an atom with a non-Coulomb central field. Part I. Theory and methods, *Math. Proc. Cambridge Philos. Soc.*, 24, pp. 89–110.
10. Fock, V. (1930). Näherungsmethode zur Lösung des quantenmechanischen Mehrkörperproblems, *Z. Phys.*, 61, pp. 126–148.
11. Slater, J. C. (1930). Note on Hartree's method, *Phys. Rev.*, 35, pp. 210–211.
12. Pratt, G. W. (1956). Unrestricted Hartree-Fock method, *Phys. Rev.*, 102, pp. 1303–1307.
13. Roothaan, C. C. J. (1951). New developments in molecular orbital theory, *Rev. Mod. Phys.*, 23, pp. 69–89.
14. Hall, G. G. (1951). The molecular orbital theory of chemical valency VIII. A method of calculating ionization potentials, *Proc. R. Soc. A*, 205, pp. 541–552.
15. Boys, S. F. (1950). Electronic wave functions I. A general method of calculation for the stationary states of any molecular system, *Proc. R. Soc. A*, 200, pp. 542–554.
16. Löwdin, P.-O. (1955). Quantum theory of many-particle systems. III. Extension of the Hartree-Fock scheme to include degenerate systems and correlation effects, *Phys. Rev.*, 97, pp. 1509–1520.
17. Allen, S. E., Mahatthananchai, J., Bode, J. W. and Kozlowski, M. C. (2012). Oxyanion steering and CH$-\pi$ interactions as key elements in an n-heterocyclic carbene-catalyzed $[4+2]$ cycloaddition, *J. Am. Chem. Soc.*, 134, pp. 12098–12103.
18. Houk, K. N., Gustafson, S. M. and Black, K. A. (1992). theoretical secondary kinetic isotope effects and the interpretation of transition state geometries. 1. The Cope rearrangement,*J. Am. Chem. Soc.*, 114, pp. 8565–8572.
19. Møller, C. and Plesset, M. S. (1934). Note on an approximation treatment for many-electron systems, *Phys. Rev.*, 46, pp. 618–622.
20. Head-Gordon, M., Pople, J. A. and Frisch, M. J. (1988). MP2 energy evaluation by direct methods, *Chem. Phys. Lett.*, 153, pp. 503–506.
21. Krishnan, R. and Pople, J. A. (1978). Approximate fourth-order perturbation theory of the electron correlation energy, *Int. J. Quantum Chem.*, 14, pp. 91–100.
22. Zubatyuk, R. I., Shishkin, O. V., Gorb, L. and Leszczynski, J. (2009). Homonuclear versus heteronuclear resonance-assisted hydrogen bonds: Tautomerism, aromaticity, and intramolecular hydrogen bonding in heterocyclic systems with different exocyclic proton donor/acceptor, *J. Phys. Chem. A*, 113, pp. 2943–2952.

23. Zubatyuk, R. I., Volovenko, Y. M., Shishkin, O. V., Gorb, L. and Leszczynski, J. (2007). Aromaticity-controlled tautomerism and resonance-assisted hydrogen bonding in heterocyclic enaminone-iminoenol systems, *J. Org. Chem.*, 72, pp. 725–735.
24. Cato, J. M. A., Majumdar, D., Roszak, S. and Leszczynski, J. (2013). Exploring relative thermodynamic stabilities of formic acid and formamide dimers — role of low-frequency hydrogen-bond vibrations, *J. Chem. Theory Comput.*, 9, pp. 1016–1026.
25. Shishkin, O., Konovalova, I., Gorb, L. and Leszczynski, J. (2009). Novel type of Mixed O–H··· N/O–H··· π hydrogen bonds: Monohydrate of pyridine, *Struct. Chem.*, 20, pp. 37–41.
26. Del Bene, J. E., Alkorta, I. and Elguero, J. (2015). Exploring the $(H_2C{=}PH_2)^+$:N-base potential surfaces: Complexes stabilized by pnicogen, hydrogen, and tetrel bonds, *J. Phys. Chem. A*, 119, pp. 11701–17110.
27. Bowling, N. P. and McMahon, R. J. (2006). Enediyne isomers of tetraethynylethene, *J. Org. Chem.*, 71, pp. 5841–5847.
28. Esselman, B. J., Emmert, F. L., Wiederhold, A. J., Thompson, S. J., Slipchenko, L. V. and McMahon, R. J. (2015). Thermal isomerizations of diethynyl cyclobutadienes and implications for fullerene formation, *J. Org. Chem.*, 80, pp. 11863–11868.
29. Gonzalez, J., Taylor, E. C. and Houk, K. N. (1992). Why are isoxazoles unreactive in diels-alder reactions? an ab initio computational study, *J. Org. Chem.*, 57, pp. 3753–3755.
30. Thomas, B. E. and Houk, K. N. (1993). Tuning Exo/endo stereoselectivity in ene reaction, *J. Am. Chem. Soc.*, 115, pp. 790–792.
31. Roos, B. O., Taylor, P. R. and Siegbahn, P. E. M. (1980). A complete active space SCF method (CASSCF) using a density matrix formulated super-CI approach, *Chem. Phys.*, 48, pp. 157–173.
32. Bernardi, F., Bottoni, A., Field, M. J., Guest, M. F., Hillier, I. H., Robb, M. A. and Venturini, A. (1988). MC-SCF study of the Diels-Alder reaction between ethylene and butadiene, *J. Am. Chem. Soc.*, 110, pp. 3050–3055.
33. Li, Y. and Houk, K. N. (1993). Diels-Alder Dimerization of 1,3-butadiene: An ab initio CASSCF study of the concerted and stepwise mechanisms and butadiene-ethylene revisited, *J. Am. Chem. Soc.*, 115, pp. 7478–7485.
34. Hrovat, D. A. and Borden, W. T. (2001). CASSCF and CASPT2 calculations on the cleavage and ring inversion of bicyclo[2.2.0]hexane find that these reactions involve formation of a common twist-boat diradical intermediate, *J. Am. Chem. Soc.*, 123, pp. 4069–4072.
35. Greer, E. M., Cosgriff, C. V. and Doubleday, C. (2013). Computational evidence for heavy-atom tunneling in the Bergman cyclization of a 10-membered-ring enediyne, *J. Am. Chem. Soc.*, 135, pp. 10194–10197.

36. Bearpark, M. J., Bernardi, F., Olivucci, M., Robb, M. A. and Smith, B. R. (1996). Can fulvene S_1 decay be controlled? A casscf study with MMVB dynamics, *J. Am. Chem. Soc.*, 118, pp. 5254–5260.
37. Doubleday, C. (1996). Lifetime of trimethylene calculated by variational unimolecular rate theory, *J. Phys. Chem.*, 100, pp. 3520–3526.
38. Doubleday, C. (1996). Tetramethylene optimized by MRCI and by CASSCF with a multiply polarized basis set, *J. Phys. Chem.*, 100, pp. 15083–15086.
39. Fantacci, S., Migani, A. and Olivucci, M. (2004). CASPT2//CASSCF and TDDFT//CASSCF mapping of the excited state isomerization path of a minimal model of the retinal chromophore, *J. Phys. Chem. A*, 108, pp. 1208–1213.
40. March, N. H. (1957). The Thomas-Fermi approximation in quantum mechanics, *Adv. Phys.*, 6, pp. 1–101.
41. Hohenberg, P. and Kohn, W. (1964). Inhomogeneous electron gas, *Phys. Rev.*, 136, pp. B864–B871.
42. Kohn, W. and Sham, L. J. (1965). Self-consistent equations including exchange and correlation effects, *Phys. Rev.*, 140, A1133–A1138.
43. Perdew, J. P. and Schmidt, K. (2001). Jacob's ladder of density functional approximations for the exchange-correlation energy, *AIP Conf. Proc.*, 577, pp. 1.
44. Cohen, A., Mori-Sánchez, P. and Yang, W. (2012). Challenges for Density Functional Theory, *Chem. Rev.*, 112, pp. 289–320.
45. Becke, A. D. (1993). A new mixing of Hartree–Fock and local density-functional theories, *J. Chem. Phys.*, 98, pp. 1372–1377.
46. Becke, A. D. (1993). Density-functional thermochemistry. III. The role of exact exchange, *J. Chem. Phys.*, 98, pp. 5648–5652.
47. Lee, C., Yang, W. and Parr, R. G. (1988). Development of the Colle-Salvetti correlation-energy formula into a functional of the electron density, *Phys. Rev. B*, 37, pp. 785–789.
48. Zhao, Y. and Truhlar, D. G. (2007). Density functionals with broad applicability in chemistry, *Acc. Chem. Res.*, 41, pp. 157–167.
49. Zhao, Y. and Truhlar, D. G. (2006). A new local density functional for main-group thermochemistry, transition metal bonding, thermochemical kinetics, and noncovalent interactions, *J. Chem. Phys.*, 125, pp. 194101–194118.
50. Zhao, Y. and Truhlar, D. G. (2006). Density Functional for Spectroscopy: no long-range self-interaction error, good performance for rydberg and charge-transfer states, and better performance on average than B3LYP for ground states, *J. Phys. Chem. A*, 110, pp. 13126–13130.
51. Zhao, Y. and Truhlar, D. G. (2008). The M06 suite of density functionals for main group thermochemistry, thermochemical kinetics, noncovalent interactions, excited states, and transition elements: Two new functionals and systematic testing of four m06-class functionals and 12 other functionals, *Theor. Chem. Acc.*, 120, pp. 215–241.

52. Fedorov, A., Moret, M. E. and Chen, P. (2008). Gas-phase synthesis and reactivity of a gold carbene complex, *J. Am. Chem. Soc.*, 130, pp. 8880–8881.
53. Valero, R., Costa, R., Moreira, I. P. R., Truhlar, D. G. and Illas, F. (2008). Performance of the M06 family of exchange-correlation functionals for predicting magnetic coupling in organic and inorganic molecules, *J. Chem. Phys.*, 128, pp. 114103–114108.
54. Benitez, D., Tkatchouk, E., Gonzalez, A. Z., Goddard III, W. A. and Toste, F. D. (2009). On the impact of steric and electronic properties of ligands on gold(i)-catalyzed cycloaddition reactions, *Org. Lett.*, 11, pp. 4798–4801.
55. Um. J. M., Gutierrez, O., Schoenebeck, F., Houk, K. N. and MacMillan, D. W. C. (2010). Nature of intermediates in organo-somo catalysis of α-arylation of aldehydes, *J. Am. Chem. Soc.*, 132, pp. 6001–6005.
56. Cheong, P. H.-Y., Paton, R. S., Bronner, S. M., Im, G.-Y. J., Garg, N. K. and Houk, K. N. (2010). Indolyne and aryne distortions and nucleophilic regioselectivites, *J. Am. Chem. Soc.*, 132, pp. 1267–1269.
57. Albrecht, L., Dickmeiss, G., Cruz-Acosta, F., Rodríguez-Escrich, C., Davis, R. L. and Jørgensen, K. A. (2012). Asymmetric organocatalytic formal [2 + 2]-cycloadditions via bifunctional H-bond directing dienamine catalysis, *J. Am. Chem. Soc.*, 134, pp. 2543–2546.
58. Hoye, T. R., Baire, B., Niu, D., Willoughby, P. H. and Woods, B. P. (2012). The hexahydro-Diels-Alder reaction, *Nature*, 490, pp. 208–212.
59. Denis, P. A. (2013). Organic chemistry of graphene: The Diels-Alder reaction, *Chem. Eur. J.*, 19, pp. 15719–15725.
60. Schreiner, P. R., Chernish, L. V., Gunchenko, P. A., Tikhonchuk, E. Y., Hausmann, H., Serafin, M., Schlecht, S., Dahl, J. E. P., Carlson, R. M. K. and Fokin, A. A. (2011). Overcoming lability of extremely long alkane carbon-carbon bonds through dispersion forces, *Nature,* 477, pp. 309–312.
61. Chai J.-D. and Head-Gordon, M. (2008). Long-range corrected hybrid density functionals with damped atom-atom dispersion corrections, *Phys. Chem. Chem. Phys.*, 10, pp. 6615–6620.
62. Chai, J.-D. and Head-Gordon, M. (2008). Systematic optimization of long-range corrected hybrid density functionals, *J. Chem. Phys.*, 128, pp. 084106.
63. Lu, T. and Wheeler, S. E. (2013). Origin of the superior performance of (thio)squaramides over (thio)ureas in organocatalysis, *Chem. Eur. J.*, 19, pp. 15141–15147.
64. Miertus, S., Scrocco, E. and Tomasi, J. (1981). Electrostatic interaction of a solute with a continuum. A direct utilization of ab initio molecular potentials for the prevision of solvent effects, *Chem. Phys.,* 55, pp. 117–129.
65. Tomasi, J., Mennucci, B. and Cammi, R. (2005). Quantum mechanical continuum solvation models, *Chem. Rev.*, 105, pp. 2999–3094.
66. Adamo, C. and Barone, V. (1999). Toward reliable density functional methods without adjustable parameters: The PBE0 model, *J. Chem. Phys.*, 110, pp. 6158–6170.

67. Perdew, J. P., Burke, K. and Ernzerhof, M. (1996). Generalized gradient approximation made simple, *Phys. Rev. Lett.*, 77, pp. 3865–3868.
68. Perdew, J. P., Burke, K. and Ernzerhof, M. (1996). Errata to generalized gradient approximation made simple, *Phys. Rev. Lett.*, 78, p. 1369.
69. Pichierri, F. and Yamamoto, Y. (2007). Mechanism and chemoselectivity of the pd(II)-catalyzed allylation of aldehydes: A density functional theory study, *J. Org. Chem.*, 72, pp. 861–869.
70. Sancho-García, J. C., Adamo, C. and Pérez-Jiménez, A. J. (2016). Describing excited states of [n]cycloparaphenylenes by hybrid and double-hybrid density functionals: From isolated to weakly interacting molecules, *Theor. Chem. Acc.*, 135, pp. 1–12.
71. Jacquemin, D., Preat, J., Wathelet, V. and Perpète, E. A. (2005). Theoretical investigation of the absorption spectrum of thioindigo dyes, *THEOCHEM*, 731, pp. 67–72.
72. Jacquemin, D., Preat, J., Wathelet, V., Fontaine, M. and Perpéte, E. A. (2006). Thioindigo dyes: Highly accurate visible spectra with TD-DFT, *J. Am. Chem. Soc.*, 128, pp. 2072–2083.
73. Sun, S. and Brown, A. (2014). Simulation of the resonance Raman spectrum for uracil, *J. Phys. Chem. A*, 118, pp. 9228–9238.
74. Fayet, G., Joubert, L., Rotureau, P. and Adamo, C. (2008). Theoretical study of the decomposition reactions in substituted nitrobenzenes, *J. Phys. Chem. A*, 112, pp. 4054–4059.
75. Ashirov, R. V., Shamov, G. A., Lodochnikova, O. A., Litvynov, I. A., Appolonova, S. A. and Plemenkov, V. V. (2008). Tetramerization of 3-methyl-cyclopropene-3-carbonitrile: A novel CN-Alder-ene reaction, *J. Org. Chem.*, 73, pp. 5985–5988.
76. Runge, E. and Gross, E. K. U. (1984). Density-functional theory for time-dependent systems, *Phys. Rev. Lett.*, 52, pp. 997–1000.
77. Jacquemin, D., Perpète, E. A., Scuseria, G. E., Ciofini, I. and Adamo, C. (2008). TD-DFT performance for the visible absorption spectra of organic dyes: Conventional versus long-range hybrids, *J. Chem. Theory Comput.*, 4, pp. 123–135.
78. Guillaumont, D. and Nakamura, S. (2000). Calculation of the absorption wavelength of dyes using time-dependent density-functional theory (TD-DFT), *Dyes Pigm.*, 46, pp. 85–92.
79. Balanay, M. P. and Kim, D. H. (2008). DFT/TD-DFT molecular design of porphyrin analogues for use in dye-sensitized solar cells, *Phys. Chem. Chem. Phys.*, 10, pp. 5121–5127.
80. Park, S. S., Won, Y. S., Choi, Y. C. and Kim, J. H. (2009). Molecular design of organic dyes with double electron acceptor for dye-sensitized solar cell, *Energy Fuels*, 23, pp. 3732–3736.
81. Shen, L., Ji, H.-F. and Zhang, H.-Y. (2005). A TD-DFT Study on triplet excited-state properties of curcumin and its implications in elucidating the photosensitizing mechanisms of the pigment, *Chem. Phys. Lett.*, 409, pp. 300–303.

82. Dumont, É. and Monari, A. (2015). Interaction of palmatine with DNA: An environmentally controlled phototherapy drug, *J. Phys. Chem. B*, 119, pp. 410–410.
83. Pople, J. A. (1953). Electron interaction in unsaturated hydrocarbons, *Trans. Faraday Soc.*, 49, pp. 1375–1385.
84. Pariser, R. and Parr, R. G. (1953). A semi-empirical theory of the electronic spectra and electronic structure of complex unsaturated molecules. I, *J. Chem. Phys.*, 21, pp. 466–471.
85. Hoffmann, R. (1963). An extended Hückel theory. I. Hydrocarbons, *J. Chem. Phys.*, 39, pp. 1397–1412.
86. Woodward, R. B. and Hoffmann, R. (1965). Stereochemistry of electrocyclic reactions, *J. Am. Chem. Soc.*, 87, pp. 395–397.
87. Pople, J. A. and Segal, G. A. (1965). Approximate self-consistent molecular orbital theory. II. Calculations with complete neglect of differential overlap, *J. Chem. Phys.*, 43, pp. S136–S151.
88. Pople, J. A., Santry, D. P. and Segal, G. A. (1965). Approximate self-consistent molecular orbital theory. I. Invariant procedures, *J. Chem. Phys.*, 43, pp. S129–S135.
89. Pople, J. A. and Segal, G. A. (1966). Approximate self-consistent molecular orbital theory. III. CNDO results for AB2 and AB3 systems, *J. Chem. Phys.*, 44, pp. 3289–3296.
90. Pople, J. A., Beveridge, D. L. and Dobosh, P. A. (1967). Approximate self-consistent molecular-orbital theory. V. Intermediate neglect of differential overlap, *J. Chem. Phys.*, 47, pp. 2026–2033.
91. Dewar, M. J. S. and Thiel, W. (1977). Ground states of molecules. The MNDO method. Approximations and parameters, *J. Am. Chem. Soc.*, 99, pp. 4899–4907.
92. Dannenberg, J. J. and Rocklin, D. (1982). A theoretical study of the mechanism of the thermal decomposition of azoalkanes and 1,1-diazenes, *J. Org. Chem.*, 47, pp. 4529–4534.
93. Dewar, M. J. S., Zoebisch, E. G., Healy, E. F. and Stewart, J. J. P. (1985). AM1: A new general purpose quantum mechanical molecular model, *J. Am. Chem. Soc.*, 107, pp. 3902–3909.
94. Sodupe, M., Oliva, A., Bertran, J. and Dannenberg, J. J. (1989). An AM1 and MNDO theoretical study of the Diels-Alder reaction between β-angelica lactone and cyclopentadiene, *J. Org. Chem.*, 54, pp. 2488–2490.
95. Dewar, M. J. S., Hwang, J. C. and Kuhn, D. R. (1991). An AM1 study of the reactions with ozone with ethylene and 2-butene, *J. Am. Chem. Soc.*, 113, pp. 735–741.
96. Nelsen, S. F. and Blomgren, F. (2001). Estimation of electron transfer parameters from AM1 calculations, *J. Org. Chem.*, 66, pp. 6551–6559.
97. Reynolds, C. H. (1988). An AM1 theoretical study of the structure and electronic properties of porphyrin, *J. Org. Chem.*, 53, pp. 6061–6064.

98. Frontera, A., Suner, G. A. and Deya, P. M. (1992). Double group-transfer reactions: A theoretical (AM1) approach, *J. Org. Chem.*, 57, pp. 6731–6735.
99. Stewart, J. J. P. (1989). Optimization of parameters for semiempirical methods II. Applications, *J. Comput. Chem.*, 10, pp. 221–264.
100. Stewart, J. J. P. (1989). Optimization of parameters for semiempirical methods I. Method, *J. Comput. Chem.*, 10, pp. 209–220.
101. Buam, D. M. L. and Lyngdoh, R. H. D. (2000). Self-associative base-pairing in some nitrogen heterocycles: A PM3 SCF-MO study, *J. Mol. Struc. (THEOCHEM)*, 505, pp. 149–159.
102. Ishihara, Y., Tanaka, T., Miwatashi, S., Fujishima, A. and Goto, G. (1994). Regioselective Friedel-Crafts acylation of 2,3,4,5-tetrahydro-1*h*-2-benzazepine and related nitrogen heterocycles, *J. Chem. Soc., Perkin Trans.*, 1, pp. 2993–2999.
103. Rescifina, A., Chiacchio, U., Corsaro, A., Piperno, A. and Romeo, R. (2011). isoxazolidinyl polycyclic aromatic hydrocarbons as DNA-intercalating antitumor agents, *Eur. J. Med. Chem.*, 46, pp. 129–136.
104. Williams, C. I., Whitehead, M. A. and Jean-Claude, B. J. (1997). A semiempirical PM3 treatment of benzotetrazepinone decomposition in acid media, *J. Org. Chem.*, 62, pp. 7006–7014.
105. Neuvonen, K., Fülöp, F., Neuvonen, H., Koch, A., Kleinpeter, E. and Pihlaja, K. (2001). Substituent influences on the stability of the ring and chain tautomers in 1,3-o, n-heterocyclic systems: Characterization by ^{13}C NMR chemical shifts, PM3 charge densities, and isodesmic reactions, *J. Org. Chem.*, 66, pp. 4132–4140.
106. Bottino, P., Dunkel, P., Schlich, M., Galavotti, L., Deme, R., Regdon, G., Bényei, A., Pintye-Hódi, K., Ronsisvalle, G. and Mátyus, P. (2012). Study on the scope of *tert*-amino effect: New extensions of Type 2 reactions to bridged biaryls, *J. Phys. Org. Chem.*, 25, pp. 1033–1041.
107. Marri, E., Pannacci, D., Galiazzo, G., Mazzucato, U. and Spalletti, A. (2003). Effect of the nitrogen heteroatom on the excited state properties of 1,4-distyrylbenzene, *J. Phys. Chem. A*, 107, pp. 11231–11238.
108. Türker, L. and Gümüş, S. (2008). Positional effect of nitrogen substitution on a certain perylene chromophobe — A semiempirical treatment, *Polycyclic Aromat. Compd.*, 28, pp. 4–14.
109. Budyka, M. F. and Oshkin, I. V. (2006). Theoretic investigation of the size and charge effects in photochemistry of heteroaromatic azides, *J. Mol. Struc. (THEOCHEM)*, 759, pp. 137–144.
110. Stewart, J. J. P. (2007). Optimization of parameters for semiempirical methods V: Modification of NDDO approximations and application to 70 elements, *J. Mol. Model.*, 13, pp. 1173–1213.
111. Stewart, J. J. P. (2013). Optimization of parameters for semiempirical methods VI: More modifications to the NDDO approximations and re-optimization of parameters, *J. Mol. Model.*, 19, pp. 1–32.

112. Rocha, G. B., Freire, R. O., Simas, A. M. and Stewart, J. J. P. (2006). RM1: A reparametrization of AM1 for H, C, N, O, P, S, F, Cl, Br, and I, *J. Comput. Chem.*, 27, pp. 1101–1111.
113. Moura, G. L. C. and Simas, A. M. (2010). Two-photon absorption by fluorene derivatives: Systematic molecular design, *J. Phys. Chem. C*, 114, pp. 6106–6116.
114. Marjanović, B., Juranić, I. and Ćirić-Marjanović, G. (2011). Revised mechanism of Boyland-Sims oxidation, *J. Phys. Chem. A*, 115, pp. 3536–3550.
115. Cojocaru, C., Rotaru, A., Harabagiu, V. and Sacarescu, L. (2013). Molecular structure and electronic properties of pyridylindolizine derivative containing phenyl and phenacyl groups: Comparison between semi-empirical calculations and experimental studies, *J. Mol. Struct.*, 1034, pp. 162–172.
116. Cardoso, C. L., Castro-Gamboa, I., Bergamini, G. M., Cavalheiro, A. J., Silva, D. H. S., Lopes, M. N., Araújo, A. R., Furlan, M., Verli, H. and Bolzani, V. d. S. (2011). An unprecedented neolignan skeleton from *Chimarrhis turbinata*. *J. Nat. Prod.*, 74, pp. 487–491.
117. Brik, M. G., Kuznik, W., Gondek, E., Kityk, I. V., Uchacz, T., Szlachcic, P., Jarosz, B. and Plucinski, K. J. (2010). Optical absorption measurement and quantum-chemical simulations of optical properties of novel fluoro derivatives of pyrazoloquinoline, *Chem. Phys.*, 370, pp. 194–200.
118. Svensson, M., Humbel, S., Froese, R. D. J., Matsubara, T., Sieber, S., Morokuma, K. (1996). ONIOM: A multilayered integrated MO + MM method for geometry optimizations and single point energy predictions. A test for Diels–Alder reactions and $Pt(P(t\text{-}Bu)_3)_2 + H_2$ oxidative addition, *J. Phys. Chem.*, 100, pp. 19357–19363.
119. Chung, L. W., Sameera, W. M. C., Ramozzi, R., Page, A. J., Hatanaka, M., Petrova, G. P., Harris, T. V., Li, X., Ke, Z., Liu, F., Li, H.-B., Ding, L. and Morokuma, K. (2015). The ONIOM method and its applications, *Chem. Rev.*, 115, pp. 5678–5796.
120. Remko, M., Walsh, O. A. and Richards, W. G. (2001). Theoretical study of molecular structure, tautomerism, and geometrical isomerism of moxonidine: Two-layered ONIOM calculations, *J. Phys. Chem. A*, 105, pp. 6926–6931.
121. Jensen, F. (2013). Atomic orbital basis set, *WIREs Comput. Mol. Sci.*, 3, pp. 273–295.
122. See: https://bse.pnl.gov/bse/portal. Accessed on October 2, 2017.
123. Davidson, E. R. and Feller, D. (1986). Basis set selection for molecular calculations, *Chem. Rev.*, 86, pp. 681–696.
124. Hehre, W. J., Stewart, R. F. and Pople, J. A. (1969). Self-consistent molecular-orbital methods. I. use of Gaussian expansions of Slater-type atomic orbitals, *J. Chem. Phys.*, 51, pp. 2657–2664.
125. Hehre, W. J., Ditchfield, R., Stewart, R. F. and Pople, J. A. (1970). Self-consistent molecular orbital methods. IV. Use of Gaussian expansions of

slater-type orbitals. Extension to second-row molecules, *J. Chem. Phys.*, 52, pp. 2769–2773.
126. Reveles, J. U. and Köster, A. M. (2004). Geometry optimization in density functional methods, *J. Comput. Chem.*, 25, pp. 1109–1116.
127. Izquierdo, M., Osuna, S., Filippone, S., Martín-Domenech, A., Solà, M. and Martín, N. (2009). H-bond-assisted regioselective (*cis-1*) intramolecular nucleophilic addition of the hydroxyl group to [60]fullerene, *J. Org. Chem.*, 74, pp. 1480–1487.
128. Delgado, J. L., Osuna, S., Bouit, P.-A., Martínez-Alvarez, R., Espíldora, E., Solà, M. and Martín, N. (2009). Competitive retro-cycloaddition reaction in fullerene dimers connected through pyrrolidinopyrazolino rings, *J. Org. Chem.*, 74, pp. 8174–8180.
129. Konev, A. S., Khlebnikov, A. F. and Frauendorf, H. (2011). Bisaziridine tetracarboxylates as building blocks in the stereoselective synthesis of c_{60}-fullerene diads and dumbbell-like bis-c_{60}-fullerene triads, *J. Org. Chem.*, 76, pp. 6218–6229.
130. Izquierdo, M., Osuna, S., Filippone, S., Martín-Domenech, A., Solà, M. and Martín, N. (2009). Regioselective intramolecular nucleophilic addition of alcohols to c_{60}: One-step formation of a *cis*-1 bicyclic-fused fullerene, *J. Org. Chem.* 74, pp. 6253–6259.
131. Ditchfield, R., Hehre, W. J. and Pople, J. A. (1971). Self-consistent molecular-orbital methods. IX. An extended guassian-type basis for molecular-orbital studies of organic molecules, *J. Chem. Phys.*, 54, pp. 724–728.
132. Binkley, J. S., Pople, J. A., Hehre, W. J. (1980). Self-consistent molecular orbital methods. 21. Small split-valence basis sets for first-row elements, *J. Am. Chem. Soc.*, 102, pp. 939–947.
133. Hehre, W. J., Ditchfield, R. and Pople, J. A. (1972). Self-consistent molecular orbital methods. XII. Further extensions of Gaussian-type basis sets for use in molecular orbital studies of organic molecules, *J. Chem. Phys.*, 56, pp. 2257–2261.
134. Krishnan, R., Binkley, J. S., Seeger, R. and Pople, J. A. (1980). Self-consistent molecular orbital methods. XX. A basis set for correlated wave functions, *J. Chem. Phys.*, 72, pp. 650–654.
135. Hariharan, P. C. and Pople, J. A. (1973). The influence of polarization functions on molecular orbital hydrogenation energies, *Theor. Chim. Acta*, 28, pp. 213–222.
136. Clark, T., Chandrasekhar, J., Spitznagel, G. W. and Schleyer, P. V. R. (1983). Efficient diffuse function-augmented basis sets for anion calculations. III. The 3-21+G basis set for first-row elements, Li-F, *J. Comput. Chem.*, 4, pp. 294–301.
137. Thomas, B. E., Evanseck, J. D. and Houk, K. N. (1993). Electrocyclic reactions of 1-substituted 1,3,5,7-octatetraenes. An ab initio molecular orbital study of torquoselectivity in eight-electron electrocyclizations, *J. Am. Chem. Soc.*, 115, pp. 4165–4169.

138. McCarrick, M. A., Wu, Y. D. and Houk, K. N. (1993). Hetero-Diels-Alder reaction transition structures: Reactivity, stereoselectivity, catalysis, solvent effects, and the exo-lone-pair effect, *J. Org. Chem.*, 58, pp. 3330–3343.
139. Hoffner, J., Schottelius, M. J., Feichtinger, D. and Chen, P. (1998). Chemistry of the 2,5-didehydropyridine biradical: Computational, kinetic, and trapping studies toward drug design, *J. Am. Chem. Soc.*, 120, pp. 376–385.
140. Navarro-Vázquez, A., Prall, M. and Schreiner, P. R. (2004). Cope reaction families: To be or not to be a biradical, *Org. Lett.*, 6, pp. 2981–2984.
141. Prall, M., Wittkopp, A. and Schreiner, P. R. (2001). Can fulvenes form from enediynes? A systematic high-level computational study on parent and benzannelated enediyne and enyne–allene cyclizations, *J. Phys. Chem. A*, 105, pp. 9265–9274.
142. Kumpf, R. A. and Dougherty, D. A. (1993). A mechanism for ion selectivity in potassium channels: Computational studies of cation-π interactions, *Science*, 261, pp. 1708–1710.
143. Mecozzi, S., West, A. P. and Dougherty, D. A. (1996). Cation-π interactions in aromatic biological and medicinal interest: Electrostatic potential surfaces as a useful qualitative guide, *Proc. Natl. Acad. Sci.*, 93, pp. 10566–10571.
144. Mecozzi, S., West, A. P. and Dougherty, D. A. (1996). Cation-π interactions in simple aromatics: Electrostatics provide a predictive tool, *J. Am. Chem. Soc.*, 118, pp. 2307–2308.
145. Gallivan, J. P. and Dougherty, D. A. (1999). Cation-π interactions in structural biology, *Proc. Natl. Acad. Sci.*, 96, pp. 9459–9464.
146. Bekele, T., Christian, C. F., Lipton, M. A. and Singleton, D. A. (2005). "Concerted" transition state, stepwise mechanism. dynamic effects in C2-C6 enyne allene cyclizations, *J. Am. Chem. Soc.*, 127, pp. 9216–9223.
147. Zeidan, T. A., Manoharan, M. and Alabugin, I. V. (2006). Ortho effect in the Bergman cyclization: Interception of *p*-benzyne intermediate by intramolecular hydrogen abstraction, *J. Org. Chem.*, 71, pp. 954–961.
148. Kraka, E. and Cremer, D. (1994). CCSD(T) Investigation of the Bergman cyclization of enediyne. Relative stability of *o*-, *m*-, and *p*-didehydrobenzene, *J. Am. Chem. Soc.*, 116, pp. 4929–4936.
149. Samanta, D., Rana, A. and Schmittel, M. (2014). Nonstatistical dynamics in the thermal Garratt–Braverman/[1,5]-H shift of one ene–diallene: An experimental and computational study, *J. Org. Chem.*, 79, pp. 8435–8439.
150. Tishchenko, O., Kryachko, E. S. and Nguyen, M. T. (2002). Low energy barriers of H-atom abstraction from phenols, *J. Mol. Struct.*, 615, pp. 247–250.
151. Chou, P.-T., Yu, W.-S., Cheng, Y.-M., Pu, S.-C., Yu, Y.-C., Lin, Y.-C., Huang, C.-H and Chen, C.-T. (2004). Solvent-polarity tuning excited-state charge coupled proton-transfer reaction in *p-N,N*-ditolylaminosalicylaldehydes, *J. Phys. Chem. A*, 108, pp. 6487–6498.
152. Alabugin, I. V., Manoharan, M., Breiner, B. and Lewis, F. D. (2003). Control of kinetics and thermodynamics of [1,5]-shifts by aromaticity:

A view through the prism of Marcus theory, *J. Am. Chem. Soc.*, 125, pp. 9329–9342.
153. Gu, J., Chen, K., Jiang, H. and Leszczynski, J. (1999). The radical transformation in artemisinin: A DFT study, *J. Phys. Chem. A*, 103, pp. 9364–9369.
154. Ortega, D. E., Gutierrez-Oliva, S., Tantillo, D. J. and Toro-Labbe, A. (2015). A detailed analysis of the mechanism of a carbocationic triple shift rearrangement, *Phys. Chem. Chem. Phys.*, 17, pp. 9771–9779.
155. Hong, Y. J., Ponec, R. and Tantillo, D. J. (2012). Changes in charge distribution, Molecular volume, accessible surface area and electronic structure along the reaction coordinate for a carbocationic triple shift rearrangement of relevance to diterpene biosynthesis, *J. Phys. Chem. A*, 116, pp. 8902–8909.
156. Hong, Y. J., Giner, J.-L. and Tantillo, D. J. (2013). Triple shifts and thioether assistance in rearrangements associated with an unusual biomethylation of the sterol side chain, *J. Org. Chem.*, 78, pp. 935–941.
157. Tantillo, D. J. (2011). Biosynthesis *via* carbocations: Theoretical studies on terpene formation, *Nat. Prod. Rep.*, 28, pp. 1035–1053.
158. Krenske, E. H., Houk, K. N. and Harmata, M. (2015). Computational analysis of the stereochemical outcome in the imidazolidinone-catalyzed enantioselective (4 + 3)-cycloaddition reaction, *J. Org. Chem.*, 80, pp. 744–750.
159. Le, H. T., Flammang, R., Gerbaux, P., Bouchoux, G. and Nguyen, M. T. (2001). Ionized phenol and its isomers in the gas phase, *J. Phys. Chem. A*, 105, pp. 11582–11592.
160. Warner, P. M. (1996). *Ab initio* calculations on heteroatomic systems using density functional theory and diffuse basis functions, *J. Org. Chem.*, 61, pp. 7192–7194.
161. Dunning, T. H. (1989). Gaussian basis sets for use in correlated molecular calculations. I. The atoms boron through neon and hydrogen, *J. Chem. Phys.*, 90, pp. 1007–1023.
162. Kendall, R. A., Dunning, T. H. and Harrison, R. J. (1992). Electron affinities of the first-row atoms revisited. Systematic basis sets and wave functions, *J. Chem. Phys.*, 96, pp. 6796–6806.
163. Woon, D. E. and Dunning, T. H. (1995). Gaussian basis sets for use in correlated molecular calculations. V. Core-valence basis sets for boron through neon, *J. Chem. Phys.*, 103, pp. 4572–4585.
164. Brayfindly, E. E., Irace, E. E., Castro, C. and Karney, W. L. (2015). Stone-Wales rearrangements in polycyclic aromatic hydrocarbons: A computational study, *J. Org. Chem.*, 80, pp. 3825–3831.
165. Irace, E. E., Brayfindly, E. E., Vinnacombe, G. A., Castro, C. and Karney, W. L. (2015). Stone–Wales rearrangements in hydrocarbons: From planar to bowl-shaped substrates, *J. Org. Chem.*, 80, pp. 11718–11725.

Chapter 3

Brief History of Applied Theoretical Organic Chemistry

Steven M. Bachrach

Department of Chemistry, Monmouth University, West Long Branch, NJ 07764-1898, USA

3.1 Introduction

Applied theoretical organic chemistry began with Hückel theory,[1] which concerns the application of quantum mechanics to conjugated π-systems. This dramatically simplified approximation to the Schrödinger equation allowed for the rationalization of the stability of aromatic compounds (those containing $4n + 2\,\pi$-electrons) and the instability of the class of conjugating cyclic molecules containing $4n\,\pi$-electrons.

With this auspicious start, organic chemists, especially physical organic chemists, have regularly employed quantum mechanics to rationalize, and even formulate predictions of, organic structures, properties, and reactions. The focus of this book is on the application of quantum mechanical computations to understanding organic chemistry, and so this chapter provides a survey of some of the seminal papers where computational chemistry was the key element of the study.

Perhaps the most consequential application of quantum theory to organic chemistry is the concept or the conservation of orbital symmetry, also known as the Woodward–Hoffmann rules.[2] Woodward and Hoffmann demonstrated that the feasibility of pericyclic reactions can be rationalized on the basis of the symmetry of the occupied orbitals, distinguishing reactions that are allowed due to conservation of symmetry and those that are forbidden due to changes in orbital symmetry. Application of the Woodward–Hoffmann rules allowed for the prediction of stereochemical outcomes. Extensions of these concepts allowed for rationalizing regioselectivity and the notion of antiaromaticity and Möbius topologies in certain allowed reactions, along with explaining the stereoselection of photochemical pericyclic reactions.

Seeman[3] recently published a brilliant historical analysis of the genesis of the Woodward–Hoffmann rules. Computations of model transition states performed by Hoffmann using extended Hückel theory (EHT) were essential toward an understanding of the role of orbital symmetry in dictating the course of pericyclic reactions. Though these computations are very rudimentary and remarkably idealized as viewed today, they nonetheless provided a firm quantum mechanical grounding for the full development of the theory of orbital conservation.[4]

Over the past decades, the incredible improvements in computing technology have rendered EHT obsolete. Even the next generation of approximate quantum mechanics (QM) technique, namely the semi-empirical methods such and modified neglect of diatomic overlap (MNDO) and Austin model 1 (AM1), are sparingly used by organic chemists (except for some specialized circumstances like enzyme activity and molecular dynamics). Therefore, this chapter will highlight computational work that employed *ab initio* techniques: either wave function methods, such as Hartree–Fock (HF), perturbation theory, and couple-cluster theory, or density functional theory (DFT). This historical overview is not designed to be comprehensive but rather to provide a sense of the scope of organic chemistry problems that have been addressed with computations and how theoretical and computational developments have shaped what problems have been examined.

3.2 Historical overview

The history of computational organic chemistry will be divided into three periods. The first period addresses the establishment of computational quantum chemistry as a legitimate discipline. The second period, from the mid-1970s through the mid-1990s, firmly establishes computational chemistry as a tool for both understanding organic chemistry and making predictions. The last period, from the mid-1990s through today, witnesses the transition of computations from specialists to the general organic chemist. For each period, a number of important applications along with technical developments that allowed for new problems to be tackled will be discussed.

3.2.1 *Early 1970s: Can quantum chemical computations be trusted?*

In the late 1950s, Roothaan[5] detailed the procedure for solving for the HF wave function for molecular systems. With Boys'[6] suggestion of using Gaussian functions as the atomic orbital basis set, all of the mathematics was in place for calculating molecular properties. There were two outstanding issues. First, computers in the 1960s were just too slow, too expensive, and lacked sufficient memory (both core and disk) to tackle any real organic molecules. Clearly, technological improvements in the computing industry were already pointing toward the day in the not too distant future when computers would be powerful enough for solving quantum chemical problems.

The second problem was not so easily redressed. Longuet–Higgins dominated the theoretical landscape of the 1960s. His notion of a proper computation was one that could be done on the proverbial back-of-an-envelope.[7a] He was certainly not alone in this attitude. Computers with their hidden workings were not to be trusted, an attitude not much different from the reception of the general mathematics community to the computational solution of the four-color map problem.[8] What was needed was a real molecule with some controversial aspect, and a molecule small enough that existing computational technology could be applied. It turned out that methylene, CH_2, was great choice.[9,10]

3.2.2 *Methylene, part 1: Structure of the triplet state*

In 1970, Bender and Schaefer were looking for a molecule to employ their "magnificent device," a state-of-the-art CI program.[7b] Herzberg, the soon-to-be Nobel laureate, had determined that triplet methylene was linear.[11] Methylene fit their size requirements and had just enough geometric complexity. Their CISD/DZ computations indicated that triplet methylene was bent, with an H–C–H angle of 135.1°.[12] By the end of the year, two ESR studies,[13,14] along with Herzberg's own reassessment of his UV data,[15] confirmed the computations. Computational chemistry had won its first major victory.

3.2.3 *Methylene, part 2: The singlet–triplet energy gap*

In 1972, two separate computational studies, by Bender *et al.*[16] and Hay *et al.*[17] concluded that triplet methylene lies about 11 kcal mol^{-1} below the singlet state. This was in agreement with most experiments and seemed to settle the matter. However, Lineberger's photoelectron spectroscopy experiment on CH_2^- indicated that the gap was much larger, almost 20 kcal mol^{-1}.[18] Goddard, Schaefer, and other researchers continued to use improving computations and always found a singlet–triplet gap of about 10–11 kcal mol^{-1}.[19–21]

A year later, Harding and Goddard[22] offered an explanation: their new computations suggested that the PES spectrum CH_2^- shows three hot bands, overestimating the energy gap. The final resolution came in the early 1980s. A far–infra-red (IR) laser magnetic resonance study indicated an energy gap of 9 kcal mol^{-1}.[23] Lineberger reinvestigated the PES of CH_2^- with a flowing afterglow instrument and observed no hot bands, and a singlet–triplet gap in complete agreement with the computations.[24] Quantum chemistry had been used to correct two significant experimental errors and thus established itself has a new useful sub-discipline of chemistry!

3.3 1970s–1990s: Expanding the scope of computations

The next phase in the history of computational organic chemistry was facilitated by two technological advances. Computers, as always, became faster and cheaper, and in the early 1980s, the development of the minicomputer, most notably, the DEC Vax series, meant that individual research groups could own their own dedicated computing facility. Second, computational chemistry software was evolving, and two general-purpose programs were in constant development and improvement: the *Gaussian* suite from Pople's group and *GAMESS* from the Gordon group. In addition to a more streamlined input using keywords to select computational options, two key developments contributed greatly toward the wider usage of computations: the ability to optimize to a transition state structure and the ability to compute analytical frequencies. The latter feature allowed for characterization of critical points and prediction of IR spectra.

During this period, computational organic chemistry was pushed forward by a number of research groups led by organic chemists, in particular: Borden, Houk, Jorgenson, Kollman, Schleyer, Streitwieser, and Wiberg. Computational chemistry was the dominant, if not exclusive, domain of these groups. A key characteristic of the research by these groups was the close association of the computations to experimental results: to offer interpretation and predictions for new experiments. The significant accomplishments of these and other research groups are too many to detail here. Instead, we provide a small sample of some of the major developments.

3.3.1 *Torquoselectivity*

The Woodward–Hoffmann rules specify a conrotatory pathway for pericyclic reactions involving $4n$ electrons and a disrotatory pathway for reactions with $4n + 2$ electrons. However, there are two possible stereoproducts for many reactions. For example, the disrotatory ring opening of *trans*-3,4-dimethylcyclobutene can proceed in two allowed ways, giving two different products (Eq. (3.1)).

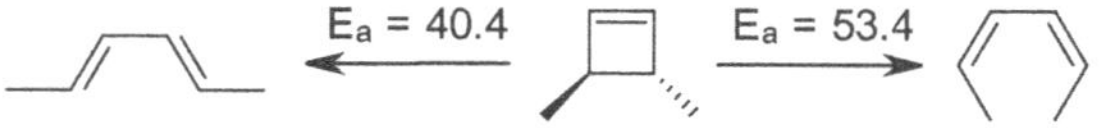

(3.1)

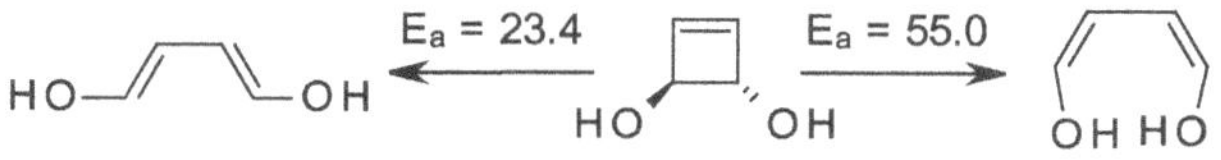

(3.2)

Houk noted that these two pathways are different and termed the preference for one path over the other *torquoselectivity*. The pathway leading to the *trans,trans* product in Eq. (3.1) is kinetically favored by 13 kcal mol^{-1} over the path leading to the *cis,cis* product.[25,26] This might be expected based on sterics; however, Houk noted that the selectivity for the diol analogue (Eq. (3.2)) is 33 kcal mol^{-1}, even though the steric interactions are likely to be similar. This led to Houk's molecular orbital (MO) model for torquoselection and the general rule which states that an electron donor group will prefer outward rotation while an acceptor group will prefer inward rotation.[27] Application of this model provides, for example, an explanation for why the ring opening of *trans*-3,4-bis(trimethylsilyl)cyclobutene (Eq. (3.3)), predominantly gives the sterically congested product.[28,29] Houk also examined torquoselectivity in electrocylic reactions involving six[30] and eight[31] π-electrons.

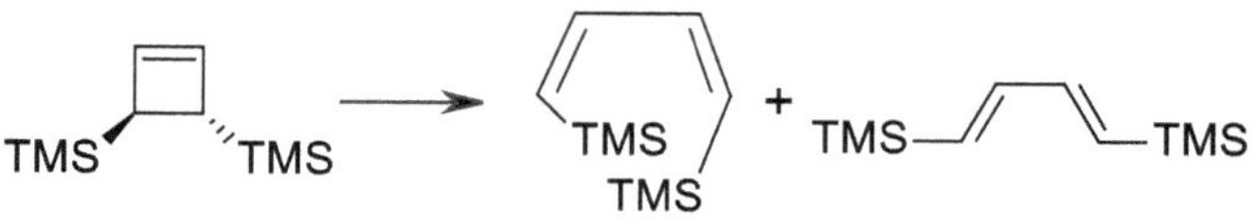

(3.3)

3.3.2 *Cope rearrangement*

Computational chemistry has been very successful in elucidating the mechanism and making stereoselectivity predictions about a wide swath of pericyclic reactions.[32] This makes the story of the mechanism of the Cope rearrangement so unusual, and so cautionary.

The Cope rearrangement can, in principle, proceed through three different pathways: (a) through a single transition state structure (i.e., a concerted reaction), (b) through initial cleavage of the C_3–C_4 bond to form two allyl radicals, or (c) through initial formation of the C_1–C_6 bond invoking the cyclohexane-1,4-diyl intermediate. Path (b) can be discounted because the energy for cleaving the bond far exceeds the activation barrier of the Cope rearrangement, and no crossover products are observed. So, the question becomes whether there is an intermediate along the Cope rearrangement pathway.

This reaction seems to be ideally suited for the complete active space self-consistent field (CASSCF) method to allow for electron configurations that describe the three possible paths to contribute to the wave function. HF, as usual, greatly overestimates the activation energy.[33] CASSCF computations located a concerted transition state and a diyl intermediate slightly lower in energy.[34–36] However, the barrier leading to the diyl and the concerted barrier are larger than 10 kcal mol^{-1}. This result was quite surprising, as it points toward the unexpected importance of dynamic correlation in this seemingly simple reaction.

The mechanism of the Cope rearrangement was only settled when multireference perturbation theory was utilized. Two different studies, one by Hrovat *et al.*[37] and the second by Dupuis *et al.*[38] revealed a single transition state structure and no diyl intermediate. Nonetheless, a multireference wavefunction is necessary to describe both a diyl and diradical component to the description of the transition state structure. The take-home message was that organic reactions can be much more complicated than anticipated, a lesson that needs to be repeatedly learned!

3.3.3 S_N2 *reaction in solution*

The potential energy surface (PES) of a typical S_N2 reaction is well known to anyone who has been through college organic chemistry: reactants move uphill over a single barrier and then down to product, as in Fig. 3.1(a). This is for a reaction in solution. The reaction in the gas phase has a different profile, with reactants first coming together to form an ion–dipole complex, which then goes through a barrier

Figure 3.1. PES for an S_N2 reaction in (a) solution-phase and (b) gas-phase.

to form an exit channel ion–dipole complex before separating into products, as shown in Fig. 3.1(b).[39]

Can computation reproduce these surfaces? The gas-phase surface for a variety of S_N2 reactions had been computed in the 1970s[40,41] but it was not until 1984 that Chandrasekhar *et al.*[42] provided the definitive computed solution-phase surface. They modeled the identity reaction $Cl^- + CH_3Cl$ at HF/6-31G* in the gas phase, generating a surface like that of Fig. 3.1(b). Next, for a series of points along the gas-phase reaction coordinate, they ran Monte Carlo simulation allowing 250 water molecules to equilibrate about the fixed geometry of the substrate. Each water molecule was treated by the TIP4P approximation, which treats every water molecule as a set of fixed charge points along with a Lennard–Jones term. The resulting PES shows no ion–dipole complexes, but rather a single barrier between reactant and product, i.e., Fig. 3.1(a). The computed barrier height (26.3 kcal mol^{-1}) nearly exactly matched the experimental value (26.6 kcal mol^{-1}).

3.3.4 *Benzyne*

The discovery of a number of natural products containing an enediyne moiety[43] spurred a large number of computations into the nature of the Bergman cyclization. The product of the Bergman cyclization is a *p*-benzyne fragment (**3**).[44] This led to a revived interest in the benzynes by both experimental and computational chemists. Of the many interesting aspects of the benzynes, we address here their heats of formation.

Collision-induced dissociation of benzyne precursors gave ΔH_f (**1**)=106, $\Delta H_f(\mathbf{2}) = 116$, and $\Delta H_f(\mathbf{3}) = 128\,\text{kcal mol}^{-1}$.[45] The value for **1** was in agreement with previous experiments.[46,47] Shortly after these experiments were reported, computations raised doubts about the values for **2** and **3**.

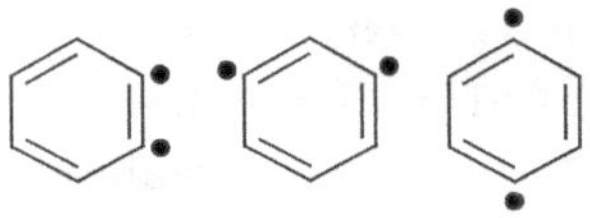

1 2 3

Computations of the benzynes are complicated by the fact that each of them requires a multiconfiguration wave function because each has (differing) biradical character. Borden's π-SDCI computation indicated that **2** is $16\,\text{kcal mol}^{-1}$ higher in energy than **1**, and **3** is another $13\,\text{kcal mol}^{-1}$ higher.[48] Squires' CCSD(T) study found a similarly large difference in the relative energies of the benzynes, much greater that the energy differences from the experiment.[49] These computations led Squires to reevaluate his experiment, recognizing that trace water might allow for the benzynes to rearrange. His later experiment confirmed the energy differences predicted by the computations.[50]

3.4 1990s–today: Quantum computations in the wider organic chemistry community

The current state of computational organic chemistry is one where computations are performed not just by specialists, i.e., chemists trained in quantum mechanics, but rather computation has become another important tool utilized by organic chemists. Just like nuclear magnetic resonance (NMR), where one does not need to be a specialist in spectroscopy to utilize and interpret NMR spectra, computation is widely used by the general organic chemist. This shift from specialist to generalist has been facilitated by a number of developments.

Computer hardware continues to improve, with desktop systems becoming ever more powerful and affordable. Serious computational hardware can now be brought into every research laboratory for a few thousand dollars and does not require specialized personnel to set-up or maintain.

Quantum chemistry software also continues to improve. In addition to new releases of *Gaussian* and *GAMESS* that have added functionality and improved performance, other codes appeared such as *MOLCAS*, *Q-CHEM*, and *Jaguar* along with open-source programs like *Psi4* and *Orca*. A great deal of effort was also spent on making these programs more user-friendly. A key development was graphical front-ends that allow users to sketch their molecules and select computational options. Important front-ends include *Spartan*, *GaussView*, *Avogadro*, *WebMO*, and *Maestro*. These graphical front-end programs make it much easier than before to create an input file and analyze the output file.

So, while computations are much more widely utilized by organic chemists today, and many nonspecialists have made important contributions using calculations, potential pitfalls still exist. Selection of the proper QM method remains a challenge for many situations, and some front-end programs can mask the dangers ahead. The next few sections describe a few major successes along with some cautionary examples.

3.4.1 *NMR*

With the development of methods for computing NMR chemical shifts by Schindler and Kutzelnigg,[51] application to organic chemistry soon followed. The first important development was the concept of nucleus-independent chemical shifts (NICS) by Schleyer *et al.*[52] Computing the chemical shift at the center of a ring provided a measure of its aromatic or antiaromatic character (see Chapter 9). This technique continues to be broadly used to assess if a molecule is aromatic, like **4**,[53] or has some antiaromatic character, like **5**.[54]

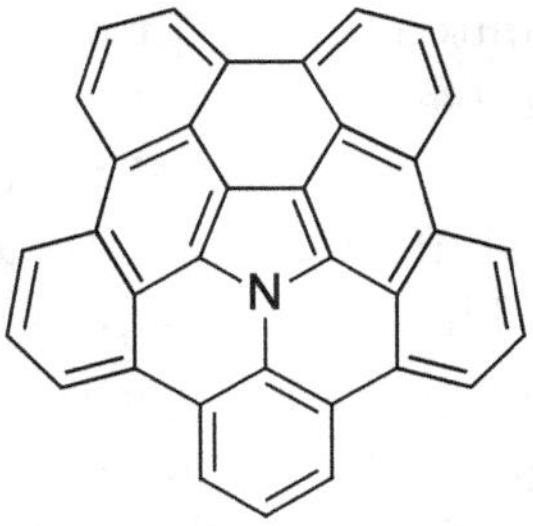

4

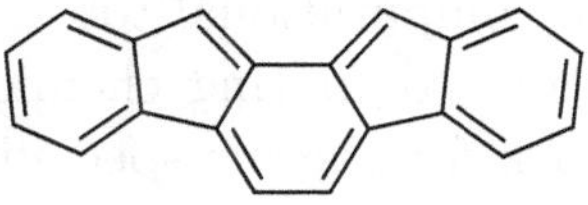

5

Few reports of applying computed NMR chemical shifts to structural problems appeared until the landmark paper by Rychnovsky[55] concerning the structure of hexacyclinol. First proposed to have structure **6**,[56] Rychnovsky was skeptical about this unusual compound. He optimized the geometry of **6** at HF/3-21G and evaluated ^{13}C chemical shifts at mPW1PW91/6-31G(d,p). These computed chemical shifts differed from the experimental values on average by 6.8 ppm, and five shifts differed by more than 10 ppm. He then proposed structure **7**, and the computed chemical shifts were in very nice agreement with the experimental values, with an average error of less than 2 ppm. This structure was subsequently synthesized by Porco *et al.*,[57] and its NMR spectra is identical to that of the natural sample.

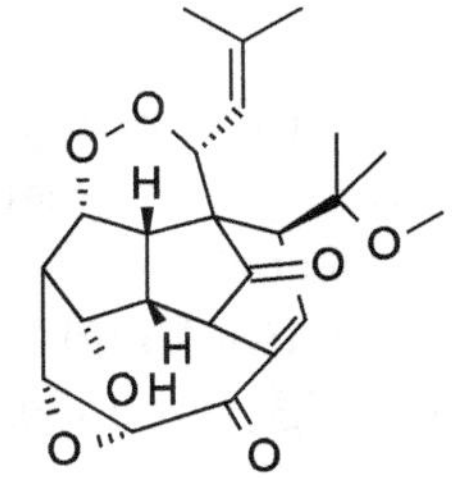

6

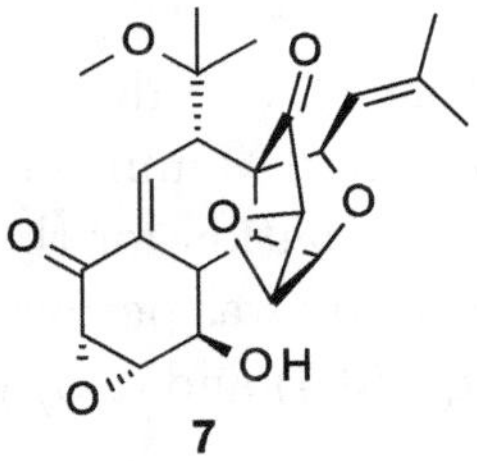

7

Following on this work, a large number of structural identification studies have made use of computed NMR spectra. Some examples include maitotoxin[58] and vannusal B,[59] both by the Nicolaou group, and citrinalin B[60] and aquatolide.[61] It is clear that computed

NMR spectra (along with other computed optical properties) can be invaluable in structure determination (see Chapter 6).[62]

Though computational tools have become easier to use over the past 20 years, computing chemical shifts may nonetheless seem a daunting task for the nonspecialist. The Hoye group has published an easy to follow set of protocols for performing a computational NMR study, including scripts to help automate the process.[63]

Computed chemical shifts often need to be scaled in some fashion to better match experiment. Jain *et al.*[64] produced a scaling procedure for both chemical shifts[64] and coupling constants,[65] and Tantillo's group provides the web resource CHEmical SHIft REpository (CHESHIRE; cheshirenmr.info) for scripts and assistance in implementing these corrections. Oftentimes, one is looking to use computed chemical shifts to discriminate between diastereomers. Smith and Goodman[66] developed the DP4 method for just such a task, and their free web app (http://www-jmg.ch.cam.ac.uk/tools/nmr/DP4/) implements this procedure.

3.4.2 *Organocatalysis*

A major emphasis of synthetic organic chemistry over the past two decades has been the development of organocatalysts. Computational chemistry provided insight into the mechanism of action for many organocatalysts. We discuss here the use of amines in asymmetric aldol reactions.

List[67] and Notz *et al.*[68] demonstrated the utility of enamines in catalyzing the aldol and Mannich reactions, with great enatioselectivity as well. The explanation for these effects was left to computation to unravel. A key study was performed by Houk and List for reactions represented in Eqs. (3.4) and (3.5) where (*S*)-proline is the catalyst.[69]

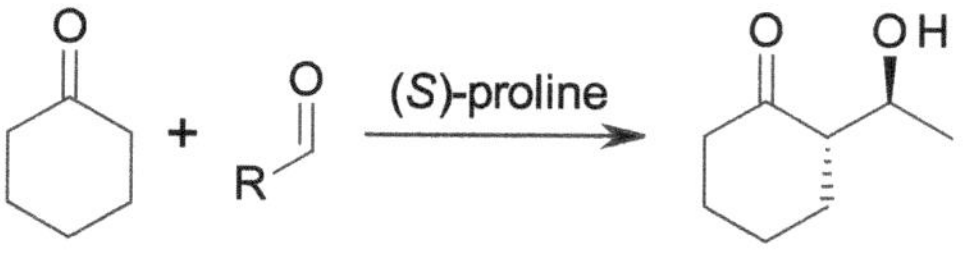

R = Ph, (3.4)

R = *i*-Pr (3.5)

The lowest energy B3LYP/6-31G* transition state structures for both of these reactions are characterized by three traits. The formation of the new C–C bond occurs with proton transfer from the carboxylic acid group to the incipient alcohol oxygen. The enamine, formed from cyclohexanone and proline, is in the *anti* orientation. The aldehyde substituent is *anti* to the enamine double bond. All of these are seen in the depiction of the transition state **8**. Computations correctly predict the major product and that reaction shown in Eq. (3.5) is more selective than that in Eq. (3.4). The features of TS **8** comprise what is now called the Houk–List model.

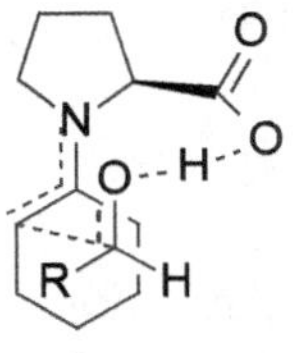

8

Computations of the intramolecular aldol reaction, the Hajos–Parrish–Wiechert–Eder–Sauer reaction, Eq. (3.6), was also examined by Bahmanyar *et al.*[70] and Clemente *et al.*[71] A number of mechanistic alternatives were studied, and the lowest energy pathway is consistent with the Houk–List model: an *anti* enamine and proton transfer concomitant with C–C bond formation. The proline catalyst is found to lower the activation energy by 11 kcal mol^{-1}.

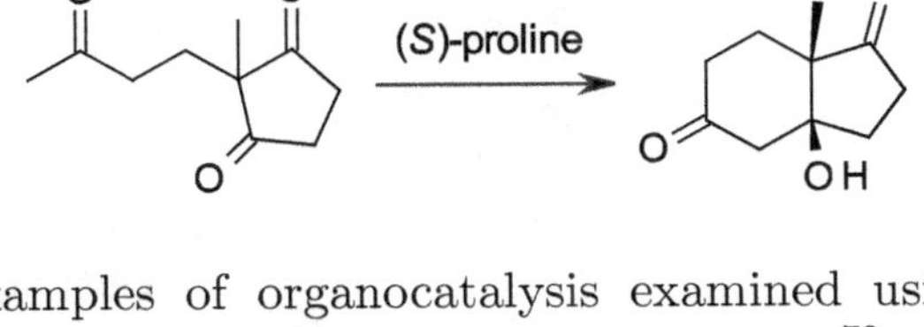

(3.6)

Other examples of organocatalysis examined using computational techniques include studies of the Mannich[72] and Michael[73] reactions and Uyeda and Jacobsen's[74] studies of urea derivatives to catalyze Claisen rearrangements and a thiourea catalyst for a hydroxyamination reaction.[75]

3.4.3 *Dispersion*

From the late-1990s into the mid-2000s, the B3LYP functional dominated computational organic chemistry, largely based on its availability within *Gaussian* and good performance for pericyclic reactions. However, in 2005 Check and Gilbert[76] noted that B3LYP underestimated the bond dissociation energy of alkanes, with increasingly poor values with increasing substitution. The following year Grimme[77] noted that B3LYP failed to properly order the relative energy of some C_8H_{18} isomers, and Schreiner noted the same problem with $C_{12}H_{12}$ isomers.[78] A study of bond separation energies for 72 hydrocarbons by Wodrich *et al.*[79] revealed poor performance not just for B3LYP but a large number of functionals.

The culprit was ultimately revealed to be inadequate treatment of medium-range correlation and dispersion (see Chapter 10).[80] A number of new functionals have now been developed that address these problems, along with adding a dispersion correction to older functionals. This points to the continuing difficulty in selecting an appropriate computational method for the problem at hand, especially given the plethora of functionals available. Using a quantum chemistry program in a black-box fashion is still not recommended!

An unexpected outcome of identifying that dispersion is missing in many functionals is just how important dispersion can be in organic chemistry.[81] We present a few examples here.

Hexaphenylethane **9** is not isolable, nor is **10**, yet the seemingly even more crowded **11** is isolable and its crystal structure has been reported. Its central C–C bond is quite long, 1.67Å, and one wonders why it should even exist! TPSS/TZV(2,2p) calculations indicate that all three analogues are unstable relative to the two radical fragments created by cleaving the central C–C bond.[82] However, when the dispersion correction is included in the computations **9** and **10** remain unstable, but **11** is stable by 14 kcal mol^{-1}. Dispersion, which is increased by the steric crowding, makes this compound stable.

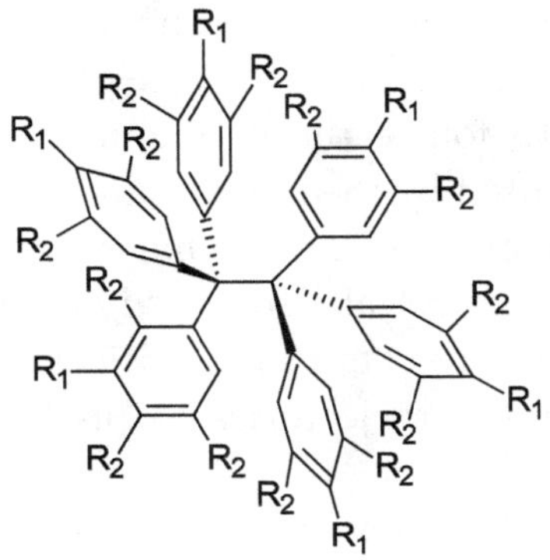

9: $R_1 = R_2 = H$
10: $R_1 =$ *t*-Bu, $R_2 = H$
11: $R_1 = R_2 =$ *t*-Bu

Schreiner has been interested in diamondoid fragments, as they exhibit unusually bonds. For example, the C–C bond joining the two cages of **12** is very long: 1.647 Å.[83] Optimization of this structure using a density functional that lacks a dispersion correction leads to a bond that is too long. However, functionals that incorporate some dispersion, either through an explicit dispersion correction (like B3LYP-D and B2PLYP-D) or with a functional that addresses mid-range or long-range correlation (like M06-2X) or both (like ωB97X-D) all provide very good estimates of this distance.

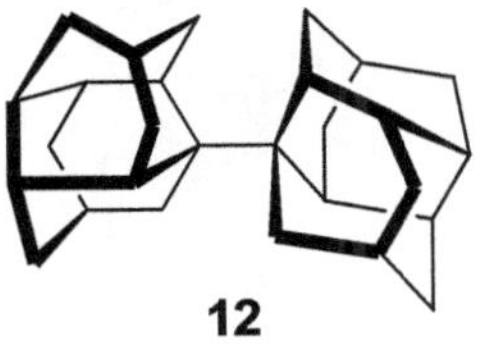

12

3.4.4 *Dynamic effects*

The last two decades has witnessed a minor revolution in how physical organic chemists think of reaction mechanisms.[84,85] Statistical theories, such as transition state theory, are based on the consideration of critical points on the PES, mainly local energy minima and transition state structures, and the reaction coordinate connecting them. What has now been observed in both experiments and computations is a much more complex situation. Reactions, often seemingly simple organic reactions, fail to follow the minimum energy path, instead following pathways dictated by the momentum the molecules carry through transition states, or crossing over flat regions, leading to nonstatistical dynamics, generally labeled as "dynamic effects" (see Chapter 12). Instead of simply computing critical points, one must compute molecular dynamics trajectories to predict reaction outcomes.

There are too many examples to explicitly cover here. Instead, we will point out a few examples. A flat energy region surrounded by a few transition states, a caldera surface, is found in a number of reactions that can go via a concerted or stepwise diradical mechanism. Examples of nonstatistical dynamics on a caldera[86] include the vinylcyclopropane to cyclopentene rearrangement,[87,88] the rearrangement of bicyclo[3.1.0]hex-2-ene,[89] and the stereomutation of cyclopropane.[90,91] These reactions express the notion of dynamic matching. Dynamic matching occurs when an entrance into the caldera matches up, through conservation of momentum, with a specific exit channel. Other examples of dynamic matching include the reaction of tetrazine with cyclopropene[92] and the substitution reaction shown in Eq. (3.7).[93]

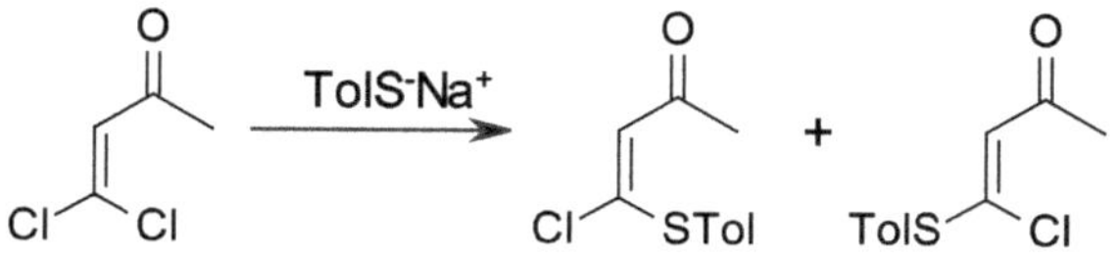

(3.7)

A second type of reaction where dynamics is important are reactions that have high barriers and after crossing this barrier, the large momentum carries the system past intervening intermediates. Examples of this include the Cope rearrangement of 1,2,6-hexatriene,[94] and the S_N2 reaction of hydroxide with methylflouride.[95]

A third category of reactions with dynamic effects are those reactions where the PES exhibits a bifurcation.[85] In these cases, a single transition state effectively leads to multiple products. Bifurcating surfaces are found in cycloadditions of ketenes[96,97] and even in some Diels–Alder reactions, such as Eq. (3.8).[98]

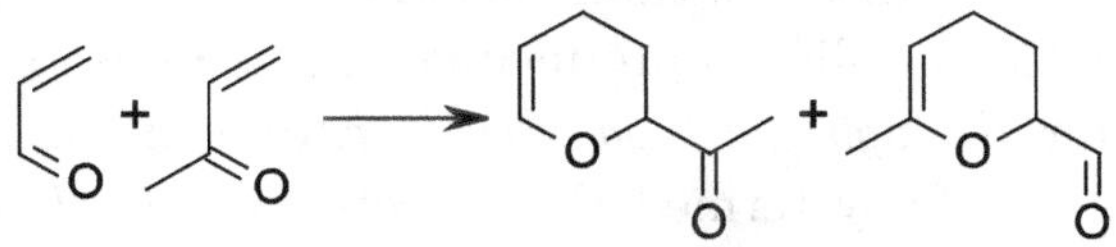

(3.8)

3.4.5 *Tunneling control*

Schreiner and Allen have recently presented a series of combined experimental and computational studies of the rearrangements of some simple carbenes. Along the way, they made two important discoveries.

In 2008, they examined the rearrangement of carbene **13** into formaldehyde.[99] The barrier for this rearrangement was computed to be almost 30 kcal mol^{-1} using the focal point method. Given this high barrier, they were astonished to find that the half-life of **13** is 2 h at 11–20 K! The only possibility was that tunneling is occurring, which was confirmed by computations using Wentzel–Kramer–Brillouin (WKB) theory. Not only did the computed half-life (122 min) agree with the experimental value, the prediction that OD-substituted **13** should have a very long half-life was also confirmed. Tunneling was then found to occur in reactions of a number of related carbenes.

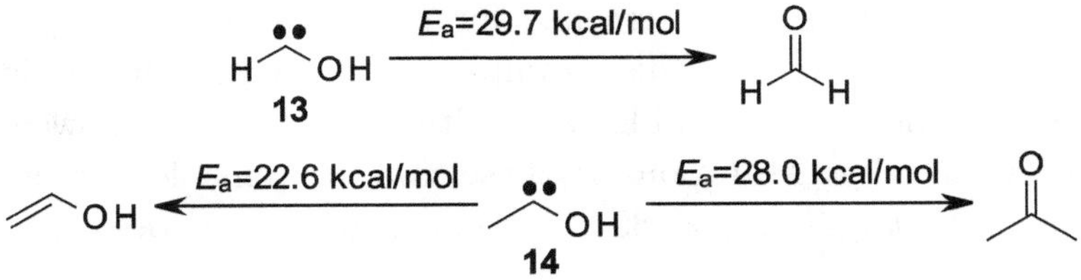

Their study of methylhydroxycarbene **14** proved to be even more interesting.[100] Focal point computations indicated that the barrier to rearrange from **14** to acetone is 28 kcal mol^{-1}, about 5.4 kcal mol^{-1} higher than the barrier for rearrangement to vinyl alcohol. However, acetone is the only observed product at low temperature. WKB computations indicated that the tunneling half-life to produce acetone is 71 min, while the half-life to produce vinyl alcohol is 190 days. Tunneling is faster *through the higher barrier*, due to the shorter path for this reaction. They termed this selectivity *tunneling control*,[101] proposing a third means for steering a reaction, in addition to kinetic and thermodynamic control.

3.4.6 *Designer enzymes*

One of the many dreams of chemists is to design an enzyme to catalyze a specific reaction, with a procedure for developing an appropriate enzyme for any given reaction. Achieving this dream remains way in the future, however, recent computational work from the Baker and Houk labs provides some glimmer of promise.

Their approach is to first design a theozyme, a model of the transition state structure for the reaction of interest complexed to catalytic groups. For example, they used **15** as the theozyme for a Diels–Alder reaction. Then, with the accompanying tyrosine and glutamine bases, this theozyme is fit into potential protein backbones using the RosettaMatch algorithm.[102] Mutations along the backbone are allowed in an attempt to optimize binding of the theozyme, using a scoring procedure. The best binders are then synthesized and screened for their ability to catalyze the Diels–Alder reaction. In this example, 84 proteins were designed, 50 were found to be soluble, and ultimately only 2 showed any catalytic activity. Mutation of these two proteins led to a best enzyme with $k_{cat}/k_{uncat} = 89$ M, demonstrating a modest catalytic effect.[103]

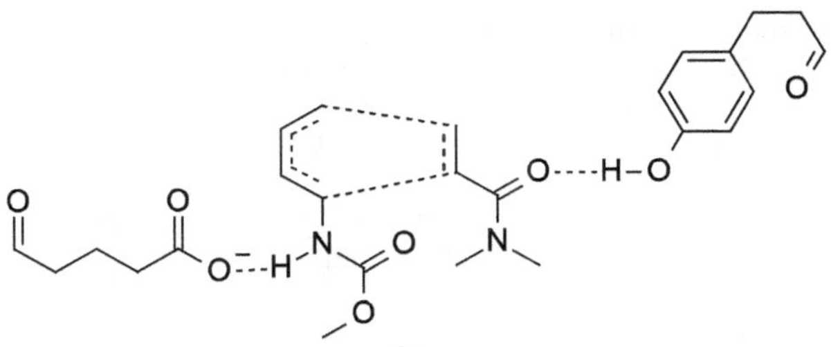

15

Other examples have shown similar performance.[104,105] Clearly, more development in this process is needed, but it certainly presents a pathway forward toward reaching the goal of designer enzymes.

3.5 Conclusion

In the four decades that have followed from Bender and Schaefer's seminal work on methylene, computational organic chemistry has grown into a respected discipline and partner with experiment in interpreting reactions and properties. As demonstrated here, computational chemistry has not just been an adjunct to experiment, but has oftentimes led to the identification of errors in experiments and has made predictions that subsequently were confirmed by experiment.

Challenges remain for computational chemists. A few areas that are likely to see significant developments in the near future[106] include dealing with large conformational and configurational spaces (see Chapter 5) and bringing some clarity to the near-Babel-like situation concerning the number of density functionals. Undoubtedly, computations will continue to bring solutions and surprises in uncovering the nature of organic chemistry.

References

1. Huckel, E. (1937). The theory of unsaturated and aromatic compounds, *Z. Elektrochem. Angew. Phys. Chem.*, 43, pp. 752–788.

2. Woodward, R. B. and Hoffmann, R. (1969). The conservation of orbital symmetry, *Angew. Chem. Int. Ed. Engl.*, 8, pp. 781–853.
3. Seeman, J. I. (2015). Woodward–Hoffmann's stereochemistry of electrocyclic reactions: From day 1 to the JACS receipt date (May 5, 1964 to November 30, 1964), *J. Org. Chem.*, 80, pp. 11632–11671.
4. Woodward, R. B. and Hoffmann, R. (1965). Stereochemistry of electrocyclic reactions, *J. Am. Chem. Soc.*, 87, pp. 395–397.
5. Roothaan, C. C. J. (1951). New developments in molecular orbital theory, *Rev. Mod. Phys.*, 23, pp. 69–89.
6. Boys, S. F. (1950). Electronic wave functions. I. A general method of calculation for the stationary states of any molecular system, *Proc. Roy. Soc.*, A200, pp. 542–554.
7. Bachrach, S. M. (2014). *Computational Organic Chemistry*, 2nd Ed. (Wiley, USA). (a) Quote from Weston Borden, p. 278; (b) Quote from Henry Schaefer III, p. 355.
8. Swart, E. R. (1980). The philosophical implications of the four-color problem, *Am. Math. Mon.*, 87, pp. 697–707.
9. Goddard, W. A., III (1985). Theoretical chemistry comes alive: Full partner with experiment, *Science*, 227, pp. 917–923.
10. Schaefer, H. F., III (1986). Methylene: A paradigm for computational quantum chemistry, *Science*, 231, pp. 1100–1107.
11. Herzberg, G. and Shoosmith, J. (1959). Spectrum and structure of the free methylene radical, *Nature*, 183, pp. 1801–1802.
12. Bender, C. F. and Schaefer, H. F., III (1970). New theoretical evidence for the nonlinearity of the triplet ground state of methylene, *J. Am. Chem. Soc.*, 92, pp. 4984–4985.
13. Bernheim, R. A., Bernard, H. W., Wang, P. S., Wood, L. S. and Skell, P. S. (1970). Electron paramagnetic resonance of triplet CH_2, *J. Chem. Phys.*, 53, pp. 1280–1281.
14. Wasserman, E., Yager, W. A. and Kuck, V. J. (1970). EPR of CH_2: A substiantially bent and partially rotating ground state triplet, *Chem. Phys. Lett.*, 7, pp. 409–413.
15. Herzberg, G. and Johns, J. W. C. (1971). On the structure of CH_2 in its triplet ground state, *J. Chem. Phys.*, 54, pp. 2276–2278.
16. Bender, C. F., Schaefer, H. F., III, Franceschetti, D. R. and Allen, L. C. (1972). Singlet-triplet energy separation, Walsh-Mulliken diagrams, and singlet d-polarization effects in methylene, *J. Am. Chem. Soc.*, 94, pp. 6888–6893.
17. Hay, P. J., Hunt, W. J. and Goddard, W. A., III (1972). Generalized valence bond wavefunctions for the low lying states of methylene, *Chem. Phys. Lett.*, 13, pp. 30–35.
18. Zittel, P. F., Ellison, G. B., O'Neil, S. V., Herbst, E., Lineberger, W. C. and Reinhardt, W. P. (1976). Laser photoelectron spectrometry of CH_2^-. Singlet-triplet splitting and electron affinity of CH_2, *J. Am. Chem. Soc.*, 98, pp. 3731–3732.

19. Harding, L. B. and Goddard, W. A., III (1977). Ab initio studies on the singlet-triplet splitting of methylene (CH_2), *J. Chem. Phys.*, 67, pp. 1777–1779.
20. Lucchese, R. R. and Schaefer, H. F., III (1977). Extensive configuration interaction studies of the methylene singlet-triplet separation, *J. Am. Chem. Soc.*, 99, pp. 6765–6766.
21. Roos, B. O. and Siegbahn, P. M. (1977). Methylene singlet-triplet separation. An ab initio configuration interaction study, *J. Am. Chem. Soc.*, 99, pp. 7716–7718.
22. Harding, L. B. and Goddard, W. A., III (1978). Methylene: Ab initio vibronic analysis and reinterpretation of the spectroscopic and negative ion photoelectron experiments, *Chem. Phys. Lett.*, 55, pp. 217–220.
23. McKellar, R. W., Bunker, P. R., Sears, T. J., Evenson, K. M., Saykally, R. J. and Langhoff, S. R. (1983). Far infrared laser magnetic resonance of singlet methylene: Singlet-triplet perturbations, singlet-triplet transitions, and the singlet-triplet splitting, *J. Chem. Phys.*, 79, pp. 5251–5264.
24. Leopold, D. G., Murray, K. K. and Lineberger, W. C. (1984). Laser photoelectron spectroscopy of vibrationally relaxed CH_2^-: A reinvestigation of the singlet-triplet splitting in merthylene, *J. Chem. Phys.*, 81, pp. 1048–1050.
25. Kirmse, W., Rondan, N. G. and Houk, K. N. (1984). stereoselective substituent effects on conrotatory electocyclic reactions of cyclobutenes, *J. Am. Chem. Soc.*, 106, pp. 7989–7991.
26. Rondan, N. G. and Houk, K. N. (1985). Theory of stereoselection in conrotatory electrocyclic reactions of substituted cyclobutenes, *J. Am. Chem. Soc.*, 107, pp. 2099–2111.
27. Dolbier, W. R., Jr., Koroniak, H., Houk, K. N. and Sheu, C. (1996). Electronic control of stereoselectivities of electrocyclic reactions of cyclobutenes: A triumph of theory in the prediction of organic reactions, *Acc. Chem. Res.*, 20, pp. 471–477.
28. Murakami, M., Miyamoto, Y. and Ito, Y. (2001). Stereoselective synthesis of isomeric functionalized 1,3-dienes from cyclobutenones, *J. Am. Chem. Soc.*, 123, pp. 6441–6442.
29. Lee, P. S., Zhang, X. and Houk, K. N. (2003). Origins of inward torquoselectivity by silyl groups and other σ-acceptors in electrocyclic reactions of cyclobutenes, *J. Am. Chem. Soc.*, 125, pp. 5072–5079.
30. Evanseck, J. D., Thomas IV, B. E., Spellmeyer, D. C. and Houk, K. N. (1995). Transition structures of thermally allowed disrotatory electrocyclizations. The prediction of stereoselective substituent effects in six-electron pericyclic reactions, *J. Org. Chem.*, 60, pp. 7134–7141.
31. Thomas, B. E., IV, Evanseck, J. D. and Houk, K. N. (1993). Electrocyclic reactions of 1-substituted 1,3,5,7-octatetraenes. An ab initio molecular orbital study of torquoselectivity in eight-electron electrocyclizations, *J. Am. Chem. Soc.*, 115, pp. 4165–4169.
32. Houk, K. N., Gonzalez, J. and Li, Y. (1995). Pericyclic transition states: Passion and punctilios, 1935–1995, *Acc. Chem. Res.*, 28, pp. 81–90.

33. Staroverov, V. N. and Davidson, E. R. (2001). The Cope rearrangement in theoretical retrospect, *J. Mol. Struct. (THEOCHEM)*, 573, pp. 81–89.
34. Osamura, Y., Kato, S., Morokuma, K., Feller, D., Davidson, E. R. and Borden, W. T. (1984). Ab initio calculation of the transition state for the cope rearrangement, *J. Am. Chem. Soc.*, 106, p. 3362.
35. Hrovat, D. A., Borden, W. T., Vance, R. L., Rondan, N. G., Houk, K. N. and Morokuma, K. (1990). Ab initio calculations of the effects of cyano substituents on the cope rearrangement, *J. Am. Chem. Soc.*, 112, pp. 2018–2019.
36. Dupuis, M., Murray, C. and Davidson, E. R. (1991). The Cope rearrangement revisited, *J. Am. Chem. Soc.*, 113, pp. 9756–9759.
37. Hrovat, D. A., Morokuma, K. and Borden, W. T. (1994). The Cope rearrangement revisited again. Results of ab initio calculations beyond the CASSCF level, *J. Am. Chem. Soc.*, 116, pp. 1072–1076.
38. Kozlowski, P. M., Dupuis, M. and Davidson, E. R. (1995). The Cope rearrangement revisited with multireference perturbation theory, *J. Am. Chem. Soc.*, 117, pp. 774–778.
39. Chabinyc, M. L., Craig, S. L., Regan, C. K. and Brauman, J. I. (1998). Gas-phase ionic reactions: Dynamics and mechanism of nucleophilic displacements, *Science*, 279, pp. 1882–1886.
40. Dedieu, A. and Veillard, A. (1972). Comparative study of Some S_n2 reactions through ab initio calculations, *J. Am. Chem. Soc.*, 94, pp. 6730–6738.
41. Keil, F. and Ahlrichs, R. (1976). Theoretical study of S_n2 reactions. Ab initio computations on HF and CI level, *J. Am. Chem. Soc.*, 98, pp. 4787–4793.
42. Chandrasekhar, J., Smith, S. F. and Jorgenson, W. L. (1984). S_n2 reaction profiles in the gas phase and aqueous solution, *J. Am. Chem. Soc.*, 106, pp. 3049–3050.
43. Nicolaou, K. C. and Dai, W.-M. (1991). Chemistry and biology of the enediyne anticancer antibiotics, *Angew. Chem. Int. Ed. Engl.*, 30, pp. 1387–1416.
44. Bergman, R. G. (1973). Reactive 1,4-dehydroaromatics, *Acc. Chem. Res.*, 6, pp. 25–31.
45. Wenthold, P. G., Paulino, J. A. and Squires, R. R. (1991). The absolute heats of formation of *o*-, *m*-, and *p*-benzyne, *J. Am. Chem. Soc.*, 113, pp. 7414–7415.
46. Guo, Y. and Grabowski, J. J. (1991). Reactions of the benzyne radical anion in the gas phase, the acidity of the phenyl radical, and the heat of formation of *o*-benzyne, *J. Am. Chem. Soc.*, 113, pp. 5923–5931.
47. Riveros, J. M., Ingemann, S. and Nibbering, N. M. M. (1991). Formation of gas phase solvated bromine and iodine anions in ion/molecule reactions of halobenzenes. Revised heat of formation of benzyne, *J. Am. Chem. Soc.*, 113, pp. 1053–1053.
48. Nicolaides, A. and Borden, W. T. (1993). CI calculations on didehydrobenzenes predict heats of formation for the *meta* and *para* isomers that are

substantially higher than previous experimental values, *J. Am. Chem. Soc.*, 115, pp. 11951–11957.
49. Wierschke, S. G., Nash, J. J. and Squires, R. R. (1993). A multiconfigurational SCF and correlation-consistent CI study of the structures, stabilities, and singlet-triplet splittings of *o*-, *m*-, and *p*-benzyne, *J. Am. Chem. Soc.*, 115, pp. 11958–11967.
50. Wenthold, P. G. and Squires, R. R. (1994). Biradical thermochemistry from collision-induced dissociation threshold energy measurements. Absolute heats of formation of *ortho*-, *meta*-, and *para*-benzyne, *J. Am. Chem. Soc.*, 116, pp. 6401–6412.
51. Schindler, M. and Kutzelnigg, W. (1982). Theory of magnetic susceptibilities and NMR chemical shifts in terms of localized quantities. II. Application to some simple molecules, *J. Chem. Phys.*, 76, pp. 1919–1933.
52. Schleyer, P. v. R., Maerker, C., Dransfeld, A., Jiao, H. and Hommes, N. J. R. v. E. (1996). Nucleus-independent chemical shifts: A simple and efficient aromaticity probe, *J. Am. Chem. Soc.*, 118, pp. 6317–6318.
53. Ito, S., Tokimaru, Y. and Nozaki, K. (2015). Benzene-fused azacorannulene bearing an internal nitrogen atom, *Angew. Chem. Int. Ed.*, 54, pp. 7256–7260.
54. Shimizu, A. and Tobe, Y. (2011). Indeno[2,1-a]fluorene: An air-stable ortho-quinodimethane derivative, *Angew. Chem. Int. Ed.*, 50, pp. 6906–6910.
55. Rychnovsky, S. D. (2006). Predicting NMR spectra by computational methods: Structure revision of hexacyclinol, *Org. Lett.*, 8, pp. 2895–2898.
56. Schlegel, B., Hartl, A., Dahse, H.-M., Gollmick, F. A., Grafe, U., Dorfelt, H. and Kappes, B. (2002). Hexacyclinol, A new antiproliferative metabolite of *Panus rudis* HKI 0254, *J. Antibiot.*, 55, pp. 814–817.
57. Porco, J. A. J., Shun Su, S., Lei, X., Bardhan, S. and Rychnovsky, S. D. (2006). Total synthesis and structure assignment of (+)-hexacyclinol, *Angew. Chem. Int. Ed.*, 45, pp. 5790–5792.
58. Nicolaou, K. C. and Frederick, M. O. (2007). On the structure of maitotoxin, *Angew. Chem. Int. Ed.*, 46, pp. 5278–5282.
59. Saielli, G., Nicolaou, K. C., Ortiz, A., Zhang, H. and Bagno, A. (2011). Addressing the stereochemistry of complex organic molecules by density functional theory-NMR: Vannusal B in retrospective, *J. Am. Chem. Soc.*, 133, pp. 6072–6077.
60. Mercado-Marin, E. V., Garcia-Reynaga, P., Romminger, S., Pimenta, E. F., Romney, D. K., Lodewyk, M. W., Williams, D. E., Andersen, R. J., Miller, S. J., Tantillo, D. J., Berlinck, R. G. S. and Sarpong, R. (2014). Total synthesis and isolation of citrinalin and cyclopiamine congeners, *Nature*, 509, pp. 318–324.
61. Lodewyk, M. W., Soldi, C., Jones, P. B., Olmstead, M. M., Rita, J., Shaw, J. T. and Tantillo, D. J. (2012). The correct structure of aquatolide — experimental validation of a theoretically-predicted structural revision, *J. Am. Chem. Soc.*, 134, pp. 18550–18553.

62. Lodewyk, M. W., Siebert, M. R. and Tantillo, D. J. (2012). Computational prediction of ^{1}H and ^{13}C chemical shifts: A useful tool for natural product, mechanistic, and synthetic organic chemistry, *Chem. Rev.*, 112, pp. 1839–1862.
63. Willoughby, P. H., Jansma, M. J. and Hoye, T. R. (2014). A guide to small-molecule structure assignment through computation of (^{1}H and ^{13}C) NMR chemical shifts, *Nat. Protoc.*, 9, pp. 643–660.
64. Jain, R., Bally, T. and Rablen, P. R. (2009). Calculating accurate proton chemical shifts of organic molecules with density functional methods and modest basis sets, *J. Org. Chem.*, 74, pp. 4017–4023.
65. Bally, T. and Rablen, P. R. (2011). Quantum-chemical simulation of ^{1}H NMR spectra. 2. Comparison of DFT-based procedures for computing proton-proton coupling constants in organic molecules, *J. Org. Chem.*, 76, pp. 4818–4830.
66. Smith, S. G. and Goodman, J. M. (2010). Assigning stereochemistry to single diastereoisomers by GIAO NMR calculation: The DP4 probability, *J. Am. Chem. Soc.*, 132, pp. 12946–12959.
67. List, B. (2004). Enamine catalysis is a powerful strategy for the catalytic generation and use of carbanion equivalents, *Acc. Chem. Res.*, 37, pp. 548–557.
68. Notz, W., Tanaka, F. and Barbas, C. F., III (2004). Enamine-based organocatalysis with proline and diamines: The development of direct catalytic asymmetric Aldol, Mannich, Michael, and Diels-Alder reactions, *Acc. Chem. Res.*, 37, pp. 580–591.
69. Bahmanyar, S., Houk, K. N., Martin, H. J. and List, B. (2003). Quantum mechanical predictions of the stereoselectivities of proline-catalyzed asymmetric intermolecular aldol reactions, *J. Am. Chem. Soc.*, 125, pp. 2475–2479.
70. Bahmanyar, S. and Houk, K. N. (2001). The origin of stereoselectivity in proline-catalyzed intramolecular aldol reactions, *J. Am. Chem. Soc.*, 123, pp. 12911–12912.
71. Clemente, F. R. and Houk, K. N. (2004). Computational evidence for the enamine mechanism of intramolecular aldol reactions catalyzed by proline, *Angew. Chem. Int. Ed.*, 43, pp. 5766–5768.
72. Bahmanyar, S. and Houk, K. N. (2003). Origins of opposite absolute stereoselectivities in proline-catalyzed direct Mannich and Aldol reactions, *Org. Lett.*, 5, pp. 1249–1251.
73. Yang, H. and Wong, M. W. (2012). (*S*)-Proline-catalyzed nitro-Michael reactions: Towards a better understanding of the catalytic mechanism and enantioselectivity, *Org. Biomol. Chem.*, 10, pp. 3229–3235.
74. Uyeda, C. and Jacobsen, E. N. (2008). Enantioselective Claisen rearrangements with a hydrogen-bond donor catalyst, *J. Am. Chem. Soc.*, 130, pp. 9228–9229.

75. Brown, A. R., Uyeda, C., Brotherton, C. A. and Jacobsen, E. N. (2013). Enantioselective thiourea-catalyzed intramolecular cope-type hydroamination, *J. Am. Chem. Soc.*, 135, pp. 6747–6749.
76. Check, C. E. and Gilbert, T. M. (2005). Progressive systematic underestimation of reaction energies by the B3LYP model as the number of C-C bonds increase: Why organic chemists should use multiple DFT models for calculations involving polycarbon hydrocarbons, *J. Org. Chem.*, 70, pp. 9828–9834.
77. Grimme, S. (2006). Seemingly simple stereoelectronic effects in alkane isomers and the implications for Kohn-Sham density functional theory, *Angew. Chem. Int. Ed.*, 45, pp. 4460–4464.
78. Schreiner, P. R., Fokin, A. A., Pascal, R. A. and deMeijere, A. (2006). Many density functional theory approaches fail to give reliable large hydrocarbon isomer energy differences, *Org. Lett.*, 8, pp. 3635–3638.
79. Wodrich, M. D., Corminboeuf, C., Schreiner, P. R., Fokin, A. A. and Schleyer, P. v. R. (2007). How accurate are DFT treatments of organic energies? *Org. Lett.*, 9, pp. 1851–1854.
80. Grimme, S. (2010). *n*-Alkane isodesmic reaction energy errors in density functional theory are due to electron correlation effects, *Org. Lett.*, 12, pp. 4670–4673.
81. Wagner, J. P. and Schreiner, P. R. (2015). London dispersion in molecular chemistry — reconsidering steric effects, *Angew. Chem. Int. Ed.*, 54, pp. 12274–12296.
82. Grimme, S. and Schreiner, P. R. (2011). Steric crowding can stabilize a labile molecule: Solving the hexaphenylethane riddle, *Angew. Chem. Int. Ed.*, 50, pp. 12639–12642.
83. Fokin, A. A., Chernish, L. V., Gunchenko, P. A., Tikhonchuk, E. Y., Hausmann, H., Serafin, M., Dahl, J. E. P., Carlson, R. M. K. and Schreiner, P. R. (2012). Stable alkanes containing very long carbon–carbon bonds, *J. Am. Chem. Soc.*, 134, pp. 13641–13650.
84. Carpenter, B. K. (2005). Nonstatistical dynamics in thermal reactions of polyatomic molecules, *Annu. Rev. Phys. Chem.*, 46, pp. 57–89.
85. Ess, D. H., Wheeler, S. E., Iafe, R. G., Xu, L., Çelebi-Ölçüm, N. and Houk, K. N. (2008). Bifurcations on potential energy surfaces of organic reactions, *Angew. Chem. Int. Ed.*, 47, pp. 7592–7601.
86. Collins, P., Kramer, Z. C., Carpenter, B. K., Ezra, G. S. and Wiggins, S. (2014). Nonstatistical dynamics on the Caldera, *J. Chem. Phys.*, 141, p. 034111.
87. Doubleday, C., Nendel, M., Houk, K. N., Thweatt, D. and Page, M. (1999). Direct dynamics quasiclassical trajectory study of the stereochemistry of the vinylcyclopropane-cyclopentene rearrangement, *J. Am. Chem. Soc.*, 121, pp. 4720–4721.

88. Doubleday, C. (2001). Mechanism of the vinylcyclopropane-cyclopentene rearrangement studied by quasiclassical direct dynamics, *J. Phys. Chem. A*, 105, pp. 6333–6341.
89. Doubleday, C., Suhrada, C. P. and Houk, K. N. (2006). Dynamics of the degenerate rearrangement of bicyclo[3.1.0]hex-2-ene, *J. Am. Chem. Soc.*, 128, pp. 90–94.
90. Doubleday, C., Jr., Bolton, K. and Hase, W. L. (1997). Direct dynamics study of the stereomutation of cyclopropane, *J. Am. Chem. Soc.*, 119, pp. 5251–5252.
91. Hrovat, D. A., Fang, S., Borden, W. T. and Carpenter, B. K. (1997). Investigation of cyclopropane stereomutation by quasiclassical trajectories on an analytical potential energy surface, *J. Am. Chem. Soc.*, 119, pp. 5253–5254.
92. Törk, L., Jiménez-Osés, G., Doubleday, C., Liu, F. and Houk, K. N. (2015). Molecular dynamics of the Diels–Alder reactions of tetrazines with alkenes and N_2 extrusions from adducts, *J. Am. Chem. Soc.*, 137, pp. 4749–4758.
93. Bogle, X. S. and Singleton, D. A. (2012). Dynamic origin of the stereoselectivity of a nucleophilic substitution reaction, *Org. Lett.*, 14, pp. 2528–2531.
94. Debbert, S. L., Carpenter, B. K., Hrovat, D. A. and Borden, W. T. (2002). The iconoclastic dynamics of the 1,2,6-heptatriene rearrangement, *J. Am. Chem. Soc.*, 124, pp. 7896–7897.
95. Sun, L., Song, K. and Hase, W. L. (2002). A S_n2 reaction that avoids its deep potential energy minimum, *Science*, 296, pp. 875–878.
96. Ussing, B. R., Hang, C. and Singleton, D. A. (2006). Dynamic effects on the periselectivity, rate, isotope effects, and mechanism of cycloadditions of ketenes with cyclopentadiene, *J. Am. Chem. Soc.*, 128, pp. 7594–7607.
97. Gonzalez-James, O. M., Kwan, E. E. and Singleton, D. A. (2012). Entropic intermediates and hidden rate-limiting steps in seemingly concerted cycloadditions. Observation, prediction, and origin of an isotope effect on recrossing, *J. Am. Chem. Soc.*, 134, pp. 1914–1917.
98. Wang, Z., Hirschi, J. S. and Singleton, D. A. (2009). Recrossing and dynamic matching effects on selectivity in a Diels-Alder reaction, *Angew. Chem. Int. Ed.*, 48, pp. 9156–9159.
99. Schreiner, P. R., Reisenauer, H. P., Pickard, F. C., IV, Simmonett, A. C., Allen, W. D., Matyus, E. and Csaszar, A. G. (2008). Capture of hydroxymethylene and its fast disappearance through tunnelling, *Nature*, 453, pp. 906–909.
100. Schreiner, P. R., Reisenauer, H. P., Ley, D., Gerbig, D., Wu, C.-H. and Allen, W. D. (2011). Methylhydroxycarbene: Tunneling control of a chemical reaction, *Science*, 332, pp. 1300–1303.
101. Ley, D., Gerbig, D. and Schreiner, P. R. (2012). Tunnelling control of chemical reactions — The organic chemist's perspective, *Org. Biomol. Chem.*, 10, pp. 3781–3790.
102. Zanghellini, A., Jiang, L., Wollacott, A. M., Cheng, G., Meiler, J., Althoff, E. A., Röthlisberger, D. and Baker, D. (2006). New algorithms and an

in silico benchmark for computational enzyme design, *Protein Sci.*, 15, pp. 2785–2794.

103. Siegel, J. B., Zanghellini, A., Lovick, H. M., Kiss, G., Lambert, A. R., St.Clair, J. L., Gallaher, J. L., Hilvert, D., Gelb, M. H., Stoddard, B. L., Houk, K. N., Michael, F. E. and Baker, D. (2010). Computational design of an enzyme catalyst for a stereoselective bimolecular Diels-Alder reaction, *Science*, 329, pp. 309–313.
104. Jiang, L., Althoff, E. A., Clemente, F. R., Doyle, L., Röthlisberger, D., Zanghellini, A., Gallaher, J. L., Betker, J. L., Tanaka, F., Barbas, C. F., Hilvert, D., Houk, K. N., Stoddard, B. L. and Baker, D. (2008). De novo computational design of retro-aldol enzymes, *Science*, 319, pp. 1387–1391.
105. Richter, F., Blomberg, R., Khare, S. D., Kiss, G., Kuzin, A. P., Smith, A. J. T., Gallaher, J., Pianowski, Z., Helgeson, R. C., Grjasnow, A., Xiao, R., Seetharaman, J., Su, M., Vorobiev, S., Lew, S., Forouhar, F., Kornhaber, G. J., Hunt, J. F., Montelione, G. T., Tong, L., Houk, K. N., Hilvert, D. and Baker, D. (2012). Computational design of catalytic dyads and oxyanion holes for ester hydrolysis, *J. Am. Chem. Soc.*, 134, pp. 16197–16206.
106. Bachrach, S. M. (2014). Challenges in computational organic chemistry, *WIREs: Comput. Mol. Sci.*, 4, pp. 482–487.

Chapter 4

Solvation

Carlos Silva López and Olalla Nieto Faza

Departmento de Química Orgánica Universidad de Vigo,
Campus Universitario, 36310 Vigo, Spain

4.1 Introduction

4.1.1 *Scope of this chapter*

The problem of simulating solvation, as we will disclose in the following lines, is extraordinarily complex. This topic alone, if throughly described, can fill enough pages to deserve a whole book on its own. This chapter therefore cannot cover the entirety of solvation modeling; actually we do not dare to aim even close to that goal. In the following lines we will cover the most common treatments of solvation applied to medium-sized molecules in quantum chemical calculations. And we will focus mostly on a qualitative and interpretative approach rather than an accurate and mathematical description of the methods described. The latter can be found in the original references that will be provided along with the text and in other volumes, which are more experiment-than layman-oriented. We hope that this chapter will be of help to those entering the field and looking for the first insights on how these methods work, rather than to those looking for the specific implementation of each method in quantum chemistry packages.

The chapter will first analyze the continuum approach of simulating solvation and explain its advantages and disadvantages. A second part will then describe discrete and pseudo-discreet methods that have been successfully applied to solve a number of solvation problems where explicit interactions between the solute and the solvent are crucial. Lastly, under the epigraph Practical Considerations, a brief assessment on a number of recipes that are currently generally accepted as a reasonable balance between cost and quality will be provided.

4.1.2 *Occurrence of gas phase vs. solution phase chemistry*

Most of the chemistry occurs in the gas phase in nature. After all vacuum dominates the vast extension of the cosmos where interstellar chemistry takes place. The study of this chemistry has therefore the advantage that it is mostly devoid of environmental effects. If we focus on the chemistry occurring closer around us, or even that is caused by us, the situation changes quite significantly. On Earth, chemistry in the gas phase is also relatively common, and many efforts have been, and still are devoted, to a better understanding, for instance, of atmospheric chemistry or fragmentation in Mass Spectrometry. However, most of the chemistry that surrounds us is carried out or takes place in a condensed phase, typically in solution phase. Just to highlight a few, most of the biochemistry occurring in the cytoplasm, all the photochemistry on the surface of the oceans, most of the industrial processes and, finally, most of the fine chemistry carried out in research laboratories is carried out under solvation conditions.

Computational chemistry aimed at understanding any of these processes therefore needs to be armed with methods that allow a description of the solvation environment. Such description is far from trivial, and countless efforts have been taken in this direction. An overview of the most important difficulties to correctly describe solvation, and a summary of popular approaches to tackle this problem in computational chemistry will be addressed in this chapter.

4.1.3 *The dimension of the problem*

To put into perspective the complexity of simulating solvation, let us make a mental exercise of imagining a random solvent and considering the perturbation that implies including a solute molecule in an otherwise perfectly pure solvent. A sample of pure solvent will feature molecules in a homogeneous but random fashion. This means that the solvent molecule-to-molecule distances will be kept constant across the sample, hence maintaining density, and the orientation of a particular solvent molecule at a specific moment in time will be impossible to anticipate. To include a solute in this dynamic lattice, we need to create first a cavity in the solvent: this implies creating more surface tension, which requires energy. Then the solute can be placed in the cavity, and both the solvent and the solute need to rearrange at several levels to achieve a new equilibrium. Moving from faster to slower, electron densities of the solvent will respond to the electron density of the solute, and vice-versa; the solute and adjacent solvent molecules may adjust their geometry to the new shortrange environment (the energy minimum of the solute in the gas phase is likely different from the energy minimum in a solvent environment, and the energy minima of solvent molecules in contact with the solute is likewise different from that in the pure solvent); and the solvent molecules may need to relocate (translate) to find new minima around the solute molecule. This relocation creates a degree of organization that will show up in the region of solvent surrounding the solute. This region is known as a solvation sphere, and the orientation of solvent molecules within this sphere becomes quite stable with respect to the solute. This is due to strong and specific interactions between the solute and the solvent around it. All these changes will propagate to the outer solvation shells, although most of them fade quickly with the distance to the solute. The new state is obviously also dynamic, and solvent molecules may exchange positions between shells. In summary, a physically meaningful representation of solvation needs to include all these effects and it has to include enough solvation shells (a droplet) to ensure that the outer molecules in the system suffer negligible effects from solute solvation. The latter means that the outer solvent molecules approximate the behavior of the pure solvent. At least, therefore, a simulation should include:

- A quantum mechanical description of the solute.
- A quantum mechanical description of the solvent molecules.
- A large number of solvation shells.
- All of the above treated dynamically over a long time sample, such that average values are representative.

If we take into account that the number of molecules needed for each solvation shell grows with the square of the radius, it becomes quickly obvious that such a system is intractable with current resources at a quantum chemical level once a handful of solvation shells are taken into account (Fig. 4.1). To complicate it further, creating a solvation model with a sphere of solvent molecules brings another artifact to the system. Molecules at the edge of the sphere are not fully solvated; they become surface solvent molecules and they will rearrange during the dynamic simulation to behave as such. The ratio of bulk vs. surface solvent molecules then becomes unrealistically small, and its effects propagate back to the inner areas of the droplet. To avoid this, we would need to use a periodic description of the system (and thus replace the spherical droplet with other shapes, like cubes or truncated octahedrons, etc.) or create potentials and apply them to the border solvent molecules to mimic their missing surroundings. An accurate representation of solvation therefore requires a very large amount of solvation molecules, quantum mechanically described, and its dynamic behavior averaged over time. In this scenario, a number of simplifications have to be imposed to turn this problem into something tractable.

From the mental exercise performed above, the duality of solvation becomes evident. This is a macroscopic reality, in which bulk effects are important but with very strong microscopic effects. In this regard, techniques make use of a bottom-up or a top-down approach, depending on whether they stress the bulk (macroscopic) side of the

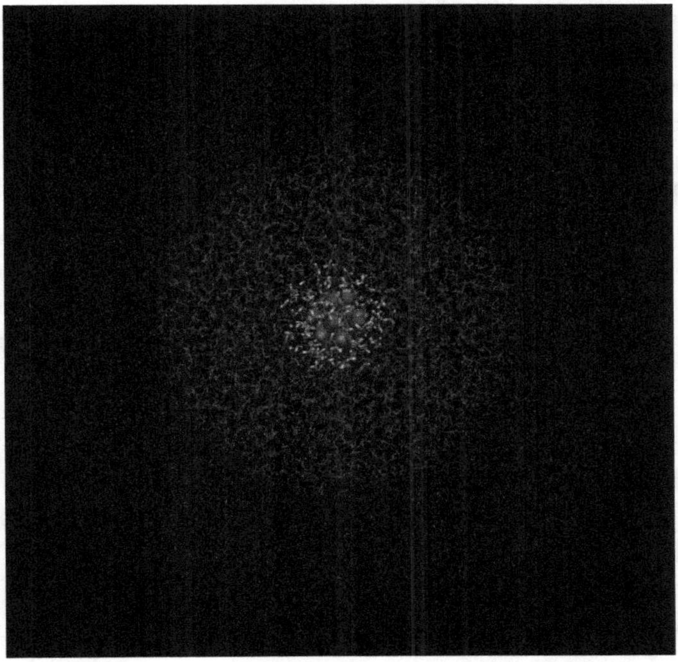

Figure 4.1. Water spheres after dynamic equilibration of 5, 10, 30 and 50 Å. They require 20, 140, 3700, and 17 500 solvent molecules, respectively.

problem or its molecular (microscopic) counterpart. Additionally, the wide variety of solvents available and their diverse array of properties (polarity, acidity/basicity, molecular size, H-bond capabilities, etc.) make it very challenging to construct a single solution protocol that provides reasonably good results for all circumstances. There is also a wide field of chemistry in solution that occurs very far from the ideal state of pure solvent and solute. Most of the biochemistry, at least all that occurs either in the intra- or the extra-cellular matrix is highly affected by other solvated species, particularly dissociated salts. For instance, these anions and cations exert extraordinary effects on bio-macromolecules featuring multiple charged sites that may participate in the chemistry object of our simulation. For these reasons, the methods and theoretical approaches to solvation are rather diverse, and they are generally constructed to simulate solvation within a confined range of scenarios.

As with any other computer-aided simulation, the methods aimed at describing solvation are more accurate the closer they get to the actual physics of the problem. However, this would imply an atomistic and fully quantum mechanics description of solvation that nowadays is intractable even for the simplest solvated molecule. The number of solvent molecules needed to obtain a fair description of this phenomenon requires very efficient methods and algorithms to keep computational cost at a reasonable level (Fig. 4.2). To achieve such efficiency, some methods even sacrifice the high level of discretization that solvation entails and model solvent as if it only had bulk properties, hence disregarding the specific interactions mentioned above. Despite the crude approximation that this implies, such methods provide reasonable results for many problems where solute–solvent specific interactions are weak. In order to gain efficiency but maintaining a discrete representation of the solvent, a fully quantum mechanics description can be replaced with empirical potentials that are much faster to compute.

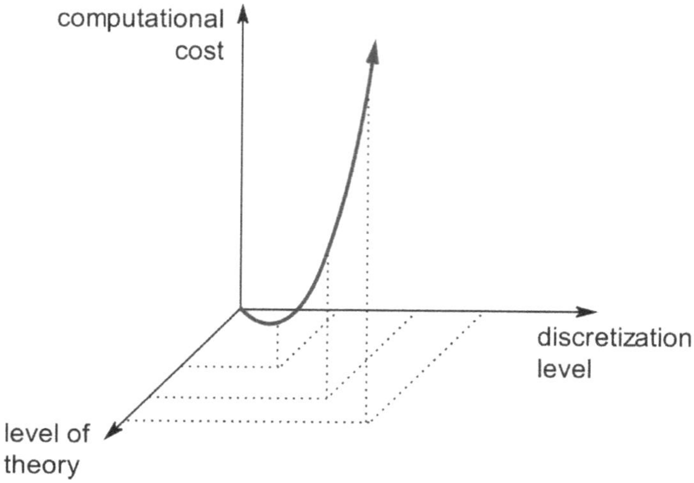

Figure 4.2. Qualitative representation of the cost of simulating solvation with respect to the level of discretization of the model employed and the theoretical level used to describe the solute molecules.

This however comes at the cost of reduced accuracy and versatility since these potentials are bound to a solvent-by-solvent recipe and a parameter set to evaluate interactions.

Very different methods are employed depending on the nature of the problem at hand and on the degree of impact expected from solvation. For instance, many small apolar molecule problems taken from organic laboratories are tackled with very simplistic solvation approaches, forsaking discretization and specific interactions without compromising much accuracy. On the other side, computational chemistry involving bio-macromolecules usually compromises on the models employed to describe the solute due to its size but they keep an atomistic description of the solvent molecules because solvation exerts an enormous effect on bio-macromolecules and a reasonable description requires the inclusion of specific solvent–solute interactions and an accurate determination of entropy.

In this chapter, we will provide an overview of the most common ways of simulating solvation, going from the simplest models in terms of level of discretization to those including an atomistic description of solvation.

4.2 Continuum methods

Continuum methods model the effect of the solvent in our system of interest by embedding this system in a continuum medium described using macroscopic properties (dielectric constant, surface tension, density, etc.). As a result, specific solute–solvent interactions stemming from the discrete molecular nature of the solvent (e.g., hydrogen bonding) are lost, although this disadvantage is offset by several advantages. The most evident of these is the reduction in computational cost, but other advantages include the possibility of taking into account long-range effects that can be lost with the cluster approach and the instantaneous response of the solvent, precluding the need of performing solvent equilibration. In replacing the molecules of solvent with a continuous dielectric, the different microstates of the solvent are averaged, and potential energy surfaces become much simpler as a result. This reduction in the number of the degrees of freedom avoids the problem of many local minima with just small changes in the solvent structure, greatly facilitating the estimation of free energies of solvated species.

The key magnitude of interest when using these methods is the free energy of solvation ΔG_{solv}, defined as the work needed to transfer one molecule of solute from the vacuum (gas phase) to the solution. This free energy can be partitioned in different terms which will be explained in Sec 4.2.8: the electrostatic interactions between solute and solvent ΔG_{el}, the energy required to generate a cavity in the solvent ΔG_{cav} and the dispersion and repulsion interactions, usually bundled in a "short-range" term (ΔG_{sr}).

$$\Delta G_{\text{solv}} = \Delta G_{\text{el}} + \Delta G_{\text{cav}} + \Delta G_{\text{sr}} \tag{4.1}$$

Under such a formulation, once ΔG_{cav} and ΔG_{sr} are calculated, solvation is essentially a problem of electrostatics: the solute is located in a cavity of a dielectric material and its electron density creates an electric field that interacts with the charges in the material (the solvent), which, on their part, alter the electronic distribution in the solute, generating a self-consistent problem that has to be solved in an iterative way (Fig. 4.3).

In the following sections, we will provide a short description of the main options available. A very comprehensive review of these techniques, including information about how they can be implemented in quantum mechanical codes, can be found in two thorough reviews by Tomasi *et al.*[1,2]

4.2.1 *Solvent as a dielectric continuum*

The main assumptions in using a dielectric continuum to model the solvent are the following:

- The solute occupies an empty cavity in the solvent.
- The solvent is a continuous dielectric characterized by a series of macroscopic parameters, such as the dielectric constant ϵ.
- The solvent is isotropic, at equilibrium at a given temperature and pressure.
- The solvent is infinite in all three spatial dimensions.

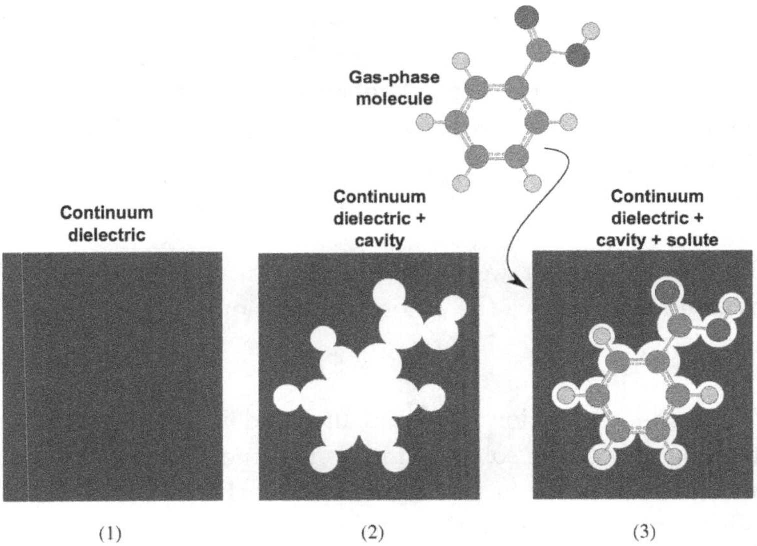

Figure 4.3. Schematic representation of the steps involved in solvation in a continuum method and the calculation of the solvation energy: (1) The reference is the solute in the gas phase and the bulk solvent in equilibrium; (2) an empty cavity is generated in the solvent of the size to accommodate the solute, at the cost of the cavitation energy ΔG_{cav}; (3) the solute is transferred into the cavity and the following main contributions to the energy have to be taken into account: ΔG_{el} (electrostatic interactions between solute and solvent) and ΔG_{sr} (dispersion and repulsion interactions between solute and solvent). If the electronic structure of the solute changes upon solvation, the corresponding term has to be added to the solvation free energy as well.

- The solution is dilute, so no interaction between two molecules of solute is considered.

When using such a model we are only interested in the effect the solvent has on the solute, forsaking all information about the structure or dynamics of the solvent. Thus, calculations are focused on the solute, and the effect of the solvent is treated as a perturbation in the Hamiltonian of the solute. Thus, the Hamiltonian in solution ($\mathbf{H}_t(\mathbf{r_m})$) can be expressed as

$$\mathbf{H}_t(\mathbf{r_m}) = \mathbf{H}_m(\mathbf{r_m}) + \mathbf{V}_{m-s}(\mathbf{r_m}, Q(\mathbf{r_m}, \mathbf{r})) \tag{4.2}$$

where $\mathbf{r_m}$ are the coordinates of the solute, $\mathbf{H}_t(\mathbf{r_m})$ is the gas-phase Hamiltonian and $\mathbf{V}_{m-s}(\mathbf{r_m},\ Q(\mathbf{r_m},\ \mathbf{r}))$ is the solute–solvent interaction potential. $Q(\mathbf{r_m}, \mathbf{r})$ is the solvent response function, which depends on the coordinates of the solute and the position in space, and can include several terms associated with different components of the interaction between the solute and solvent. In most methods, this response is electrostatic in nature and consists of the polarization of the medium with respect to the field induced by the charge density of the solute.

In this context, if we are using a quantum description of the solute, we need to solve the following equation:

$$\mathbf{H}_t|\Psi\rangle = E|\Psi\rangle \tag{4.3}$$

Since the interaction potential included in $\mathbf{H}_t$ depends on the electron density of the solute, which is not known until this equation is solved, the problem has to be solved iteratively.

The key term in computing solvation energies or energies in solution phase is electrostatic in nature, and the classical formalism of the Poisson equation is used to calculate it. To derive this equation, we use the macroscopic formalism of electrostatics, which, completely in line with our separation between a solute made of discrete particles and a solvent that is described as a continuum medium characterized by a series of averaged quantities, makes a difference between free charges (in this case they would correspond to the charges in the solute) and bound charges or polarization charges, the latter describing the response of the continuum to the electric field exerted by the former. Gauss's law (one of Maxwell's equations) for such a system would be stated (we are going to use the SI in this first part of the chapter) as

$$\nabla \cdot \mathbf{D}(\mathbf{r}) = \rho(\mathbf{r}) \tag{4.4}$$

In this context, the electric displacement field is defined as

$$\mathbf{D}(\mathbf{r}) = \varepsilon_0 \mathbf{E}(\mathbf{r}) + \mathbf{P}(\mathbf{r}) \tag{4.5}$$

where $\mathbf{E}(\mathbf{r})$ is the electric field and $\mathbf{P}(\mathbf{r})$ is the polarization (the density of dipole moments induced in the material by the electric field). The relationship between the polarization and the electric field that creates it is summarized in the electric susceptibility χ, through the expression

$$\mathbf{P}(\mathbf{r}) = \chi\varepsilon_0\mathbf{E}(\mathbf{r}) \tag{4.6}$$

Together with the relationship between $\bar{\varepsilon}$, the permittivity of the medium (a measure of how a dielectric medium is polarized by an electric field), and the electric susceptibility χ,

$$\varepsilon = (1 + \chi)\varepsilon_0 \tag{4.7}$$

These equations result in the following expression for the electric displacement

$$\mathbf{D}(\mathbf{r}) = \bar{\varepsilon}\mathbf{E}(\mathbf{r}) \tag{4.8}$$

$\bar{\varepsilon}$, which is a tensor in the more general case, becomes a scalar in isotropic media. In what are called nonlinear media, the relationship between polarization and the electric field is nonlinear (the susceptibility depends on the electric field itself) and usually a Taylor expansion in $\mathbf{E}$ is used to describe it. Also in the most general case, the polarization is not instantaneous upon changes in the electric field, and a complex permittivity is used in the frequency domain after a Fourier transform.

However, the dielectric media most commonly used for the representation of solvents can be considered to be linear, isotropic with instantaneous response, so that ε is a scalar quantity. The combination of the previous definitions with Gauss's law, lead to the following form for the Poisson equation:

$$\nabla[\varepsilon(\mathbf{r}) \cdot \nabla\phi(\mathbf{r})] = -\rho(\mathbf{r}) \tag{4.9}$$

This equation can be further simplified if the medium is considered to be homogeneous (ε does not depend on the position):

$$\nabla^2\phi(\mathbf{r}) = -\frac{\rho(\mathbf{r})}{\varepsilon} \tag{4.10}$$

In some cases, especially in the simulation of biological systems, where physiological concentrations of salts are relatively high and often determine structure and function, the solution is composed of not only a molecule of solute in an infinite bath of solvent molecules but also a cloud of mobile ions, whose movement can be described using Boltzmann statistics. In this case, a term is added to Poisson equation to account for the potential generated by these ions, leading to what is called the Poisson–Boltzmann equation

$$\nabla[\varepsilon(\mathbf{r}) \cdot \nabla\phi(\mathbf{r})] - \lambda(\mathbf{r}) \sum_{i=1}^{n} c_i q_i e^{\frac{q_i \phi(\mathbf{r})}{k_B \mathbf{T}}} = -\rho(\mathbf{r}) \quad (4.11)$$

where λ is a switching function which takes the value 0 in the solute region (impenetrable to ions) and 1 in the solvent area; c_i is the density of ions i with charge q_i; k_B is the Boltzmann constant; and T the temperature.

In the case of a 1:1 salt, this equation becomes

$$\nabla[\varepsilon(\mathbf{r}) \cdot \nabla\phi(\mathbf{r})] - \lambda(\mathbf{r})(\mathbf{r})\kappa(\mathbf{r})^2 \frac{k_B T}{q} \sinh\left(\frac{q\phi(\mathbf{r})}{k_B T}\right) = -\rho(\mathbf{r}) \quad (4.12)$$

where κ^2 is the Debye–Hückel parameter and I the ionic strength, and it can be further simplified using a Taylor expansion of the sinh term if the electrostatic potential is considered to be weak (at low ionic strength), leading to what is known as the linearized Poisson–Boltzmann equation

$$\nabla[\varepsilon(\mathbf{r}) \cdot \nabla\phi(\mathbf{r})] - \lambda(\mathbf{r})\kappa(\mathbf{r})^2\phi(\mathbf{r}) = -\rho(\mathbf{r}) \quad (4.13)$$

This linearized version has several mathematical advantages over the nonlinear form, especially the fact that the electric fields are superimposable (the field due to a series of charges is the sum of the field of the individual charges) and that the expression for the electrostatic energy of the charge distribution has the same functional form as a system in vacuum or in a dielectric (the electrostatic potential is not proportional to the charge in the full Poisson–Boltzmann equation).

The Poisson–Boltzmann equation (Eq. (4.11)) becomes the Poisson equation (Eq. (4.9)) in the absence of mobile ions, which turns into Coulomb's law when $\varepsilon(\mathbf{r})$ is constant through space.

The solution of the nonlinear Poisson–Boltzmann equation (Eq. (4.12)) cannot be obtained analytically, even for simple surfaces such as a sphere, although there are a variety of numerical algorithms that can be applied to find it. Solving the linearized Poisson–Boltzmann equation (Eq. (4.13)) is simpler and can be done using finite difference, boundary element, or finite element methods, among others.[3] In most instances of simulation of molecular properties or reactivity in chemistry, however, the presence of ionic species in the solution phase is not considered, so Eq. (4.9) is the one that needs to be solved to account for the electrostatic part of solvation. We will focus our discussion in the most common approaches for solving this problem.

When using continuum solvent methods, the dielectric constant usually assumes two different values: 0 (sometimes another low value in the simulation of biomolecular systems) inside the cavity and a higher value in the solvent region. The molecular surface (the frontier between solute and solvent regions) displays a discontinuity in $\varepsilon(\mathbf{r})$.

The electrostatic contribution to the interaction between solute and solvent can be expressed as the following integral inside the cavity (where all the charge density is supposed to be)

$$W = \int_{V_{\text{int}}} \rho(\mathbf{r})\phi(\mathbf{r})dV \tag{4.14}$$

If the solvent is assumed not to affect the charge distribution in the solute, the electrostatic contribution to the solvation free energy (the work done to bring the charge distribution of the solute into the solvent cavity vs. the work done to assemble this same charge distribution in vacuum) is then

$$\Delta G_{\text{pol}} = -\frac{1}{2}\int \rho(\mathbf{r})[\phi_{\text{sol}}(\mathbf{r}) - \phi_{\text{gas}}(\mathbf{r})]d\mathbf{r} \tag{4.15}$$

4.2.2 *The Kirkwood–Onsager model*

The Kirkwood–Onsager model is based on the Born model for obtaining the free energy of solvation of a monoatomic ion. The ion is modeled by a conducting sphere, so that the potential is zero inside it and $\phi(\mathbf{r}) = -q/(4\pi\varepsilon|\mathbf{r}|)$ outside, and its charge is uniformly spread on the surface with a charge density of $\rho(\mathbf{s}) = q/(4\pi a^2)$, where $(\mathbf{s})$ is a point on the surface and a is the radius of the sphere.

Integrating on the surface of the sphere ($|\mathbf{r}| = a$), we get that

$$G = -\frac{1}{2}\int \frac{q}{4\pi a^2}\left(-\frac{q}{4\pi\varepsilon a}\right) d\mathbf{s} = \frac{q^2}{8\pi\varepsilon a} \tag{4.16}$$

so that the polarization free energy is

$$\Delta G_{\text{pol}} = -\frac{1}{2}\left(\frac{1}{\varepsilon_0} - \frac{1}{\varepsilon}\right)\frac{q^2}{8\pi\alpha} \tag{4.17}$$

The Onsager model also uses a spherical cavity, but describes the charge distribution in the solute through a point dipole μ with a polarizability α in its center. It is the origin of the term *reaction field* often used when referring to continuum methods. The solute polarizes the dielectric, which in its turn generates a reaction field that polarizes the solute. The expression for this reaction field *the electric field which acts upon the dipole as a result of electric displacements induced by its own presence*[4] in atomic units is

$$\phi = \frac{2(\varepsilon - 1)}{2\varepsilon + 1}\frac{\mu}{a^3} \tag{4.18}$$

so that the polarization energy for a dipole is

$$\Delta G_{\text{pol}} = -\frac{(\varepsilon - 1)}{2\varepsilon + 1}\frac{\mu^2}{a^3} \tag{4.19}$$

When the variation of the dipole moment as a result of the effect of the solvent field is taken into account through α, the polarization energy is

$$\Delta G_{\text{pol}} = -\frac{\varepsilon - 1}{2\varepsilon + 1}\frac{\mu^2}{\alpha^3}\left[1 - \frac{\varepsilon - 1}{2\varepsilon + 1}\frac{2\alpha}{\alpha^3}\right]^{-1} \tag{4.20}$$

Extensions to this work have been done on two fronts: (a) replacement of spherical cavities by ellipsoids or cavities of an arbitrary shape and (b) replacement of the charged sphere or point dipole to a higher-order multipole expansion, giving rise to multipole expansion continuum solvation models.

If the point dipole in the Kirkwood–Onsager method is replaced by an arbitrary charge distribution inside the sphere, the resultant potential can be expressed through a multipolar expansion:

$$\Delta G_{\text{pol}} = -\frac{1}{2}\sum_{l=0}^{L}\sum_{m=-l}^{l}\sum_{l'=0}^{L}\sum_{m'=-l'}^{l'} M_l^m f_{ll'}^{mm'} M_{l'}^{m'} \tag{4.21}$$

where M_l^m are the multipoles of the solute charge distribution and the f, called reaction field factors, have expressions that depend on the dielectric constant and the geometrical parameters of the cavity.

Although the expressions are straightforward and their computational implementation is simple, a spherical cavity does not describe well most molecules, and the choice of radius is then problematic, leading to unphysical results. Extension of this model to ellipsoidal cavities can solve some of these problems, but most modern implementations of multipolar expansion methods use cavities of arbitrary shape, such as those used in polarizable continuum model (PCM), and multicenter multipolar expansions. In this more general case, Eq. (4.21) becomes

$$\Delta G_{\text{pol}} = \sum_{I,J} M_l^m(I) f_{ll'}^{mm'}(I,J) M_{l'}^{m'}(J) \tag{4.22}$$

where the reaction factors f no longer have an analytical expression and have to be calculated numerically.

The Born or Onsager models could be considered versions of this general method with a single center expansion truncated at the zeroth order in the first case and with only the dipole term different from zero in the second.

In computational chemistry, the Onsager method (spherical cavity, molecular dipole plus global charge in the case of an ion) was very popular in the past, since the analytical formulation was easily implemented in computational codes and the resultant calculations were fast. The volume of the solute was calculated in the gas phase as the volume inside a given contour of the density and used to estimate the radius of the cavity. This method is very sensitive to the radius of the cavity and somewhat less to the dielectric constant of the solvent (ε). The main limitations lay in the reduction of the charge distribution of the solute to just a dipole moment, neglecting all higher multipoles, and the use of a spherical cavity, which usually does not reproduce well the shape of the solute, but it can provide meaningful results for compact molecules with large dipoles that are sensitive to the molecular structure.

4.2.3 *The Generalized Born model*

The Generalized Born model considers that the solute is formed by a medium of dielectric constant ε_{in} with a charge q_i associated to each atom i, which is then embedded in the solvent, modeled as a dielectric characterized by ε_{out}.

It can be formulated as either a multicenter multipolar expansion truncated at the first term (charge) with centers of expansion at every nucleus, or a solution to the Poisson equation using Green's functions.[5,6]

The first approach leads to the most common expression for the electrostatic free energy

$$\Delta G_{\text{el}} = \sum_{ij} \Delta_{ij}^{\text{GB}} = -\frac{1}{2}\left(\frac{1}{\varepsilon_{\text{in}}} - \frac{1}{\varepsilon_{\text{out}}}\right)\sum_{ij} \frac{q_i q_j}{f^{\text{GB}}(r_{ij}, R_i, R_j)} \qquad (4.23)$$

where q_i is the charge of atom i, the r_{ij} are the distances between each pair of atoms i and j and $1/f^{\text{GB}}$ is an empirical Coulomb operator applied to all atom pairs. The most common expression for it is

$$f^{\text{GB}} = [r_{ij}^2 + R_i R_j \exp(-\gamma r_{ij}^2 / R_i R_j)]^{1/2} \qquad (4.24)$$

with γ usually being 1/4, although other values have been used (0 would be the exact value in a spherical cavity). This parameter is useful in somewhat reproducing the "dampening" of the interactions between charges in regions of the cavity that are far from spherical, due to the field lines having to cross regions of high ε_{out}, not just ε_{in}

as in the sphere. R_i is the *effective Born radius* of atom i, a quantity that depends on the atom's degree of exposure to the surface between solute and solvent.

If the Green function is used to solve the Poisson equation for such a problem, we get

$$\nabla[\varepsilon(\mathbf{r}) \cdot \nabla\mathbf{G}(\mathbf{r_i}, \mathbf{r_j})] = -4\pi\delta(\mathbf{r_i} - \mathbf{r_j}) \tag{4.25}$$

$$\mathbf{G}(\mathbf{r_i}, \mathbf{r_j}) = \frac{1}{|\mathbf{r_i} - \mathbf{r_j}|} + \mathbf{F}(\mathbf{r_i}, \mathbf{r_j}) \tag{4.26}$$

a solution composed on a Coulombic term and a reaction field $F(\mathbf{r_i}, \mathbf{r_j})$ due to the polarization charges induced in the solute/solvent boundary, which has to satisfy the Laplace equation $\nabla^2 F\ (\mathbf{r_i},\ \mathbf{r_j})$. This can be used to obtain the electrostatic free energy as

$$\Delta G_{\text{el}} = \frac{1}{2}\sum_{ij} q_i q_j F(\mathbf{r_i}, \mathbf{r_j}) \tag{4.27}$$

with the most general form[6] for $F\ (\mathbf{r_i},\ \mathbf{r_j})$ being

$$F(\mathbf{r_i}, \mathbf{r_j}) \approx F_{\text{GB}}(DG_{ii}^{\text{el}}, \Delta G_{jj}^{\text{el}}, r_{ij}) \tag{4.28}$$

where DG are self-energy contributions to the electrostatic part of the solvation free energy of charges q_i and q_j in the molecule. In calculating these terms, a simplified expression for $F\ (\mathbf{r_i},\ \mathbf{r_j})$ is used. The choice of simplified model is a spherical cavity, for which an analytical expression can be derived, leading to Eq. (4.28).

In the case of a single, spherical ion of radius R_i, Eq. (4.23) becomes

$$\Delta G_{ii}^{\text{el}} = -\frac{1}{2}\left(\frac{1}{\varepsilon_{\text{in}}} - \frac{1}{\varepsilon_{\text{out}}}\right)\frac{q_i^2}{R_i} \tag{4.29}$$

The problem with these equations is that for calculating ΔG_{el}, we need a R_i for every atom i in the solute. There are different approaches to calculating these effective Born radii, based on integral approaches (the volume integral of the energy density of a Coulomb field) or using surface integrals. The key step is first calculating the

$\Delta G_{\rm el}$ terms, which represent the work of transferring charge i from a medium characterized by $\varepsilon_{\rm in}$ (or vacuum) to its position in the solute cavity surrounded by the solvent but without any of the other charges. From these, the R_i are calculated, so that they can be used to obtain the $f^{\rm GB}(\mathbf{r_i}, \mathbf{r_j})$ terms and $\Delta G_{\rm el}$.

The main approximations introduced with this method have to do with the functional form of Eqs. (4.23) and (4.24) (associated with the method itself), the definition of the solvent cavity and the approximations used to obtain the effective Born radii, the latter being the most critical. Onufriev *et al.*[7] have shown by using the generalized Born model with effective radii calculated through the Poisson equation ("perfect radii"), that the model gives a very good approximation to the Poisson equation results, although this evaluation of the R_i is costly. The definition of the solute/solvent boundary affects the effective radii through the integrals used to calculate them, so it is not trivial, as well. Different approaches are used between the two extremes of a solute region limited to the van der Waals spheres corresponding to each atom and a solute region defined by a molecular surface comprising the interstitial space. A better description of a molecular system in the latter approach comes with the higher cost of calculating integrals in complex volumes. The kind of equilibrium point is found for these approximations defines different flavors of the method.

This method can be used for simulations where interactions are described through a force field or in quantum mechanical calculations. In the former case, the atomic charges are those defined by the force field for each atom type; in the latter, the assignation of partial charges to atoms has to be done using any of the different partition schemes of the electron density (Mulliken charges, CM4, etc.).

4.2.4 *The polarizable continuum method*

The main advantage of the PCM with respect to the previously described methods lies in its ability to describe any kind of solute, or several interacting solutes, independently of its shape or charge distribution. The key is the definition of the solute boundary, allowing for arbitrary shapes, and the way the polarization of the solvent is treated.

The Poisson equation is solved for a potential that can be separated into two terms: the potential generated by the solute (ϕ_M) and the potential generated by the polarization of the dielectric (ϕ_{pol}).

$$\phi(\mathbf{r}) = \phi_M(\mathbf{r}) + \phi_{\text{pol}}(\mathbf{r}) \tag{4.30}$$

Two sets of boundary conditions are invoked for this electrostatic potential: the first relate to its behavior at infinity and the second at the surface (ϕ has to be continuous at Γ, and the component of the field normal to the surface, discontinuous).

$$\lim_{r \to \infty} r\phi(\mathbf{r}) = 0 \tag{4.31}$$

$$\lim_{r \to \infty} r^2 \nabla\phi(\mathbf{r}) = 0 \tag{4.32}$$

$$\phi_{\text{in}}(\mathbf{s}) = \phi_{\text{out}}(\mathbf{s}) \tag{4.33}$$

$$\frac{\partial \phi_{\text{in}}(\mathbf{s})}{\partial \mathbf{n}} = \varepsilon \frac{\partial \phi_{\text{out}}(\mathbf{s})}{\partial \mathbf{n}} \tag{4.34}$$

These continuity/discontinuity boundary conditions at the surface Γ result in a polarization surface density charge that is the cornerstone of the method.

PCM is a general formulation of the electrostatics problem in continuum methods based in the representation of the solvent reaction potential ($\phi_{\text{pol}}(\mathbf{r}) = (\phi_\sigma(\mathbf{r}))$ using an apparent surface charge density $\sigma(\mathbf{s})$ spread on the cavity surface Γ ($\mathbf{s}$ is a position vector on the surface). The reaction potential arising from such a charge distribution is:

$$\phi_\sigma(\mathbf{r}) = \int_\Gamma \frac{\sigma(\mathbf{s})}{|\mathbf{r} - \mathbf{s}|} ds \tag{4.35}$$

The electrostatic contribution to the free energy of solvation is then the interaction of this reaction potential generated by the apparent surface charges (ASC), with the charges in the solute ρ.

The key issues in this method, which are going to define its different versions, are: the definition of these apparent surface charges, the building of the cavity surface and how the integration of Eq. (4.35) on this complex surface is approached.

The surface is usually built as a superposition of overlapping spheres corresponding to the atoms in the solute, sometimes with some smaller spheres added to eliminate crevasses (van der Waals, solvent-accessible surface, or solvent-excluding surface). Different set of radii can be chosen, which have been optimized and scaled for specific applications. Another option is to use an isodensity surface of the solute at a given ρ.

The surface is discretized in a set of small finite elements (tesserae), and the surface density charge is discretized as a set of point charges (q_i) located at each of these surface elements. The tesserae are sufficiently small that the charge density can be considered constant in its whole surface A_i, so that Eq. (4.35) becomes:

$$\phi(\mathbf{r}) \simeq \sum_i \frac{\sigma(\mathbf{s_i})A_i}{|\mathbf{r} - \mathbf{s_i}|} = \sum_i \frac{q_i}{|\mathbf{r} - \mathbf{s_i}|} \tag{4.36}$$

Since the potential is needed to define the values of these surface charges, an iterative procedure is used to obtain the reaction potential.

This discretization of the surface, needed to carry out the numerical integration of the PCM equations, was one of the main caveats of the method. It often resulted in discontinuities and an energy functional that is not smooth with respect to the variables from which it is constructed. As the solute atoms are displaced, the spheres that define the cavity are also moved, and some new grid points can appear or disappear as the surface changes. As a result, the exploration of potential energy surfaces or trajectories in solution was not always reliable. Well known are the problems in reaching convergence in geometry optimizations, very large errors in vibrational frequency calculations and unstable molecular dynamics trajectories due to the errors in the evaluation of gradients. The work by York and Karplus[8] on the development of a smooth-COSMO method, however, was rediscovered a decade after its publication and incorporated in a more robust implementation of the PCM model.[9,10] Its most important feature is the use of a switching function around each atom that modulates the turning on and off of surface grid points.

In its most general formulation, PCM adopts the form of a set of N linear equations (N being the number of tesserae) that has to be solved using the boundary elements method (BEM). The key step in this BEM approach involves recasting the differential Poisson equation as an integral equation defined on the solute/solvent boundary surface and an integral relating the boundary solution to the solution at points in the domain.

$$\mathbf{kq} = \mathbf{Rf} \tag{4.37}$$

In these equations $\mathbf{q}$ is a vector formed by the surface charges at the discretization points (q_i), $\mathbf{f}$ are the values of the electrostatic quantity (the normal component of the electric field E_n for classical D-PCM and the electrostatic potential ϕ for the other methods) of the solute at the discretization points, and the matrices $\mathbf{K}$ and $\mathbf{R}$ depend on the "flavor" of PCM used (see Table 4.1).

The matrix $\mathbf{S}$ represents the potential at the surface and $\mathbf{D}^*$ the electric field normal to the surface generated by the surface charges. The interpretation of $\mathbf{D}$ is less straightforward, but it can be thought of as the integral of the normal field on the surface. In a general formulation of PCM, where the mathematical formalism is independent of how the cavity is defined, these matrices correspond to the discretization of the following operators:

$$\mathbf{S}\sigma(\mathbf{s}) = \int_\Gamma \frac{\sigma(\mathbf{s}')}{|\mathbf{s} - \mathbf{s}'|} d^2\mathbf{s}' \tag{4.38}$$

Table 4.1. Definition of the matrix operators in $\mathbf{Kq} = \mathbf{Rf}$.[11]

Methods	Refs.	$\mathbf{K}$	$\mathbf{R}$	$\mathbf{f}$	f_ε
D-PCM	[12]	$\mathbf{A}^{-1} - \frac{f_\varepsilon}{2\pi}\mathbf{D}^*$	$-f_\varepsilon \frac{1}{2\pi}$	$E_\mathbf{n}^\rho(\mathbf{s})$	$\frac{\varepsilon-1}{\varepsilon+1}$
COSMO	[13]	$\mathbf{S}$	$-f_\varepsilon \mathbf{I}$	$\phi_\rho(\mathbf{s})$	$\frac{\varepsilon-1}{\varepsilon+x}$
C-PCM	[14]	$\mathbf{S}$	$-f_\varepsilon \mathbf{I}$	$\phi_\rho(\mathbf{s})$	$\frac{\varepsilon-1}{\varepsilon}$
IEF-PCM	[15]	$\mathbf{S} - \frac{f_\varepsilon}{2\pi}\mathbf{DAS}$	$-f_\varepsilon \left(\mathbf{I} - \frac{1}{2\pi}\mathbf{DA}\right)$	$\phi_\rho(\mathbf{s})$	$\frac{\varepsilon-1}{\varepsilon+1}$
SS(V)PE	[11]	$\mathbf{S} - \frac{f_\varepsilon}{4\pi}(\mathbf{DAS} + \mathbf{SAD}^*)$	$-f_\varepsilon \left(\mathbf{I} - \frac{1}{2\pi}\mathbf{DA}\right)$	$\phi^\rho(\mathbf{s})$	$\frac{\varepsilon-1}{\varepsilon+1}$

$$\mathbf{D}^*\sigma(\mathbf{s}) = \int_\Gamma \left[\frac{\partial}{\partial \mathbf{n_s}} \frac{1}{|\mathbf{s} - \mathbf{s}'|} \sigma(\mathbf{s}') d^2\mathbf{s}'\right] \tag{4.39}$$

$$\mathbf{D}\sigma(\mathbf{s}) = \int_\Gamma \left[\frac{\partial}{\partial \mathbf{n_{s'}}} \frac{1}{|\mathbf{s} - \mathbf{s}'|} \sigma(\mathbf{s}') d^2\mathbf{s}'\right] \tag{4.40}$$

A is a diagonal matrix with the surface areas of the tesserae in the diagonal, and **I** is a unit matrix. For more information about how to build these matrices, the reader is referred to the original literature or the reviews in Refs. [1, 10, 13].

Nowadays, PCM is an umbrella term that includes a large family of ASC methods, whose main characteristics are a realistic definition of the solute/solvent boundary and the use of a nontruncated solution of the electrostatic problem (the definition of the potential in Eq. (4.35) is exact if the surface density charge is correctly chosen). Thus, the original PCM of Tomasi has received the name D-PCM to set it apart from the other versions. A modification of D-PCM was generated so that a boundary based on isodensity surfaces could be used, resulting in the I-PCM method (Isodensity-PCM). Further extension of this method resulted in SCI-PCM (self-consistent isodensity), where the surface is allowed to vary between SCF iterations.[14]

Up until this moment, we have considered that all the charge of the solute lies inside the cavity. This assumption, correct for force field representations of the solute (its electrostatic properties are represented by a series of discrete charges), is however flawed for quantum mechanical solutes, where the continuous charge density extends to infinity. The charge in the solute residing outside the cavity, named *escaped charge*, generates a potential affecting the electrostatic term of the solvation energy, a problem that can be addressed in different ways. The simplest way is implemented in D-PCM; the escaped charge is eliminated by distributing a compensating charge, of the same magnitude and opposite sign to it, over the surface, called *renormalization of the apparent charge.*

Another problem with D-PCM is that it requires the explicit evaluation of $E^\rho_\mathbf{n}(\mathbf{s})$ (the component of the electric field generated by the solute normal to the surface), which is more expensive and more numerically unstable than the evaluation of $\phi^{\rho}(\mathbf{s})$ (the potential). The electric field is also one order of magnitude more sensitive than the potential to the escaped charge.

The conductor-like screening model (COSMO) was independently developed by Klamt and Schüürmann.[15] Instead of the exact dielectric boundary condition, this model uses the much simpler condition of a conductor ($\varepsilon = \infty$). One of its main advantages is the reliance on ϕ^{ρ} instead of $E^{\rho}_{\mathbf{n}}$ and its implicit treatment of the escaped charge. Although its discussion could be easily included in this section, since it shares the general formalism of other PCMs, we preferred to devote a single section to it, together with the C-PCM, to account for this singularity.

The SS(V)PE model (Surface and Simulation of Volume Polarization for Electrostatics) was developed by Chipman[16] while studying alternative boundary conditions for PCM models that improved the treatment of the escaped charge by taking into account the volume polarization it causes outside the cavity. In this method, which also avoids the use of $E^{\rho}_{\mathbf{n}}$, the volume polarization is simulated through an additional volume polarization charge density, that needs to be incorporated to the usual surface polarization in the calculation of the apparent surface. It is more complex numerically than COSMO, but it does not need to incorporate any empirically derived quantity.

Cancès and Menucci[17] developed another version of PCM, the IEF-PCM (Integral Equation Formalism), using a general framework that, in its more complex version, allows even the description of anisotropic dielectrics. The simplified version, which also avoids the use of $E^{\rho}_{\mathbf{n}}$, has been shown to be equivalent to SS(V)PE.

The IEF-PCM/SS(V)PE approach affords not only an exact treatment of the surface polarization (surface charges) but also an approximate treatment of the volume polarization due to the escaped charge. It is nowadays considered the best available boundary condition among the PCM models since it can be formulated independently of the way the cavity is defined, is robust with respect to the escaped charge, and is computationally efficient. First and second derivatives can be calculated analytically, easily providing energies, geometries, and vibrational frequencies in solution, together with magnetic and some spectroscopic properties.[10] It can be considered a general case of the other PCM methods, and is equivalent to C-PCM or COSMO at high ε values.

4.2.5 *Conductor-like screening model*

The COSMO represents a slightly different way of solving the continuum problem. In COSMO, all the essential ideas behind PCM apply. The molecule is surrounded by a surface that represents the frontier between the solute electron density and the solvent. This boundary can be built in as many ways as in the PCM approach, but often times is constructed by addition of atomic spheres of parametrized radii centered on the nuclear positions. This surface, unlike in PCM, is assumed to be a perfect conductor, not a dielectric surface. A conductor mirrors the electrostatic potential of the solute and therefore completely screens its partial charges. The effect of this surface on the electron density of the solute is solved variationally and then the resulting interaction energy scaled by a factor of $f_\varepsilon = (\varepsilon - 1)/(\varepsilon + x)$ to account for the finite permittivity, where Klamt proposed in his original description of COSMO a value of x for neutral and $x = 0$ for charged solutes (see Table 4.1).[18]

The better performance of COSMO related to the effect of the escaped charge and the use of the potential instead of the electric field as **f**, prompted the development of GCOSMO by Truong and Stefanovic[19] and of the C-PCM model by Barone and Cossi[20] and Cossi *et al.*,[21] implementing the conductor boundary condition in D-PCM. The main difference with COSMO is the choice of x as 0 instead of 1/2.[20] Both scaling factors have been justified on the basis of analytic solutions of simpler models. The choice of x is not very relevant for solvents with high dielectric constants, but it is worth devoting some time to it when dealing with low ε solvents.

Another difference and a considerable advantage of COSMO is that Klamt improved its implementation to take into account the outlying charge which results in lower artifacts due to this escaped charge. This correction is however, nonvariational and is computed after self-consistency is achieved in the computation of the wave function.[22,23] For interested readers, a very detailed and more technical discussion on continuum methods has been recently composed by Herbert and Lange.[24]

4.2.6 *Minnesota solvation models*

Truhlar and Cramer have also devoted substantial effort into the construction and improvement of a continuum scheme that can afford accurate solvation energies. For the last two decades these efforts have crystallized in a series of SMx methods, where SM stands for *solvent model* and the x is a numeral that stands for the version (sometimes reflecting the year of publication).[25] These series of models are also based upon the ideas of the continuum dielectric surrounding the solute molecule. The SMx models rely on the generalized Born equation to solve the electrostatic problem augmented with geometry-dependent atomic surface tensions to capture nonelectrostatic effects (cavitation, dispersion, exchange repulsion, solvent structure). The SMx methods also rely on a cavity built through the addition of overlapping spheres with specific radii. In the last two versions of this model, significant differences have arisen. SM8 uses a partial charge scheme (CM4) that roughly resembles the Mulliken approach and may become unstable with respect to the basis set size. This implementation was therefore rather restrictive in terms of the levels of theory available.[26] With SM12 this limitation has been removed since the new implementation employs a charge scheme derived from the molecular electrostatic potential (specifically they consider ChElPG and Merz–Kollman–Singh charges) and therefore it is much more stable with respect to the basis set size.[27] SM12 has also been parametrized with an extensive set of solvents for which experimental data is available and it is therefore more versatile than its older siblings.

A different approach by the same authors pivots around the idea of avoiding the use of partial charges altogether. This method is named SMD, where D stands for the electron density (of the solute) and is deemed *universal*, meaning that it can be applied to any charged or uncharged solute in any solvent for which a few key magnitudes are known (dielectric constant, refractive index, bulk surface tension, and acidity and basicity parameters).[28] This method is enjoying extensive use in current computational studies since it provides reasonably accurate free energies of solvation and has been implemented in many of the most popular quantum chemistry packages. According to the developers, SMD achieves mean unsigned errors of 0.6–1.0 kcal/mol in the solvation free energies of some tested neutral solutes and mean unsigned errors of 4 kcal/mol on average for ions.

4.2.7 *Construction of the cavity*

The construction of the cavity is a key aspect of PCM that can significantly affect the computed properties of the system. This is because it affects not only the electrostatic part of the reaction potential but also other components of the solvation free energy.

There are different approaches for generating this solute/solvent boundary, some of which we will discuss here (for more details, the reader is directed to the original literature and/or the manuals of the codes being used).

As briefly commented in the previous section, I-PCM and SCI-PCM define the cavity as the surface enclosing most of the electronic density of the solute, using a certain isovalue of the electronic density to define the solute/solvent boundary. One of the advantages of SCI-PCM is that as the density changes along the SCF iterations, the cavity changes with it.

Most PCMs, however, use atom-centered spheres as the basis of their cavity. A sphere with a defined radius is assigned to each atom, the van der Waals surface being just the boundary of this set of spheres (no other spheres are used to smooth the surface). Then a spherical probe of radius r_p representing one molecule of the solvent is run over these overlapping spheres so that two surfaces can be defined as a result (see Fig. 4.4): (a) The *solvent-excluded surface*

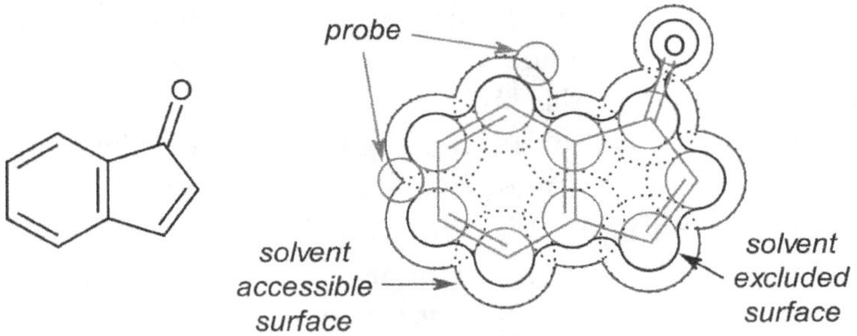

Figure 4.4. Solvent-accessible surface and solvent-excluded surface, two different ways of building the solute/solvent boundary in PCM.

(or Connolly surface) is the surface delimited by the trajectory of the closest point of the probe as it rolls along the atomic spheres. (b) The *solvent-accessible surface* is defined by the trajectory of the center of this same probe and is equivalent to a van der Waals surface where the atomic radii have been increased by r_p.

There are two components to the solvent-excluded surface: the contact surface, where the probe directly touches the van der Waals spheres, and the surface formed by the probe when it is touching more than one atom. As a result, the surface includes the van der Waals volume and interstitial volume. This volume is smaller than that enclosed by the solvent-accessible surface, although both are successfully used in different applications.

A second concern in PCM is how to discretize this surface in order to assign charges to the surface elements and to calculate the appropriate integrals. The most modern PCM approaches involve the use of atom-centered Lebedev grids, with the quadrature weights used to define the surface area associated with each point.[9,10] The surface charge density is then represented as the sum of a set of spherical Gaussian functions centered at each discretization point (akin to the expansion of the solute's density in a set of basis functions used in quantum methods). These methods also incorporate, as previously discussed, the use of switching functions to turn on and off new grid points as the geometry of the solute changes, in order to achieve smooth energies.

Except in the case of isodensity-based approaches for the building of the solute/solvent boundary, atomic radii are the most important variable implicated in the building of the cavity, and their choice can deeply impact the solvation energies obtained. There are many options available, but the most commonly used in most modern applications are Bondi van der Waals radii scaled by a factor of 1.2 (used in Q-CHEM or GAMESS).[29] This factor can be changed in most codes, since the 1.2 value might not be optimal.

Other popular options are the UFF (Universal Force Field)[30] van der Waals radii, with a scale factor of 1.1 (the default in Gaussian09). The United Atom approach uses UFF radii but does not provide a separated sphere for hydrogens, embedding them in the sphere corresponding to the heavy atom to which they are attached. There are specific versions of these radii specially parametrized for the HF/6-31G(d) level of theory (UAHF) or the PBE1PBE/6-31G(d) level of theory (UAKS).

In most codes, atomic radii, or even the whole surface, can be given as input, so that the user can control the shape and size of the solute/solvent boundary.

4.2.8 *Other contributions to the solvation free energy*

Up until now, we have only discussed the electrostatic interactions between solute and solvent, neglecting all other interactions affecting the process of bringing a solute from the gas phase to a cavity in the solvent.

The solvation free energy, defined as the free energy involved in such a process would be defined as follows:

$$\Delta G_{\text{solv}} = \Delta G_{\text{el}} + \Delta G_{\text{cav}} + \Delta G_{\text{sr}} + p\Delta V \tag{4.41}$$

where ΔG_{solv} is the energy of solvation, ΔG_{el} is the electrostatic interactions between solute and solvent, ΔG_{cav} is the free energy involved in the formation of the cavity in the solute at equilibrium, and ΔG_{sr} is a short-range, free-energy term that includes the repulsive and attractive van der Waals interactions between solute and solvent. If the structure of the solute or/and its partition function for nonelectronic degrees of freedom change in solution from their gas-phase values, an additional terms need to be included to account for this: ΔG_{e}, the difference between the electronic energy of the isolated molecule and the electronic energy computed in solution, and ΔG_{tm}, a libration term that reflects that the accessible volume for the solute in solution does not correspond to that in the gas phase.

The cavitation energy is a term that comprises the energetic penalty for creating the cavity in terms of surface tension and the entropy loss involved in the reorganization of the solvent around the solute.

It is usually calculated using Pierotti's scaled particle theory,[31] where a cavitation free energy is associated with every atom through a term representing the cavitation free energy of a sphere with radius r_i in a solvent composed of spheres of radius r_s at a given temperature, pressure and density ($\Delta G_{\mathrm{cav},i}$):

$$\Delta G_{\mathrm{cav}} = \sum_{i}^{N} \frac{A_i}{4\pi r_i^2} \Delta G_{\mathrm{cav},i} \tag{4.42}$$

where A_i is the surface exposed to the solvent of the sphere associated to atom i, with radius r_i.

The calculation of dispersion is much more complicated, since it is a purely quantum mechanical effect that is very difficult to model when there are no explicit solvent molecules. In classical approaches to it, which are usually based on polarizabilities, it can be difficult to separate dispersion from the actual electrostatic interactions.

When cavitation and dispersion free energies are calculated explicitly, they are often quite large and of opposite signs. That results in a convenient cancellation of errors but makes their validation complicated since there are no experimental values to compare with them.

As a result, ΔG_{cav} and ΔG_{sr} are often evaluated together using an empirical function based on the type of atom and its exposed surface area (A_i), through an expression similar to:

$$\Delta G_{\mathrm{cav}} + \Delta G_{\mathrm{sr}} = \sum_{i}^{N} A_i \sigma_i \tag{4.43}$$

where σ_k is an empirically determined "surface tension" or atomic solvation parameter.

Although the expressions for cavitation, dispersion, and other terms in the solvation energy are based on physical principles and models, they are usually somewhat parametrized to improve the agreement with experimentally measured quantities. This parametrization results in them implicitly including contributions that compensate for the errors in the evaluation of the electrostatic terms, so, as a result, ideally they should only be used in conjunction with the method for which they have been developed.

4.2.9 *Nonequilibrium properties*

The most obvious applications of solvation models using a continuum dielectric are the simulation of chemical processes in condensed phases and the estimation of free energies of solvation. Both of these applications, and many others that the reader may consider, involve a solvent surrounding the solute molecules in a situation of chemical and physical equilibrium. This is, in the first place, why most of these methods are intertwined variationally within the self-consistent field procedure to obtain the wave function. However, there are some notable exceptions for which a physically meaningful description of the process requires solvent at a nonequilibrium state. For example, let us assume we are interested in obtaining vertical excitations of a solute. The initial state is the solute molecule in equilibrium with its solvent environment and this is the presumed situation for most solvent simulations. The final state involves a solute molecule in an excited state and the solvent molecules around this solute will need to respond to such a change in its electron distribution. These solvent molecules will respond rapidly via polarizing the electron density, but there is a slow component of the response that involves reorientation (translation and rotation of the solvent molecules around the solute) that will not occur before the solute relaxes back to the ground state. This situation therefore involves nonequilibrium solvation: the geometries of the solvent around the solute are held fixed in the photophysical process and they optimally solvate the initial but not the final state. This effect destabilizes the final electronic state of the vertical excitation and therefore implies a blue-shift with respect to fully equilibrating the solvation environment in both states. Some solvation models based on the dielectric continuum are able to simulate this situation. This is achieved by requiring two different values of the dielectric constant for the solvent under simulation: ε and ε_∞. The first value represents the static dielectric constant of the solvent whereas the second is its dynamic (or optical) dielectric constant. In nonequilibrium simulations, only the optical dielectric constant is employed and allowed to screen the charges of the solute.

More complex examples featuring nonequilibrium solvation are simulating a fluorescence phenomenon or a photochemical reaction. Both imply excitation and de-excitation steps that must occur under a solvent not fully equilibrated in the arriving electronic state (the excited state in the excitation step and the ground state in the de-excitation step) and an adiabatic relaxation (in the fluorescence phenomenon) or an adiabatic chemical step (in the photochemical reaction) that take place at rates that allow the solvent to fully adjust to the solute changes (see Fig. 4.5).

4.3 Discrete and pseudo-discrete methods

In a number of chemical problems, the role of solvation is such that it cannot be reasonably described with a continuum approach. More often than not solvent molecules establish strong interactions with the solute at specific sites. This situation is more common the more polar the solvent is, and it becomes impossible to overlook when the solvent is capable of establishing hydrogen bonds. A *tour de force* approach to avoid this problem is to engulf the solute molecule

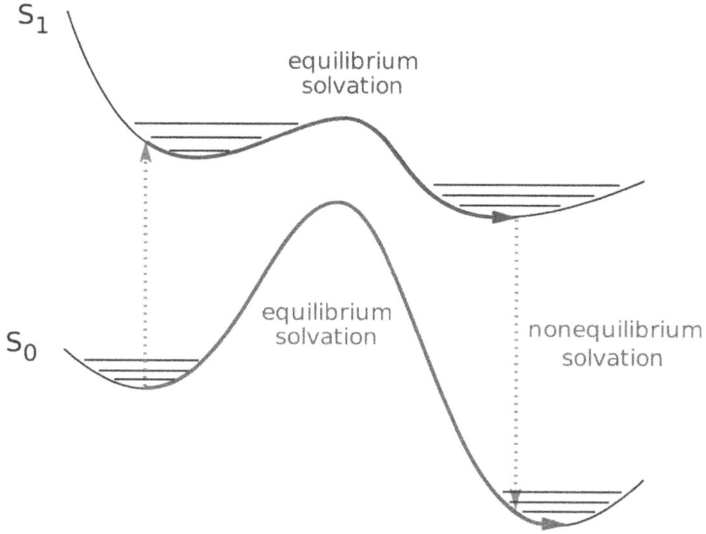

Figure 4.5. Steps of nonequilibrium and equilibrium solvation during: an adiabatic potential energy surface exploration (thermal reaction, bottom) and a nonadiabatic process (photochemical reaction, top).

in a droplet of solvent molecules, but such a solution becomes excruciatingly expensive even for medium-sized solute molecules as it requires extensive sampling until average physical properties converge. Balanced strategies between a simplistic continuum description of the solvent and its atomistic representation, however, exist. The idea behind microsolvation is that, in numerous circumstances, a reasonably accurate representation of solvation only needs to include some explicit solvent molecules at those sites where specific interactions are expected.

These approaches can be quite efficient at capturing most of the effects created by explicit solute–solvent interactions provided that an adequate number and location of solvent molecules is employed. Additionally, when this number becomes large, semi-discrete techniques are available that allows the substitution of solvent molecules with potentials. These approaches will be discussed below.

4.3.1 *Microsolvation*

Microsolvation implies the use of a small number of solvent molecules around a solute to simulate the effect of solvation on the latter. These techniques may be used in conjunction with a continuum dielectric surrounding both the solute and the explicit solvent molecules (see Fig. 4.6). The choice of whether to combine these approaches is problem-based. When the goal is an improved representation of the solvation effects with respect to using a continuum dielectric alone, the combined approach is quite reasonable. In some instances, however, a bottomup approach is carried out, in which the solute molecule is considered as is in a vacuum environment, and then solvent molecules are added sequentially to explore when a chemical change occurs or at which size the solvent cluster converges into the fully solvated solution. A typical example of this approach is the investigation of the structures of aminoacids. In the gas phase, an amino acid is more stable in its nonionized form; however, in water solution, the zwitterionic structure is much more stable. Sequentially adding water molecules to a gas-phase aminoacid molecule in both forms helps determine the critical amount of solvation that allows the zwitterion to coexist with the nonionized form.[32,33] This kind of study can be extended not only to different protonation patterns but

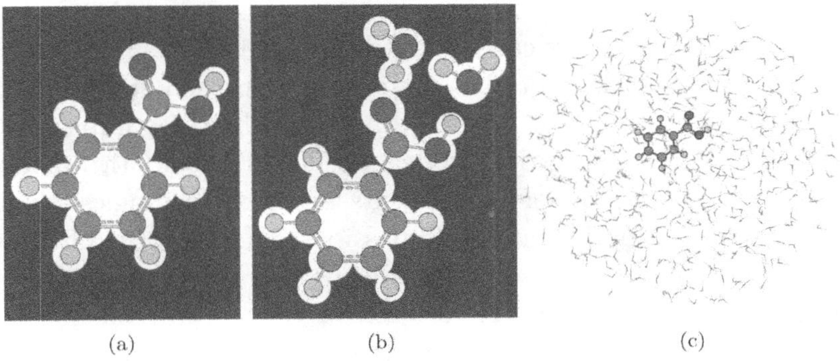

Figure 4.6. Illustration of the solvation environment when simulating benzoic acid with a continuum dielectric (a), two molecules of microsolvation and a continuum dielectric (b) or a 15$\AA$ droplet of explicit water molecules (c).

also to variations in the conformational space, the structure of hydrogen bonds (intra- vs. inter-molecular), or the complexation pattern around metal centers, for instance.[34–37] In this latter approach, using a dielectric continuum would only obscure the results since it diminishes the perturbation applied to a vacuum surrounded molecule when transformed into the microsolvated one.

There are a few methodological advantages of microsolvation beyond those inherent to an explicit description of the noncovalent interactions between solute and solvent molecules. Microsolvation does not impose restrictions on the methodology available to tackle the chemical problem of interest, other than the increased size of the system to calculate. Therefore, all the different levels of theory available to the gas phase molecule are available to the solvated one. Additionally, for a desired level of theory, all the property evaluation techniques and the energy derivatives that are available in vacuum are also available under a microsolvation scheme. For instance, depending on the implementation of a continuum method, like PCM or COSMO, calculation of NMR chemical shifts may be available in the gas phase, but not in the solution phase. A more critical magnitude is the gradient and the second derivatives of the energy with respect to the atomic displacement. The first step of almost every computational simulation in chemistry involves optimization of the geometry of the system under study. This process is relatively straightforward when we can compute first derivatives. Furthermore, when these simulations involve computing transition states, second derivatives are also rather useful. Unfortunately, some implementations of continuum solvation models do not offer smooth functions for the energy and therefore gradients become meaningless, which render geometry optimizations impossible. The use of explicit solvent molecules in a microsolvation scheme may flatten the potential energy surface and make the location of certain stationary points challenging, but they are at least algorithmically possible. Luckily, a few modern implementations of continuum methods circumvent these difficulties and make gradients and second derivatives available for those levels of theory that provide them in the gas phase.

Microsolvation also incurs in a number of difficulties that are not found in the continuum approaches. Let us consider a solute surrounded by a small number of solvent molecules to rescue relevant explicit solute–solvent interactions. Even the smallest possible solute (for instance, a simple alkaline earth cation like Mg^{2+}) requires a few solvent molecules to fill just the first solvation shell (six water molecules are needed for Mg^{2+} in water, for instance). The nonbonding interactions between each of the solvent molecules and the solute are relatively weak and the molecules around the solute are therefore labile. Under these circumstances, the configuration of the solvent around the solute is not very strictly imposed, and small changes in their relative positions involve just a small energy penalty. This creates a multiple minimum problem that is mostly dependent on the number of solvation molecules included in the microsolvation scheme and it becomes quickly intractable to simple optimization techniques. As the number of solvent molecules increase, the representation of the solvation spheres around the solute is improved, at the cost of complicating the potential energy surface and requiring an automatic search engine to locate a representative set of solute–solvent configurations. Simulations including anything but just a few solvent molecules become quickly very complex, and more efficient descriptions of the discrete solvent molecules are made necessary.

4.3.2 *Effective fragment potentials*

The microsolvation scheme is rather efficient at capturing the explicit interactions between solute and solvent at a relatively moderate computational cost. This is the case when dealing with small- to medium-sized molecules with only one or two sites to saturate with explicit solvent molecules. However, as the system under study gets larger or simply more complex, the number of explicit solvent molecules needed to capture these effects escalates very quickly. This is a two-fold problem: on the one hand, the cost of each energy evaluation, even with the most cost-efficient methods scales, roughly, with the second power of the number of electrons included in the quantum mechanics system; and on the other hand, the presence of multiple solvent molecules aggravates the local minimum problem. Putting it simply, the first problem means that all the calculations are significantly more expensive and the second means that, as the system grows, the topology of the potential energy surface becomes more complex and manually looking for an energy minimum is no longer a valid approach; an automatic statistical sampling is needed. To this end, some approximations have been developed aimed at drastically reducing the cost of the calculations while trying to retain most of the quantum mechanics in the description of the system.

One such approximation is the Effective Fragment Potential (EFP). The idea behind this methodology is to replace the explicit solvent molecules that do not establish new covalent bonds or break existing ones along the chemical process under study with potentials. EFPs are analytic functions easily computed, and they replace all the electrons of the solvent molecules they replace. These potentials mimic the behavior of the electron density of the solvent molecules, and they interact directly with the wave function of the solute. The resulting system has the same (or a very similar) complex potential energy surface, with the associated multiple close lying minima and very often still needs an automatic statistical sampling algorithm to locate a representative collection of energy minima, but the drastically reduced cost of the energy evaluations due to the much smaller number of electrons makes the simulation viable.[38–40]

There are currently two versions of the EFP model. The EFP1 has a simpler mathematical expression: it introduces Coulombic, induction, and repulsive interactions via one-electron terms in the *ab initio* Hamiltonian and is parametrized for water as a solvent and with single determinant methods in mind (HF and density functionals). This flavor or EFP is therefore usable but rather limited. The second version, EFP2, is significantly more versatile: it may be applicable to any species (including water) and it includes terms describing not only repulsion but also charge transfer and dispersion.[41,42] The EFP2 method relies on a parameter-free scheme for all the interaction terms described earlier, this means that effective fragment potentials can be constructed *ad hoc* and relatively easily for any species that we would like to model (Fig. 4.7).

Current improvements to the EFP scheme include the possibility of wrapping the solute and the EFP with a dielectric continuum (using the PCM), thus rescuing also the bulk effect of the solvent. Analytic gradients have been coded for this efficient combination of microsolvation and a continuum model, so geometry optimizations are therefore also possible.[43,44] Recent developments on the EFP scheme oriented to biomolecular problems have allowed one to partition a large system into a quantum mechanics fragment and the remaining portion can be described with EFP even if there is a covalent bond between these two moieties.[45] The implementation of this boundary between a QM and an EFP section and its applications is, however, out of the scope of this chapter.

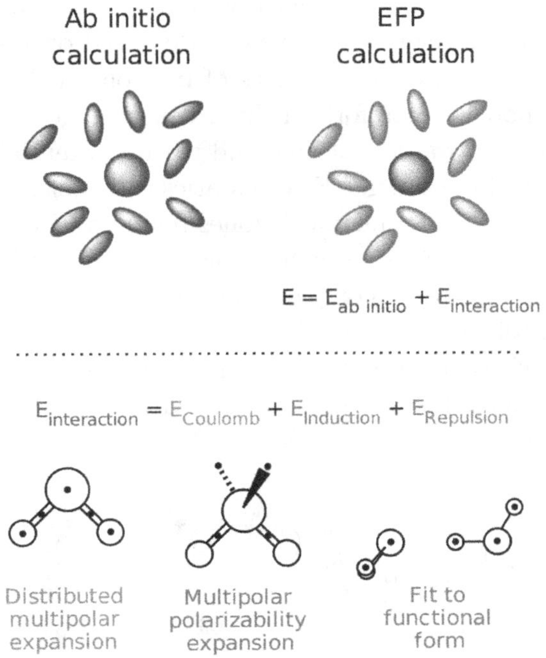

Figure 4.7. Schematic of the construction of the EFP.

4.3.3 *Water models for large-scale solution-phase simulations*

Some chemical problems, particularly when macromolecules are involved, require an enormously large number of water molecules to fully solvate the system under investigation. Most of these simulations also require a discrete description of solvent since entropic effects are fundamental for a correct free energy evaluation. In such cases, even the EFPs may become prohibitively expensive and simpler and faster potentials substituting solvent molecules are needed. In these simulations, particularly when bio-macromolecules are involved, the ubiquitous solvent is water and, for this reason, a number of highly efficient models have been developed for water simulation in large-scale problems. These models are often coupled with a molecular mechanics force field to also allow for a very efficient description of the solute, but this is not strictly necessary, and most water models can be combined with other levels of theory. Most water molecules developed up to date are based on rigid structures in an attempt to reduce the degrees of freedom in the solvated system. These models, therefore, include constant geometric parameters to define each water molecule and then a potential that mimics the nonbonding interactions. Electrostatics is modeled via Coulomb's law and dispersion via a Lennard–Jones potential. Depending on the complexity of this potential, water molecules are classified into categories. When the potential only depends on partial charges centred at the nuclear positions the model is termed "three-site"; if additional (nonnuclear) sites are used to construct the nonbonding interaction potential, the model is termed "n-site," where n is the total number of points employed (see Fig. 4.8).

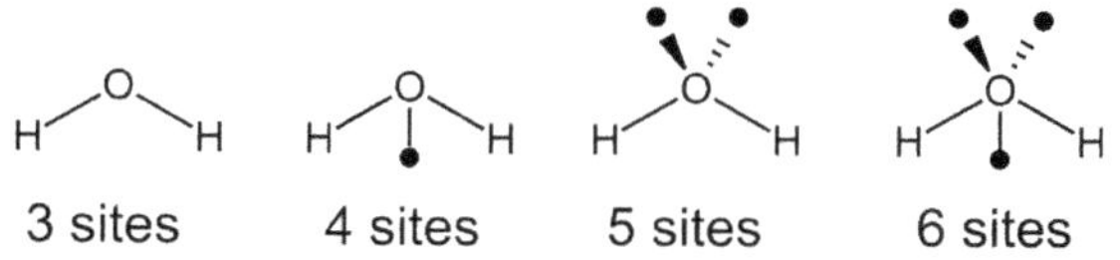

Figure 4.8. Classification of water models depending on the number of sites employed to construct the electrostatic and dispersion terms.

Three-site models include partial charges only at the nuclear positions and probably the most popular model in this set is TIP3P by Jorgensen *et al.*[46] This method has found prolific application and can be found implemented in most molecular dynamics packages.

The four-site models add a dummy atom with a partial charge in the bisection of the HOH bond angle. This dummy atom is therefore mass-less and it is only employed to improve the electrostatic description of the water molecule via an extra pair of parameters (the amount of charge displaced to this fourth point and its exact location with respect to the oxygen nucleus). The first water model developed was a four-site one created by Bernal and Fowler (BF) but its parametrization was somehow deficient.[47] The four-site model by Jorgensen (TIP4P) significantly improves upon both the BF and TIP3P models and it is ubiquitous in biomolecular simulations.[48] Further improvements have been applied on this model, like TIP4P/Ice to be used in ice simulations or TIP4P-Ew to be used with Ewald sums when running simulations with periodic boundary conditions. The earlier water models were unpolarizable, later developments included polarizability to some extent in two ways: (1) The mass-less site can feature fluctuating charge instead of a fixed value (as in TIP4P-FQ[49] for instance) and (2) The mass-less site (in these models named Drude particle) instead of being located at a fixed position is attached to the oxygen by a harmonic spring-like function (4,184 kJ mol^{-1} $Å^{-2}$ (as in the SWM4-DP model, where the name stands for *simple water model with four sites and Drude polarizability*).

Five- and six-site models contain dummy atoms located at the lone-pair positions in a tetrahedral representation of water. These models are therefore the only ones including off-plane partial charges which can account for the corresponding off-plane terms in the polarizability tensor of water. Additionally, the presence of more parameters allows for a better mathematical fit and improved results on physical magnitudes of both solid and liquid water. For instance, these models are the only ones capable of correctly reproducing the temperature of the maximum of density of water (277 K). In the five-site set, the Jorgensen model is also the most used (TIP5P and TIP5P-E for periodic boundary conditions).[50,51] Six-site models are currently not very popular but a couple of examples are available, such as the broad scope SWM6 and a six-site model focused on the simulation of ice/water mixtures.[52,53] New developments in this later scheme are, however, rather promising.[54]

4.3.4 *Computing weak association constants in solution phase*

Some important chemical processes occurring in solution phase involve a small energy exchange. Examples of such processes may be host–guest recognition/association in organic substrates or, more biologically oriented, ligand–receptor binding. In these processes, the entropic contributions stemming from solvent microstates are essential and, therefore, an adequate account of the physics involved requires a discrete representation of the solvent molecules. Some efficient approaches to simulate solution phase chemistry with discrete solvation have been provided before. Most of them, however, would fail to provide reasonable binding constants. The main problem is that the energy exchanged in these equilibria is often times an order of magnitude smaller than the energy difference between two microstates of the solvent. A stochastic simulation would require untractably long runs to tighten convergence such that the binding energy becomes relevant with respect to the noise created by the different solvation configurations.

In this context, the binding equilibrium between an arbitrary host and its guest is represented in a straightforward way as

$$\text{Host} + \text{Guest} \rightleftharpoons \text{Host} - \text{Guest} \tag{4.44}$$

and the free energy of binding would be expressed as

$$\Delta G_{\text{bind}}^{\text{solv}} = \Delta G_{H-G}^{\text{solv}} - \Delta G_{H}^{\text{solv}} - \Delta G_{G}^{\text{solv}} \tag{4.45}$$

However, as we mentioned above, in a simulation trying to directly describe this chemistry, the contributions from the solvent configurations would obscure the energy exchanged in the host–guest

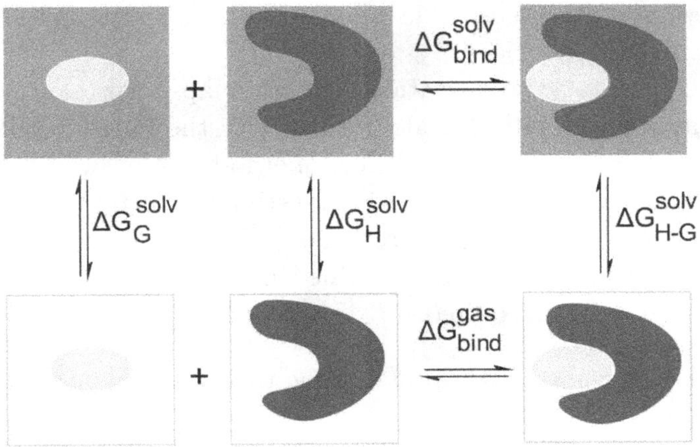

Figure 4.9. Thermodynamic cycle representing the MM-PBSA or MM-GBSA approaches.

binding process. A thermodynamic cycle like that represented in Fig. 4.9 is instead employed.

In the MM-PBSA and MM-GBSA approaches,[55] standing for Molecular Mechanics Poisson–Boltzmann (Generalized Born) Surface Area, the energy terms associated with the thermodynamic cycle described in Fig. 4.9 are approximated as:

$$\Delta G_{\text{bind}}^{\text{solv}} \approx \langle \Delta G_{H-G}^{\text{solv}} \rangle - \langle \Delta G_{H}^{\text{solv}} \rangle - \langle \Delta G_{G}^{\text{solv}} \rangle \tag{4.46}$$

where values in brackets are averages obtained over an ensemble of snapshots on which a Poisson–Boltzmann or a Generalized Born formulation is used to account for the free energies of solvation as follows:

- A solution-phase dynamics trajectory is obtained for the host–guest complex.
- Solution-phase dynamics trajectories are obtained for the host and the guest molecules (this is usually only performed if the conformational shape of either of these species is expected to change significantly from the free to the bound state, otherwise all the data is extracted from a single trajectory of the host–guest complex).

- A representative ensemble of snapshots is extracted from the dynamics trajectories.
- The electrostatics of solvation for each snapshot is captured with a continuum solvation model (either with the Poisson–Boltzmann equation or the Generalized Born method).
- An empirical correction is used to estimate the hydrophobic contributions to the solvation free energy.[56]
- Entropy contributions are obtained in the gas phase via normal mode analysis, if required.

Each of the averages approximating the free energy terms in solution phase is computed as

$$\langle \Delta_X^{\text{solv}} \rangle = \langle E_{MM}^{\text{gas}} \rangle + \langle \Delta G_{\text{solv}} \rangle \langle E_{\text{LCPO}} \rangle - \langle T S_{MM}^{\text{gas}} \rangle \tag{4.47}$$

where ΔG_{solv} is the solvation free energy computed at each snapshot with the Generalized Born or the Poisson–Boltzmann methods and LCPO is the "linear combinations of pairwise overlaps" that is often employed to obtain the hydrophobic correction. If entropic contributions are computed via normal mode analysis (usually in the gas phase) the corresponding molecular mechanics TS_{MM} terms are also appended.

This methodology has been applied successfully to a number of host–guest association processes, both in the organic chemistry field and, more broadly, in biomolecular ligand–receptor and protein–protein complexes.[55,57–61] Recently more exotic interactions, like those being established between carbon nanoparticles and proteins, have also been described through this hybrid discrete-continuum approach.[62]

4.4 Practical considerations

In the simulation of chemical problems involving medium-sized molecules the general advice with regard to solvation models runs fairly parallel to common sense. If your system of interest is fairly apolar and devoid of functionality that could interact strongly with the chosen solvent, a continuum method seems like a good and generally accepted choice to simulate solvation effects in a cost-efficient manner. The usual protocol not very long ago was obtaining optimized geometries in the gas phase with a reasonably good level of theory and a single point calculation including a continuum model to account for solvation effects on the electronic energy. Given the new advances in the implementation of continuum methods, it is hard to find an excuse for this approach nowadays since geometry optimization and property evaluation within the continuum model is possible at a very affordable cost. The errors associated with the evaluation of solvation via single point calculations can be large since the stationary points on the gas phase and solution phase potential energy surfaces are often displaced with respect to each other. A very simple illustration on this kind of error is provided in Fig. 4.10.

Optimization within a continuum solvent and using a modern density functional, for instance, with a sufficiently large basis set is therefore strongly advised. This should be the lowest standard for production-quality calculations on medium-sized molecules for which specific solute to solvent interactions are not expected.

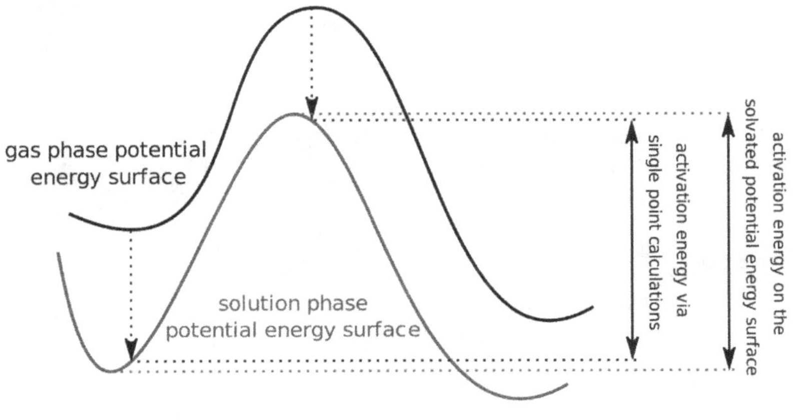

Figure 4.10. Activation barrier computed via single point corrections for solvation with a continuum model and the actual barrier on the solvated potential energy surface using the same solvation model.

In the case of highly polar molecules and/or solvents, and particularly when hydrogen bonds or other strong noncovalent interactions are expected, a microsolvation approach is necessary. This strategy oftentimes does not need to involve many solvent molecules. In many cases, an educated guess in which the most salient sites in terms of noncovalent interactions are saturated with one or two solvent molecules affords very good results. Once the specific interactions are captured with explicit solvent molecules, the bulk effect can be entrusted to the continuum model. Actually the strong effect of only one or two solvent molecules around a strongly interacting solute has been captured even experimentally.[63–66] One example is the observation via microwave spectroscopy of changes to the tautomerization equilibrium of some substrates when one and two water molecules are added (see Fig. 4.11).[67,68] Also, the microsolvated environment of amino acids in proteins has been successfully probed even experimentally via amino acid analogues that feature several proton acceptors and a unique spectroscopic signature for protonation at each of these acceptors.[69]

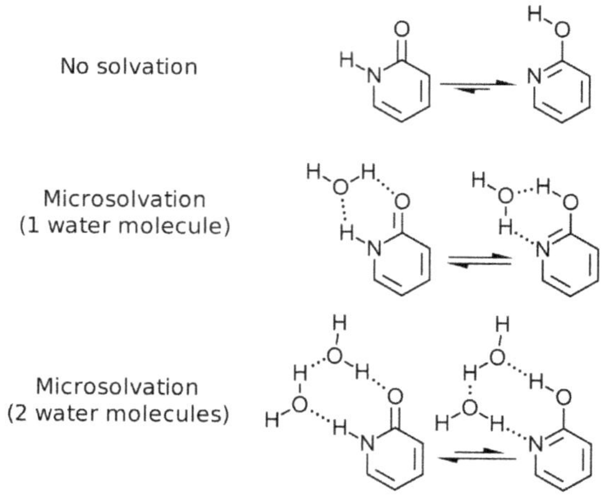

Figure 4.11. The presence of one or two water molecules affects the tautomeric equilibrium of 2-hydroxypyridine/2-pyridone, which is shifted to the enol form for the bare molecules but to the keto tautomer for the hydrated forms.

More complex problems in which the solvent is expected to have a very active role may need a more extensive microsolvation droplet. Chemical problems requiring such an approach are relatively easy to identify since they are affected or even governed by strong noncovalent interactions. Some notorious examples in this group are the conformational analysis of oligopeptides in water, relative pK_a evaluation or long-range transprotonation processes common in some polyfunctionalized molecules and in proton-relay mechanisms. This approach requires careful handling and most often implies the use of a random sampling engine (like a Monte Carlo calculation) to ensure the location of a representative number of local minima and, ideally, the global minimum. Cost-effective approaches, like the use of effective fragment potentials, are in these cases strongly advised since the computational cost of extensively sampling the phase space of a molecule surrounded by a large number of quantum-mechanically described solvent molecules is enormous.

References

1. Tomasi, J., Mennucci, B. and Cammi, R. (2005). Quantum mechanical continuum solvation models, *Chem. Rev.*, 105, pp. 2999–3094.
2. Tomasi, J. and Persico, M. (1994). Molecular interactions in solution: An overview of methods based on continuous distributions of the solvent, *Chem. Rev.*, 94, pp. 2027–2094.
3. Lu, B. Z., Zhou, Y. C., Holst, M. J. and McCammon, J. A. (2008). Recent progress in numerical methods for the Poisson-Boltzmann equation in biophysical applications, *Commun. Comput. Phys.*, 3, pp. 973–1009.
4. Onsager, L. (1936). Electric moments of molecules in liquids, *J. Am. Chem. Soc.*, 58, pp. 1486–1493.
5. Still, W. C., Tempczyk, A., Hawley, R. C. and Hendrickson, T. (1990). Semianalytical treatment of solvation for molecular mechanics and dynamics, *J. Am. Chem. Soc.*, 112, pp. 6127–6129.
6. Sigalov, G., Scheffel, P. and Onufriev, A. (2005). Incorporating variable dielectric environments into the generalized Born model, *J. Chem. Phys.*, 122, p. 094511.
7. Onufriev, A., Case, D. A. and Bashford, D. (2002). Effective Born radii in the generalized Born approximation: The importance of being perfect, *J. Comput. Chem.*, 23, pp. 1297–1304.
8. York, D. M. and Karplus, M. (1999). A smooth solvation potential based on the conductor-like screening model, *J. Phys. Chem.*, 103, pp. 11060–11079.

9. Lange, A. W. and Herbert, J. M. (2010). Polarizable continuum reaction-field solvation models affording smooth potential energy surfaces, *Phys. Chem. Lett.*, 1, pp. 556–561.
10. Scalmani, G. and Frisch, M. J. (2010). Continuous surface charge polarizable continuum models of solvation. I. General formalism, *J. Chem. Phys.*, 132, p. 114110.
11. Chipman, D. M. and Dupuis, M. (2002). Implementation of solvent reaction fields for electronic structure, *Theor. Chem. Acc.*, 107, pp. 90–102.
12. Miertuŭs, S., Scrocco, E. and Tomasi, J. (1981). Electrostatic interaction of a solute with a continuum. A direct utilization of ab initio molecular potentials for the prevision of solvent effects, *Chem. Phys.*, 55, pp. 117–129.
13. Chipman, D. M. (2002). Comparison of solvent reaction field representations, *Theor. Chem. Acc.*, 107, pp. 80–89.
14. Foresman, J. B., Keith, T. A., Wiberg, K. B., Snoonian, J. and Frisch, M. J. (1996). Solvent effects. 5. Influence of cavity shape, truncation of electrostatics, and electron correlation on ab initio reaction field calculations, *J. Phys. Chem.*, 100, pp. 16098–16104.
15. Klamt, A. and Schüürmann, G. (1993). COSMO: A new approach to dielectric screening in solvents with explicit expressions for the screening energy and its gradient, *J. Chem. Soc., Perkin Trans.*, 2, pp. 799–805.
16. Chipman, D. M. (2000). Reaction field treatment of charge penetration, *J. Chem. Phys.*, 112, pp. 5558–5564.
17. Cancès, E. and Mennucci, B. (2001). The escaped charge problem in solvation continuum models, *J. Chem. Phys.*, 115, pp. 6130–6135.
18. Klamt, A., Moya, C. and Palomar, J. (2015). A comprehensive comparison of the IEFPCM and SS(V)PE continuum solvation methods with the COSMO approach, *J. Chem. Theory Comput.*, 11, pp. 4220–4225.
19. Truong, T. N. and Stefanovich, E. V. (1995). A new method for incorporating solvent effect into the classical ab initio molecular orbital and density functional theory frameworks for arbitrary shape cavity, *Chem. Phys. Lett.*, 240, pp. 253–260.
20. Barone, V. and Cossi, M. (1998). Quantum calculation of molecular energies and energy gradients in solution by a conductor solvent model, *J. Phys. Chem. A*, 102, pp. 1995–2001.
21. Cossi, M., Rega, N., Scalmani, G. and Barone, V. (2003). Energies, structures, and electronic properties of molecules in solution with the C-PCM solvation model, *J. Comput. Chem.*, 24, pp. 669–681.
22. Klamt, A. and Jonas, V. (1996). Treatment of the outlying charge in continuum solvation models, *J. Chem. Phys.*, 105, pp. 9972–9981.
23. Baldridge, K. and Klamt, A. (1997). First principles implementation of solvent effects without outlying charge error, *J. Chem. Phys.*, 106, pp. 6622–6633.

24. Herbert, J. M. and Lange, A. W. (2016). The polarizable continuum model for (bio)molecular electrostatics: Basic theory and recent advances for macromolecules and simulations. In *Many-Body Effects and Electrostatics in Multi-Scale Computations of Biomolecules*, eds. Q. Cui, P. Ren, and M. Meuwly, Chapter 1 (Pan Stanford, Singapore), pp. 1–54.
25. Cramer, C. J. and Truhlar, D. G. (2006). SMx continuum models for condensed phases. In *Trends and Perspectives in Modern Computational Science*, Lecture Series on Computer and Computational Sciences, eds. G. Maroulis and T. E. Simos, Vol. 6 (Brill/VSP, Leiden) pp. 112–139.
26. Cramer, C. J. and Truhlar, D. G. (2008). A universal approach to solvation modeling, *Acc. Chem. Res.*, 41, pp. 760–768.
27. Marenich, A. V., Cramer, C. J. and Truhlar, D. G. (2013). Generalized Born Solvation Model SM12, *J. Chem. Theory Comput.*, 9, pp. 609–620.
28. Marenich, A. V., Cramer, C. J. and Truhlar, D. G. (2009). Universal solvation model based on solute electron density and on a continuum model of the solvent defined by the bulk dielectric constant and atomic surface tensions, *J. Phys. Chem. B*, 113, pp. 6378–6396.
29. Bondi, A. (1964). Van der Waals volumes and radii, *J. Phys. Chem.*, 68, pp. 441–451.
30. Rappe, A. K., Casewit, C. J., Colwell, K. S., III, W. A. G. and Skiff, W. M. (1992). UFF, A full periodic table force field for molecular mechanics and molecular dynamics simulations, *J. Am. Chem. Soc.*, 114, pp. 10024–10035.
31. Pierotti, R. A. (1976). A scaled particle theory of aqueous and nonaqueous solutions, *Chem. Rev.*, 76, pp. 717–726.
32. Bachrach, S. M. (2008). Microsolvation of glycine: A DFT study, *J. Phys. Chem. A*, 112, pp. 3722–3730.
33. Mullin, J. M. and Gordon, M. S. (2009). Alanine: Then there was water, *J. Phys. Chem. B*, 113, pp. 8657–8669.
34. Bachrach, S. M. (2014). Microsolvation of 1,4-butanediol: The competition between intra- and intermolecular hydrogen bonding, *J. Phys. Chem. A*, 118, pp. 1123–1131.
35. Wu, X., Tan, K., Tang, Z. and Lu, X. (2014). Hydrogen bonding in microsolvation: Photoelectron imaging and theoretical studies on $Au_x(H_2O)_n$ and $Au_x(CH_3OH)_n$ (x = 1, 2; n = 1, 2) complexes, *Phys. Chem. Chem. Phys.*, 16, pp. 4771–4777.
36. Zeng, Z., Hou, G.-L., Song, J., Feng, G., Xu, H.-G. and Zheng, W.-J. (2015). Microsolvation of $LiBO_2$ in water: Anion photoelectron spectroscopy and ab initio calculations, *Phys. Chem. Chem. Phys.*, 17, pp. 9135–9147.
37. Larrucea, J., Rezabal, E., Marino, T., Russo, N. and Ugalde, J. M. (2010). Ab initio study of microsolvated Al3-aromatic amino acid complexes, *J. Phys. Chem. B*, pp. 114, 9017–9022.

38. Jensen, J. H., Day, P. N., Gordon, M. S., Basch, H., Cohen, D., Garmer, D. R., Kraus, M. and Stevens, W. J. (2009). Effective fragment method for modeling intermolecular hydrogen-bonding effects on quantum mechanical calculations. Chapter 9, pp. 139–151.
39. Day, P. N., Jensen, J. H., Gordon, M. S., Webb, S. P., Stevens, W. J., Krauss, M., Garmer, D., Basch, H. and Cohen, D. (1996). An effective fragment method for modeling solvent effects in quantum mechanical calculations, *J. Chem. Phys.*, 105, pp. 1968–1986.
40. Chen, W. and Gordon, M. S. (1996). The effective fragment model for solvation: Internal rotation in formamide, *J. Chem. Phys.*, 105, pp. 11081–11090.
41. Gordon, M. S., Freitag, M. A., Bandyopadhyay, P., Jensen, J. H., Kairys, V. and Stevens, W. J. (2001). The effective fragment potential method: A QM-based MM approach to modeling environmental effects in chemistry, *J. Phys. Chem. A*, 105, pp. 293–307.
42. Gordon, M. S., Slipchenko, L., Li, H. and Jensen, J. H. (2007). The effective fragment potential: A general method for predicting intermolecular interactions. *Annu. Rep. Comput. Chem.*, 3, pp. 177–193.
43. Bandyopadhyay, P., Gordon, M. S., Mennucci, B. and Tomasi, J. (2002). An integrated effective fragment-polarizable continuum approach to solvation: Theory and application to glycine, *J. Chem. Phys.*, 116, pp. 5023–5032.
44. Li, H. and Gordon, M. S. (2007). Polarization energy gradients in combined quantum mechanics, effective fragment potential, and polarizable continuum model calculations, *J. Chem. Phys.*, 126, p. 124112.
45. Kairys, V. and Jensen, J. H. (2000). QM/MM boundaries across covalent bonds: A frozen localized molecular orbital-based approach for the effective fragment potential method, *J. Phys. Chem. A*, 104, pp. 6656–6665.
46. Jorgensen, W. L., Chandrasekhar, J., Madura, J. D., Impey, R. W. and Klein, M. L. Comparison of simple potential functions for simulating liquid water, *J. Chem. Phys.*, 79, http://aip.scitation.org/doi/abs/10.1063/1.445869.
47. Bernal, J. D. and Fowler, R. H. A theory of water and ionic solution, with particular reference to hydrogen and hydroxyl ions, *J. Chem. Phys.*, 1, http://aip.scitation.org/doi/abs/10.1063/1.1749327.
48. Jorgensen, W. L. and Madura, J. D. (1985). Temperature and size dependence for Monte Carlo simulations of TIP4P water, *Mol. Phys.*, 56, pp. 1381–1392.
49. Rick, S. W. Simulations of ice and liquid water over a range of temperatures using the fluctuating charge model, *J. Chem. Phys.*, 114, http://aip.scitation.org/doi/abs/10.1063/1.1336805.
50. Mahoney, M. W. and Jorgensen, W. L. A Five-site model for liquid water and the reproduction of the density anomaly by rigid, nonpolarizable potential functions, *J. Chem. Phys.*, 112, http://aip.scitation.org/doi/abs/10.1063/1.481505.
51. Rick, S. W. A reoptimization of the five-site water potential (TIP5P) for use with Ewald sums, *J. Chem. Phys.*, 120, http://aip.scitation.org/doi/abs/10.1063/1.1652434.

52. Yu, W., Lopes, P. E. M., Roux, B. and MacKerell, A. D. Six-site polarizable model of water based on the classical Drude oscillator, *J Chem. Phys.*, 138, http://aip.scitation.org/doi/abs/10.1063/1.4774577.
53. Nada, H. and van der Eerden, J. P. J. M. An intermolecular potential model for the simulation of ice and water near the melting point: A six-site model of H_2O, *J. Chem. Phys.*, 118, http://aip.scitation.org/doi/abs/10.1063/1.1562610.
54. Tröster, P., Lorenzen, K. and Tavan, P. (2014). Polarizable six-point water models from computational and empirical optimization, *J. Phys. Chem. B*, 118, pp. 1589–1602.
55. Kollman, P. A., Massova, I., Reyes, C., Kuhn, B., Huo, S., Chong, L., Lee, M., Lee, T., Duan, Y., Wang, W., Donini, O., Cieplak, P., Srinivasan, J., Case, D. A. and Cheatham, T. E. (2000). Calculating structures and free energies of complex molecules: Combining molecular mechanics and continuum models, *Acc. Chem. Res.*, 33, pp. 889–897.
56. Weiser, J., Shenkin, P. S. and Still, W. C. (1999). Approximate atomic surfaces from linear combinations of pairwise overlaps (LCPO), *J. Comput. Chem.*, 20, pp. 217–230.
57. Bea, I., Gotsev, M. G., Ivanov, P. M., Jaime, C. and Kollman, P. A. (2006). Chelate effect in cyclodextrin dimers: A computational (MD, MM/PBSA, and MM/GBSA) study, *J. Org. Chem.*, 71, pp. 2056–2063.
58. Martins, S. A., Perez, M. A. S., Moreira, I. S., Sousa, S. F., Ramos, M. J. and Fernandes, P. A. (2013). Computational alanine scanning mutagenesis: MM-PBSA vs TI, *J. Chem. Theory Comput.*, 9, pp. 1311–1319.
59. Xu, L., Sun, H., Li, Y., Wang, J. and Hou, T. (2013). Assessing the performance of MM/PBSA and MM/GBSA methods. 3. The impact of force fields and ligand charge models, *J. Phys. Chem. B*, 117, pp. 8408–8421.
60. Leonis, G., Steinbrecher, T. and Papadopoulos, M. G. (2013). A contribution to the drug resistance mechanism of darunavir, amprenavir, indinavir, and saquinavir complexes with HIV-1 protease due to flap mutation I50V: A systematic MM-PBSA and thermodynamic integration study, *J. Chem. Inf. Model.*, 53, pp. 2141–2153.
61. Ylilauri, M. and Pentikãinen, O. T. (2013). MMGBSA as a tool to understand the binding affinities of filamin-peptide interactions, *J. Chem. Inf. Model.*, 53, pp. 2626–2633.
62. Calvaresi, M., Bottoni, A. and Zerbetto, F. (2015). Thermodynamics of binding between proteins and carbon nanoparticles: The Case of C60@Lysozyme, *J. Phys. Chem. C*, 119, pp. 28077–28082.
63. Alonso, J. L., Cocinero, E. J., Lesarri, A., Sanz, M. E. and López, J. C. (2006). The glycine-water complex, *Angew. Chem. Int. Ed.*, 45, pp. 3471–3474.
64. Blanco, S., López, J. C., Lesarri, A. and Alonso, J. L. (2006). Microsolvation of formamide: A rotational study, *J. Am. Chem. Soc.*, 128, pp. 12111–12121.
65. Evangelisti, L. and Caminati, W. (2010). Internal dynamics in complexes of water with organic molecules. Details of the internal motions in tert-butylalcohol-water, *Phys. Chem. Chem. Phys.*, 12, pp. 14433–14441.

66. Lin, W., Wu, A., Lu, X., Tang, X., Obenchain, D. A. and Novick, S. E. (2015). Internal dynamics in the molecular complex of CF_3CN and H_2O, *Phys. Chem. Chem. Phys.*, 17, pp. 17266–17270.
67. Mata, S., Cortijo, V., Caminati, W., Alonso, J. L., Sanz, M. E., López, J. C. and Blanco, S. (2010). Tautomerism and microsolvation in 2-hydroxypyridine/2-pyridone, *J. Phys. Chem. A*, 114, pp. 11393–11398.
68. Lopez, J. C., Sanchez, R., Blanco, S. and Alonso, J. L. (2015). Microsolvation of 2-azetidinone: A model for the peptide group-water interactions, *Phys. Chem. Chem. Phys.*, 17, pp. 2054–2066.
69. Shen, J.-Y., Chao, W.-C., Liu, C., Pan, H.-A., Yang, H.-C., Chen, C.-L., Lan, Y.-K., Lin, L.-J., Wang, J.-S., Lu, J.-F., Chun-Wei Chou, S., Tang, K.-C. and Chou, P.-T. (2013). Probing water micro-solvation in proteins by water catalysed proton-transfer tautomerism, *Nat. Commun.*, 4, Article No. 2611.

Chapter 5

Conformational Searching for Complex, Flexible Molecules

Alexander C. Brueckner, O. Maduka Ogba, Kevin M. Snyder, H. Camille Richardson and Paul Ha-Yeon Cheong

Department of Chemistry, Oregon State University, 153 Gilbert Hall, Corvallis, Oregon 97333, USA

5.1 Introduction

Conformational analysis is essential to the study of organic reactions using theory. In this chapter, we examine the practical considerations behind software-guided conformational searches using molecular mechanics, analyze the strengths and weaknesses of specific parameters, and show examples involving both ground- and transition-state conformational searches. Finally, we provide a general summary of suggestions.

5.2 General steps for all conformational searches

The convenience of modern computer hardware and software means that most of the chore of performing a conformational search falls on the computer. However, there are certain aspects, particularly the initial input structure, analysis, and software parameters, which must be carefully considered by the user. The first two aspects are beyond the scope of this work, but this chapter will provide practical analyses and suggestions of the conformational search software parameters.

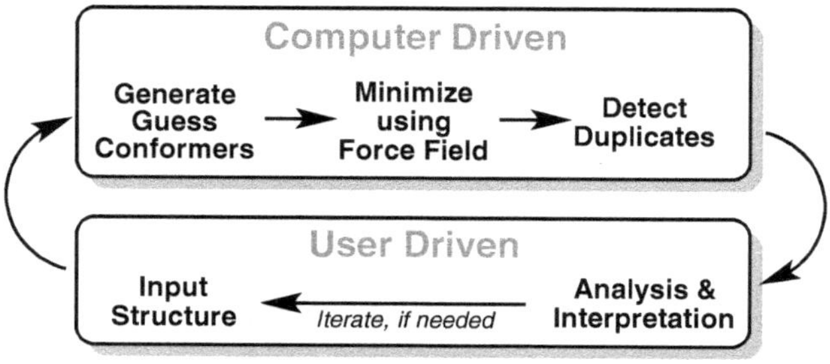

Figure 5.1. General steps for all conformational searches.

All conformational search programs in general operate as follows: (1) *guess conformer generation*, in which the program generates new conformers based on the given input structure; (2) *minimization*, in which the software uses the user-selected molecular mechanics force field to optimize the geometry of the guess structure to its closest energy minimum on the potential energy surface; and (3) *detection of duplication*, where the software decides whether the new structure is unique or a duplicate. This overall process is repeated until specified criteria are met: e.g., all new guess conformations are duplicates of previously found conformations. (Fig. 5.1). In cases where possible, we recommend using high-accuracy quantum mechanics methods to refine the lowest energy conformers.

5.2.1 *Guess conformer generation algorithms*

Saunders *et al.*[1] published a seminal work on conformational searches in 1990, in which they reported the importance of guess conformer generation algorithms. The two major approaches they discussed and are still operative today are: (1) **systematic** searches, which generate new guess conformers by changing the structure at regular increments (e.g., generating a new structure for every 15° rotation around every C–C bond), whereas in contrast, (2) **stochastic** searches generate new conformers by using random increments. What Saunders *et al.* discovered was that an exhaustive systematic search with *infinitesimally small increments* will most likely be exhaustive, i.e., find all conformations. However, given a finite number of guess conformers and computer resources, stochastic searches were more effective in locating conformers.

Modern approaches have focused on developing and improving on stochastic methods, the simplest being a *Cartesian stochastic search.* In this method, an atom is randomly displaced, and the resulting structure minimized. This is an inefficient way to generate new guess conformations, as a change in the XYZ coordinate of an atom may not correspond to a change in molecular conformation. In *Monte Carlo Multiple-Minimum* (MCMM)[2] searches, bond torsions are rotated by random increments to generate new structures. *Low-Mode Sampling* (LMOD)[3] and *Large-Scale Low-Mode Sampling* (LLMOD)[4] searches change the geometry along low-frequency vibrational modes in order to locate new conformers. This method is not designed to examine conformations that span across high-energy barriers for interconversion but is effective at exhaustively searching the local conformational space. Hybrid methods can be found in some programs, such as the *Mixed Torsional/Low-Mode Sampling* (MTLMS), as implemented in the Schrödinger *MacroModel* suite.[5] In redoing the classic Saunders–Houk conformational search experiments, we have discovered that LSLMS and MCMM are the most time-efficient, and MCMM gave the most number of conformers (Fig. 5.2). However, MTLMS and LMS gave the most number of lowest energy conformers, despite the greater computational expense. It is crucial to note that the global minima found by the different algorithms may not be identical — it is imperative that the user visualizes and analyzes the results, particularly for conformationally flexible molecules where the local minima may be just as important as the identity of the global.

5.2.2 *Minimization: Choice of molecular mechanics force field*

The second step of the conformational search process is the guess structure minimization using a molecular mechanics force field. These methods consist of a set of Newtonian equations, force constants, and predetermined parameters for describing bonding and nonbonding

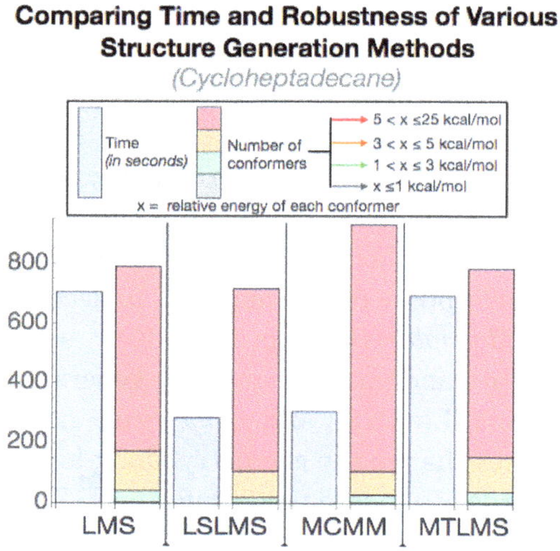

Figure 5.2. Comparing the efficiencies of the guess conformer generation algorithms implemented in Schrödinger MacroModel for cycloheptadecane. Numbers of unique conformers generated within 1, 3, 5, and 25 kcal/mol with respect to the lowest energy conformer are shown for each algorithm.

interactions that are used to calculate energies, forces, and structures. These methods are empirical in nature, derived from experimental data or high-level *ab initio* calculations, and usually parameterized for specific atom types. We focus on some of the most common force fields parameterized for organic molecules: (1) *Assisted Model Building with Energy Refinement* (AMBER*)[6] for proteins and nucleic acids; (2) Molecular Mechanics 3 (MM3*)[7] for aliphatic hydrocarbons; (3) *Merck Molecular Force Field* (MMFFs)[8] for general organic molecules; and (4) *Optimized Potentials for Liquid Simulations* (OPLS)[9] for organic molecules and peptides. In evaluating these force field methods for Mandelalide A, it is clear that there is no shortage of unique conformers (Fig. 5.3). This is true for most organic molecules — the choice of the force field is not qualitatively as critical as the user analysis that follows the conformational search, so long as the force field method is parametrized for the system of interest. In our experience, the most reliable way of verifying the quantitative quality of conformational searches of complex organic

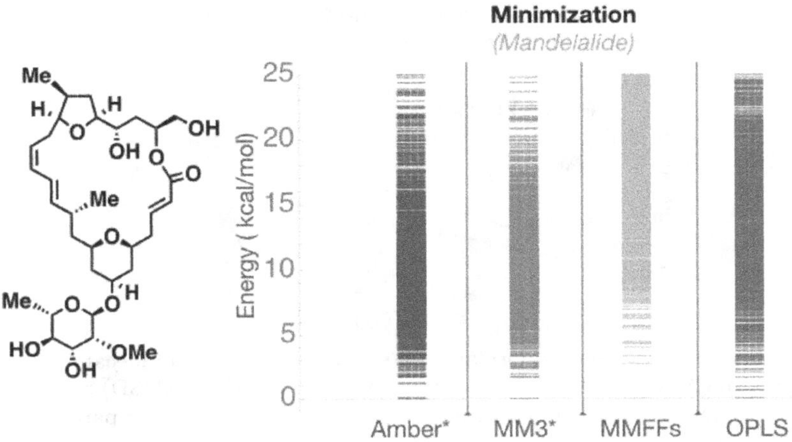

Figure 5.3. Comparing the conformer energy distributions using different force field structure minimization methods, holding all other parameters constant, for Mandelalide A.

molecules is by comparing one or two key structural parameters of the global minimum with experiments. For example, in mandelalide, we have compared trans-annular ROESY NMR correlations and distances with the minimum generated by different force field methods to identify the best method for follow-on work.

5.2.3 *Duplicates detection*

The last step of the conformational search process is duplicates detection. The algorithm calculates the root-mean-square deviation (RMSD) or maximum atom deviation (MAD) of the minimized guess conformation and compares them to all other previously found conformers. Based on a user-defined cutoff, the software determines if the new candidate is a duplicate or a new, unique conformation. If a stringent cutoff is set, a large number of redundant conformers may be saved. With a loose cutoff, conformers that are unique may be considered duplicates. In practice, the default cutoff of 0.5 Å RMSD found in many programs are sufficient to give a representative sample of the conformational space (Fig. 5.4). However, for large molecules, RMSD may no longer be a meaningful measure of

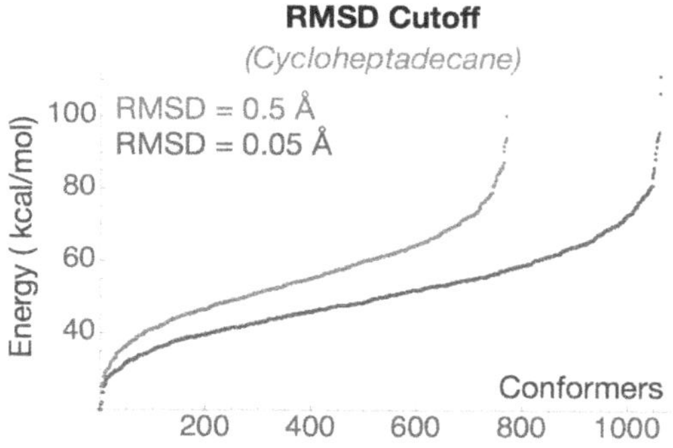

Figure 5.4. Comparing conformational searches of cycloheptadecane using stringent (0.05 Å) vs. loose (0.5 Å) root-mean-square deviation (RMSD) cutoff for categorizing new structures as duplicates or unique, holding all other parameters constant.

uniqueness because significant changes in conformations in a small part of the molecule could be drowned out by the rest of the molecule that has not changed. We strongly recommend using MAD, as this is much more sensitive.

5.3 Other critical parameters

5.3.1 *Minimization parameter: Nonbonding cutoff*

Nonbonding interactions decay remarkably slowly (inverse of the square of the distance) and can be costly to compute for large systems. Historically, these efficiency considerations were clearly important for practicality reasons when computing resources were scarce, and cutoffs were used to speed up the computations of nonbonding interactions. However, on modern computers, we recommend against distance cutoffs. On a modern machine, we find that the efficiency gained is minimal (~20% faster), and, more importantly, however, more low-energy unique conformers were found when nonbonding cutoffs were omitted (Fig. 5.5). In a recent study of the natural product Coibamide,[10] the ground-state conformation of the natural product as confirmed by NMR could only be found when cutoffs were not used.

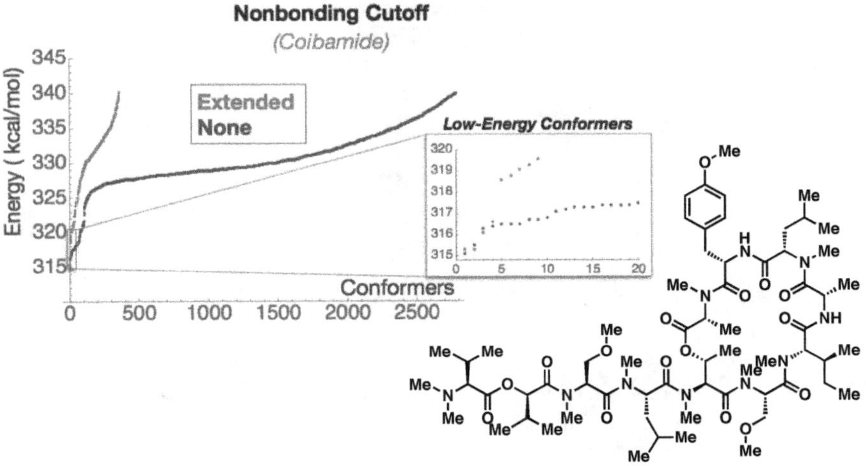

Figure 5.5. The effect of using "Extended" vs. "None" nonbonding cutoff algorithms in generating conformers of Coibamide, holding all other parameters constant.

5.3.2 *Guess conformer generation parameter: Maximum number of steps*

The maximum number of guess conformers to be generated during a conformational search is a critical parameter. Too few, and one risks missing the global minimum (Fig. 5.6); too many, and efficiency suffers. Considering that force fields are "rapid" methods, in the sense that each structure may be computed in seconds, we strongly suggest that the user errs on the side of caution. In our experience, 10,000 seems to be enough for most organic molecules up to ~200 atoms.

5.4 How do I know my conformational search was done correctly?

For simple organic molecules, the suggestions in the previous sections will most likely prove sufficient. For complex systems (e.g., >100 atoms), it will be prudent to compare to some experimental structural observables (NMR, X-ray, etc.) on a related model system. If the experiments are not a possibility, we recommend running the

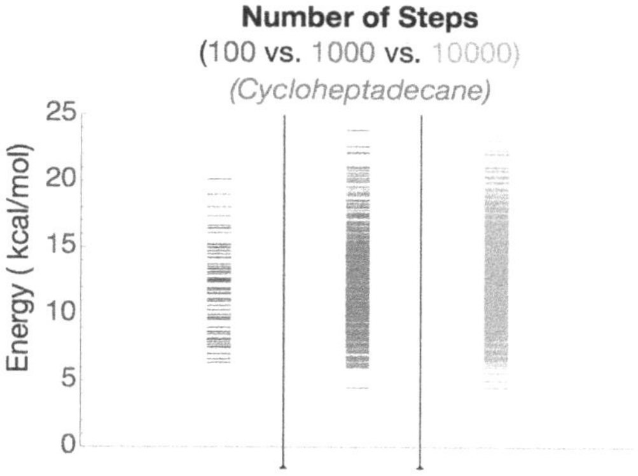

Figure 5.6. The effect of varying the maximum number of steps (100, 1,000, and 10,000) allowed in generating conformers of cycloheptadecane, holding all other parameters constant.

conformational search in an iterative manner, feeding the global minimum from the previous run as the initial input structure of subsequent runs, until the global minimum is consistent across multiple runs.

We show several case studies in the rest of the chapter that shows the parameters that have proven effective, the results of the conformational search, and the discovered chemistry.

5.5 Case studies: Ground-state conformational searches

5.5.1 *Case study 1: Effect of macrocycle substituents on conformational flexibility of Mandelalide A*

Mandelalide A is a marine natural product with high cytotoxicity to human NCI-H460 lung cancer cells (IC_{50}, 12 nM) and mouse Neuro-2A neuroblastoma cells (IC_{50}, 29 nM), originally isolated by Sikorska *et al.*[11] In this case study, we look at the effects of Mandelalide A substituents on the conformational flexibility. Specifically, we ran conformational searches with and without the methyl and methoxy substituents on the macrocycle (highlighted in pink, Fig. 5.7). The unsubstituted macrocycle was considerably more

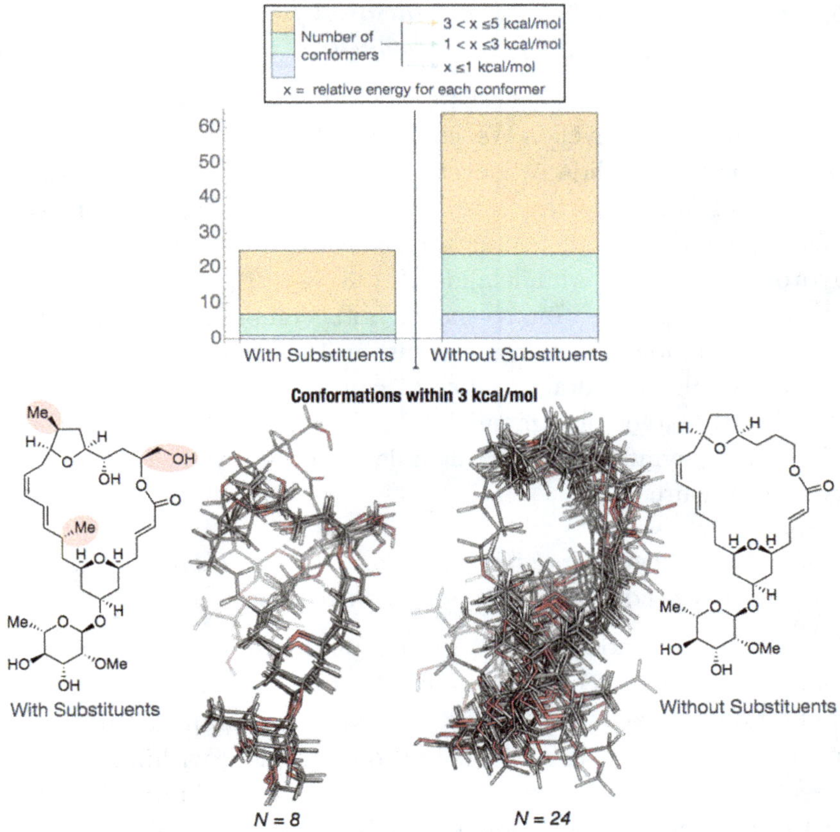

Figure 5.7. Effects of Mandelalide A substituents on the conformational flexibility.

Parameters for Conformational Search in Figure 5.7
Structure Generation Method: MTLMS
Minimization Force Field: MMFFs
MAD Duplication Cutoff: 0.1 Å
Electrostatic Cutoff: None
Max. Number of Steps: 10,000
Solvent: None

flexible (i.e., yielded a larger number of low-energy conformers within 3 kcal/mol) than the parent compound (24 vs. 8, respectively). The overlay of these conformers graphically shows the dramatic conformational rigidification effects of the macrocycle substituents.

5.5.2 *Case study 2: Conformational flexibility of acyclic vs. cyclic polyketides*

One often assumes that acyclic compounds are more conformationally flexible than cyclic. We compared the conformational search results for Mandelalide A, a cyclic polyketide, and Tautomycetin, an acyclic polyketide, to test this assumption (Fig. 5.8). Tautomycetin was chosen because it contains roughly the same number of non-hydrogen atoms as Mandelalide A (43 vs. 42), and similar array of functional groups. Interestingly, both compounds yield roughly the same number of conformers within 3 kcal/mol (8 vs. 9, respectively). While the total number of conformers for tautomycetin is significantly larger than mandelalide, the unsubstituted mandelalide (Fig. 5.7) is significantly more flexible than the acyclic Tautomycetin by any measure.

5.5.3 *Case study 3: Effect of Ca^{2+} counterion on the conformations of ionomycin*

The antibiotic ionomycin[12] is an ionophore for Ca^{2+} ions. We ran conformational searches on ionomycin with and without the Ca^{2+} counterion to evaluate its effect in structural organization of ionomycin. An overlay of the low-energy conformations within 3 kcal/mol reveals drastic differences in conformational flexibility (Fig. 5.9). In the presence of Ca^{2+}, ionomycin is remarkably rigid, but without the organizing counterion, ionomycin is more flexible, with no persistent conformational motif within 3 kcal/mol of the global minimum.

5.6 Transition state conformational searches

5.6.1 *Case study 4: Proline catalyzed intermolecular aldol transition states*

In a seminal report, Houk and List[13] reported the density functional theory (DFT) study of an intermolecular proline-catalyzed asymmetric intermolecular aldol reaction (Fig. 5.10). This led to the development of the Houk–List proline stereoselectivity model which serves

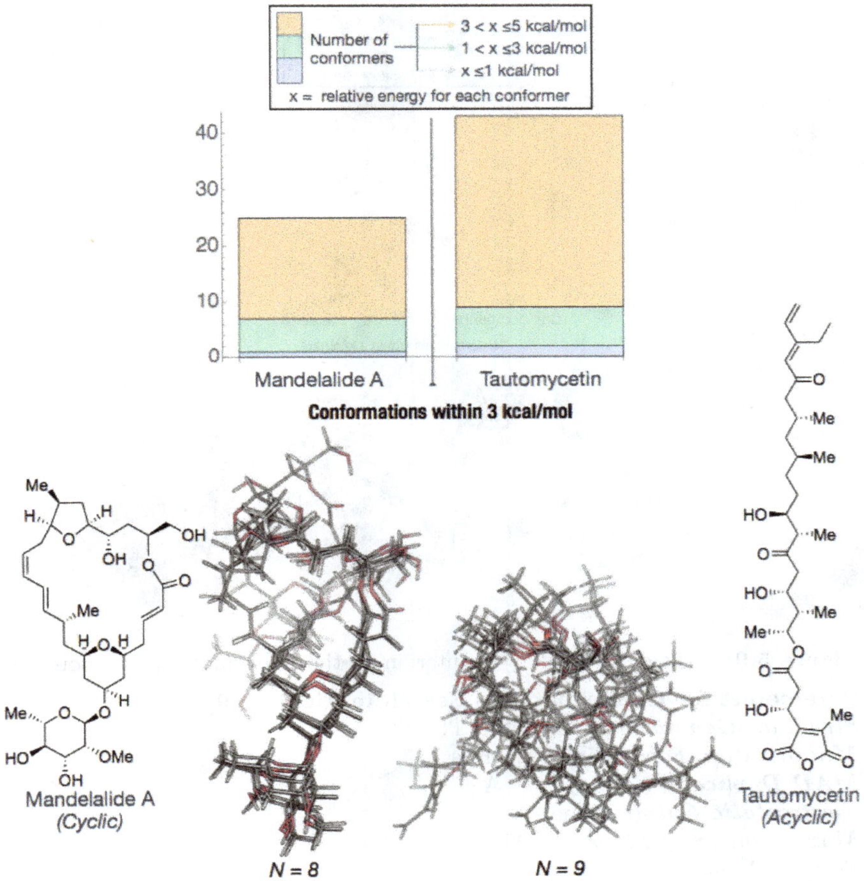

Figure 5.8. Comparing the conformational flexibility of cyclic Mandelalide A and acyclic tautomycetin.

Parameters for Conformational Search in Figure 5.8
Structure Generation Method: MTLMS
Minimization Force Field: MMFFs
MAD Duplication Cutoff: 0.1 Å
Electrostatic Cutoff: None
Max. Number of Steps: 10,000
Solvent: None

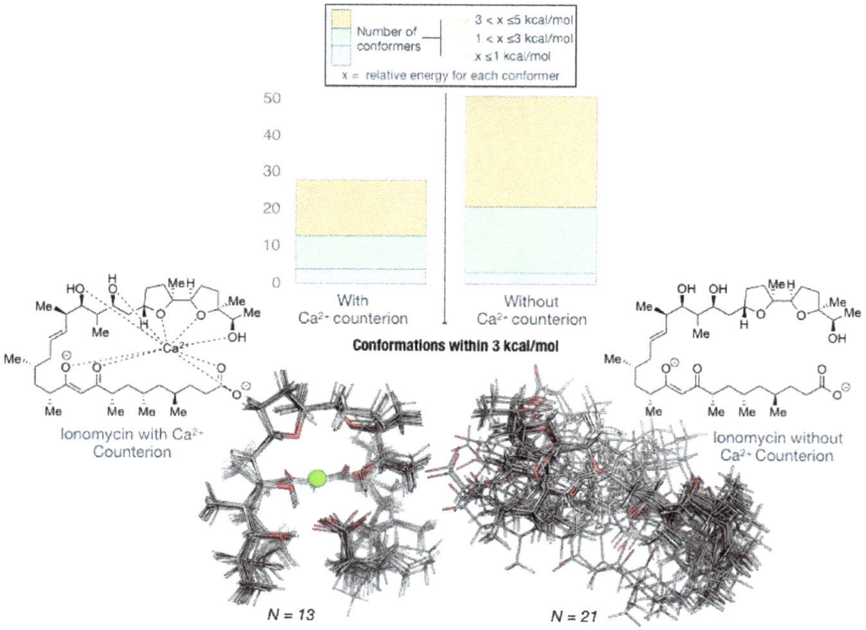

Figure 5.9. The effect of Ca^{2+} counterion on the conformations of ionomycin.

Parameters for Conformational Search in Figure 5.9
Structure Generation Method: MTLMS
Minimization Force Field: MMFFs
MAD Duplication Cutoff: 0.1 Å
Electrostatic Cutoff: None
Max. Number of Steps: 10,000
Solvent: None

as the general template for explaining many organocatalytic reactions beyond proline.[13] Moreover, it showed the potential of DFT in predicting stereoselectivities of organic reactions. In this section, we will address protocols for performing transition-state conformational searches, using the proline–aldol reaction.

Transition state searches: Since almost all publicly available force fields do not treat transition states, special considerations must be made in performing a transition-state conformational search. The bond-forming/breaking atoms must remain fixed (or constrained)

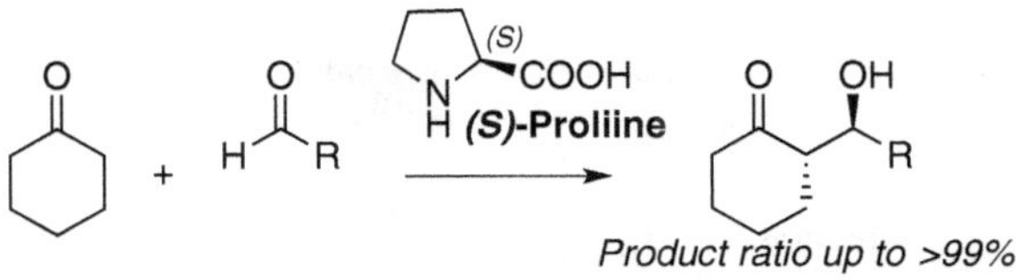

Figure 5.10. Proline-catalyzed intermolecular aldol reaction.

during the conformational search at an appropriate distance. This can readily be adapted from literature precedence. For example, the key C–C bond formation in the aldol reaction is set at 2.1 Å in this case study — approximately at the DFT computed TS bond distances in the Houk–List study.[14]

In general, we recommend fixing the key bond-forming/breaking atoms and constraining bond distances for key nonbonding interactions (e.g., hydrogen bonding between proline carboxylic acid and electrophile carbonyl in the proline–aldol case). The difference between fixing and constraining is subtle but significant. In the former, atoms will not change locations, but in the latter, there is some give. It is critical to not have more than two atoms fixed in any conformational searches as this dramatically limits the program's ability to sample the large-scale changes in the conformational space.

The conformational search results for the four diastereomeric transition states (Fig. 5.11) are shown in Fig. 5.12. The conformational search:

(1) predicts the right major product. There is a distinct energetic difference among all four diastereomers, with *anti-re* predicted as the lowest energy.
(2) wrongly predicts that the *syn-si is more stable than anti-si and syn-re*. The DFT results by Houk and List show that, in fact, *syn-si* is least stable.
(3) reveals different "families" of related conformations. In each transition state, there are several energy plateaus. The lowest energy plateau is consistent with structures in which the cyclohexanone

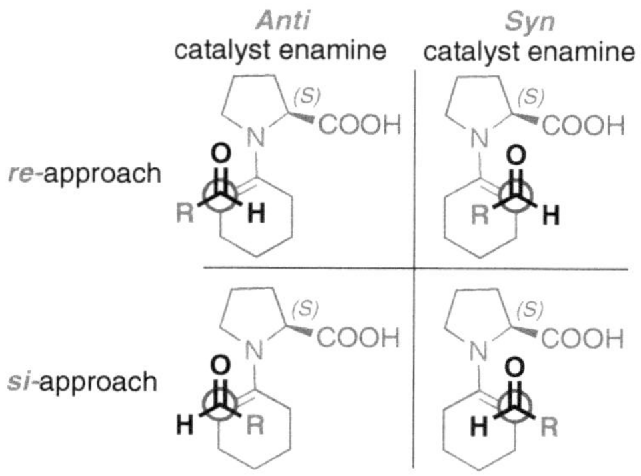

Figure 5.11. There are four possible diastereomeric transition states in the key C–C bond-forming step between the catalyst enamine and the benzaldehyde. Specifically, the catalyst enamine can be in *anti* vs. *syn* configuration and can attack the *re* vs. *si* face of the benzaldehyde.

is in a chair conformation. The next plateau consists of structures with a twist-boat cyclohexanone. Lastly, the structures in the highest energy plateau exhibit a boat cyclohexanone.

5.6.2 *Case study 5: Substrate-binding transition states in the organocatalyzed desymmetrization of meso-1,2-diols*

In this final case study, we conduct a conformational search for a transition state where multiple mechanisms are possible but are not energetically separated like in the proline case above. We specifically perform a conformational search of the initial binding of the substrate alcohol to the catalyst oxazolidine (Fig. 5.13). The forming C–O bond was fixed at 1.7 Å, and the chloride-providing base-catalysis was constrained at 1.3 ± 0.1 Å.

The conformational search gave 27 conformers within 3 kcal/mol of the lowest energy structure. A detailed look at the conformers reveals four distinct families based on unique hydrogen-bonding patterns (Fig. 5.14). However, unlike in the proline example, the plot here revealed there was no energetic distinction between these hydrogen-bonding families. This leads to the hypothesis that

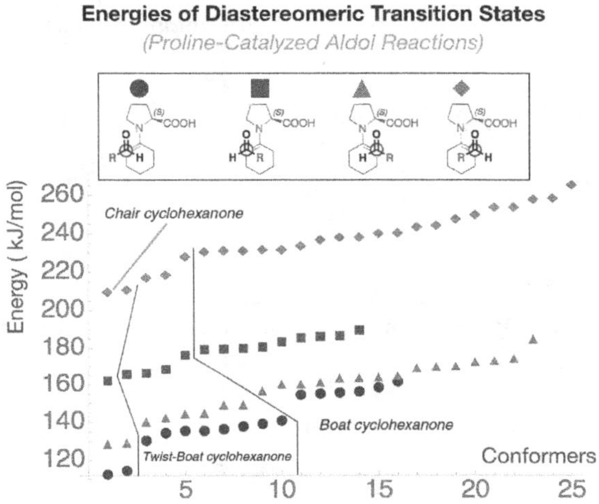

Figure 5.12. Conformational search results of the four diastereomeric C–C bond-forming proline–aldol transition states.

Parameters for Conformational Search in Figure 5.12
Structure Generation Method: MTLMS
Minimization Force Field: MMFFs
MAD Duplication Cutoff: 0.5 Å
Electrostatic Cutoff: None
Max. Number of Steps: 10,000
Solvent: None
Fixed Atoms: The two carbons involved in the bond-forming process (denoted in the Newman projection in Figs. 5.9 and 5.10). Fixed at 2.1 Å.
Constrained Bond Distance: The proline carboxylate H and aldehyde carbonyl O at *2.1 Å* with a force constant of 100 kJ/mol.

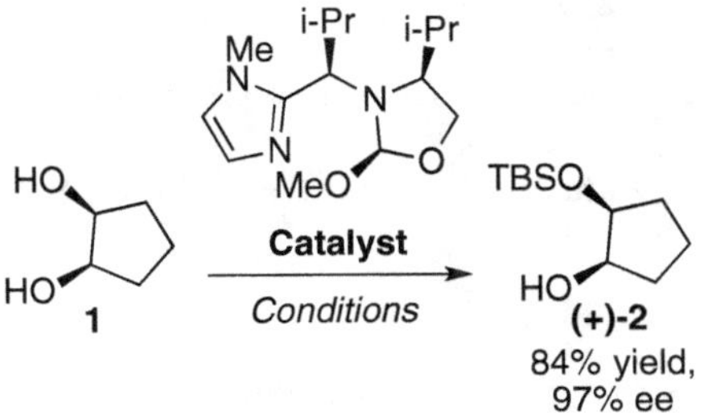

Figure 5.13. Organocatalyzed desymmetrization of *meso*-1,2-diols.

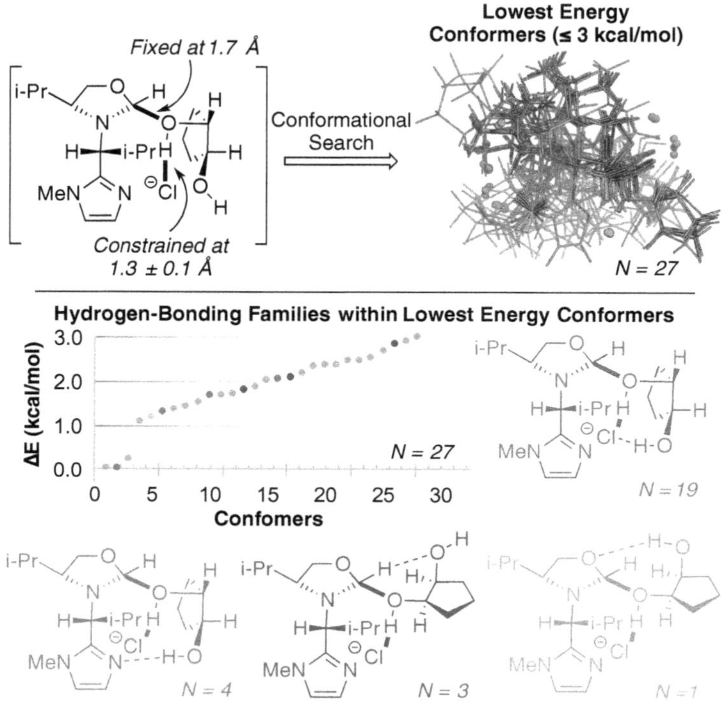

Figure 5.14. Conformational search results of the substrate-binding transition state for the reaction in Fig. 5.13.

Parameters for Conformational Search in Figure 5.14
Structure Generation Method: MTLMS
Minimization Force Field: MMFFs
MAD Duplication Cutoff: 0.1 Å
Electrostatic cutoff: None
Max. Number of Steps: 10,000
Solvent: None
Fixed Atoms: The carbon and oxygen atoms involved in the bond-forming process. Fixed at 1.7 Å.
Constrained Bond Distance: The substrate alcohol H and chloride ion were constrained at 1.3 ± 0.1 Å.

structural features other than hydrogen bonding (e.g., torsions, van der Waals interactions, etc.) may be more energetically important in this transformation.

5.7 Concluding remarks/practical advice

In summary, we have briefly outlined the conformational search process and important parameters to consider for an effective search. We have also provided five practical case studies (three ground-state and two transition-state cases) that demonstrate the conformational search of complex, flexible molecules. There remain significant challenges in the efficient analyses of large conformational spaces (>100), for which the best tools and techniques are as yet unclear, and remains the domain of active research and experts. However, the discussions of this chapter should prove sufficient for the identification of the most stable conformers.

References

1. Saunders, M., Houk, K. N., Wu, Y.-D., Still, W. C., Lipton M., Chang, G. and Guida, W. C. (1990). Conformations of cycloheptadecane. A comparison of methods for conformational searching, *J. Am. Chem. Soc.*, 112, pp. 1419–1427.
2. Chang, G., Guida, W. C. and Still, W. C. (1989). An internal coordinate Monte Carlo method for searching conformational space, *J. Am. Chem. Soc.*, 111, pp. 4379–4386.
3. Kolossváry, I. and Guida, W. C. (1996). Low mode search. An efficient, automated computational method for conformational analysis: Application to cyclic and acyclic alkanes and cyclic peptides, *J. Am. Chem. Soc.*, 118, pp. 5011–5019.
4. Kolossváry, I. and Keseru, G. M. (2001). Hessian-free low-mode conformational search for large-scale protein loop optimization: Application to c-*jun* N-terminal kinase JNK3, *J. Comp. Chem.*, 22, pp. 21.
5. MacroModel (2016). Version 11.2 (Schrödinger, LLC, New York, NY).
6. Weiner, P. K. and Kollman, P. A. (1981). AMBER: Assisted model building with energy refinement. A general program for modeling molecules and their interactions, *J. Comp. Chem.*, 2, pp. 287–303.
7. Allinger, N. L., Yuh, Y. H. and Lii J.-H. (1989). Molecular mechanics. The MM3 force field for hydrocarbons, *J. Am. Chem. Soc.*, 111, pp. 8551–8566.
8. Halgren, T. A. (1996). Merck molecular force field. I. Basis, form, scope, parameterization, and performance of MMFF94, *J. Comp. Chem.*, 17, pp. 490–519.
9. Jorgensen, W. L. and Tirado-Rives, J. (1988). The OPLS potential functions for proteins. Energy minimizations for crystals of cyclic peptides and crambin, *J. Am. Chem. Soc.*, 110, pp. 1657–1666.

10. Snyder, K. M., Sikorska, J., Ye, T., Fang, L., Su, W., Carter, R. C., McPhail, K. L. and Cheong, P. H.-Y. (2016). Towards theory driven structure elucidation of complex natural products: Mandelalides and Coibamide A, *Org. Biomol. Chem.*, 14, pp. 5826–5831.
11. Sikorska, J., Hau, A. M., Anklin, C., Parker-Nance, S., Davies-Colemani, M. T., Ishmael, J. E. and McPhail, K. L. (2012). Mandelalides A-D, cytotoxic macrolides from new lissoclinum species of South African turnicate, *J. Org. Chem.*, 77 (14), pp. 6066–6075.
12. Lui, C. and Hermann, T. E. (1978). Characterization of ionomycin as a calcium ionophore, *J. Biol. Chem.*, 253, pp. 5892–5894.
13. Bahmanyar, S., Houk, K. N., Martin, H. J. and List, B. (2003). Quantum mechanical predictions of the stereoselectivities of proline-catalyzed asymmetric intermolecular aldol reactions, *J. Am. Chem. Soc.*, 125, pp. 2475–2479.
14. Cheong, P. H.-Y., Legault, C. Y., Um, J. M., Çelebi-Ölçüm, N. and Houk, K. N. (2011). Quantum mechanical investigations of organocatalysis: Mechanisms, reactivities, and selectivities, *Chem. Rev.*, 111, pp. 5042–5137.

Chapter 6

NMR Prediction

Kelvin E. Jackson and Robert S. Paton

Chemistry Research Laboratory, University of Oxford, Mansfield Road, Oxford OX1 3TA, UK

6.1 Introduction

The importance of Nuclear Magnetic Resonance (NMR) spectroscopy in organic chemistry is unquestionable, not only for structure elucidation but also in the quantitative analysis of selectivities and conformational or chemical equilibria. It is therefore unsurprising that intense effort has been expended in developing computational simulations of NMR spectra to be able to faithfully reproduce experimental measurements to the point where reliable predictions are possible. In this chapter, we discuss the computation of NMR spectroscopic observables predominantly through the perspective of electronic structure theory. This is motivated by the numerous observations we and others have made that quantum mechanical calculations of these observables can be of high-enough accuracy for their application to solve organic and analytical problems. Here we discuss the theoretical and practical considerations pertinent to the prediction of chemical shifts, coupling constants, and intramolecular contacts with quantum chemistry. Several applications from the chemical literature serve to highlight both current "best practice" in terms of computational methodology and the predictive power of computational NMR prediction.

6.2 Shielding tensor prediction

Most chemists will be familiar with the idea of the NMR chemical shift: that magnetically active nuclei have different resonant frequencies in a magnetic field, and that these frequency differences result from variations in the distribution of electrons around the nuclei. The computation of chemical shifts therefore relies on an accurate description of the electron distribution in atoms and molecules, and upon describing the response of the electron density to a magnetic field. The complete theory of quantum mechanical NMR calculations extends far beyond the scope of this chapter. We focus our discussion on those aspects germane to the practical calculation of NMR properties using widely available electronic structure packages. Good introductions to the theory of NMR computations can be found in textbooks[1] and in reviews.[2]

6.2.1 *Theoretical background*

The NMR shielding, σ, can be expressed as a static second derivative of the total energy with respect to two perturbations, the magnetic field, B, and the nuclear magnetic moment, μ (see Eq. (6.1))

$$\sigma = \left.\frac{\partial^2 E}{\partial B \partial \mu}\right|_0 \tag{6.1}$$

The introduction of a magnetic field into electronic structure calculations leads to the so-called *gauge problem*. In essence, it is necessary to define an arbitrary coordinate origin for the vector potential associated with the magnetic field — the dependence on this origin vanishes from the NMR properties when they are calculated exactly using an (infinitely) large basis set. Inevitably this is not practical, introducing a dependence of the calculated shielding on the choice of gauge. Today, the most widely implemented solution to this problem is the use of a basis set composed from gauge-including atomic orbitals (GIAOs) for the computation of NMR parameters.[3] In the GIAO approach, the conventional basis functions are multiplied by a field-dependent factor, which has the effect of eliminating any origin-dependence, even for approximate molecular orbitals (MOs) and finite basis sets. Alternative schemes have been developed, such as individual gauges for localized orbitals (IGLO) methods[4]; however, GIAO calculations converge faster with the basis-set size and so probably represent the best available method.[5]

The familiar NMR chemical shift values, δ, are obtained (experimentally and computationally) from the shielding tensors, σ, by Eq. (6.2)

$$\delta = \sigma_{\text{ref}} - \sigma \tag{6.2}$$

where σ_{ref} represents the shielding tensor of a suitably defined reference compound (e.g., $SiMe_4$). It follows that shielding tensors and chemical shifts share the same units, most often quoted in parts per million (ppm) which is therefore the format adopted by most computer programs. The shielding tensors computed by the GIAO method result from the sum of two parts, corresponding to diamagnetic and paramagnetic shielding terms. The first of these terms depends solely on the unperturbed electron density, whereas the second is due to the density under the influence of the magnetic field. The paramagnetic shielding term is influenced by both occupied and virtual (i.e., unoccupied) molecular orbitals.

Through implementation of the GIAO method in electronic structure calculations, it is possible to obtain shielding tensors (and therefore, trivially, the NMR chemical shifts) using several different levels of theory for any set of nuclear coordinates. Standard software packages have implemented the evaluation of the static second-derivative of the energy with Hartree–Fock (HF), density functional theory (DFT), and correlated wave function theory (WFT) methods. We will therefore consider the various modeling decisions that can be made to ensure accurate results. These practical considerations are outlined in the following sections.

6.2.2 *Effects of structure and conformation*

Molecular shape and conformation affects the distribution of the electron density and therefore the computed chemical shifts as well. Rigid molecule (i.e., one which is described by a single conformation) geometry optimization using either *ab initio* or DFT calculations yield structures generally considered suitable for the computation of chemical shifts. For example, a comparison of experimental bond distances and angles obtained with those optimized with HF and DFT (using BLYP and B3LYP functionals) shows small average errors for all three methods in the region of 0.02 Å and 0.5°, respectively.[6] Using these optimized geometries to compute shielding constants with GIAO-B3LYP/6-311++G(d,p) calculations provides ^{13}C chemical shifts to within 6 ppm mean absolute deviation (MAD) with respect to experiment. In fact, using HF geometries gave higher accuracy than DFT structures by around 2–3 ppm. Given small differences in optimized structures obtained using different DFT functionals, a wide range has been employed for optimization prior to NMR calculations. Nevertheless, in studying the ^{13}C and ^{1}H NMR spectrum of the 1-adamantyl cation, Harding *et al.*[7] discovered that the chemical shifts were sensitive to the level of theory employed for geometry optimization, and that DFT geometries gave inferior results when compared with correlated WFT (Fig. 6.1). The root-mean-square error (RMSE) compared to experiment for ^{13}C shifts was 12.8 ppm with B3LYP geometries, 6.7 ppm for MP2 geometries, and could be reduced to 4.8 ppm RMSE with CCSD(T)/cc-pVTZ optimizations. Evidently, in this case, ^{13}C and ^{1}H chemical shifts are acutely sensitive to the geometry used. On the other hand, structures obtained from molecular mechanics (MM) calculations with the MMFF force field have also been used to compute chemical shifts

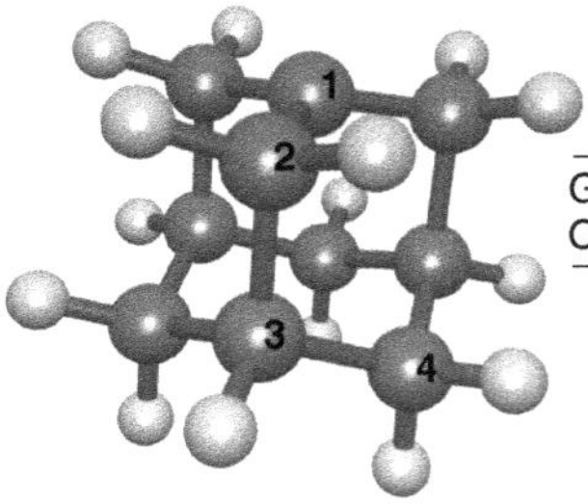

	δ (ppm)	
	GAIO-CCSD(T)/qz2p CCSD(T)/cc-pVTZ	expt (T = 193 K)
C1	300.4	300
C2	69.7	65.7
C3	93.0	86.8
C4	36.7	34.5

Figure 6.1. Chemical shifts of the C_{3v}-symmetric 1-adamantyl cation.[7]

in a subsequent quantum mechanics (QM) calculation with a high-enough accuracy to assign the structure of a natural product.[8] On such occasions though, it is usually necessary to perform an empirical correction to the shielding tensors to facilitate direct comparison with experimental chemical shifts.

For flexible organic molecules with multiple conformations, it is necessary to consider how this will affect the chemical shift. As conformational interconversion is generally rapid on the NMR timescale, the observed chemical shift will be a weighted average over all of the individual conformations. NMR calculations for individual conformations must be performed separately, and the chemical shifts averaged to produce an ensemble average value for each position. Assuming a thermal equilibrium between the conformers, the Boltzmann weighting factors for each conformer (w_i) can be obtained from the relative Gibbs energies (Eq. (6.3))

$$w_i = \frac{e^{-G_i/RT}}{\sum_i^N e^{-G_i/RT}} \tag{6.3}$$

The sum of the Boltzmann-weighted shielding tensors obtained separately for each conformation is then comparable to the experimental, time-averaged value. In practice, it is often expedient to assume that the entropy and volume associated with each conformation is effectively identical, because this enables relative energies to be used in Eq. (6.3) (rather than Gibbs energies) and avoids the need to perform several frequency calculations. For molecules with only a handful of conformations, it may be feasible to manually optimize and perform the NMR calculations. In more flexible cases, this may be a more daunting proposition and it will become desirable to automate the task of conformational searching (see Chapter 5): this is possible using numerous programs, although inevitably an inexpensive computational method (e.g., MM) must be used. The low-energy conformations are selected and reoptimized at a higher level of theory or may be used directly to compute shielding tensors. Willoughby

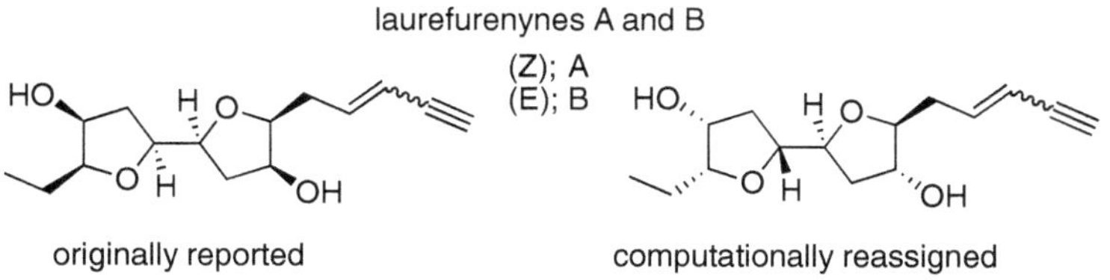

Figure 6.2. Stereostructure reassignment of laurefurenyene, a flexible 2,2′-bifuranyl natural product, based on Boltzmann-averaged ^{13}C and ^{1}H computed chemical shifts.

et al.[9] have produced a helpful workflow to assist in carrying out this sequence of steps for flexible organic molecules.

In predicting the ^{13}C shielding tensors for the natural product laurefurenyne (Fig. 6.2), it was necessary to consider 32 different diastereomers, each of which has several freely rotatable single bonds, resulting in the population of several conformations at room temperature.[10] It was necessary to carry out a 10,000 step Monte Carlo Multiple Minimum (MCMM) conformational search with the MMFF force field to generate possible conformers. The low-energy structures within 10 kJ/mol of the global energy minimum (a larger window of 20 kJ/mol was used in testing and did not result in any improvement in the results) were reoptimized with DFT (in this case with ωB97XD/6-31G(d)). This process can be automated through the use of, e.g., Python scripts.[9,10] The effect of using the MM structures directly for the computation of shielding tensors was also evaluated. In this scenario, the single point energy obtained from the DFT computation of the shielding tensor was used to generate the Boltzmann factor for conformational averaging. For the 2,800 low-energy conformations, the cost of DFT optimizations and NMR calculations was 18,500 CPU hours vs. 3,800 CPU hours when omitting the DFT optimizations. The results from both approaches were qualitatively similar, and both predicted the correct identity of the natural product on the basis of ^{13}C calculations. However, the range between best and worst structures was smaller for the MM structures, indicating that the structural assignment of laurefurenyne (Fig. 6.2) could not be made with the same confidence as from DFT optimizations. Additionally, ^{1}H results favored the incorrect structure from the MM structures, whereas DFT optimizations gave the correct laurefurenyne diastereomer.

Aside from the issue of conformational isomerism, we should also acknowledge that the use of static molecular geometries is to neglect the effects of internal rotation and vibration upon chemical shifts.[11] Shielding tensors depend upon internal motions in a nonuniform way, which are themselves fundamentally anharmonic. Although these effects are often assumed to be negligible, or at least can be systematically corrected through empirical corrections, it was shown recently that errors in computed chemical shifts can be reduced by taking an average value from multiple snapshots produced from 25 (quasi-classical) molecular dynamics trajectories, sampling every eighth point.[12] The dynamic contribution to the shielding tensor may be evaluated using a relatively low QM level, and added to a static evaluation of shielding tensor at a demanding level of theory. This approach leads to excellent agreement between calculated ^{13}C and ^{1}H chemical shifts for a set of small organic molecules, with mean errors of 0.5 and 0.02 ppm, respectively, and can be applied to natural products with similar levels of performance.

6.2.3 *Level of theory for shielding tensor calculations*

As with many other applications in electronic structure theory, one is faced with a choice over the level of theory and an associated basis set in evaluating of the nuclear shielding tensors. The shieldings are evaluated computationally as tensors, and the *isotropic* value is obtained as the average of the diagonal elements (Eq. (6.4))

$$\sigma = \begin{pmatrix} \sigma_{xx} & \sigma_{xy} & \sigma_{xz} \\ \sigma_{yx} & \sigma_{yy} & \sigma_{yz} \\ \sigma_{zx} & \sigma_{zy} & \sigma_{zz} \end{pmatrix}, \quad \sigma_{\text{iso}} = \frac{\sigma_{xx} + \sigma_{yy} + \sigma_{zz}}{3} \tag{6.4}$$

It is this isotropic shielding tensor that is comparable to the isotropic chemical shift values obtained for organic molecules in solution. As discussed earlier in the chapter, quantum mechanical calculations of nuclear shielding depend on both the occupied MOs (the diamagnetic shielding) and unoccupied MOs (the paramagnetic shielding). Despite the fact that virtual orbital energies computed by DFT are systematically too low, it has found significant popularity in evaluation of isotropic shielding tensors. The convergence of DFT-computed properties with respect to basis set size typically occurs more quickly than for correlated WFT–based methods, making calculations tractable for fairly large organic molecules consisting of hundreds of atoms. GIAO NMR calculations are compatible with both pure and hybrid density functionals employing the local density approximation (LDA), generalized gradient approximation (GGA), and higher derivatives such as meta-GGAs.

In terms of computational accuracy, the relative merits of different functionals is dependent on whether one considers the "raw" shielding tensors directly outputted or chemical shifts, which are obtained from comparison to a reference compound or through a scaling process discussed in Sec. 6.2.5. Isotropic shielding tensors obtained from calculations at the extrapolated CCSD(T)/aug-cc-pCV[TQ]Z level of theory have been used to benchmark HF and DFT values for ^{1}H, ^{7}Li, ^{13}C, ^{14}N, ^{17}O, ^{19}F, ^{27}Al, ^{31}P, and ^{33}S nuclei in a set of small molecules with seven atoms or fewer.[13] Of the various density functionals considered (LDA, BLYP, B3LYP, PBE, KT2, B97, PBE0, CAM-B3LYP, all with the aug-cc-pVQZ basis set) there was a slight improvement over HF-computed shielding tensors, but the DFT shieldings were of lower accuracy than CCSD. The authors noted a general improvement on average upon moving from LDA to GGA functionals, and a further modest improvement with hybrid functionals, with DFT generally underestimating the shielding constants. For cases where HF performs badly, such as ^{14}N shieldings, this is inherited by hybrid DFT functionals, which perform worse than pure functionals. Notably, the use of orbital-dependent functionals as in the optimized effective potential method (OEP) showed the most consistent improvements when compared with conventional DFT implementations. Standard databases of shielding constants have also been used to show that the KT2 functional that was specifically designed for NMR applications, and the SSB-D functional, outperform PBE and OPBE for ^{13}C and ^{1}H tensors.[14]

Comparisons of experimental chemical shift values with DFT results have led to different recommendations depending upon whether these values are obtained from the shielding tensors. For example, the RMS error for DFT computed ^{1}H chemical shifts across 80 small organic molecules is minimized to 0.13 ppm by B3LYP and TPSS functionals.[15] A bespoke hybrid GGA functional, WP04, specifically parameterized for ^{1}H chemical shift also fares well. For this comparison, the shifts were obtained from a scaling process (see below) rather than by comparison to a reference compound. In the case of ^{13}C chemical shifts (using a TMS reference) OPBE and OPW91 gave the lowest MAEs, and outperformed B3LYP.[16] Semi-empirical corrections to standard density functionals have been developed to correct the virtual orbital energies: Magyarfalvi and Pulay[17] has shown that a uniform shift of 0.025 Hartree applied to virtual orbitals generated using BLYP/B3LYP/OLYP functionals was optimal in reducing the errors with respect of CCSD(T) shielding tensors for a set of small molecules. Shifting the energy level of unoccupied molecular orbitals affects the paramagnetic component of the computed shielding tensors.

6.2.4 *Basis sets for shielding tensor calculations*

Various studies have examined the effect of varying basis set size upon computed ^{1}H and ^{13}C isotropic shielding tensors with the familiar Pople (e.g., 6-31G(d), etc.) and Dunning correlation-consistent (e.g., cc-pVnZ) families of basis sets. For example, shielding constants evaluated for ^{13}C and ^{1}H nuclei exhibit a linear relationship between basis set size and error (a larger basis set leads to smaller errors) when investigated with several density functional methods.[18] Basis sets specifically designed to help reproduce NMR data, named pcJ-n and pcS-n ($n = 0, 1, 2$, etc.), also have been developed by Jensen.[19] These basis sets incorporate "tighter" basis functions closer to the nucleus and tend to outperform standard basis sets of an equivalent size for

NMR applications. As with computed energies and other properties, extrapolation of shielding tensors to the complete basis set limit is achievable, as has been demonstrated for the *cis*-N-amide proton in formamide with a variety of levels of theory.[20] There appears to be a modest benefit to using larger basis sets to ensure convergence of computed shielding tensors, such that 6-311+G(2d,p) has been recommended by Lodweyk *et al.*[21] However, additional corrections (such as employing different reference molecules or scaling) are commonly employed in the conversion of shielding tensors into predicted chemical shifts, such that any improvements associated with using even larger basis sets are likely to be very small. Accordingly, the effects of basis set size upon ^{1}H chemical shifts (as opposed to shielding tensors) do not show systematic improvement with increasing basis set size.[22] It has been observed that increasing basis set size can in fact worsen computed ^{13}C chemical chemical shifts, although the increase in error is usually only by a few tenths of a ppm for ^{13}C and a few hundredths of a ppm for ^{1}H.[23]

6.2.5 *Chemical shift prediction*

Based on Eq. (6.1), isotropic chemical shift values for a nucleus of interest may be obtained from the difference in shielding tensor relative to a reference compound. As tetramethylsilane ($SiMe_4$) defines 0 ppm for both ^{1}H and ^{13}C chemical shifts, this is the most obvious choice of reference molecule for both sets of nuclei. Nevertheless, for calculations of ^{13}C NMR, Sarotti and Pellegrinet[24] have proposed the use of more than one reference molecule depending on the formal hybridization of the carbon atom of interest: methanol serves as the reference for sp^3-carbons and benzene for sp and sp^2-carbons. This so-called multi-standard (MTSD) approach led to average errors in chemical shift predictions of around half those obtained using a single (TMS) reference molecule for all carbon atoms for a set of 50 organic compounds. For example, at the B3LYP/6-311+G(2d,p) level, MAD in chemical shift was reduced from 5.4 ppm (with a TMS standard) to 2.4 ppm with the MTSD approach. The MSTD approach has also been developed for ^{1}H nuclei using the same two reference compounds, benzene (sp and sp^{2} C–H protons) and methanol (sp^{3} C–H protons). For 66 rigid compounds with experimental ^{1}H chemical shifts, the MADs ranged from 0.10 to 0.18 ppm (MTSD), whereas the corresponding values for TMS were between 0.10 and 0.33 ppm.[25]

In fact, it is also possible to derive chemical shift values without using any reference compound whatsoever. For example, using a collection of known experimental chemical shifts, it is possible to define a linear relationship with respect to the corresponding isotropic shielding tensors for the same compounds obtained at a given level of theory. In the ideal case, for ^{1}H or ^{13}C nuclei, this linear relationship would recover the reference shielding tensor value for TMS and have a slope of one. However, the actual value of slope and intercept obtained from a linear regression in this way can be used to derive (linearly) scaled chemical shift values for which the systematic error has been averaged out in the scaling process. This approach avoids specific errors that may be obtained for a reference compound. Migda and Rys[26] noted that TMS-referenced chemical shifts obtained from ^{13}C isotropic shielding constants resulted in MADs two to three times greater than from using a linear regression. Linear-scaling requires a substantial amount of reliable experimental and computational data for which the scaling factors and intercepts have been determined. As these values depend on the method used for shielding tensor calculation, it is important that they have been determined at the appropriate level of theory/basis set combination. Thankfully, Tantillo and colleagues have developed an online resource, the CHEmical SHIft REpository (CHESHIRE) to provide this information for several methods and basis sets.[27] Scaling factors and intercepts have been obtained for different functionals used from both optimization and NMR calculation (including B3LYP, PBE0, mPW1PW91, M06), performed with different solvation models and sizes of basis set. These were obtained for a training set of small molecules and subsequently tested against another set to produce statistical estimates of the performance of each methodology in addition to the scaling factors that can be used by others. These are applicable to ^{1}H chemical shift calculations, and to ^{13}C, provided a halogen is not directly attached: in this case, relativistic effects cause shielding of the carbon nucleus. Braddock and Rzepa[28] have proposed a systematic correction to ^{13}C-Br nuclei of 12–14 ppm to account for spin-orbit effects, which has been deployed in the successful reassignment of halogenated natural products.

6.3 Coupling constant prediction

Coupling is responsible for the splitting observed in experimental NMR spectra in all but the most simple of molecules and is the result of interactions between neighboring, nonequivalent, NMR-active nuclei and surrounding electrons.[29] The geometric dependence of spin-spin coupling makes this particularly useful in the assignment of relative stereochemistry and conformation. The magnitude of coupling (the coupling constant), typically quoted in Hertz, depends on a number of factors, including bond and dihedral angles, bond distances, hybridization and neighboring group participation (substituent effects). For analysis of organic structures, coupling constants for vicinally coupled protons (those separated by three bonds), $^3J_{\text{H-H}}$, are perhaps most routinely used. The magnitude of these values can be predicted using the Karplus equation:

$$J_{\text{H-H}} = \text{A} + \text{B}\cos\phi + \text{C}\cos 2\phi \qquad (6.5)$$

where ϕ is the dihedral angle between H–C–C–H and A, B, and C are empirically derived constants dependent on the atoms and substituents in the system.[30] In this regard, the straightforward prediction of coupling constants can be performed using the dihedral angles of an optimized molecular geometry obtained at any level of theory. However, it is also possible to compute coupling constants using quantum chemistry.[31] There are four contributions to coupling constants, Fermi contact (FC) and spin-dipolar (SD) terms, arising from the interaction of electron spin with the nuclear magnetic field, and paramagnetic spin-orbit (PSO) and diamagnetic spin-orbit (DSO) terms arising from the interaction of nuclear magnetic moments with the magnetic moment generated from the movement of electrons. The weights of these individual contributions to the coupling constant between nuclei vary depending on the character of the nuclei and the extent of their separation (geminal, vicinal, etc.), although usually the FC term is the major contribution among the four, and with standard basis sets, it has the largest error.[32]

Nuclear coupling constants can be calculated using HF, however doing so leads to great inaccuracies caused by triplet instabilities.[33] Kohn–Sham DFT, on the other hand, suffers to a significantly lower extent from such inaccuracies and therefore the majority of large organic molecule quantum chemical coupling constant assignment is performed using this method. Interestingly, modern hybrid density functionals with HF contributions have been shown to result in high levels of agreement with experimentally determined NMR spectra — often with a similar level of accuracy to more computationally demanding *ab initio* methods.[34] Basis sets too play an important role in accurately predicting NMR coupling constants because the Fermi contact term is proportional to electron density. For this reason, bespoke basis sets have been developed specifically for computational NMR prediction, in which core shells are decontracted and additional tight functions are added, with the remaining contributions assessed using standard basis sets.[32,35] For ${}^{1}J_{CC}$ and ${}^{3}J_{CC}$ coupling, studies on strychnine show that it is only necessary for the FC term, whereas the remaining terms can be computed without decontraction to give MAEs of 1.0 and 4.0 Hz, respectively.[36] In the case of proton–proton coupling, computation of the FC term alone using compact 1s basis functions and scaling the result by 0.9155 is sufficient to give a mean error of 0.5 Hz across 61 coupling constants.[37]

6.4 Organic (stereo)-structure prediction

NMR predictions have become a valuable tool in the structure elucidation of organic structures; notably the structures of several natural products have been predicted or revised following the computation of chemical shifts and/or coupling constants, which have been validated experimentally.[38,39] In most cases, this involves the consideration of flexible structures, for which Bifulco *et al.*[40,41] has shown that conformational searching and Boltzmann averaging of chemical shifts is a viable approach to accurate predictions. The use of DFT-based chemical shift calculations to assign the structure and stereochemistry of natural products has rapidly expanded since Rychnovsky's reassignment of 2006.[42] mPW1PW91/6-31G(d,p) GIAO NMR predictions were found to correlate well with the chemical shifts of related natural products whose structures had been confirmed previously, which led to the reassignment of hexacyclinol from an endoperoxide structure (Fig. 6.3) to a diepoxide which has since been confirmed by total synthesis. The structure of aquatolide, a sesquiterpene natural product, was confirmed following DFT predictions of ^{13}C and ^{1}H chemical shift, and ^{1}H–^{1}H coupling constants for the postulated structures.[43] Large differences of 24 ppm (^{13}C) and 1.3 ppm (^{1}H) between calculated and measured chemical shifts were found using the originally proposed ladderane structure. In contrast, the computed chemical shifts for an alternative skeleton based on biosynthetic considerations were in much more satisfactory agreement with the experiment. The computed coupling constants predicted a large $^{4}J_{H-H}$ coupling constant of 6.8 Hz, also in agreement with experiment, and the structural revision was subsequently confirmed by X-ray crystallography. The above examples are relatively inflexible molecules, however, related computational approaches have been successfully

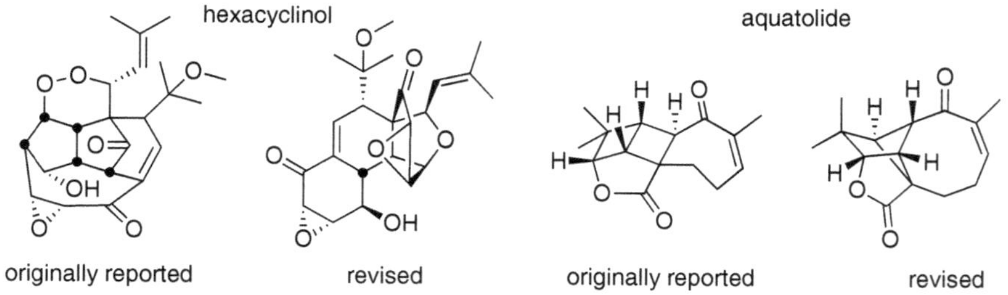

Figure 6.3. Computational structural revisions of rigid natural products based on NMR predictions.[42,43]

explored for more flexible natural products. In 2012, two synthetic groups independently verified the predicted structure of the natural product elatenyne.[44] Originally, the structure had been assigned as a bicyclic molecule with two tetrahydropyran rings, however, the reassignment confirms the presence of tetrahydrofuran rings instead (Fig. 6.4). The presence of flexible side-chains, five-membered rings, and a rotatable inter-ring C–C bond led to a number of conformers accessible, for each of the 32 possible diastereomers of the natural product.

Goodman and Burton used GIAO chemical shift calculations to predict the correct relative stereochemistry, having the lowest MAE with respect to experiment.[8] MM conformational searches were used to generate the structures, and scaled B3LYP shielding tensors were computed without DFT optimizations. Laurefurenyne, has a related bifuranyl skeleton to elatenyne and has similarly undergone a structural revision based on NMR calculations of ^{13}C and ^{1}H chemical shifts.[10] In this case, geometries obtained from a MM conformational search were optimized at the ωB97XD/6-31G(d) level of theory, and GIAO-mPW1PW91/6-311G(d,p) shielding tensors were scaled based on the experimental data.

In comparing computed with experimental chemical shift values, various statistical metrics may be used to assign the most likely structure, such as the average error or linear correlation coefficients. Particularly when comparing different stereostructures, the

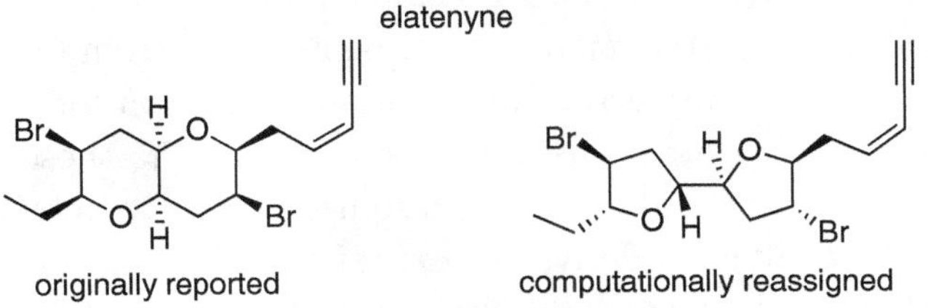

Figure 6.4. Computational structural revisions of the flexible natural product elatenyne based on chemical shift predictions.[45]

difference between correct and incorrect structures may be relatively small. Smith and Goodman[45] have introduced the DP4 metric to assign probabilities to the likelihood of a computational NMR-based structural assignment being correct. This metric relies on prior knowledge of the underlying distribution of errors ($\Delta\delta$) in scaled ^{13}C and ^{1}H chemical shifts for a large collection of correctly assigned structures. These errors were found to be distributed according to a Student's T-distribution. From this fact, a conditional probability can be assigned to a set of errors in chemical shifts obtained for a new assignment, enabling a confidence to be assigned to the likelihood of the assignment being correct. Goodman's original data set used MMFF geometries, with linearly scaled GIAO B3LYP/6-31G(d) chemical shifts to obtain the shape and width of the error distributions. Similar distributions have been observed for alternative DFT methods, such that the original DP4 parameters are transferable. Nonetheless, a modified probability (DP4+) has also been developed, which allows unscaled chemical shifts and higher levels of theory to be used.[46]

In addition to the use of chemical shifts, the nuclear Overhauser effect (nOe) has also been used to compare calculated with experimental interproton distances as a tool for structural assignment. Chini *et al.*[47] used quantitative nOe measurements to probe relative stereochemistry. The polyhydroxylated steroid conicasterol F lacks protons at either end of the epoxide, which makes experimental assignment of stereochemistry difficult. An MMFF conformational search using molecular dynamics and Monte Carlo simulations was used to generate low-energy structures that were optimized at the mPW1PW91/6-31G(d) level. DFT and experimental interproton distances enable the correct structure to be assigned from two possibilities: the mean absolute error and standard deviation for structure X is 3.0% and 2.6%, whereas for its diastereoisomer Y, these values are 7.8% and 5.9%. Butts *et al.*[48] have also used the nOe between H11b–H23b in strychnine to derive an experimental distance of 3.49 Å. This is shorter than expected from the crystallographic structure (Fig. 6.5).[48] Two conformers in solution contribute to the observed

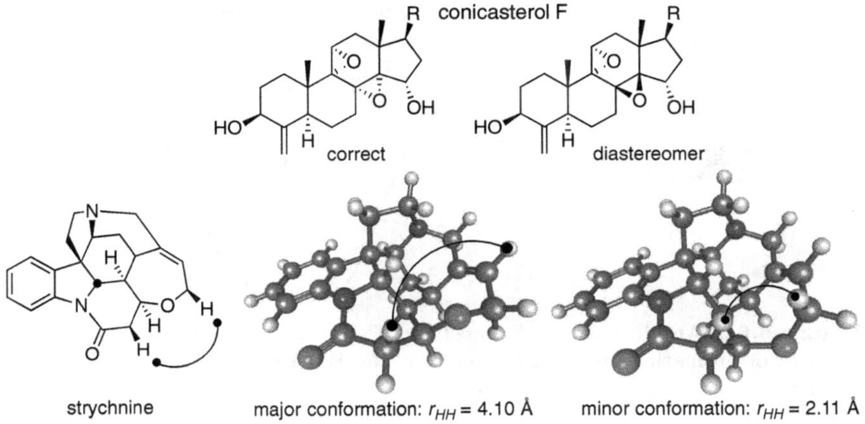

Figure 6.5. Comparison of computed and experimental *n*Oe-derived interproton distances have been used to assign relative stereochemistry and molecular conformation.[47,48]

nOe: the LCCSD(T0)/cc-pVTZ energy difference of 9.4 kJ/mol corresponds to a 2.2% population of the minor conformer. The key distance in each conformer is 4.10 and 2.11 Å, giving a weighted average interproton distance of 3.60 Å, which is comparable to the experimental distance.

6.5 Nuclear-independent chemical shift calculations

Unlike experimental measurements of chemical shift, it is possible to compute shielding tensors, and hence chemical shift values not just at nuclear positions but also at any point in space applying the definition in (Sec. 6.1). This has been termed a nucleus-independent chemical shift (NICS) calculation and allows for a computational assessment of the magnetic environment at any point in space, in or around a chemical system. The technique functions by calculating the isotropic shielding tensor at a given point, defined by the addition of a ghost or dummy atom in the program input file, in much the same way as for a typical nucleus (Fig. 6.6).

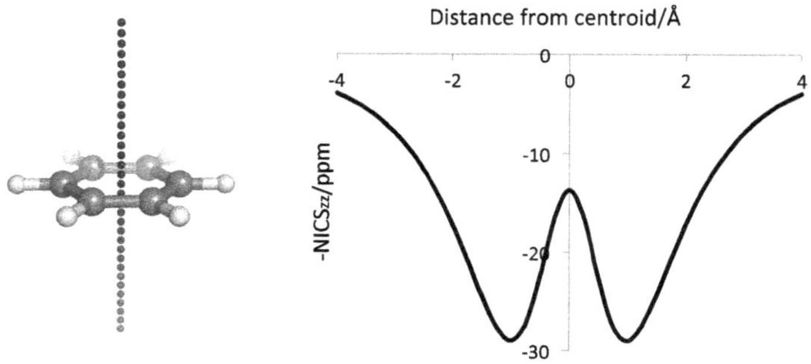

Figure 6.6. An illustration of "ghost atom" positioning and dissected $NICS_{zz}$ analysis of aromatic benzene performed at the B3LYP/6-311++G(d,p) level of theory.[53]

As a computational chemistry tool, NICS calculations have been implemented in a number of ways and their rise in popularity has been facilitated in part by their incorporation in many of the standard quantum chemical programs.[49,50] Schleyer *et al.*[51] pioneered their use in the assessment of the aromatic nature of chemical systems. In this work, the calculation of magnetic shieldings at the centroids of ring systems (defined as the nonweighted mean of the heavy atom coordinates), defined as NICS(0), could be used as a criterion of aromaticity/anti-aromaticity (see Chapter 9). This is because the characteristic electronic ring-currents in aromatic systems (e.g., benzene) give rise to shielding at their center, leading to negative values of NICS(0), whereas for anti-aromatic systems (e.g., cyclobutadiene), the opposite effect is observed and positive NICS(0) values are obtained. The results from NICS calculations were shown to be a more conclusive metric for aromaticity than previously used, diamagnetic susceptibility exaltation (Λ) values,[52] which have a greater dependence on ring size and require calibration.

Morao *et al.*[54] demonstrated the use of NICS calculations in describing the aromatic nature of transition structures (TS) in 1,3-dipolar cycloadditions. Due to the asymmetric nature of these systems, they instead opted to use the ring critical point of electron density for the location of the NICS(0) probe, rather than the ring centroid. Their use of NICS calculations in determining the extent of π-aromaticity in these systems led to the conclusion that only in-plane aromaticity was present in these reactions. Following these early studies, it was suggested that the use of standard NICS values was limited due to contamination by in-plane tensor components, which are not necessarily related to aromaticity.[55,56] This led to variations of NICS being proposed, including dissected NICS where the σ and π components are distinguished,[57] and $NICS_{zz}$, where the zz component of the NICS tensor is separated as this has been found to be particularly sensitive to π-delocalization.[58] NICS calculations may be run at points along the vector which is perpendicular to the ring plane, and which passes through the centroid of ring systems, to generate variable distance NICS values; these yield greater information about the topology of orbitals and allows for greater distinction to be made between ring systems.[59] The most primitive of these types of calculations is termed NICS(1), where a single NICS probe is positioned 1 Å above or below the aromatic ring[60] — in more sophisticated versions of this, multiple probes are positioned along the plane orthogonal vector at shorter-spaced intervals. This type of NICS analysis has been used to evaluate aromaticity, antiaromaticity, and nonaromaticity in a wide range of chemical systems. More recently, variable distance $NICS_{zz}$ calculations have been used to reveal the aromatic properties of organic and inorganic molecules, intermediates, ions, and transition state (TSs).[61] In some cases, they have resulted in the discovery of previously overlooked connections between different classes of organic reactions; Gilmore *et al.*[62] have shown the cyclic nature of electron delocalization in anionic 5-*endo-dig* ring closing TSs, leading to the assertion that these cyclizations are in fact "aborted" sigmatropic shifts. The characteristics of the TS in 5-*endo-trig* cyclizations, formally disfavored according to Baldwin's rules, have also been studied with NICS calculations.[63] The lack of aromatic ring currents observed in these ring-closing TSs indicated an absence of pericyclic character, which had been previously suggested for this class of cyclization.[64]

6.6 Practical considerations

Optimized structures of organic molecules tend to be relatively insensitive to the level of theory employed. Most applications discussed above used DFT with a medium-sized basis set, such as 6-31G(d), or larger. For geometry optimizations, standard functionals such as B3LYP have been put to good use, either in the gas phase or with an implicit solvation treatment. Where accurate relative energetics are important, for example, in Boltzmann weighting of a conformational ensemble, it may be necessary to use energetics generated using a larger basis set, or with *ab initio* methods.

The calculation of shielding constants requires a reference molecule (TMS), or alternatively, can be performed by scaling the shielding tensors. Here there is an advantage to using tried and tested methods such as mPW1PW91, PBE0, and B3LYP with a 6-311+G(2d,p) basis set since tabulated scaling coefficients have been determined previously (http://cheshirenmr.info) for data sets of organic molecules. Further improvements in accuracy are likely by using an ensemble of geometries generated from molecular dynamics trajectories instead of a single static structure. However, for organic structure elucidation, it is possible to discriminate between correct and incorrect structures on the basis of chemical shifts (^{13}C and ^{1}H) from standard geometry optimized structures. For stereostructure assignment, the use of the DP4 statistical metric enables one to assign confidence levels to an assignment, which can be performed through a webserver (http://www-jmg.ch.cam.ac.uk/tools/nmr/DP4/).

The generation of NICS values at selected positions in and around a chemical system is straightforward and can be performed with most popular quantum chemistry packages. It has been shown that NICS calculations are relatively insensitive to the levels of theory employed, and as a result, the B3LYP functional with a medium-sized basis set such as 6-311G(d,p) has been commonly used. This is because NICS calculations are localized and often situated distant from chemical nuclei. The placement of "ghost atoms" at which these calculations are run can be easily automated through simple scripts, allowing for rapid and precise positioning of these points in free space.

References

1. Kaupp, M., Buhl, M. and Malkin, V. G. (2004). *Calculation of NMR and EPR Parameters: Theory and Applications*, 1st Ed. (Wiley-VCH: Verlag GmbH & Co, KGaA).
2. Schreckenbach, G. and Ziegler, T. (1998). Density functional calculations of NMR chemical shifts and ESR g-tensors, *Theor. Chem. Acc.*, 99, pp. 71–82.
3. Ditchfield, R. (1974). Self-consistent perturbation theory of diamagnetism. I. A gauge-invariant LCAO (Linear Combination of Atomic Orbitals) method for NMR chemical shifts, *Mol. Phys.*, 27, pp. 789–807.
4. Schindler, M. and Kutzelnigg, W. (1982). Theory of magnetic susceptibilities and NMR chemical shifts in terms of localized quantities. II. Application to some simple molecules, *J. Chem. Phys.*, 76, pp. 1919–1933.
5. Wolinski K., Hinton, J. F. and Pulay, P. (1990). Efficient implementation of the gauge-independent atomic orbital method for NMR chemical shift calculations, *J. Am. Chem. Soc.*, 112, pp. 8251–8260.
6. Zhang, Y., Wu, A., Xu, X. and Yan, Y. (2007). Geometric dependence of the B3LYP-predicted magnetic shieldings and chemical shifts, *J. Phys. Chem. A*, 111, pp. 9431– 9437.
7. Harding, M. E., Gauss, J. and Schleyer, P. v. R. (2011). Why benchmark-quality computations are needed to reproduce 1-adamantyl cation NMR chemical shifts accurately, *J. Phys. Chem. A*, 115, pp. 2340–2344.
8. Smith, S. G., Paton, R. S., Burton, J. W. and Goodman, J. M. (2008). Stereostructure assignment of flexible five-membered rings by GIAO ^{13}C NMR calculations: Prediction of the stereochemistry of elatenyne, *J. Org. Chem.*, 73, pp. 4053–4062.
9. Willoughby, P. H., Jansma, M. J. and Hoye, T. R. (2014). A guide to small-molecule structure assignment through computation of (^{1}H and ^{13}C) NMR chemical shifts, *Nat. Protoc.*, 9, pp. 643–660.
10. Shepherd, D. J., Broadwith, P. A., Dyson, B. S., Paton, R. S. and Burton, J. W. (2013). Structure reassignment of laurefurenynes A and B by computation and total synthesis, *Chem. Eur. J.*, 19, pp. 12644–12648.
11. Auer, A. A., Gauss, J. and Stanton, J. F. (2003). Quantitative prediction of gas-phase ^{13}C nuclear magnetic shielding constants, *J. Chem. Phys.*, 118, pp. 10407–10417.
12. Kwan, E. E. and Liu, R. Y. (2015). Enhancing NMR prediction for organic compounds using molecular dynamics, *J. Chem. Theory Comput.*, 11, pp. 5083–5089.
13. Teale, A. M., Lutnaes, O. B., Helgaker, T., Tozer, D. J. and Gauss, J. (2013). Benchmarking density-functional theory calculations of NMR shielding constants and spin–rotation constants using accurate coupled-cluster calculations, *J. Chem. Phys.*, 138, pp. 024111.

14. Armangué, L., Solà, M. and Swart, M. (2011). Nuclear shieldings with the SSB-D functional, *J. Phys. Chem. A*, 115, pp. 1250–1256.
15. Jain, R., Bally, T. and Rablen, P. R. (2009). Calculating accurate proton chemical shifts of organic molecules with density functional methods and modest basis sets, *J. Org. Chem.*, 74, pp. 4017–4023.
16. Wu, A., Zhang, Y., Xu, X. and Yan, Y. (2007). Systematic studies on the computation of nuclear magnetic resonance shielding constants and chemical shifts: The density functional models, *J. Comput. Chem.*, 28, pp. 2431–2442.
17. Magyarfalvi, G. and Pulay, P. (2003). Assessment of density functional methods for nuclear magnetic resonance shielding calculations, *J. Chem. Phys.*, 119, pp. 1350–1357.
18. Giesen, D. J. and Zumbulyadis, N. (2002). A hybrid quantum mechanical and empirical model for the prediction of isotropic ^{13}C shielding constants of organic molecules, *Phys. Chem. Chem. Phys.*, 4, pp. 5498–5507.
19. Jensen, F. (2008). Basis set convergence of nuclear magnetic shielding constants calculated by density functional methods, *J. Chem. Theory Comput.*, 4, pp. 719–727.
20. Christensen, A. S., Sauer, S. P. A. and Jensen, J. H. (2011). Definitive benchmark study of ring current effects on amide proton chemical shifts, *J. Chem. Theory Comput.*, 7, pp. 2078–2084.
21. Lodewyk, M. W., Siebert, M. R. and Tantillo, D. J. (2012). Computational prediction of ^{1}H and ^{13}C chemical shifts: A useful tool for natural product, mechanistic, and synthetic organic chemistry, *Chem. Rev.*, 112, pp. 1839–1862.
22. Jain, R., Bally, T. and Rablen, P. R. (2009). Calculating accurate proton chemical shifts of organic molecules with density functional methods and modest basis sets, *J. Org. Chem.*, 74, pp. 4017–4023.
23. Gauss, J. (1999). Effects of electron correlation in the calculation of nuclear magnetic resonance chemical shifts, *J. Chem. Phys.*, 99, pp. 3629–3643.
24. Sarotti, A. M. and Pellegrinet, S. C. (2009). A multi-standard approach for GIAO ^{13}C NMR calculations, *J. Org. Chem.*, 74, pp. 7254–7260.
25. Sarotti, A. M. and Pellegrinet, S. C. (2012). Application of the multi-standard methodology for calculating ^{1}H NMR chemical shifts, *J. Org. Chem.*, 77, pp. 6059–6065.
26. Migda, W. and Rys, B. (2004). GIAO/DFT evaluation of ^{13}C NMR chemical shifts of selected acetals based on DFT optimized geometries, *Magn. Reson. Chem.*, 42, pp. 459–466.
27. Chemical Shift Repository with Coupling Constants Added Too. See: http://cheshirenmr.info.
28. Braddock, D. C. and Rzepa, H. S (2008). Structural reassignment of obtusallenes V, VI, and VII by GIAO-based density functional prediction, *J. Nat. Prod.*, 71, pp. 728–730.

29. Hahn, E. L. and Maxwell, D. E. (1952). Spin echo measurements of nuclear spin coupling in molecules, *Phys. Rev.*, 88, pp. 1070–1084.
30. Karplus, M. (1963). Vicinal proton coupling in nuclear magnetic resonance, *J. Am. Chem. Soc.*, 85, pp. 2870–2871.
31. Helgaker, T., Jaszuński, M. and Pecul, M. (2008). The quantum-chemical calculation of NMR indirect spin–spin coupling constants, *Prog. Nucl. Magn. Reson. Spectrosc.*, 53, pp. 249–268.
32. Deng, W., Cheeseman, J. R. and Frisch, M. J. (2006). Calculation of nuclear spin–spin coupling constants of molecules with first and second row atoms in study of basis set dependence, *J. Chem. Theory Comput.*, 2, pp. 1028–1037.
33. Lutnæs, O. B., Helgaker, T. and Jaszuński, M. (2010). Spin–spin coupling constants and triplet instabilities in Kohn–Sham theory, *Mol. Phys.*, 108, pp. 2579–2590.
34. Helgaker, T., Lutnæs, O. B. and Jaszuński, M. (2007). Density-functional and coupled-cluster singles-and-doubles calculations of the nuclear shielding and indirect nuclear spin–spin coupling constants of o-benzyne, *J. Chem. Theory Comput.*, 3, pp. 86–94.
35. Jensen, M. (2006). The basis set convergence of spin—spin coupling constants calculated by density functional methods, *J. Chem. Theory Comput.*, 2, pp. 1360–1369.
36. Williamson, R. T., Buevich, A. V. and Martin, G. E. (2012). Experimental and theoretical investigation of $^{1}J_{CC}$ and $^{n}J_{CC}$ coupling constants in strychnine, *Org. Lett.*, 14, 5098–5101.
37. Bally, T. and Rablen, P. R. (2011). Quantum-chemical simulation of ^{1}H NMR spectra. 2. Comparison of DFT-based procedures for computing proton-proton coupling constants in organic molecules, *J. Org. Chem.*, 76, pp. 4818–4830.
38. Tantillo, D. J. (2013). Walking in the woods with quantum chemistry — applications of quantum chemical calculations in natural products research, *Nat. Prod. Rep.*, 30, pp. 1079–1086.
39. Di Micco, S., Chini, M. G., Riccio, R. and Bifulco, G. (2010). Quantum mechanical calculation of NMR parameters in the stereostructural determination of natural products, *Eur. J. Org. Chem.*, pp. 1411–1434.
40. Use of DFT calculations of NMR chemical shifts for structure deter- mination/confirmation was pioneered by Bifulco: Barone, G., Gomez-Paloma, L., Duca, D., Silvestri, A., Riccio, R. and Bifulco, G. (2002). Structure validation of natural products by quantum-mechanical GIAO calculations of ^{13}C NMR chemical shifts, *Chem. Eur. J.*, 8, pp. 3233–3239.
41. Bifulco, G., Dambruoso, P., Gomez-Paloma, L. and Riccio, R. (2007). Determination of relative configuration in organic compounds by NMR spectroscopy and computational methods, *Chem. Rev.* 107, pp. 3744–3779.
42. Rychnovsky, S. D. (2006). Predicting NMR spectra by computational methods: Structure revision of hexacyclinol, *Org. Lett.*, 8, pp. 2895–2898.

43. Lodewyk, M. W., Soldi, C., Jones, P. B., Olmstead, M. M., Rita, J., Shaw, J. T. and Tantillo, D. J. (2012). The correct structure of aquatolide — Experimental validation of a theoretically-predicted structural revision, *J. Am. Chem. Soc.*, 134, pp. 18550–18553.
44. Dyson, B. S., Burton, J. W., Sohn, T.-I., Kim, B., Bae, H. and Kim, D. J. (2012). Total synthesis and structure confirmation of elatenyne: Success of computational methods for NMR prediction with highly flexible diastereomers, *J. Am. Chem. Soc.*, 134, pp. 11781–11790.
45. Smith, S. G. and Goodman, J. M. (2010). Assigning stereochemistry to single diastereoisomers by GIAO NMR calculation: The DP4 probability, *J. Am. Chem. Soc.*, 132, pp. 12946–12959.
46. Grimblat, N., Zanardi, M. M. and Sarotti, A. M. (2015). Beyond DP4: An improved probability for the stereochemical assignment of isomeric compounds using quantum chemical calculations of NMR shifts, *J. Org. Chem.*, 80, pp. 12526–12534.
47. Chini, M. G., Jones, C. R., Zampella, A., D'Auria, M. V., Renga, B., Fiorucci, S., Butts, C. P. and Bifulco, G. (2012). Quantitative NMR-derived interproton distances combined with quantum mechanical calculations of ^{13}C chemical shifts in the stereochemical determination of conicasterol F, a nuclear receptor ligand from *Theonella swinhoei, J. Org. Chem.*, 77, pp. 1489–1496.
48. Butts, C. P., Jones and C. R., Harvey, J. N. (2011). High precision NOEs as a probe for low level conformers—a second conformation of strychnine, *Chem. Commun.*, 47, pp. 1193–1195.
49. Schleyer, P. v. R., Jiao, H. J., Hommes, N., Malkin, V. G. and Malkina, O. L. (1997). An evaluation of the aromaticity of inorganic rings: Refined evidence from magnetic properties, *J. Am. Chem. Soc.*, 119, pp. 12669–12670.
50. Chen, Z., Wannere, C. S., Corminboeuf, C., Puchta, R. and Schleyer, P. v. R. (2005). Nucleus-independent chemical shifts (NICS) as an aromaticity criterion, *Chem. Rev.*, 105, pp. 3842–3888.
51. Schleyer, P. v. R., Maerker, P., Dransfeld, C., Jiao, A. H. and van Eikema Hommes, N. J. R. (1996). Nucleus-independent chemical shifts: A simple and efficient aromaticity probe, *J. Am. Chem. Soc.*, 118, pp. 6317–6318.
52. Dauben Jr., H. J., Wilson, J. D. and Laity, J. L. (1998). Diamagnetic susceptibility exaltation as a criterion of aromaticity, *J. Am. Chem. Soc.*, 90, pp. 811–813.
53. Jackson, K. E. (2016). Development and application of computational tools to analyse selectivity in organic chemistry, DPhil Thesis, The University of Oxford, pp. 138–144.
54. Morao, I., Lecea, B. and Cossío, F. P. (1997). In-plane aromaticity in 1,3-dipolar cycloadditions, *J. Org. Chem.*, 62, pp. 7033–7036.
55. Lazzeretti, P. (2000). Ring currents, *Prog. Nucl. Magn. Reson. Spectrosc.*, 36, pp. 1–88.
56. Lazzeretti, P. (2004). Assessment of aromaticity *via* molecular response properties, *Phys. Chem. Chem. Phys.*, 6, pp. 217–223.

57. Heine, T., Schleyer, P. v. R., Corminboeuf, C., Seifert, G., Reviakine, R. and Weber, J. (2003). Analysis of aromatic delocalization: Individual molecular orbital contributions to nucleus-independent chemical shifts, *J. Phys. Chem. A*, 107, pp. 6470–6475.
58. Corminboeuf, C., Heine, T., Seifert, G., Schleyer P. v. R. and Weber, J. (2004). Induced magnetic fields in aromatic [n]-annulenes – interpretation of NICS tensor components, *Phys. Chem. Chem. Phys.*, 6, pp. 273–276.
59. Stanger, A. (2006). Nucleus-independent chemical shifts (NICS): Distance dependence and revised criteria for aromaticity and antiaromaticity, *J. Org. Chem.*, 71, pp. 883–893.
60. Schleyer, P. v. R., Manoharan, M., Wang, Z. X., Kiran, B., Jiao, H. J., Puchta, R. and Hommes, N. (2001). Dissected nucleus-independent chemical shift analysis of π-aromaticity and antiaromaticity, *Org. Lett.*, 3, pp. 2465–2468.
61. Jiménez-Halla, J. O. C., Matito, E., Robles, J. and Solà, M. (2006). Nucleus-independent chemical shift (NICS) profiles in a series of monocyclic planar inorganic compounds, *J. Organomet. Chem.* 691, pp. 4359–4366.
62. Gilmore, K., Manoharan, M., Wu, J. I.-C., Schleyer, P. v. R. and Alabugin, I. V. (2012). Aromatic transition states in non-pericyclic reactions: Anionic 5–endo cyclizations are as aborted sigmatropic shifts, *J. Am. Chem. Soc.*, 134, pp. 10584–10594.
63. Johnston, C. P., Kothari, A., Sergeieva, T., Okovytyy, S. I., Jackson, K. E., Paton, R. S. and Smith, M. D. (2015). Catalytic enantioselective synthesis of indanes by a cation-directed 5-*endo-trig* cyclization, *Nat. Chem.*, 7, pp. 171–177.
64. Peng, Q. and Paton, R. S. (2016). Catalytic control in cyclizations: From computational mechanistic understanding to selectivity prediction, *Acc. Chem. Res.*, 49, pp. 1042–1051.

Chapter 7

Energy Decomposition Analysis and Related Methods

Israel Fernández

Departamento de Química Orgánica I
Facultad de Ciencias Químicas
Universidad Complutense de Madrid, Spain

7.1 Introduction

The nature of the chemical bond has been (and still is) a fundamental topic in chemistry. Since the groundbreaking heuristic theory of chemical bonding by Lewis,[1] where the concept of the octet rule and the grouping of electrons into lone and bonding pairs were suggested, an impressive number of theories and concepts have been developed to gain more insight into the bonding situation of molecules, spanning from relatively simple diatomic species to more complex organometallic or supramolecular systems.[2]

Among the different approaches to analyze the chemical bond, a particular group of bonding schemes is focused on the decomposition of the bond energy into chemically meaningful contributions. Such decomposition schemes include, to cite a few methods, the symmetry adapted perturbation theory (SAPT) scheme,[3] the block localized wavefunction energy decomposition (BLW-ED) method,[4] the absolutely localized molecular orbitals energy decomposition analysis (ALMO-EDA)[5] and the natural energy decomposition analysis (NEDA) theory.[6] In this chapter, we focus on the energy decomposition analysis (EDA) method, which was independently developed by Morokuma[7] and by Ziegler and Rauk.[8] This method, also known as extended transition-state (ETS) scheme, has been shown to give important information about the nature of the bonding in main-group compounds and transition-metal complexes.[9–12] Herein, we demonstrate the performance of the EDA method in rationalizing the nature of chemical bonding in organic chemistry by presenting illustrative applications of this approach. In addition, we also introduce a recent extension of the method where the EDA is combined with the natural orbital for chemical valence (NOCV) method.[13]

7.2 The EDA and EDA–NOCV methods

7.2.1 *Basics of the energy decomposition analysis method*

The EDA method is a powerful tool that bridges the gap between elementary quantum mechanics and a conceptually simple interpretation of the nature of the chemical bond. Within this partitioning scheme, the energy lowering associated with the formation of a chemical bond is divided into three different terms, which can be interpreted in a physically meaningful way. Although such division can be considered as arbitrary, it is mathematically unambiguously defined, and the three terms may be identified not only with classical chemical concepts but also with quantum chemical terms.

The focus of the bonding analysis is the instantaneous interaction energy, ΔE_{int}, of a bond $A-B$, which is the energy difference between the molecule and the fragments in the electronic reference state and frozen geometry of the compound AB. As mentioned above, EDA divides the bond formation between the interacting fragments into three steps that can be interpreted in a plausible way.

In the first step, the fragments, A and B, with their geometries frozen as in AB, are computed individually in appropriately selected spin states (which may not be the ground states) and are then superimposed with unrelaxed electron densities at the geometry of AB to give $A'B'$. This gives the quasi-classical electrostatic interaction[14] between the unperturbed charge distributions of the prepared fragments, ΔE_{elstat}, as the energy difference between the original AB and $A'B'$ (Eq. (7.1)).

$$\Delta E_{\text{elstat}} = \sum_{\alpha \in A} \sum_{\beta \in B} \frac{Z_\alpha Z_\beta}{R_{\alpha\beta}} + \int dr V_B(r) \rho_A(r) + \int dr V_A(r) \rho_B(r) + \iint dr_1 dr_2 \frac{\rho_A(r_1) \rho_B(r_2)}{r_{12}} \quad (7.1)$$

The term ΔE_{elstat} is usually attractive because the total nuclear–electron attraction is, in most cases, larger than the sum of the nuclear–nuclear and electron–electron repulsions. Despite that, the resulting wavefunction for this modified $A'B'$ species violates the Pauli principle because electrons with same spin from two different fragments may occupy the same spatial region. This situation is rectified by enforcing the Kohn–Sham determinant on the superimposed fragments to obey the Pauli principle by antisymmetrization and renormalization of the $A'B'$ wavefunction, thereby removing electron density, particularly from the AB bonding region where the overlap of the frozen densities is large. This step gives the Pauli repulsion term, ΔE_{Pauli}, which comprises the repulsive orbital interactions between closed shells and is responsible for any steric repulsion between molecular fragments. The stabilizing orbital interaction term, ΔE_{orb}, is calculated in the final step of the energy partitioning analysis when the molecular orbitals relax to their optimal form. The ΔE_{orb} term is always attractive as the total wavefunction is optimized during its calculation. This term therefore accounts for charge transfer (interaction between occupied orbitals on one fragment with unoccupied orbitals on the other, including HOMO–LUMO interactions), polarization (empty-occupied orbital mixing on one fragment due to the presence of another fragment) and electron-pair bonding (the stabilization arising from the formation of the electron-pair bonding configuration in which the bonding combination between the SOMOs is formed and doubly occupied).[15] Finally, the interactions due to dispersion forces may also be added in the ΔE_{disp} term.

The sum of these contributions gives the total interaction energy between fragments A and B (Eq. (7.2))

$$\Delta E_{\text{int}} = \Delta E_{\text{elstat}} + \Delta E_{\text{Pauli}} + \Delta E_{\text{orb}}(+\Delta E_{\text{disp}}) \tag{7.2}$$

Moreover, the orbital term ΔE_{orb} can be further decomposed into contributions by the orbitals belonging to different irreducible representations of the point group of the interacting system (Eq. (7.3)). This allows us to quantitatively estimate the intrinsic strength of orbital interactions from orbitals having σ, π, δ symmetry.

$$\Delta E_{\text{orb}} = \sum_{\Gamma} \Delta E_{\Gamma} \tag{7.3}$$

The interaction energy, ΔE_{int}, can be used to calculate the bond dissociation energy, D_{e}, by adding the so-called preparation energy, ΔE_{prep}, which is the energy necessary to promote the fragments from their equilibrium geometry to the geometry they adopt in AB (Eq. (7.4)). This term is often referred to also as the distortion or strain energy. The advantage of using ΔE_{int} instead of D_{e} is that the instantaneous electronic interaction of the fragments gets analyzed, which yields a direct estimate of the energy components. At variance, D_{e} values are usually affected by the geometrical and, possibly, the electronic relaxation of the fragments. For instance, there are molecules (such as high-energy materials and explosives) where the fragments A and B are lower in energy than $A{-}B$, which makes the D_{e} values useless for estimating the $A{-}B$ bond strength.

$$\Delta E(= -D_{\text{e}}) = \Delta E_{\text{prep}} + \Delta E_{\text{int}} \tag{7.4}$$

At this point, we would like to highlight the physical meaning and the relevance of the energy terms involved in the EDA. In particular, during the calculation of ΔE_{elstat}, the electron-density distribution of the fragments does not consider the polarization of the charge distribution that arises in chemical interactions. The polarization is only taken into account during the final step of the EDA, in which stabilization due to polarization and relaxation is completely included in the ΔE_{orb} term. Therefore, the calculated electrostatic interaction energy, ΔE_{elstat}, of the unpolarized fragments is not the same as the total potential energy change of bond formation because the final electron density differs from the electronic density that results from superposition of the two fragment densities. This means that the ΔE_{elstat} term contains only quasi-classical electrostatic interactions coming from the frozen electron densities of the fragments, whereas the ΔE_{orb} term contains electrostatic attractions resulting from quantum interference, potential energy changes due to polarization and relaxation and kinetic energy contributions. In addition, the ΔE_{Pauli} term also contains potential energy contributions because electronic charge is moved from the area of overlap of the fragments to nearer the nuclei. The increase in the total energy due to the ΔE_{Pauli} term comes from the kinetic energy of the electrons, which is much higher when they are closer to the nuclei. For this reason, the ΔE_{Pauli} term is sometimes called kinetic repulsion.[11,12]

For further details of the theoretical background and applications of the EDA method, we refer readers to the review articles and book chapters that were published recently.[9–12,16,17]

7.2.2 *The EDA–NOCV method*

The NOCV method[13] is a useful scheme that links the concepts of bond-order, bond-orbitals and charge rearrangement with the deformation density. Within this scheme, a few eigenfunctions of the deformation density matrix (ΔP) describe bond formation of the molecules from fragments in a compact form. The corresponding eigenvalues can be used as valence indices as well as a measure of the change in the density associated with bond formation. Despite that, this scheme does not provide information about the energetics associated with the charge rearrangement. For this reason, Mitoraj co-workers efficiently combined this scheme with the EDA method described above.[18]

The focus of the NOCV method is the deformation density $\Delta\rho(r)$, which corresponds to the difference between the densities of the fragments before and after bond formation. $\Delta\rho(r)$ is expressed as a sum of pairs of complementary orbitals (ψ_{-k}, ψ_k) corresponding to eigenvalues (ν_{-k}, ν_k) with the same absolute value but opposite in sign (Eq. (7.5)):

$$\Delta\rho^{\mathrm{orb}}(r) = \sum_{k=1}^{N/2} v_k[-\Psi_{-k}^2(r) + \Psi_k^2(r)] = \sum_{k=1}^{N/2} \Delta\rho_k(r) \qquad (7.5)$$

The complementary pairs of NOCV define the channels for electron charge transfer between the molecular fragments. Therefore, Eq. (7.5) makes it possible to express the total charge deformation $\Delta\rho(r)$ associated with the bond formation in terms of pairwise charge contributions $\Delta\rho_k(r)$ coming from particular pairs of NOCV-orbitals.

The combination of the NOCV and EDA schemes leads to the following expression for the ΔE_{orb} term (Eq. (7.6)):

$$\Delta E_{\mathrm{orb}} = \sum_{k=1}^{N/2} v_k[-F_{-k,-k}^{\mathrm{TS}} + F_{k,k}^{\mathrm{TS}}] = \sum_{k=1}^{N/2} \Delta E_k^{\mathrm{orb}} \qquad (7.6)$$

where $F_{-k,-k}^{\mathrm{TS}}$ and $F_{k,k}^{\mathrm{TS}}$ are diagonal transition-state (TS) Kohn–Sham matrix elements corresponding to NOCVs with the eigenvalues $\nu_{-\mathrm{k}}$ and ν_{k}, respectively. Please note that here the term "TS" refers to the charge density which is intermediate between the density of the final molecule AB and the superimposed fragment densities of A and B.

Therefore, the difference between the EDA and the EDA–NOCV method lies only in the expression for the orbital interaction term ΔE_{orb}. Typically, the $\Delta E_k^{\mathrm{orb}}$ term of a particular type of bond is assigned by visual inspection of the shape of the deformation density, $\Delta\rho_{\mathrm{k}}$. At variance with the EDA method, the EDA–NOCV scheme provides both qualitative ($\Delta\rho_{\mathrm{orb}}$) and quantitative ($\Delta E_{\mathrm{orb}}$) information about the individual strength of orbital interactions in chemical bonds, even in molecules with C_i symmetry.

7.3 Representative applications of the EDA and EDA–NOCV methods in organic chemistry

7.3.1 *Bonding in ethane and related species*

We first discuss the bonding situation of the D_{3d} staggered conformation of ethane focusing on the C–C single bond.[19,20] In this particular case, the EDA is carried out with C_{3v} symmetry using CH_3• radicals (whose unpaired electron is formally in a $sp^3(\sigma)$-hybridized orbital) as fragments. This means that there are orbitals with $a_1(\sigma), a_2(\delta)$ and $e(\pi)$ symmetry. The EDA data (Table 7.1) indicate that the covalent character of the C–C bond is, not surprisingly, quite significant as the orbital term becomes the major contributor to the total interaction energy (58.4%). Despite that, the ionic character of the C–C bond in ethane is far from negligible in view of the high contribution of the electrostatic term (41.6%). As expected, the partitioning of the ΔE_{orb} term in contributions coming from the different irreducible representations of the C_{3v} point group clearly confirms the σ-bond character of the C–C bond in view of the contribution of the $\Delta E_\sigma(a_1)$ term (94.6%). Despite this, the π-bonding (measured by the $\Delta E_\pi(e)$ term) in ethane is not negligible. Indeed, the value of $\Delta E_\pi = -10.0$ kcal/mol, which reflects the presence of π molecular orbitals in ethane, can be considered as a direct estimate of the hyperconjugation in this system.[20–22]

Similar EDA results, i.e., C–C bonds described mainly as σ-covalent bonds, can be found for symmetrical substituted ethanes X_3C–CX_3 where X = CH_3, SiH_3, F, Cl.[20] Table 7.1 also shows that the hyperconjugation in these species is stronger than in the parent ethane, particularly for X = SiH_3, Cl. Interestingly, the compounds with the strongest hyperconjugation, $(SiH_3)_3C$–$C(SiH_3)_3$ and Cl_3C–CCl_3, actually have the longest C–C bonds. This is a remarkable result because a short bond length is often taken as evidence for hyperconjugation. This finding indicates that other effects, such as the influence of the electronegativity of the substituent on the σ-bonding and electrostatic interactions, play a role that hampers a direct correlation between hyperconjugation and bond distances.

Table 7.1. EDA results (BP86/TZ2P level, in kcal/mol) of symmetrically substituted ethanes $X_3C–CX_3$.[c]

	$X_3C–CX_3$						
	X = H	X = CH_3	X = SiH_3	X = F	X = Cl	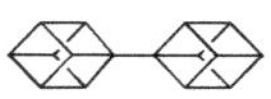	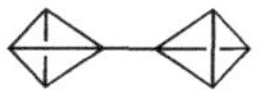
ΔE_{int}	−114.8	−93.2	−89.0	−92.4	−70.2	−105.2	−142.6
ΔE_{Pauli}	200.8	253.6	226.0	254.9	296.3	344.1	225.2
ΔE_{elstat}[a]	−131.3	−163.5	−131.4	−151.3	−157.0	−206.9	−144.0
	(41.6%)	(47.2%)	(41.7%)	(43.6%)	(42.8%)	(46.1%)	(39.2%)
ΔE_{orb}[a]	−184.2	−183.2	−183.6	−196.1	−209.5	−242.3	−223.8
	(58.4%)	(52.8%)	(58.3%)	(56.4%)	(57.2%)	(53.9%)	(60.8%)
$\Delta E_{\sigma}(a_1)$[b]	−174.3	−170.9	−165.6	−183.0	−188.2	−221.6	−196.7
	(94.6%)	(93.3%)	(90.2%)	(93.4%)	(89.8%)	(91.5%)	(87.9%)
$\Delta E_{\delta}(a_2)$[b]	0.0	−0.6	−0.5	−0.2	−0.4	−0.2	0.0
	(0.0%)	(0.3%)	(0.3%)	(0.1%)	(0.2%)	(0.1%)	(0.0%)
$\Delta E_{\pi}(e)$[b]	−10.0	−11.6	−17.5	−12.9	−20.9	−20.5	−27.1
	(5.4%)	(6.4%)	(9.5%)	(6.6%)	(10.0%)	(8.5%)	(12.1%)
ΔE_{prep}	21.8	30.2	39.7	5.3	17.2	7.8	6.4
$\Delta E(= -D_e)$	−93.8	−63.0	−49.3	−87.1	−53.0	−97.4	−136.2
r(C–C)/Å	1.532	1.591	1.612	1.565	1.593	1.476	1.425

Notes: [a]The percentage values in parentheses give the contribution to the total attractive interactions $\Delta E_{elstat} + \Delta E_{orb}$.
[b]The percentage values in parentheses give the contribution to the total orbital interactions ΔE_{orb}.
[c]The EDA was carried out using C_{3v} symmetry. Values taken from Ref. [20].

Strikingly, strong hyperconjugation is found in cubylcubane and in tetrahedranyltetrahedrane, which actually exhibit the shortest C–C bonds (1.476 and 1.425 Å, respectively, see Table 7.1). In fact, the computed ΔE_π values of cubylcubane (−20.5 kcal/mol) and tetrahedranyltetrahedrane (−27.1 kcal/mol) are much higher than in the related $Me_3C{-}CMe_3$ (−11.6 kcal/mol). Although the short C–C bond distance in tetrahedranyltetrahedrane has been explained in terms of the high percentage s-character of the C–C bond molecular orbital,[23] it can also be suggested that these relativity short C–C bond distances are at least partially caused by π-bonding contributions.

Unsymmetrically substituted ethanes, $X_3C{-}CY_3$, have also been considered.[20] Not surprisingly, the highest hyperconjugation strengths have been found for $(H_3Si)_3C{-}CF_3$ and $(H_3Si)_3C{-}CCl_3$ (ΔE_π values of −24.1 and −29.7 kcal/mol, respectively). Interestingly, these compounds do not exhibit short C–C bond distances (1.512 and 1.523 Å, respectively) which suggests, once again, that one should be extremely cautious when correlating the strength of hyperconjugation and bond lengths.

7.3.2 *From ethane to acetylene: Bonding in $H_3C{-}CH_3$, $H_2C{=}CH_2$ and $HC{\equiv}CH$*

The EDA method was particularly helpful in rationalizing the bonding situation of molecules having multiple bonds.[19] Of course, when comparing ethane with ethylene and acetylene, the selected fragments for the EDA are different in each case. Whereas two methyl radicals (doublets) are used for ethane, two H_2C triplet carbenes, where one unpaired electron is in a $sp^2(\sigma)$-hybridized orbital and the other in the $p_z(\pi)$ C-atomic orbital, must be used for ethylene. Two CH quartets are used as fragments for the EDA involving acetylene.

From the data in Table 7.2, it becomes clear that the total interaction energy increases from ethane (ΔE_{int} = −114.8 kcal/mol) to ethylene (ΔE_{int} = −191.1 kcal/mol) and to acetylene (ΔE_{int} = −280.0 kcal/mol), which nicely agrees with the corresponding C–C bond strengths (measured by the respective bond dissociation

Table 7.2. EDA results (BP86/TZ2P level, in kcal/mol) of ethane, ethylene and acetylene.

	$H_3C–CH_3$	$H_2C{=}CH_2$	$HC{\equiv}CH$
Symmetry	D_{3d}	D_{2h}	$D_{\infty h}$
ΔE_{int}	−114.8	−191.1	−280.0
ΔE_{Pauli}	200.8	281.9	255.4
ΔE_{elstat} [a]	−131.3	−181.6	−147.5
	(41.6%)	(38.4%)	(27.6%)
ΔE_{orb} [a]	−184.2	−291.4	−387.9
	(58.4%)	(61.6%)	(72.4%)
ΔE_{σ} [b]	−174.3	−212.2	−215.5
	(94.6%)	(72.8%)	(55.6%)
ΔE_{π} [b]	−10.0	−79.2	−172.4
	(5.4%)	(27.2%)	(44.4%)
ΔE_{prep}	21.8	12.9	32.8
ΔE (= $-D_e$)	−93.8	−178.2	−247.2
r(C–C)/Å	1.532	1.333	1.206

Notes: [a]The percentage values in parentheses give the contribution to the total attractive interactions $\Delta E_{elstat} + \Delta E_{orb}$.
[b]The percentage values in parentheses give the contribution to the total orbital interactions ΔE_{orb}. Values taken from Ref. [19].

energies, D_e). Not surprisingly, the contribution of the orbital interactions to the C=C double bond is higher than to the C–C single bond. Similarly, the orbital term is also higher in acetylene than in ethylene, which confirms that the covalent character of the C–C bond steadily increases from ethane to acetylene. Similarly, the total π-bonding (measured by the ΔE_{π} term) also follows the same trend. For instance, the π-bond strength in $H_2C{=}CH_2$ ($\Delta E_{\pi} = -79.2$ kcal/mol) amounts to 27.2% of the total orbital interactions, while the σ contribution ($\Delta E_{\sigma} = -212.2$ kcal/mol) is 72.8% of ΔE_{orb}. This means that the π-bond strength in ethylene has approximately one-third the strength of the σ bond.

The above EDA results suggest that the much higher interaction energy of acetylene compared with that of ethylene ($\Delta\Delta E_{int} =$ 88.9 kcal/mol) comes almost exclusively from the increase in the strength of the orbital interactions ($\Delta\Delta E_{orb} = 96.5$ kcal/mol). This is mainly the result of π-bonding interactions in acetylene.

Indeed, whereas the σ-bond strength is rather similar in both species ($\Delta E_\sigma = -212.2$ and -215.5 kcal/mol), the π-bonding in acetylene ($\Delta E_\pi = -172.4$ kcal/mol) is nearly twice as strong as in ethylene ($\Delta E_\pi = -79.2$ kcal/ mol). This result agrees with our chemical intuition because acetylene possesses a degenerate π orbital that has two components, while the π orbital of ethylene has only one. The calculated data therefore suggest that each of the π-orbital components of acetylene has about the same strength as the π orbital in ethylene.

7.3.3 *Direct estimate of π-conjugation with the EDA*

7.3.3.1 *Conjugation in 1,3-butadiene vs. 1,3-butadiyne and polyenes*

Conjugation, the interaction of one p-orbital with another across an intervening σ-bond, can be measured as the difference in energy between a conjugated molecule and its hypothetical energy (virtual state) if the entire contribution stemming from conjugation could be accounted for and excised.[24] It is important, however, to recognize that conjugation and other organic chemistry concepts such as hyperconjugation or aromaticity are just bonding models and, therefore, they are only virtual thermodynamic quantities that cannot be measured experimentally. Thus, we are dealing with terms (or "fuzzy" concepts[25]) that do not possess a precise meaning (i.e., conjugation is explained in MO theory with $\pi - \pi^*$-interactions, while the same phenomenon is explained in VB theory in terms of resonance structures). In fact, these heuristic bonding models, which are extremely useful and profusely used by organic chemists, have been called *unicorns of chemical bonding models*, because everyone seems to know what they mean but, like the mythical animal, nobody has ever seen one because they are not experimentally observable.[26]

The most traditional approaches to quantitatively estimate π-conjugation use the reaction energies of isodesmic reactions or the stepwise heats of hydrogenation of the conjugated molecule, following the initial suggestion by Kistiakowsky[27] and Conant and Kistiakowsky.[28] However, both approaches suffer from the problem that the difference between the conjugated molecule and the reference system comprises not only alterations of the π-bonding, but also changes in other parts of the system. Following Kistiakowsky's method, it has been reported that the conjugative stabilization in 1,3-butadiene may be estimated as the difference between the first and second hydrogenation energies ($\Delta\Delta E = 3.7$ kcal/mol, Fig. 7.1).[29] Using the same approach, it was concluded that there is no conjugative stabilization in 1,3-butadiyne because the energies for the hydrogenation of the two degenerate triple bonds are nearly the same ($\Delta\Delta E = 0.2$ kcal/mol, Fig. 7.1). This result is highly surprising because one would expect that the strength of conjugation in 1,3-butadyine should be at least twice as strong as the conjugation in 1,3-butadiene. A similar methodology was then applied to different polyacetylenes, leading to the same conclusion, i.e., no conjugative stabilization in these species.[30]

It therefore seems clear that the (hyper)conjugative strengths estimated by means of the latter procedures are not always reliable because they are often contaminated by flaws such as hyperconjugation, strain and Coulomb repulsion imbalance — especially in charged systems — or uncompensated van der Waals attractions.[31–33] In contrast, the EDA method makes it possible to directly estimate π-conjugation using only the molecule of interest without recourse to external reference systems. Indeed, the ΔE_π term provides a quantitative estimation of the stabilization that arises

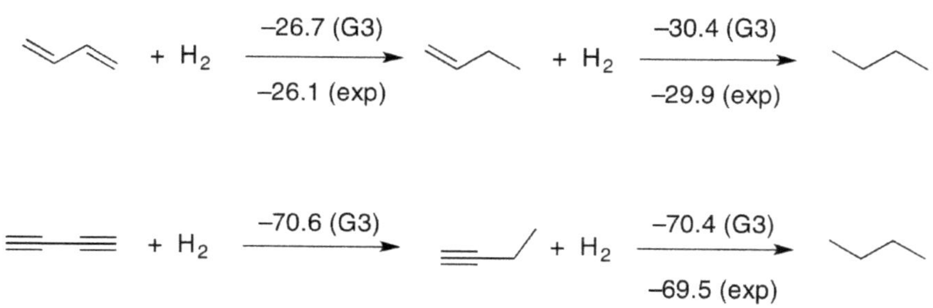

Figure 7.1. Experimental and computed hydrogenation heats (in kcal/mol) of 1,3-butadiene and 1,3-butadiyene. Energy values taken from Ref. [29].

from the mixing of occupied π-orbitals with vacant π^*-orbitals in molecules having, at least, a mirror plane (i.e., C_s symmetry).

The EDA method was then used to resolve the controversy about the reported lack of conjugation in 1,3-butadiyne.[21,22] To this end, the central C–C single bond of 1,3-butadiene and 1,3-butadiyne was analyzed. Thus, the interacting fragments are calculated in the electronic doublet state, with the corresponding unpaired electron in a σ-orbital (i.e., two $H_2C{=}CH\bullet$ radicals for 1,3-butadiene and two $HC{\equiv}C\bullet$ radicals for 1,3-butadiyne). As expected, the C–C bond in 1,3-butadiyne exhibits a much larger interaction energy (ΔE_{int} = −176.0 kcal/mol) than in 1,3-butadiene (ΔE_{int} = −128.5 kcal/mol, Table 7.3). Although the electrostatic character of the former bond is smaller (33.9%) than that of the latter (42.8%), the

Table 7.3. EDA results (BP86/TZ2P level, in kcal/mol) of 1,3-butadiene, 1,3-butadiyne and polyenes.

			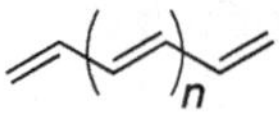			
	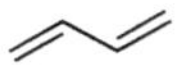		$n = 1$	$n = 2$	$n = 3$	$n = 4$
ΔE_{int}	−128.5	−176.0	−278.9	−429.9	−581.3	−733.1
ΔE_{Pauli}	268.4	161.8	501.0	737.2	976.1	1216.8
ΔE_{elstat}[a]	−169.9 (42.8%)	−114.6 (33.9%)	−340.2 (43.6%)	−521.5 (43.9%)	−686.0 (44.0%)	−860.4 (44.1%)
ΔE_{orb}[a]	−227.0 (57.2%)	−223.3 (66.1%)	−439.7 (56.4%)	−654.6 (56.1%)	−871.4 (56.0%)	−1089.5 (55.9%)
ΔE_{σ}[b]	−207.5 (91.4%)	−178.3 (79.8%)	−398.3 (90.6%)	−589.8 (90.1%)	−782.1 (89.7%)	−975.1 (89.5%)
ΔE_{π}[b]	−19.5 (8.8%)	−45.0 (20.2%)	−41.4 (9.4%)	−64.9 (9.9%)	−89.3 (10.2%)	−114.4 (10.5%)
ΔE_{prep}	13.0	5.4	17.4	22.7	27.8	32.9
ΔE (= $-D_e$)	−115.5	−170.6	−261.1	−407.2	−553.5	−700.2
r(C–C)/Å	1.453	1.361	1.444	1.441	1.439	1.438
BLW[c]	12.7	27.8				

Notes: [a]The percentage values in parentheses give the contribution to the total attractive interactions $\Delta E_{elstat} + \Delta E_{orb}$.
[b]The percentage values in parentheses give the contribution to the total orbital interactions ΔE_{orb}.
[c]BLW values taken from Ref. [34]. Values taken from Refs. [21, 22].

largest contribution to the C–C attraction in both molecules comes from the orbital term ΔE_{orb}. Interestingly, although the σ-bonding is stronger in both molecules, the ΔE_{π} values in Table 7.3 indicate that the conjugative stabilization in 1,3-butadiyne is slightly more than twice that calculated for *trans*-1,3-butadiene. This is a reasonable result, because, according to our chemical intuition, the conjugation of the two π-systems in 1,3-butadiyne should be at least twice as strong as the conjugation in 1,3-butadiene where there is only one π-system. This is nicely reflected in the remarkable computed shortening of the central C–C bond distance in 1,3-butadiyne compared to 1,3-butadiene (1.361 vs. 1.453 Å).

A similar result was obtained by means of the BLW method,[34] which uses a completely different partitioning scheme. In short, the BLW method[35,36] first constructs the wavefunction of an electron-localized reference Lewis structure (i.e., a resonance structure in which conjugative interactions are disabled) prior to self-consistent optimization of the block-localized MOs. The impact of the (hyper)conjugative interactions is evaluated in terms of delocalization energies by comparing the energy of the lowest energy resonance structure with that of the standard (i.e., canonical) structure of the molecule under investigation. For the particular case of 1,3-butadiene and 1,3-butadiyne, the π-conjugation strength was computed to be 12.7 and 27.8 kcal/mol, respectively,[34] which confirms that the conjugative stabilization in 1,3-butadiyne is nearly twice as strong as that in 1,3-butadiene.

Conjugation has been invoked to explain many properties of conjugated polyenes such as their excitation energies and absorption spectra.[37] These properties traditionally have been related to the steady decrease of the HOMO–LUMO gap when going from one polyene to another. The EDA results for the homologous series of five conjugated polyenes from *trans*-1,3-butadiene to all-*trans*-1,3,5,7,9,11-dodecahexaene show that the increase of ΔE_{π} has very similar values as the increase of the overall interaction energy ΔE_{int}.[20,22] This is a hint why the properties above can be modeled by considering only the π-electrons.[38] Although the total

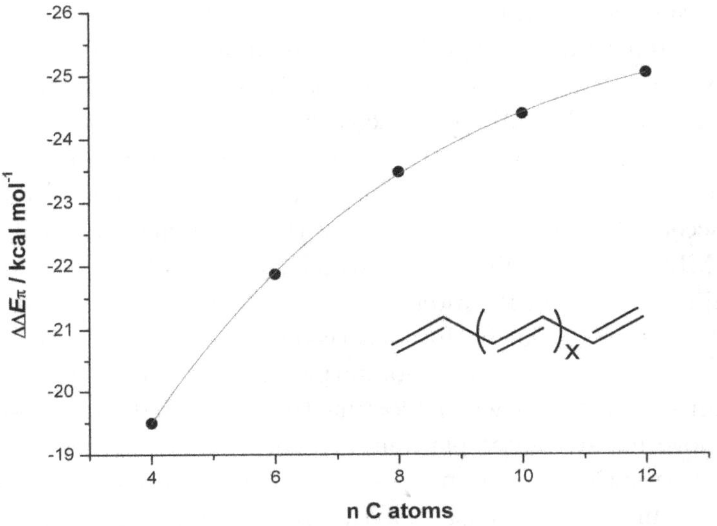

Figure 7.2. Computed $\Delta\Delta E_\pi$ vs. the number of carbon atoms in polyenes. Values taken from Ref. [20].

π-conjugation increases linearly with the number of C=C moieties in the molecule, the change in the π-conjugation from one polyene to another (measured by $\Delta\Delta E_\pi$) follows an increase, converging toward a final value which appears to be reached after 6 to 8 C=C moieties, as shown in Fig. 7.2.

7.3.3.2 *Correlation between the computed EDA derived π-conjugation strength and experimental data*

The EDA method is not only useful to quantify the conjugative strength of small π-conjugated molecules (see above) but can also be successfully applied to systems having extended π-conjugation. Among them, donor-substituted cyanoethynylethenes (CEEs) constitute a versatile class of modular building blocks that combines the scaffolding capacity of tetraethynylethenes (TEEs) with the electron-accepting properties of tetracyanoethylenes (TCNEs). This family of organic materials, prepared by Moonen and co-workers,[39] can function as electronic acceptor groups for nonlinear optical applications when substituted with dimethylaniline (DMA) donor groups.

It has been proposed that the extent of π-conjugation in these systems can be experimentally estimated by means of different experimental techniques, such as X-ray crystallography, cyclic voltammetry and NMR spectroscopy.[39] In fact, the conjugation effectiveness is nicely revealed by large changes in the first oxidation and reduction potentials of CEEs with respect to the isolated donor and acceptor moieties, as well as by the differences in the ^{13}C and ^{1}H NMR chemical shifts. Interestingly, the trend in the computed π-conjugation of these systems (measured by the ΔE_{π} term of the EDA) is rather similar to that observed by the available experimental data.[40] For that reason, it is not surprising that excellent linear relationships were found when plotting the calculated ΔE_{π} values vs. the above experimental parameters. For instance, a nearly perfect linear correlation was found between the ΔE_{π} values and the ^{13}C-NMR chemical shifts (δ_{ipso}) of the aryl carbon atoms *ipso* to the NMe_2 group of the donor DMA group (correlation coefficient of 0.99 and standard deviation of 0.43, see Fig. 7.3). These results confirm

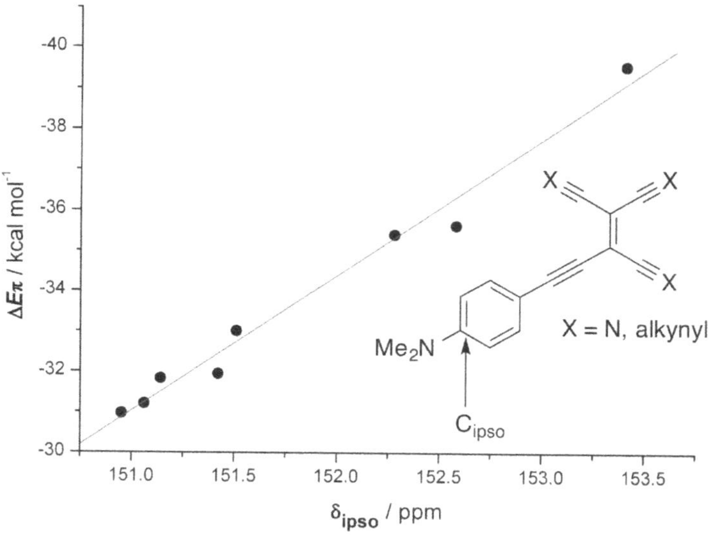

Figure 7.3. Plot of the ΔE_{π} values vs. the ^{13}C NMR chemical shift of the aryl carbon *ipso* to the NMe_2 group. Values taken from Ref. [40].

the ability of the EDA method to directly estimate the conjugative strength, even in large systems with extended π-conjugation.[40,41]

Similar correlations were found between the computed ΔE_{π} values and the Hammett σ-substituent constants.[42] The latter empirical parameters are still widely used in organic chemistry for discussing substituent effects. It is well known that the original σ_m and σ_p values reflect the extent to which substituents in a *meta* or *para* position on a phenyl ring interact with a reaction site through a combination of resonance and field/inductive effects.[43,44] However, the greatest incidence of failures of the simple Hammett equation were encountered when substituents capable of accepting or donating a pair of electrons are in direct conjugation with the reaction center (an effect known as *through resonance*). To overcome this problem, Brown and co-workers proposed an improved scale of σ constants (σ^+ or σ^-) that directly reflects the (hyper)conjugative effect of a given substituent.[45] The good performance of the EDA method to estimate π-conjugation is reflected in the linear relationships found when plotting these experimental parameters vs. the ΔE_{π} values computed either for p-substituted benzyl cations (ΔE_{π} vs. σ_p^+, correlation coefficient of 0.97, see Fig. 7.4) or p-substituted benzyl anions (ΔE_{π} vs. σ_p^-, correlation coefficient of 0.98).[42] Both correlations confirm that the values calculated for the intrinsic π-conjugation given by the EDA are reasonable for interpreting the chemical properties of a molecule, and may even be used to semi-quantitatively predict Hammett constants for hitherto unknown substituents.[42]

7.3.4 *Aromaticity strength with the EDA*

7.3.4.1 *Aromaticity in benzene, heterocycles and related compounds*

The fuzzy nature of aromaticity is clearly reflected in the good number of methods proposed to qualitatively and also quantitatively assess the aromatic character of a system (see Chapter 9).[46–48] Among them, those methods based on the so-called energetic criterion, if well-defined, are often preferred because the stability of a

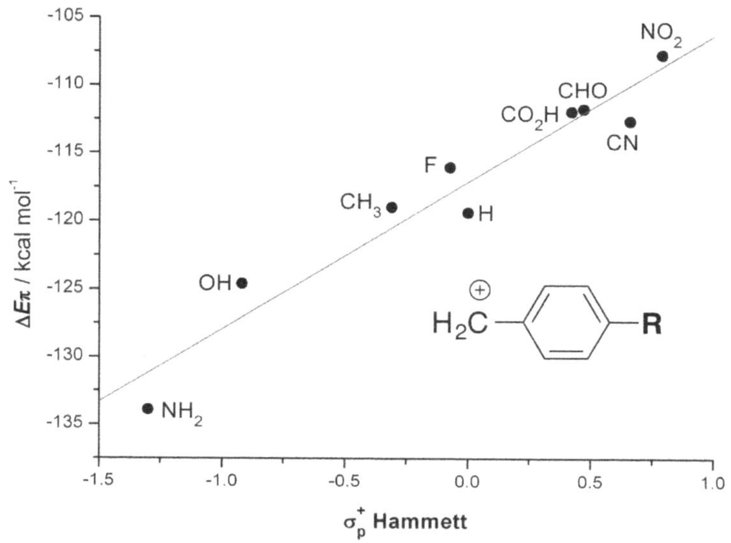

Figure 7.4. Plot of the ΔE_π values in *para*-substituted benzylic cations vs. Hammett–Brown constants for *para* substituents. Values taken from Ref. [42].

cyclic π-conjugated compound with respect to an acyclic compound is considered by many to be the primary quantity defining aromaticity. Taking this into account, the aromatic stabilization energies (ASE) in aromatic compounds, i.e., the extra stabilization energy due to aromaticity, can be computed just by comparing their π-cyclic conjugation strength with the π-conjugation of acyclic reference systems. Therefore, these ASE values can be easily estimated with the EDA method as the difference between the total amount of conjugative stabilization of single bonds in cyclic unsaturated systems and that amount in acyclic systems with the same number of single bonds according to Eq. (7.7)

$$\mathrm{ASE} = \Delta E_\pi(\text{cyclic}) - \Delta E_\pi(\text{acyclic}) \qquad (7.7)$$

Table 7.4 gathers the EDA derived ASE values[22,49] for typical Hückel 6π-aromatic and heteroaromatic compounds as well as for cyclobutadiene, the archetypal 4π-antiaromatic species.[50] Benzene was homolytically fragmented into three C_2H_2 fragments

Table 7.4. EDA results (BP86/TZ2P level, in kcal/mol) for (anti)aromatic compounds.[a]

Cyclic compound		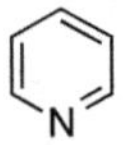	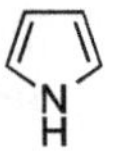	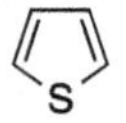		
ΔE_π	−107.7	−113.5	−98.0	−77.6	−77.7	−9.6
Acyclic reference	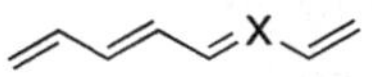		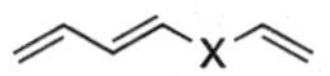			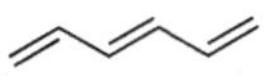
	X = CH	X = N	X = NH	X = S	X = O	
ΔE_π	−65.2	−67.8	−76.9	−55.7	−61.5	−41.5
ASE	**42.5**	**45.7**	**21.1**	**21.9**	**16.2**	**−31.9**

Note: [a]Values taken from Ref. [49].

in open-shell singlet states. Although these species possess broken symmetry, this fragmentation scheme does not affect the relaxation of the π-orbitals and, therefore, does not impact the relevant ΔE_π term. Note that although the acyclic reference system for benzene, namely 1,3,5,7-octatetraene, has more π electrons (eight instead of six in benzene), it was chosen because it presents the same number of C–C interactions. Using this approach, an ASE value of 42.5 kcal/mol was computed for benzene, which confirms that the strength of π-conjugation is much higher in the cyclic compound than in the corresponding acyclic reference species (Table 7.4). As expected, positive values were also found for the heteroaromatic compounds, whereas a highly negative ASE value (ASE = −31.9 kcal/mol) was calculated for cyclobutadiene. Therefore, this EDA-based approach constitutes a useful energy scale for assessing the strength of aromaticity, which can be used to classify compounds according to their relative ASE values. Thus, compounds with ASE > 0 can be considered as aromatic (the highest ASE values were computed for benzene and pyridine, ca. 45 kcal/mol), whereas antiaromatic compounds exhibit highly negative ASE values. Small positive ASE values are typical for homoaromatic compounds, while small negative ASE values are indicative of homoantiaromaticity.[49]

This approach has been particularly useful for quantitatively assessing the aromatic nature of systems where the application of the very popular methods for quantifying aromaticity of a molecule (e.g., Nucleus Independent Chemical Shift, NICS,[51] or magnetic susceptibility anisotropy, $\Delta\chi$,[52] calculations) fail or have severe limitations. For example, the EDA method has been successfully applied to metallabenzenes,[53,54] the organometallic analogs of benzene where a CH group is replaced by an isolobal transition metal fragment,[55,56] neutral exocyclic cyclopropenes $(HC)_2C{=}E$,[57] group 14 homologues of the cyclopropenylium cation,[58] and hyperaromatic planar cyclopolyenes.[59] It should be pointed out that in these challenging systems, the ASE values provided by the EDA method are again in nice agreement with those values obtained by applying the BLW scheme and dissected NICS values.

Pierrefixe and Bickelhaupt[60,61] used a different fragmentation scheme to rationalize the origins of the D_{6h}-delocalized and D_{2h}-localized equilibrium geometries of benzene and cyclobutadiene, respectively. In benzene, the first, third and fifth CH groups are treated as one fragment and the remaining CH groups (second, fourth and sixth) constitute the second fragment. This scheme results in two fragments, each forming a regular triangle having nine unpaired electrons (six σ type and three π type, see Fig. 7.5). In this case, the interaction energy between these triangular fragments covers the

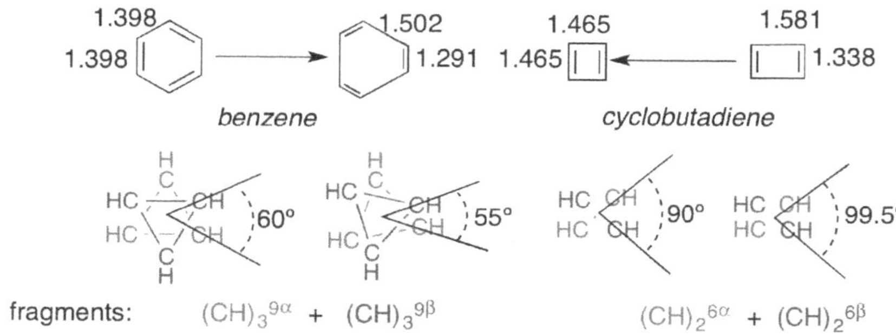

Figure 7.5. Equilibrium geometries of benzene (D_{6h}) and cyclobutadiene (D_{2h}) and respective distorted geometries (top) and schematic representation of the two fragments used for the EDA calculations (bottom, see Ref. [60]).

bond formation of all σ and π bonds. By rotating the two fragments with respect to each other, the effect of the geometrical distortion from a D_{3h}-localized molecule, having three shorter C=C double bonds and three longer C–C single bonds, to the D_{6h}-delocalized structure, with six equal C=C bond distances, can be studied. Similarly, the distortion from the D_{2h}-localized to the D_{4h}-delocalized system can be analyzed for cyclobutadiene. In both cases, it was found that the σ-electron system tends to produce delocalized structures. In the case of the antiaromatic cyclobutadiene, all π-overlap effects unidirectionally favor localization of the double bonds and overcompensate the gain of σ interaction upon delocalization, which prevents cyclobutadiene from becoming delocalized. In contrast, in the aromatic system benzene, the π system emerges from a subtle interplay of counteracting overlap effects and is, therefore, not stabilizing enough to overcome the delocalizing σ system.[60] This finding is in nice agreement with the previous report by Shaik and Hiberty, who elegantly demonstrated using the VB model[62] that it is the σ system that enforces the D_{6h}-delocalized structure of benzene upon the π system, which intrinsically strives for localized double bonds.[63]

7.3.5 *Hydrogen bonding with the EDA–NOCV method*

The nature of hydrogen bonding has also attracted considerable attention from both experimental (by means of NMR, IR or Compton profile anisotropy techniques) and theoretical points of view.[64] Several energy decomposition schemes also have been applied to rationalize the hydrogen bond formation that arises from the interaction of fragments containing the X′ and XH moieties involved in the X′ ··· HX bond.[65–68] However, most of these approaches provide a quantitative description of the bonding, with practically no direct information on the qualitative nature of bonding.

Using the EDA–NOCV method, Kovács and co-workers[19] and Kurczab and co-workers[69] were able to analyze the different components of the hydrogen bonding in several systems including biologically relevant species such as DNA base pairs. Different types of hydrogen bonds were considered, namely OH··· O, NH··· O and

Table 7.5. EDA–NOCV results (BP86/TZ2P level, in kcal/mol) of salicylic acid dimer, formamide dimer and adenine–thymine pair.

	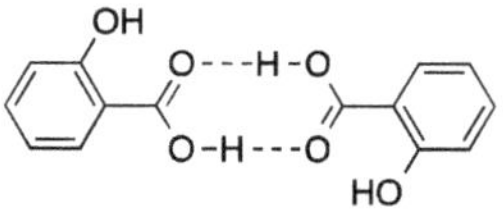	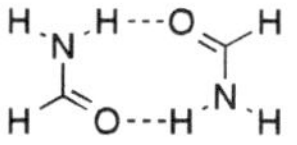	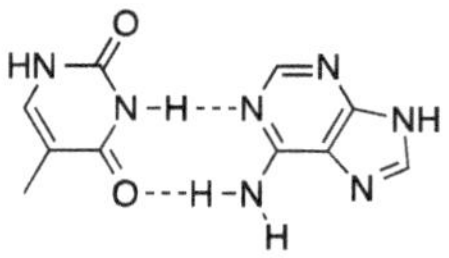
ΔE_{int}	−15.2	−13.6	−13.0
$\Delta E_{\mathrm{Pauli}}$	48.2	28.4	38.7
$\Delta E_{\mathrm{elstat}}$ [a]	−35.8 (52.6%)	−26.0 (59.5%)	−31.9 (59.2%)
ΔE_{Orb} [a]	−32.2 (47.4%)	−17.7 (40.5%)	−22.0 (40.8%)
ΔE_{σ}	−25.7 (OH···O)	−13.6 (NH···O)	−12.8 (NH···N) −5.3 (NH···O)
ΔE_{π}	−2.8	−1.8	−2.5
ΔE_{prep}	4.6	1.7	2.2

Notes: [a]The percentage values in parentheses give the contribution to the total attractive interactions $\Delta E_{\mathrm{elstat}} + \Delta E_{\mathrm{orb}}$. Values taken from Ref. [69].

NH···N. From the data in Table 7.5, it becomes evident that both the electrostatic and orbital attractions contribute to the total interaction to nearly the same extent (the $\Delta E_{\mathrm{elstat}}$ term being slightly larger).

The EDA–NOCV method makes it possible to further decompose the orbital term into contributions coming from different deformation densities. For instance, three deformation density components ($\Delta\rho_{\sigma 1}$, $\Delta\rho_{\sigma 2}$ and $\Delta\rho_{\pi}$) contribute to the bonding in the adenine–thymine system (Fig. 7.6). The energetically most important contributions ($\Delta E_{\sigma 1} = -12.8$ kcal/mol and $\Delta E_{\sigma 2} = -5.3$ kcal/mol) involve the donation of electron density from the doubly occupied lone pairs of the nitrogen (from adenine) or oxygen (from thymine) atoms to the unoccupied σ^* molecular orbitals of the N–H bonds. As shown in Fig. 7.6, the interaction involving LP(N_{adenine}) → σ^*(NH_{thymine}) is stronger than that involving LP(O_{thymine}) → σ^*(NH_{adenine}), which agrees with the higher donor ability of nitrogen. Moreover, the deformation density $\Delta\rho_{\pi}$ illustrates the minor participation of π-type orbitals in the adenine-thymine bonding (corresponding

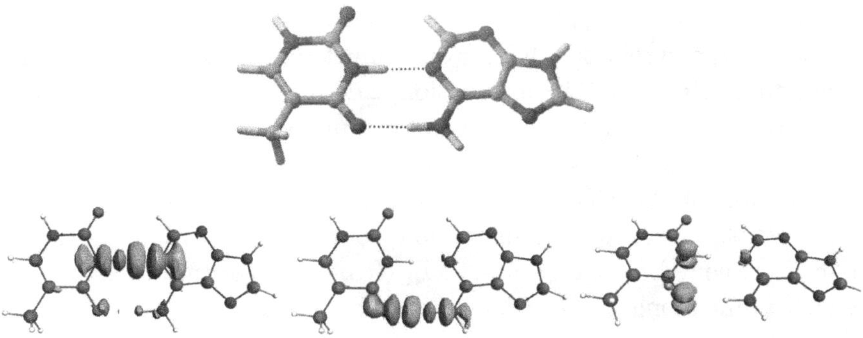

Figure 7.6. Bonding in the adenine–thymine pair. The contours show the EDA–NOCV deformation densities describing the NH···N ($\Delta\rho_{\sigma 1}$) and NH···O(=C) ($\Delta\rho_{\sigma 2}$) hydrogen bonds as well as the π-resonance ($\Delta\rho_{\pi}$). Values taken from Ref. [69].

$\Delta E_{\pi} = -2.6$ kcal/mol) and accounts for the so-called Resonance Assisted Hydrogen Bonding (RAHB)[70] in this particular system. In general, it was found that the stabilization coming from $\Delta\rho_{\sigma}$ is ca. four times stronger than that coming from $\Delta\rho_{\pi}$ (whose strength is in the -1.8 to -6.4 kcal/mol range).[69]

7.3.6 *Analyzing chemical reactions: The activation strain model (ASM) of reactivity*

The ASM[71–74] is an application of the EDA not only on the equilibrium geometries but also on TS structures as well as nonstationary points along the reaction coordinate under consideration. Within this model, also known as *distortion–interaction model* (see Chapter 13), the potential energy surface $\Delta E(\zeta)$ can be decomposed into two contributions along the reaction coordinate ζ: the strain (also called distortion) energy $\Delta E_{\text{strain}}(\zeta)$, which is associated with the structural deformation that the reactants undergo, plus the interaction $\Delta E_{\text{int}}(\zeta)$ between these increasingly deformed reactants (Eq. (7.8)). This equation is essentially the same as Eq. (7.4), which means that the ΔE_{prep} and ΔE_{strain} are equivalent terms.

$$\Delta E(\zeta) = \Delta E_{\text{strain}}(\zeta) + \Delta E_{\text{int}}(\zeta) \tag{7.8}$$

The strain $\Delta E_{\text{strain}}(\zeta)$ is determined by the rigidity of the reactants and the extent to which groups must reorganize in a particular reaction mechanism. The interaction $\Delta E_{\text{int}}(\zeta)$ between the reactants depends on their electronic structure and on how they are mutually oriented as they approach each other. The latter can, as in the application of the EDA on ground states, be separated into Pauli, electrostatic, orbital and distortion contributions. It is the interplay between $\Delta E_{\text{strain}}(\zeta)$ and $\Delta E_{\text{int}}(\zeta)$ that determines if and at which point along ζ a barrier arises. For instance, at the TS structure, the reaction profile reaches its maximum, therefore satisfying $d\Delta E_{\text{strain}}(\zeta)/d\zeta = -d\Delta E_{\text{int}}(\zeta)/d\zeta$.

The ASM of reactivity has allowed us to gain more insight into the physical factors that control how the activation barriers arise in different fundamental processes. For instance, the ASM has been successfully applied to elimination and nucleophilic substitution reactions, pericyclic reactions and transition-metal-mediated reactions. The reader is referred to the recent reviews by van Zeist and Bickelhaupt,[71] Fernández and Bickelhaupt[72] and Fernández[73,74] for further discussion of these applications. Herein, we have selected the application of the combination of the ASM/EDA methods to understand the complete regioselectivity of the cycloaddition reactions involving C_{60} fullerene.

C_{60} fullerene exhibits two types of C–C bonds, namely the pyracylenic type [6,6]-bond, where two six-membered rings are fused, and the corannulenic [5,6]-bond, which corresponds to the ring junction between a five- and a six-membered ring. In general, cycloaddition reactions in empty fullerenes show a large (or exclusive) preference for [6,6]- over [5,6]-bonds.[75] However, the origin of this experimentally well-established [6,6]-regioselectivity has not been completely explained until a recent ASM study of the Diels–Alder cycloaddition reaction between the parent C_{60} fullerene and cyclopentadiene.[76] The reaction profile computed for this particular transformation (Fig. 7.7) clearly indicates that the experimentally observed complete [6,6]-regioselectivity takes place under both kinetic and thermodynamic control, in view of the considerably higher activation energy and

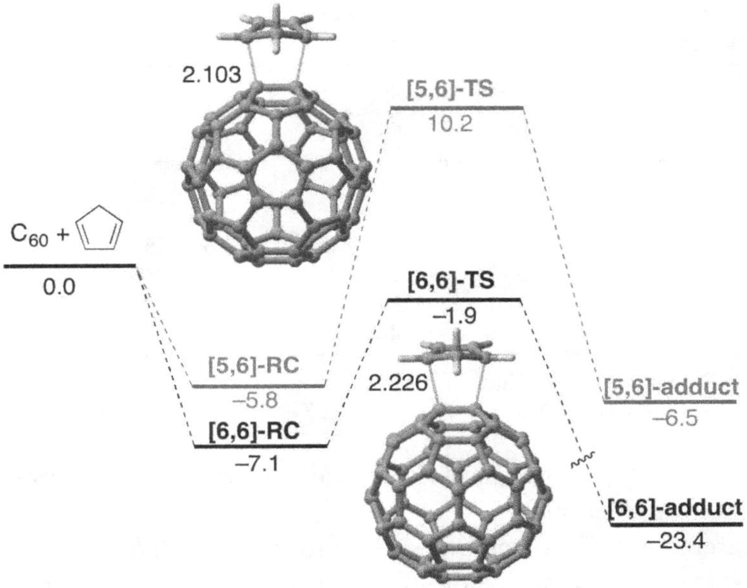

Figure 7.7. Computed reaction profile (BP86-D3/TZVP//RI-BP86-D3/def2-SVP level) for the Diels–Alder reaction between C_{60} and cyclopentadiene. Relative energies and bond distances are given in kcal/mol and angstroms, respectively. Values taken from Ref. [76].

less exothermic reaction energy computed for the formation of the corresponding [5,6]-cycloadduct.

From the data in Fig. 7.7, it becomes evident that the [6,6] reaction pathway, which proceeds through an earlier TS, is favored along the entire reaction coordinate. For instance, the initial reactant complex **[6,6]-RC** lies 1.3 kcal/mol below its [5,6] counterpart. According to the EDA method (Table 7.6), this is due to a higher electrostatic attraction, a higher orbital interaction, and more stabilizing dispersion forces. The latter term, ΔE_{disp}, clearly constitutes the major contributor to the total interaction between C_{60} and cyclopentadiene (45–50%), which agrees with the van der Waals nature of these reactant complexes and highlights the importance of including dispersion corrections in the calculations involving C_{60}.[77] At variance, this term only contributes ca. 10% to total attractions in the corresponding TS

Table 7.6. EDA results (BP86-D3/TZ2P//RI-BP86-D3/def2-SVP level, in kcal/mol) of key species in the cycloaddition between C_{60} and cyclopentadiene.

	[6,6]-pathway		[5,6]-pathway	
	[6,6]-RC	[6,6]-TS	[5,6]-RC	[5,6]-TS
ΔE_{int}	−7.2	−21.3	−5.7	−21.3
ΔE_{Pauli}	13.2	109.7	12.1	150.4
ΔE_{elstat}[a]	−5.9 (29%)	−53.5 (41%)	−5.1 (28%)	−71.2 (41%)
ΔE_{orb}[a]	−5.4 (26%)	−62.2 (48%)	−4.1 (23%)	−85.1 (50%)
ΔE_{disp}[a]	−9.1 (45%)	−15.3 (11%)	−8.7 (49%)	−15.5 (9%)
$r(\text{C}\cdots\text{C})/\text{Å}$	3.117	2.226	3.198	2.103

Notes: [a]The percentage values in parentheses give the contribution to the total attractive interactions $\Delta E_{\text{elstat}} + \Delta E_{\text{orb}} + \Delta E_{\text{disp}}$. Values taken from Ref. [76].

structures, where both the electrostatic and orbital terms become the dominant attractions.

Interestingly, the computed TS interactions, ΔE_{int}, are exactly the same (−21.3 kcal/mol, Table 7.6) for the [6,6]- and [5,6]-pathways, which might suggest that the reaction profiles of the two pathways differ solely due to strain. However, from the activation strain diagrams depicted in Fig. 7.8, which show the reaction profiles from the initial reactant complexes to the corresponding transition states, it is clear that the strain energy is rather similar along both reaction pathways (ΔE_{strain} is even slightly less destabilizing for the [5,6]-pathway). At variance, the stronger interaction between the reactants along the entire reaction coordinate in the [6,6]-reaction pathway is the major factor controlling the regioselectivity of the process. As a consequence, the [6,6]-TS structure is reached earlier than in the [5,6]-pathway and therefore has a lower strain, which, in turn, is translated into a much lower activation barrier.

The further partitioning of the interaction energy by means of the EDA method indicates that the computed difference in the ΔE_{int} term comes almost exclusively from the difference in the orbital interactions (measured by the ΔE_{orb} term), as the other terms (namely, Pauli repulsion, electrostatic and dispersion interactions) are nearly identical for both reaction paths along the entire reaction coordinate (see Fig. 7.9). For instance, at the same C···C distance of 2.35 Å,

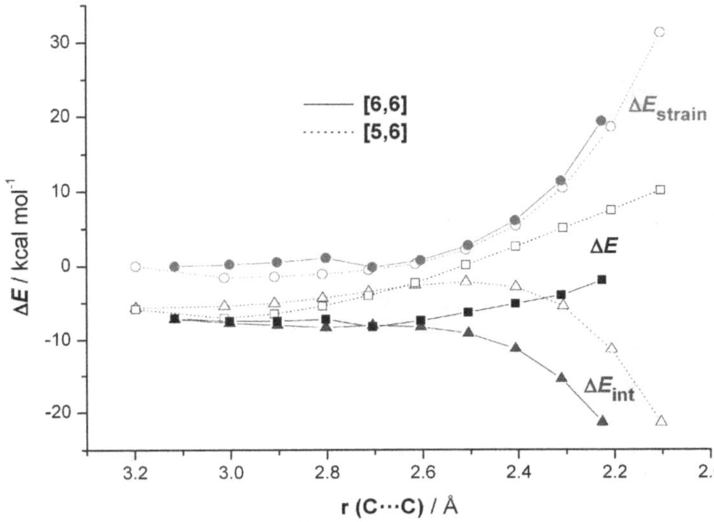

Figure 7.8. Activation-strain analysis of the [4+2]-cycloaddition reaction between C_{60} and cyclopentadiene along the reaction coordinate projected onto the forming C$\cdots$C bond. Solid lines refer to the [6,6]-pathway, whereas dotted lines to the [5,6]-pathway. All data have been computed at the BP86-D3/TZ2P//RI-BP86-D3/def2-SVP level. Values taken from Ref. [76].

$\Delta E_{\text{orb}} = -35.6$ kcal/mol for the [5,6]-pathway, whereas a lower (i.e., more stabilizing) value of $\Delta E_{\text{int}} = -41.7$ kcal/mol was computed for the [6,6]-path. The stronger orbital interactions along the [6,6]-pathway are the result of a more effective ⟨HOMO-(cyclopentadiene)| LUMO(C_{60})⟩ overlap, because the (three-fold degenerate) LUMO of C_{60} has the appropriate π^* character on [6,6] but not on [5,6] bonds.[76]

The combination of the ASM and EDA methods also has been quite useful for rationalizing the reactivity and selectivity of similar processes involving endohedral fullerenes of the type $Ng_2@C_{60}$ (Ng = He–Xe)[78] and $Sc_3N@C_{78}$[79] and related bowl-shaped polycyclic aromatic hydrocarbons.[80]

7.4 Practical considerations

The quantitative energy decomposition scheme and the EDA–NOCV approach described above have been implemented in the Amsterdam

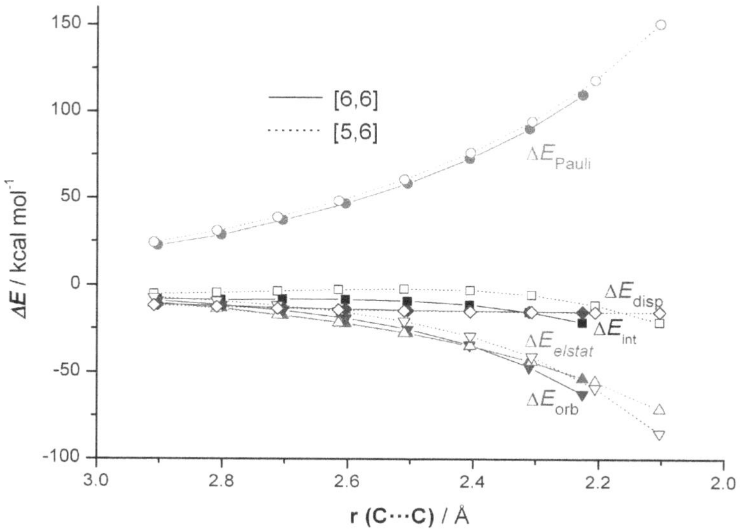

Figure 7.9. Decomposition of the interaction energy for the reaction between C_{60} and cyclopentadiene along the reaction coordinate projected onto the C···C bond length. Solid lines refer to the [6,6]-pathway, whereas dotted lines to the [5,6]-pathway. All data have been computed at the BP86-D3/TZ2P//RI-BP86-D3/def2-SVP level.

density functional (ADF)[11,81] program package. Although the selected examples in this chapter were computed with ADF, the EDA can be also performed in GAMESS[82,83] and AOMix[84] programs. In addition, the EDA calculations are usually carried out within the framework of the density functional theory (DFT) and, therefore, the corresponding EDA results strongly depend on the choice of the selected functional. In this sense, most of the reported analyses either involving organic or organometallic systems have been carried out using the BP86 functional, which has been shown to provide reasonable results.[16,17] In principle, EDA calculations can also be performed using wavefunction-based methods. However, this is only possible at the Hartree–Fock level as no EDA for correlated wavefunctions has been developed so far.

Regarding the inclusion of dispersion interactions in the analyses, note that if an explicit correction term for dispersion interaction is employed (for instance, the D3 correction suggested by Grimme[85,86]), the EDA results will remain unaltered and the dispersion correction will appear as the extra ΔE_{disp} term (see Eq. (7.2)). However, the EDA results will change significantly if the dispersion interactions constitute a part of the selected functional.

It becomes obvious that the computed EDA results strongly depend upon the choice of the electronic state of the reference fragments (e.g., singlet vs. triplet, ionic vs. covalent, electron-sharing vs. donor–acceptor fragments). In many situations, especially when the EDA is used to find the best description of a bonding situation, the choice of the electronic state of the fragments is not always clear. In general, the best description of a bonding situation is one that provides the weakest orbital attractions, because in this case the separated fragments are closest to their situation in the molecule. In contrast, large repulsive values for orbital interactions typically indicate an incorrect setup of the fragments for the EDA calculation.

As a representative example, we present the bonding analysis of the HF molecule. Two possible fragmentations can be envisaged, namely the H–F bond is broken homolytically (covalent bond with H• and F• as fragments) or, alternatively, a heterolytic fragmentation of the H–F bond (ionic bond with charged H^+ and F^- as fragments). From the data in Table 7.7, the heterolytic rupture seems to be a

Table 7.7. EDA results (BP86/TZ2P level, in kcal/mol) for HF.

	H• and F• fragments	H^+ and F^- fragments
ΔE_{int}	−145.5	−381.0
$\Delta E_{\mathrm{Pauli}}$	181.3	0.0
$\Delta E_{\mathrm{elstat}}{}^{\mathrm{a}}$	−54.6 (16.7%)	−241.5 (63.4%)
$\Delta E_{\mathrm{orb}}{}^{\mathrm{a}}$	−272.2 (83.3%)	−139.4 (36.6%)
$\Delta E_{\sigma}{}^{\mathrm{b}}$	−266.4 (97.9%)	−113.7 (81.5%)
$\Delta E_{\pi}{}^{\mathrm{b}}$	−5.7 (2.1%)	−25.8 (18.5%)

Notes: [a]The percentage values in parentheses give the contribution to the total attractive interactions $\Delta E_{\mathrm{elstat}} + \Delta E_{\mathrm{orb}}$.
[b]The percentage values in parentheses give the contribution to the total orbital interactions ΔE_{orb}.

better description of the bond as the corresponding orbital term is weaker ($\Delta E_{\text{orb}} = -139.4$ kcal/mol vs. $\Delta E_{\text{orb}} = -272.2$ kcal/mol). Despite this, we point out that ΔE_{orb} provides only a guideline to the best fragmentation, but the fragmentation that is chosen in the end should be determined by our chemical intuition and the particular system which is being analyzed. In the particular case of H–F, an H^{+}+ F^{-} fragmentation makes no sense as we are considering a gas phase environment where only homolytic bond breaking would occur. If some (polar) solvent would be included, this fragmentation would be the more reasonable choice.

7.5 Summary and conclusion

By selecting illustrative examples from organic chemistry, we show in this chapter that the EDA method constitutes a useful tool to gain insight into the nature of the chemical bond. The EDA makes it possible to rationalize, both qualitatively and quantitatively, chemical concepts such as the multiplicity or the covalent/ionic nature of a bond and bonding models such as (hyper)conjugation and aromaticity. The EDA can therefore be considered as a bridge connecting classical heuristic bonding models with a rigorous quantum chemical approach. In addition, this robust approach, as well as the EDA–NOCV extension, can be applied successfully not only to equilibrium geometries but also to TS structures and nonstationary points along the reaction coordinate under consideration. Therefore, in combination with the ASM of reactivity, the EDA constitutes a powerful method to understand chemical reactivity, which is crucial for the design of novel and/or more efficient transformations.

Acknowledgements

We are grateful for financial support from the Spanish MINECO-FEDER (Grants CTQ2013-44303-P and CTQ2014-51912-REDC) and Fundación BBVA (Convocatoria 2015 de Ayudas Fundación BBVA a Investigadores y Creadores Culturales).

References

1. Lewis, G. N. (1916). The atom and the molecule, *J. Am. Chem. Soc.*, 38, pp. 762–785.
2. Frenking, G. and Shaik, S. (eds.) (2014). *The Chemical Bond* (Wiley-VCH: Weinheim).
3. Jeziorski, B., Moszynski, R. and Szalewicz, K. (1994). Perturbation theory approach to intermolecular potential energy surfaces of van der Waals complexes, *Chem. Rev.*, 94, pp. 1887–1930.
4. Mo, Y., Gao, J. and Peyerimhoff, S. D. (2000). Energy decomposition analysis of intermolecular interactions using a block-localized wave function approach, *J. Chem. Phys.*, 112, pp. 5530–5538.
5. Khaliullin, R. Z., Cobar, E. A., Lochan, R. C., Bell, A. T. and Head-Gordon, M. (2007). Unravelling the origin of intermolecular interactions using absolutely localized molecular orbitals, *J. Phys. Chem. A*, 111, pp. 8753–8765.
6. Schenter, G. K. and Glendening, E. D. (1996). Natural energy decomposition analysis: The linear response electrical self energy, *J. Phys. Chem.*, 100, pp. 17152–17156.
7. Morokuma, K. (1971). Molecular orbital studies of hydrogen bonds, *J. Chem. Phys.*, 55, pp. 1236–1244.
8. Ziegler, T. and Rauk, A. (1977). On the calculation of bonding energies by the Hartree–Fock Slater method, *Theor. Chim. Acta*, 46, pp. 1–10.
9. Lein, M. and Frenking, G. (2005). The nature of the chemical bond in the light of an energy decomposition analysis. In: eds. Dykstra, C. E., Frenking, G., Kim, K. S. and Scuseria, G. E., *Theory and Applications of Computational Chemistry: The First 40 Years*, Chapter 13 (Elsevier: Amsterdam), pp. 291–372.
10. Frenking, G., Wichmann, K., Fröhlich, N., Loschen, C., Lein, M., Frunzke, J. and Rayón, V. M. (2003). Towards a rigorously defined quantum chemical analysis of the chemical bond in donor–acceptor complexes, *Coord. Chem. Rev.*, 238–239, pp. 55–82.
11. te Velde, G., Bickelhaupt, F. M., Baerends, E. J., van Gisbergen, S. J. A., Fonseca Guerra, C., Snijders, J. G. and Ziegler, T. (2001). Chemistry with ADF, *J. Comput. Chem.*, 22, pp. 931–967.
12. Bickelhaupt, F. M. and Baerends, E. J. (2000). Kohn-Sham density functional theory: Predicting and understanding chemistry, *Rev. Comp. Chem.*, 15, pp. 1–86.
13. Mitoraj, M. and Michalak, A. (2007). Natural orbitals for chemical valence as descriptors of chemical bonding in transition metal complexes, *J. Mol. Model.*, 13, pp. 347–355.
14. This term was first suggested by Ruedenberg. See: Ruedenberg, K. (1962). The physical nature of the chemical bond, *Rev. Mod. Phys.*, 34, pp. 326–376.
15. Bickelhaupt, F. M., Nibbering, N. M. M., van Wezenbeek, E. M. and Baerends, E. J. (1992). Central bond in the three $CN^{\bullet}$ dimers NC–CN, CN–CN and CN–NC: Electron pair bonding and Pauli repulsion effects, *J. Phys. Chem.*, 96, pp. 4864–4873.

16. Frenking, G. and Bickelhaupt, F. M. (2014). The EDA perspective of chemical bonding. In: eds. Frenking, G. and Shaik, S., *The Chemical Bond. Fundamental Aspects of Chemical Bonding*, Chapter 4 (Wiley-VCH: Weinheim), pp. 121–157.
17. von Hopffgarten, M. and Frenking, G. (2012). Energy decomposition analysis, *WIREs Comput. Mol. Sci.*, 2, pp. 43–62.
18. Mitoraj, M. P., Michalak, A. and Ziegler, T. (2009). A combined charge and energy decomposition scheme for bond analysis, *J. Chem. Theory Comput.*, 5, pp. 962–975.
19. Kovács, A., Esterhuysen, C. and Frenking, G. (2005). The nature of the chemical bond revisited: An energy-partitioning analysis of nonpolar bonds, *Chem. Eur. J.*, 11, pp. 1813–1825.
20. Fernández, I. and Frenking, G. (2006). Direct estimate of the strength of conjugation and hyperconjugation by the energy decomposition analysis method, *Chem. Eur. J.*, 12, pp. 3617–3629.
21. Cappel, D., Tüllmann, S., Krapp, A. and Frenking, G. (2005). Direct estimate of the conjugative and hyperconjugative stabilization in diynes, dienes, and related compounds, *Angew. Chem. Int. Ed.*, 44, pp. 2–5.
22. Fernández, I. (2014). Direct estimate of conjugation, hyperconjugation, and aromaticity with the energy decomposition analysis method. In: eds. Frenking, G. and Shaik, S., *The Chemical Bond. Chemical Bonding Across the Periodic Table*, Chapter 12 (Wiley-VCH: Weinheim), pp. 357–381.
23. Tanaka, M. and Sekiguchi, A. (2005). Hexakis(trimethylsilyl)tetrahedranyltetrahedrane, *Angew. Chem. Int. Ed.*, 44, pp. 5821–5823.
24. Jarowski, P. D., Wodrich, M. D., Wannere, C. S., Schleyer, P. v. R. and Houk, K. N. (2004). How large is the conjugative stabilization of diynes? *J. Am. Chem. Soc.*, 126, pp. 15036–15037.
25. Gonthier, J. F., Steinmann, S. N., Wodrich, M. D. and Corminboeuf, C. (2012). Quantification of "fuzzy" chemical concepts: A computational perspective, *Chem. Soc. Rev.*, 41, pp. 4671-4687.
26. Frenking, G. and Krapp, A. (2007). Unicorns in the world of chemical bonding models, *J. Comput. Chem.*, 28, pp. 15–24.
27. Kistiakowsky, G. B., Ruhoff, J. R., Smith, H. A. and Vaughan, W. E. (1936). Heats of organic reactions. IV. Hydrogenation of some dienes and of benzene, *J. Am. Chem. Soc.*, 58, pp. 146–153.
28. Conant, J. B. and Kistiakowsky, G. B. (1937). Energy changes involved in the addition reactions of unsaturated hydrocarbons, *Chem. Rev.*, 20, pp. 181–194.
29. Rogers, D. W., Matsunaga, N., Zavitsas, A. A., McLafferty, F. J. and Liebman, J. F. (2003). The conjugation stabilization of 1,3-butadiyne is zero, *Org. Lett.*, 5, pp. 2373–2375.
30. Rogers, D. W., Matsunaga, M., McLafferty, F. J., Zavitsas, A. A. and Liebman, J. F. (2004). On the lack of conjugation stabilization in polyynes (polyacetylenes), *J. Org. Chem.*, 69, pp. 7143–7147.

31. Mo, Y. and Schleyer, P. v. R. (2006). An energetic measure of aromaticity and antiaromaticity based on the Pauling–Wheland resonance energies, *Chem. Eur. J.*, 12, pp. 2009–2020.
32. George, P., Trachtman, M., Bock, C. W. and Brett, A. M. (1975). An alternative approach to the problem of assessing stabilization energies in cyclic conjugated hydrocarbons, *Theor. Chim. Acta*, 38, pp. 121–129.
33. George, P., Trachtman, M., Brett, A. M. and Bock, C. W. (1977). Comparison of various isodesmic and homodesmotic reaction heats with values derived from published ab initio molecular orbital calculations, *J. Chem. Soc., Perkin Trans.*, 2, pp. 1036–1047.
34. Wodrich, M. D., Wannere, C. S., Mo, Y., Jarowski, P. D., Houk, K. N. and Schleyer, P. v. R. (2007). The concept of protobranching and its many paradigm shifting implications for energy evaluations, *Chem. Eur. J.*, 13, pp. 7731–7744.
35. Mo, Y. (2011). Rotational barriers in alkanes, *WIREs Comput. Mol. Sci.*, 1, pp. 164–171.
36. Mo, Y. (2014). The block-localized wavefunction (BLW) perspective of chemical bonding. In: eds. Frenking, G. and Shaik, S., *The Chemical Bond. Fundamental Aspects of Chemical Bonding*, Chapter 6 (Wiley-VCH: Weinheim), pp. 199–231.
37. Carey, F. A. and Sundberg, R. J. (2000). *Advanced Organic Chemistry, Part A: Structure and Mechanism*, 4th edn. (Plenum: New York).
38. Reichardt, C. (2003). *Solvents and Solvent Effects in Organic Chemistry*, 3rd edn. (Wiley-VCH: Weinheim).
39. Moonen, N. N. P., Pomerantz, W. C., Gist, R., Boudon, C., Gisselbrecht, J.-P., Kawai, T., Kishioka, A., Gross, M., Irie, M. and Diederich, F. (2005). Donor-substituted cyanoethynylethenes: π-conjugation and band-gap tuning in strong charge-transfer chromophores, *Chem. Eur. J.*, 11, pp. 3325–3341.
40. Fernández, I. and Frenking, G. (2006). π-Conjugation in donor-substituted cyanoethynylethenes: An EDA study, *Chem. Commun.*, pp. 5030–5032.
41. For a related study on bis(gem-diethynylethene) compounds, see: Fernández, I. and Frenking, G. (2007). EDA study of π-conjugation in tunable bis(gem-diethynylethene) fluorophores, *J. Org. Chem.*, 72, pp. 7367–7372.
42. Fernández, I. and Frenking, G. (2006). Correlation between Hammett substituent constants and directly calculated π-conjugation strength, *J. Org. Chem.*, 71, pp. 2251–2256.
43. Hammett, L. P. (1935). Some relations between reaction rates and equilibrium constants, *Chem. Rev.*, 17, pp. 125–136.
44. Hammett, L. P. (1970). *Physical Organic Chemistry*, 2nd edn. (McGraw-Hill: New York).
45. Okamoto, Y. and Brown, H. C. (1957). A quantitative treatment for electrophilic reactions of aromatic derivatives, *J. Org. Chem.*, 22, pp. 485–494.
46. Shaik, S., Shurki, A., Danovich, D. and Hiberty, P. C. (2001). A different story of π-delocalization: The distortivity of π-electrons and its chemical manifestations, *Chem. Rev.*, 101, pp. 1501–1540.

47. Gleiter, R. and Haberhauer, G. (2012). *Aromaticity and Other Conjugation Effects* (Wiley-VCH: Weinheim).
48. Schleyer, P. v. R., Wu, J. I., Cossío, F. P. and Fernández, I. (2014). Aromaticity in transition structures, *Chem. Soc. Rev.*, 43, pp. 4909–4921.
49. Fernández, I. and Frenking, G. (2007). Direct estimate of conjugation and aromaticity in cyclic compounds with the EDA method, *Faraday Discuss.*, 135, pp. 403–421.
50. Bally, T. (2006). Cyclobutadiene, the antiaromatic paradigm? *Angew. Chem. Int. Ed.*, 45, pp. 6616–6619.
51. Chen, Z., Wannere, C. S., Corminboeuf, C., Putcha, R. and Schleyer, P. v. R. (2005). Nucleus-independent chemical shifts (NICS) as an aromaticity criterion, *Chem. Rev.*, 105, pp. 3842–3888.
52. Flygare, W. H. (1974). Magnetic interactions in molecules and an analysis of molecular electronic charge distribution from magnetic parameters, *Chem. Rev.*, 74, pp. 653–687.
53. Fernández, I. and Frenking, G. (2007). Aromaticity in metallabenzenes, *Chem. Eur. J.*, 13, pp. 5873–5884.
54. Fernández, I., Frenking, G. and Merino, G. (2015). Aromaticity of metallabenzenes and related compounds, *Chem. Soc. Rev.*, 44, pp. 6452–6463.
55. Landorf, W. C. and Haley, M. M. (2006). Recent advances in metallabenzene chemistry, *Angew. Chem. Int. Ed.*, 45, pp. 3914–3936.
56. Dalebrook, A. F. and Wright, L. J. (2012). Metallabenzenes and metallabenzenoids, *Adv. Organomet. Chem.*, 60, pp. 93–177.
57. Wang, Y., Fernández, I., Duvall, M., Wu, J. I.-C., Li, Q., Frenking, G. and Schleyer, P. v. R. (2010). Consistent aromaticity evaluations of methylenecyclopropene analogues, *J. Org. Chem.*, 75, pp. 8252–8257.
58. Fernández, I., Duvall, M., Wu, J. I.-C., Schleyer, P. v. R. and Frenking, G. (2011). Aromaticity in group 14 homologues of the cyclopropenylium cation, *Chem. Eur. J.*, 17, pp. 2215–2224.
59. Fernández, I., Wu, J. I. and Schleyer, P. v. R. (2013). Substituent effects on "hyperconjugative" aromaticity and antiaromaticity in planar cyclopolyenes, *Org. Lett.*, 15, pp. 2990–2993.
60. Pierrefixe, S. C. A. H. and Bickelhaupt, F. M. (2007). Aromaticity: Molecular-orbital picture of an intuitive concept, *Chem. Eur. J.*, 13, pp. 6321–6328.
61. Pierrefixe, S. C. A. H. and Bickelhaupt, F. M. (2008). Aromaticity and antiaromaticity in 4-, 6-, 8-, and 10-membered conjugated hydrocarbon rings, *J. Phys. Chem. A*, 12, pp. 12816–12822.
62. Shaik, S., Danovich, D., Wu, W. and Hiberty, P. C. (2014). The valence bond perspective of the chemical bond. In: eds. Frenking, G. and Shaik, S., *The Chemical Bond. Fundamental Aspects of Chemical Bonding*, Chapter 5 (Wiley-VCH: Weinheim), pp. 159–198.
63. Shaik, S.S. and Hiberty, P.C. (1985). When does electronic delocalization become a driving force of molecular shape and stability? 1. The aromatic sextet, *J. Am. Chem. Soc.*, 107, pp. 3089–3095.

64. Hirao, H. and Wang, X. (2014). Hydrogen bonding. In: eds. Frenking, G. and Shaik, S., *The Chemical Bond. Chemical Bonding Across the Periodic Table*, Chapter 17 (Wiley-VCH: Weinheim), pp. 501–521.
65. Guerra, C. F., Bickelhaupt, F. M., Snijders, J. G. and Baerends, E. J. (1999). The nature of the hydrogen Bond in DNA base pairs: The role of charge transfer and resonance assistance, *Chem. Eur. J.*, 5, pp. 3581–3594.
66. Glendening, E. D. (2005). Natural energy decomposition analysis: Extension to density functional methods and analysis of cooperative effects in water clusters, *J. Phys. Chem. A*, 109, pp. 11936–11940.
67. Pendás, A. M., Blanco, M. A. and Francisco, E. (2006). The nature of the hydrogen bond: A synthesis from the interacting quantum atoms picture, *J. Chem. Phys.*, 125, pp. 184112–184120.
68. Wolters, L. P. and Bickelhaupt, F. M. (2012). Halogen bonding versus hydrogen bonding: A molecular orbital perspective, *ChemistryOpen*, 1, 96–105.
69. Kurczab, R., Mitoraj, M. P., Michalak, A. and Ziegler, T. (2010). Theoretical analysis of the resonance assisted hydrogen bond based on the combined extended transition state method and natural orbitals for chemical valence scheme, *J. Phys. Chem. A*, 114, pp. 8581–8590.
70. Gilli, P., Bertloasi, V., Ferretti, V. and Gilli, G. (1994). Evidence for resonance-assisted hydrogen bonding. 4. Covalent nature of the strong homonuclear hydrogen bond. Study of the O-H—O system by crystal structure correlation methods, *J. Am. Chem. Soc.*, 116, pp. 909–915.
71. van Zeist, W.-J. and Bickelhaupt, F. M. (2010). The activation strain model of chemical reactivity, *Org. Biomol. Chem.*, 8, pp. 3118–3127.
72. Fernández, I. and Bickelhaupt, F. M. (2014). The activation strain model and molecular orbital theory: Understanding and designing chemical reactions, *Chem. Soc. Rev.*, 43, pp. 453–4967.
73. Fernández, I. (2014). Understanding trends in reaction barriers. In: ed. Pignataro, B., *Discovering the Future of Molecular Sciences*, Chapter 7 (Wiley-VCH: Weinheim), pp. 165–187.
74. Fernández, I. (2014). Combined activation strain model and energy decomposition analysis methods: A new way to understand pericyclic reactions, *Phys. Chem. Chem. Phys.*, 16, pp. 7662–7671.
75. Hirsch, A. (1994). *The Chemistry of the Fullerenes.* (Thieme: Stuttgart).
76. Fernández, I., Solá, M. and Bickelhaupt, F. M. (2013). Why do cycloaddition reactions involving C60 prefer [6,6] over [5,6] bonds? *Chem. Eur. J.*, 19, pp. 7416–7422.
77. Osuna, S., Swart, M. and Solá, M. (2011). Dispersion corrections essential for the study of chemical reactivity in fullerenes, *J. Phys. Chem. A*, 115, pp. 3491–3496.
78. Fernández, I., Solà, M. And Bickelhaupt, F. M. (2014). Origin of reactivity trends of noble gas endohedral fullerenes $Ng_2@C_{60}$ (Ng = He to Xe), *J. Chem. Theory Comput.*, 10, pp. 3863–3870.

79. Bickelhaupt, F. M., Solá, M. and Fernández, I. (2015). Understanding the reactivity of endohedral metallofullerenes: C_{78} versus $Sc_3N@C_{78}$, *Chem. Eur. J.*, 21, pp. 5760–5768.
80. García-Rodeja, Y., Solá, M., Bickelhaupt, F. M. and Fernández, I. (2016). Reactivity and selectivity of bowl-shaped polycyclic aromatic hydrocarbons: Relationship to C_{60}, *Chem. Eur. J.*, 22, pp. 1368–1378.
81. ADF Molecular Modeling Suite, Scientific Computing and Modeling. http://www.scm.com.
82. Schmidt, M. W., Baldridge, K. K., Boatz, J. A., Elbert, S. T., Gordon, M. S., Jensen, J. H., Koseki, S., Matsunaga, N., Nguyen, K. A., Su, S., Windus, T. L., Dupuis, M. and Montgomery, J. A. (1993). General Atomic and Molecular Electronic Structure System, *J. Comput. Chem.*, 14, pp. 1347–1363.
83. GAMESS program. http://www.msg.ameslab.gov/gamess/.
84. S. I. Gorelsky, AOMix: Program for Molecular Orbital Analysis, University of Ottawa, 2013. http://www.sg-chem.net/.
85. Grimme, S., Antony, J., Ehrlich, S. and Krieg, H. J. (2010). A consistent and accurate ab initio parametrization of density functional dispersion correction (DFT-D) for the 94 elements H-Pu, *Chem. Phys.*, 132, pp. 154104–154119.
86. Grimme, S. (2014). Dispersion interaction and chemical bonding. In: eds. Frenking, G. and Shaik, S., *The Chemical Bond. Chemical Bonding Across the Periodic Table*, Chapter 16 (Wiley-VCH: Weinheim), pp. 477–499.

Chapter 8

Systems with Extensive Delocalization

L. Zoppi and K. K. Baldridge

Department of Chemistry, University of Zurich, Winterthurerstrasse 190, Zurich, CH-8057, Switzerland

8.1 Introduction

The strong potential for new technological applications incorporating π-conjugated aromatic molecules has created an emergence of research areas in both academia and industry with promise for development of large-scale, low-cost, flexible electronic devices.[1] Organic molecules with semiconducting properties, such as aromatic hydrocarbons, tend to have high polarizabilities, small band-gaps, and delocalized π networks that support mobile charge carriers.[2] To utilize organic materials successfully in technologies such as organic light-emitting diodes (OLEDs),[1a,3] organic photovoltaic cells (OPVCs),[4] and organic field-effect transistors (OFETs),[5] material design depends on access to functional building blocks that manifest a desired property. A key design principle for practical applications lies in the understanding and control of the solid-state packing and intermolecular interactions of individual π molecular units of designed fragments.[6] As conductor dimensions approach the nanoscale, design principles focus on producing molecules with tunable functionality to enable control of charge transfer (CT) properties at the molecular scale. However, the ability to fully control the electronic character of organic fragments still remains a formidable challenge and a critical need has emerged for improved understanding of electronic transport phenomena in these systems.

The conventional mechanism for charge transport phenomena in conjugated materials involves the overlap of delocalized π molecular orbitals that extend through a plane and facilitate the transport. The possibility of confining the electronic states in one or more directions, according to morphology and dimensionality,[7] together with low dielectric screening of organic materials leads to the formation of electronic excitations in the form of Coulomb-bound electron–hole pairs — excitons[8] — that are responsible for the majority of the characteristic optoelectronic properties in organic-based materials in general.

Photoexcitation of an organic semiconductor causes an electron to be promoted to the conduction band, leaving a "hole" behind in the valence band. Depending on the polarizability of the material, excitons may be strongly bound and localized on a single subunit (Frenkel exciton) or weakly bound and delocalized over many sub-units (Mott–Wannier exciton).[8] The former situation is common in organic materials where, due to the low dielectric constants of organic materials ($\varepsilon \approx 3 - 4$), electron–hole pairs with significant binding energies ($\approx$1 eV) can be formed.[8] When excitons are generated by direct photoexcitation of the material, they can later dissociate into free carriers (unbound electrons and holes) and collected at opposite electrodes (photovoltaics).[9] Conversely, when negative (electron) and positive (holes) charge carriers are injected into an organic material from opposite electrodes, recombination of oppositely charged carriers can occur resulting in a photo-emission process (electroluminescence).[3c]

The ability to explicitly design π-conjugated systems with tailored nanoscale properties specialized for components in nanoscale electronics and optoelectronic devices is of great interest,[10] particularly in view of recent abilities to construct single-molecule junctions and measure their transport properties.[11] Fundamental device technology based on properties inherent in single organic molecules offers, in principle, unlimited possibilities for technological development since diverse electronic features can be tailored with adept chemical design and synthesis.[12] However, despite being able to control molecular design through chemical synthesis, the ability to actually guide the electrical transport properties of single molecules via external means is still an onerous undertaking.

Toward this end, the extended family of curved aromatics compounds based on the smallest bowl-shaped fullerene fragment, corannulene, $C_{20}H_{10}$ (Fig. 8.1),[13] have been key targets of interest in several directions in materials design.[14] The intriguing electronic properties of these bowl fragments can be ascribed to the local curvature of the parent molecule (Fig. 8.1), which imparts a substantial dipole[15] to the structure, and can be modulated through functionalization. Such structural diversification offers a means of tailoring the electronic properties (e.g., ionization potential [IP], electron affinity [EA], dipoles, polarizability) as well as crystal packing motifs.[3b,16]

The ability to design and tune specific structural features provides an array of materials for testing and developing prototypes with unique electronic properties. When assembled in the solid state, these compounds provide an array of structures supported in varying complex environments, which can be exploited as active molecular layers in optoelectronic applications,[3b,14b,17] aggregated as monolayers on metallic surfaces for work-function engineering,[14c,18] or, as single molecules in junctions for transport processes.[16b,19] Many functionalized derivatives display columnar packing motifs in the

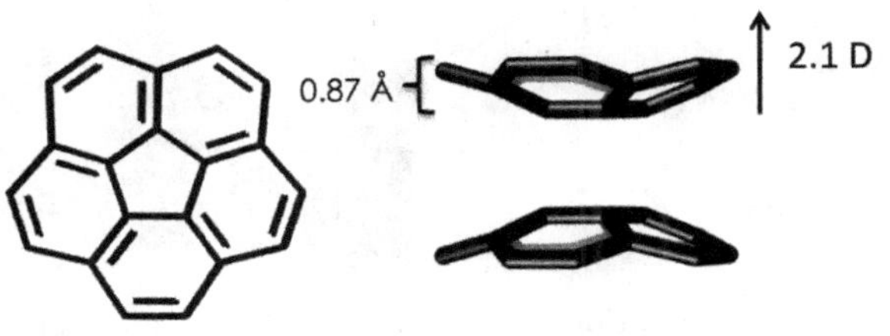

Figure 8.1. Corannulene structure indicating molecular bowl depth and intrinsic dipole.

crystal, where the approaching π surfaces within the stacks define a preferential channel for charge transport along the direction of the stack propagation.[3b,16a] The strong electronic accepting ability of the parent molecule, which can accommodate up to four electrons in the two degenerate LUMOs,[14c] the evidence for CT states in the crystal,[3b] and the special delocalized nature of molecular orbitals SAMOs,[20] impart effective charge transport properties to resulting materials when molecular units are assembled in quantum nanostructures or solids (Fig. 8.2).

Functionalized derivatives have been shown to work as emitting components in OLEDs,[17] as acceptor materials in organic solar cells (OPVs),[14b] and as active elements in organic field-effect transistor (OFETs) devices.[21] Such studies have motivated more thorough theoretical investigation into the electronic properties, CT excitations, and mechanistic aspects of various functionalized derivatives to facilitate the understanding of the processes involved, as well as to optimize materials properties and device morphologies that can maximize absorption (emission) efficiencies, electron and hole transport, and charge collection (charge recombination).

In any of these complex devices, critical questions of the optical excitation and CT phenomena connected to the individual molecules, the material, as well as the functional organic interfaces arise. Of paramount importance is control of the energy-level alignments between different materials at interfaces, which determine the energy

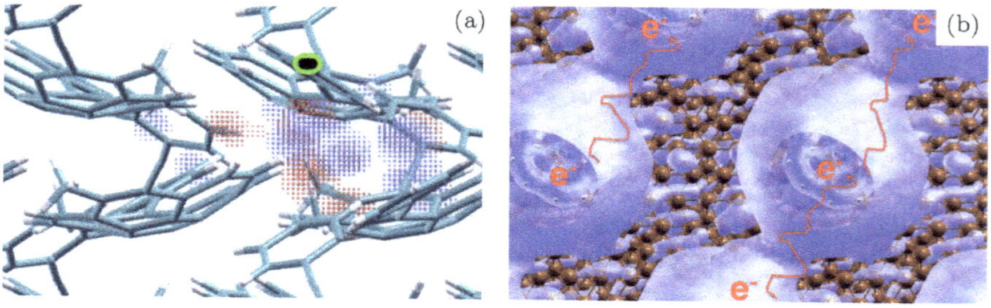

Figure 8.2. (a) CT excitation in 1,6-diphenylethynyl-2,5-dimethylcorannulene crystal.[3b] (b) Super-atomic-molecular-orbitals (SAMOs) for $C_{20}H_{10}$ molecules supported on a Cu(111) surface.[20]

barrier for electron (holes) crossing between different materials and, in the ultimate analysis, the charge injection efficiency of a device.[22] In order to capture such detail, effective computational methodology must be able to provide accurate predictions of quantities such as IP, EA, the transport gap (Δ(IP–EA)), and the exciton binding energy (inferred from the difference between the transport and optical absorption gaps). These processes are illustrated in Fig. 8.3.

It is well known that density functional theory (DFT), within standard local and semilocal approximations to the exchange-correlation potential, will not quantitatively describe excitation properties of molecular solids, and in particular, transport and optical gaps.[23] From a theoretical point of view, accurate description of electronic excitation processes in general represents a major challenge, and several approaches are being developed to address this problem as conveyed in the literature. Among the possible strategies, constrained DFT (CDFT)[24] represents one approach to the

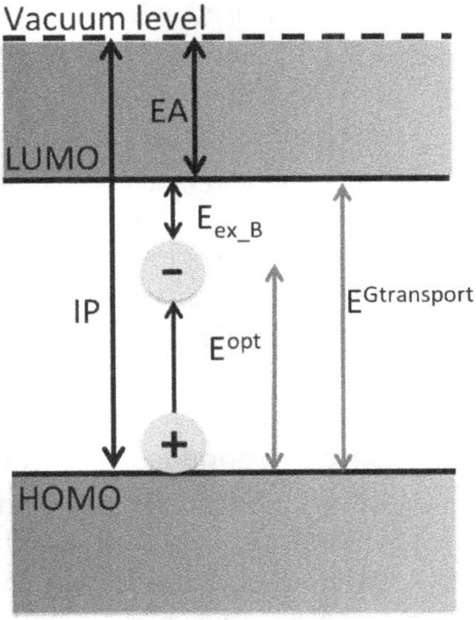

Figure 8.3. Schematic representation of an exciton indicating the IP, EA, the transport gap ($E^{Gtransport} = IP - EA$), the exciton binding energy E_{ex_B}, and the optical gap ($E^{opt} = E^{Gtransport} - E_{ex\_B}$).

problem. In the CDFT methodology, one defines an appropriate density functional together with a specific constraint on the charge density, thereby forcing localization of charge onto a particular fragment in a molecule. While CDFT methodologies[24] can be computationally as efficient as nonconstrained DFT, and have been shown to provide in specific cases quite reasonable results,[25] such methods do not take into account either the perturbing field generated by the light or the interaction between the electron and the hole. When an electron–hole couple is created by promotion of an electron into an unoccupied state, it is not possible to represent the interaction of the two charge carriers because the Hamiltonian depends only on the electronic density of the occupied states.

A second approach is the use of standard time-dependent DFT (TD-DFT).[26] TD-DFT is able to qualitatively describe optical excitations with strong spatial overlap between involved occupied and unoccupied states. However, this method completely fails in describing weakly overlapping electron–hole pairs due to the lack of direct (attractive) interaction between electron and hole. Only if the underlying density functional is fully nonlocal would TD-DFT be able to account for electron–hole direct interactions.[26b]

An alternative framework, considered state-of-the-art for characterizing excited states, involves solution of the Bethe–Salpeter equation (BSE),[26a] which belongs to the many-body perturbation theory (MBPT) class of methodologies. This method is typically calculated within the GW approximation, as GW-BSE.[26a] Such combined methodology is frequently used for investigations of bulk systems,[27] molecular crystals,[28] and one-dimensional (1D) extended systems (e.g., nanowires).[29] Within this formalism, neutral excitation energies and optical spectra are obtained from solution of the BSE, which, using the GW data, accounts for electron–hole interactions.[26a] BSE has a similar structure as single configuration interaction, except that the electron–hole interaction includes the information on the dielectric response of the electronic gas.[30] As documented in the literature,[10b,10c] a proper description of electronic structure within MBPT approaches is essential for accurately describing excited state properties of polyaromatic hydrocarbons. Key examples include studies involving polymeric systems and oligomeric chains, for example, polyacetylene, which was considered a protoype material for a long time,[31] and pentacene, one of the most investigated systems.[10c] For assemblage of donor–acceptor systems, such as C_{60}/C_{70} as an acceptor and an organic molecule as a donor (pentacene or porphyrin), electron–hole interactions are a key component for the accurate predictability of the excitation properties.[32]

The inherent challenge for first-principles–based methodology is not only the establishment of a reliable pallet of methods but also validation with corresponding benchmark protocols for predictions. Ideally, methods should be equally accessible across a broad class of systems with comparable benchmark accuracy, including bulk materials, interfaces, and nanostructures. In this respect, the extended family of curved carbon π systems offers the prospect of a tunable spectrum of structure and electronic properties in connection with their morphology and dimensionality, as manifested through zero-dimensional (0D) molecules, three-dimensional (3D) molecular crystals, 3D organic–metallic surface interfaces, and 1D organic nanojunctions (Fig. 8.4).

This chapter offers discussion of the current state-of-the-art electronic structure approaches for the prediction of excitation properties and CT/charge transport phenomena of complex materials and interfaces, with illustration of these capabilities from a series of applications involving curved aromatics systems. In particular, the focus centers on key topics in energy-related phenomena, with a particular emphasis on design aspects of functional materials and comparison with experimental data. Importantly, the need for clarification of several "unknowns" in experimental investigation of materials phenomenon has stimulated and enabled important theoretical developments and extensions of standardized methods for this particularly challenging class of systems, pushing forward the predictive capability of general schemes.

Four illustrative cases are discussed: (1) superatom states of hollow molecules (SAMOs); (2) CT excitations in molecular

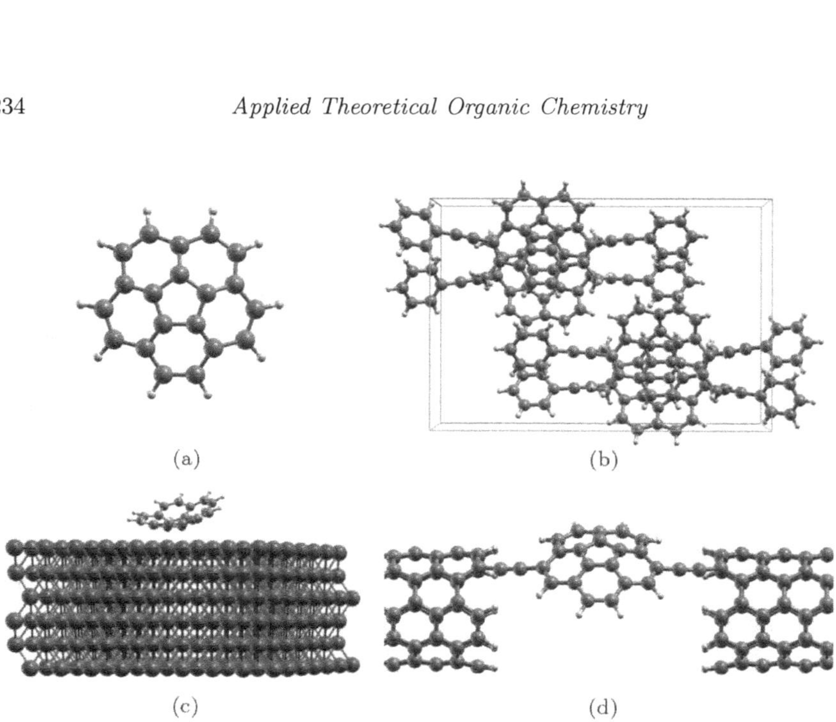

Figure 8.4. (a) $C_{20}H_{10}$ molecular structure (0D); (b) 1,6-diphenylethynyl-2,5-dimethylcorannulene corannulene crystal (3D); (c) $C_{20}H_{10}$ molecules arranged on a Cu(111) surface (3D); (d) $C_{20}H_{10}$ molecule assembled into a molecular junction (1D).

crystals/organic interfaces; (3) WF modification at organic metal interfaces; and (4) electronic transport through a molecular nanojunction.

8.2 Super-atomic-molecular-orbitals

As 0D constructs, curved and spheroid aromatic compounds are ideal model systems for exploring nanomaterial properties. Interest in their use as active elements in nanoscale electronic devices[20] and electron acceptor components of organic solar cells[33] increases the need for better understanding of their fundamental electronic properties. Most importantly, materials based on curved or hollow aromatic complexes enable exploitation of a unique mechanism of intermolecular charge transport distinct from the conventional mechanism involving tightly bound π molecular orbital overlap.

Key to electron transport routes is evidence of a characteristic set of diffuse molecular orbitals called SAMOs.[34] These orbitals were first investigated in the closed hollow aromatic molecule, C_{60}.[34,35] SAMOs are virtual orbitals that arise from the central potential of the hollow molecular core, evoking well-defined hydrogen-like s, p, and d orbital angular momentum shapes, as illustrated in Fig. 8.5 for C_{60}. As observed, SAMO orbitals extend well beyond the more tightly bound π orbitals. These orbitals are distinct from conventional Rydberg type orbitals.[35] In SAMO, electron correlation effects inside the closed or curved cavity manifest the additional shallow attractive potential, resulting in significant electron density inside the cavity. In the case of Rydberg-type orbitals, the extended orbital only sees the core as a point charge.[35]

The significance of SAMOs lies in the possibility for exploitation of the nearly free conducting channels, which arise when the molecular units are assembled in series in quantum nanostructures or solids.[20] Such channels have been observed in low-temperature scanning tunneling microscopy experiments (LT-STM) for C_{60} molecules assembled on noble metal surfaces.[34] Unfortunately, prospects for exploiting SAMOs in practical applications must be tempered by the fact that typically these orbitals are unoccupied. In the case of C_{60}, the SAMOs lie several electron volts above the lowest unoccupied molecular orbital (LUMO) and therefore are difficult to exploit for this purpose (Fig. 8.5).[34]

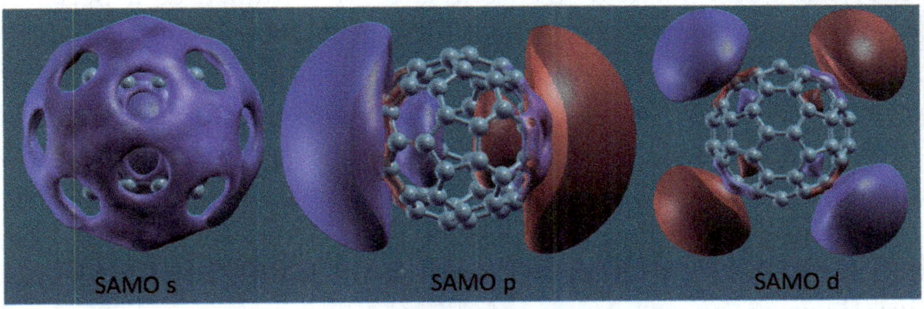

Figure 8.5. Single-point DFT calculation illustrating the 3D representation of SAMOs: typical s, p, and d-like symmetric shape in C_{60}.

The physical origin of SAMOs is ascribed to many-body screening and polarization effects, typical of a polarizable assembly (e.g., graphene) that undergoes a topological distortion, such as wrapping or rolling into a nanotube or fullerene.[36] At a solid–vacuum interface, these many-body interactions give rise to a series of degenerate image potential (IP) states in the near-surface region on both sides of a graphene sheet, which float above and below the molecular plane and undergo free motion parallel to it.[36] Topological distortion of the molecular sheet breaks symmetry, lowering/raising the energy of the IP states on the concave/convex side of the resulting material, revealing SAMOs.[35]

Theoretical investigation of SAMOs requires methods that properly include polarization and correlation effects, which are crucial for describing IP states. These states are typically poorly described with conventional DFT approaches[37] due to the approximate nature of the exchange and correlation (XC) potential.[38] In order to capture these effects, the choice of the methodology falls on MBPT[26a] electronic structure approaches within the GW scheme.[39] In this approximation,[26a,39] the self-energy, Σ, is the product of a single-particle Green function, G, and a nonlocal and dynamically screened Coulomb potential, W, ($\Sigma = iGW$). Due to the high computational demands of fully self-consistent GW (scGW), a range of perturbative GW schemes, from non-self-consistent to partially self-consistent have emerged.[40] The lowest rung in this hierarchy, standard in practical calculations, is a non-self-consistent scheme, i.e., G_0W_0[40,41] where the quasi-particle (QP) energies are obtained as a perturbative first-order correction to the DFT eigenvalues. It is important to note that GW theory is an approximation for GWΓ theory (former work of Hedin and Lunquist).[39] It is not straightforward that full scGW could provide better QP energies than any perturbative GW scheme.[42]

The use and appropriateness of the GW method for the prediction of IP states has raised some skepticism in the QM community. However, as well documented in the literature, we also strongly believe that the GW methodology is suitable for description of many-body screening and long-range polarization effects, both crucial for treating IP states.[43] Further criticism could be raised in general concerning the use of DFT wavefunctions for the purpose of SAMO description. However, differences between the Kohn and Sham and real QP wavefunction come from the incorrect long-range behavior of approximate exchange potentials in conventional DFT schemes, which scarcely affects the GW-corrected QP energies.[44]

Taking a slightly different view from typically studied spheroid fullerenes, the present section details theoretical investigation of the electronic properties of SAMOs in a series of curved aromatic constructs, $C_{20}H_{10}$, $C_{30}H_{10}$, $C_{40}H_{10}$, $C_{50}H_{10}$, with focus on their suitability for applications in molecular circuits (electron transport). In order to carry out the calculations with a reliable outcome, a customized hybrid methodology was developed, using a standard B97D/Def2-TZVPP[45] DFT (e.g., in GAMESS)[46] for full optimization and Hessian characterization of structures, followed by a plane-wave DFT formalism (in Quantum-ESPRESSO)[47] within MBPT[26a] in the GW approximation[39] (SAX).[48] The resulting calculated SAMOs for corannulene are shown in Fig. 8.6. As have been demonstrated for the closed-shaped hollow C_{60} structure, corannulene is also observed to manifest the characteristic diffuse molecule-centered hydrogenic-like s, p, and d shapes.

To be relevant for charge transport, the SAMO-derived states need to be either the first unoccupied state or cross the Fermi level

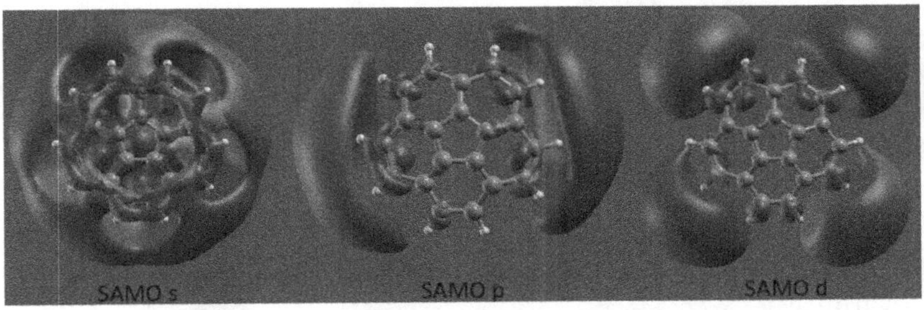

Figure 8.6. DFT-LDA 3D representation of SAMOs illustrating the typical s, p, and d-like symmetric shapes in corannulene, $C_{20}H_{10}$.

Table 8.1. $\Delta E_{\text{SAMO-LUMO}}$ predictions for $C_{20}H_{10}$ comparing different levels of theory.

	$\Delta E_{\text{SAMO-LUMO}}$(eV)		
SAMO	DFT(LDA)	DFT(PBE_0)	G_0W_0@LDA
s	2.3	1.8	0.3 eV
p	2.5	2.0	0.4 eV
d	2.8	2.3	0.5 eV

of the material such that occupation of the delocalized SAMOs can be enhanced. In terms of design, therefore, controlling the energy difference between the LUMO and SAMO of lowest energy becomes quite important. Calculated results for the smallest in the series of bowl structures, $C_{20}H_{10}$, is reported in Table 8.1. The results show gas-phase predictions from both G_0W_0 as well as DFT for $\Delta E_{\text{SAMO-LUMO}}$ As for C_{60}, the SAMOs for $C_{20}H_{10}$ are also revealed in the unoccupied part of the spectrum in a simple DFT calculation (here, LDA or PBE_0^{49}). As discussed, inclusion of many-body effects at the G_0W_0-level provide a more accurate positioning of SAMOs with respect to the simple DFT predictions. Initialization of the GW calculation from a standard L-DFT PZ functional[50] (G_0W_0@LDA) results in a SAMO level of s-type symmetry that corresponds to the first unoccupied level after the LUMO. Notably, the $\Delta E_{\text{SAMO-LUMO}}$ energy gap for the SAMO orbital of s-type symmetry in $C_{20}H_{10}$ is predicted to be 0.3 eV, nearly an order of magnitude smaller than in C_{60} (calc. 2.4 eV).[14a] The proximity of LUMO with respect to the delocalized SAMOs in $C_{20}H_{10}$ suggests possibilities for exploiting SAMO-mediated electron transport in a material using these bowl fragments as functional units.

Consideration of electronic and transport properties across the full series of buckybowls of increased curvature, $C_{20}H_{10}$–$C_{50}H_{10}$, becomes of interest based on the results for the smallest of the series shown in Fig. 8.7.

The presence of strong intrinsic molecular dipoles manifested by the curvature in these molecular fragments, together with the increasingly large polarizable surface of π electron density, are fundamental properties of interest for potentially exploiting such systems for material devices.[19] The increasing bowl depth and change in curvature across the series has revealed interesting and systematic trends in the structure and various electronic properties toward a tube structure (Fig. 8.7).[19] It is of interest, therefore, to see if the same trends can be observed in SAMO-related electronic properties.

As shown in previous work for shallow molecular bowls,[19] dipoles induced across the relatively large surface area of the cap component (pentagon) in corannulene may become comparable or larger than the intrinsic molecular dipole (perpendicular to the cap). On the other hand, for deeper bowls, such an effect is not so apparent, reflecting the increased conjugated area of the belt region while approaching what one might find in a tube-like structure. To understand the implications of the structural and electronic property variation from the shallow bowl of $C_{20}H_{10}$ to the tube-like structure of $C_{50}H_{10}$ on the associated SAMO electronic properties, G_0W_0 calculations were used to determine SAMO states across the full series $C_{20}H_{10}$–$C_{50}H_{10}$. The G_0W_0@LDA predicted SAMO–LUMO gap energy values for the set are summarized in Table 8.2 and the trends visualized in Fig. 8.8.

Comparison of the $\Delta E_{\mathrm{SAMO-LUMO}}$ energy gaps shows a sharp increase from $C_{20}H_{10}$ to $C_{30}H_{10}$, followed by a more modest increase

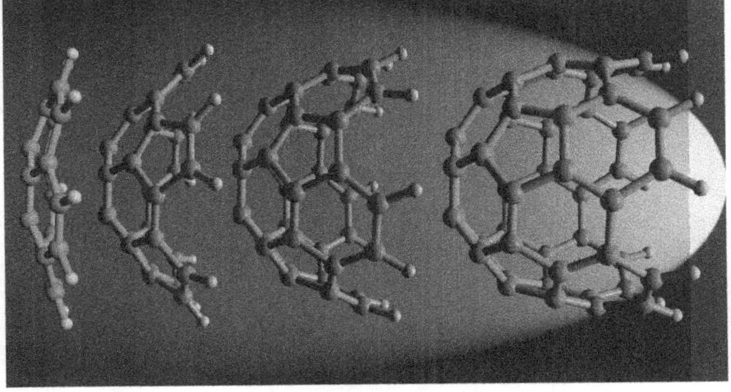

Figure 8.7. Curved aromatic bowl constructs of increasing size and depth.

Table 8.2. $\Delta E_{\text{SAMO-LUMO}}$ predictions for $C_{20}H_{10}$, $C_{30}H_{10}$, $C_{40}H_{10}$, $C_{50}H_{10}$ at the G_0W_0@LDA level of theory.

	$\Delta E_{\text{SAMO-LUMO}}$ (eV)		
System	s	p	d
$C_{20}H_{10}$	0.3	0.4	0.5
$C_{30}H_{10}$	1.3	1.9	1.7
$C_{40}H_{10}$	2.0	2.2	2.6
$C_{50}H_{10}$	1.6	2.1	2.5

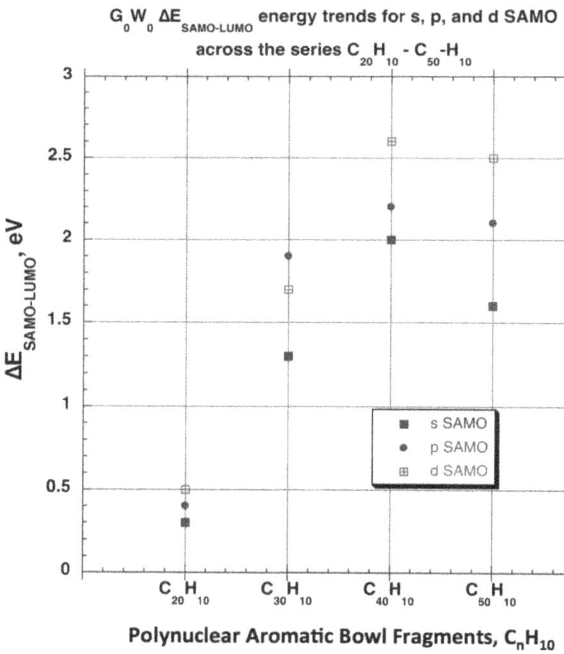

Figure 8.8. Trends in G_0W_0@LDA predicted $\Delta E_{\text{SAMO-LUMO}}$ across the series of aromatic bowl constructs of increasing size and bowl depth, $C_{20}H_{10}$, $C_{30}H_{10}$, $C_{40}H_{10}$, $C_{50}H_{10}$, in eV.

to $C_{40}H_{10}$, and then a significant drop for the more tube-like structure, $C_{50}H_{10}$ (Fig. 8.9). For the lowest energy s-type SAMO, the gap relative to $C_{20}H_{10}$ (0.3 eV) increases by 1.0 eV for $C_{30}H_{10}$, by 1.7 eV for $C_{40}H_{10}$, and then levels off to an increase of 1.3 eV for $C_{50}H_{10}$. Importantly, in the case of the smallest bowl, $C_{20}H_{10}$, the $\Delta E_{\mathrm{SAMO-LUMO}}$ energy gap is still one order of magnitude lower than the next higher analogue, $C_{30}H_{10}$. SAMOs of p- and d-type symmetry are significantly higher in energy than the LUMO (from 1.7 eV up to 2.6 eV) and do not show promise for enhancing the occupation of the delocalized orbitals.

It is also of interest to observe where the analogous results for C_{60} lie in the series in terms of "curvature" and resulting $\Delta E_{\mathrm{SAMO-LUMO}}$. The G_0W_0@LDA predicted trends across the full set of buckybowls, $C_{20}H_{10}$, $C_{30}H_{10}$, $C_{40}H_{10}$, $C_{50}H_{10}$, as well as C_{60}, are shown in Fig. 8.9. One finds that the value of $\Delta E_{\mathrm{SAMO-LUMO}}$ for C_{60} of 2.4 eV is still on the increase, in trend with the bowl structures shown, before the

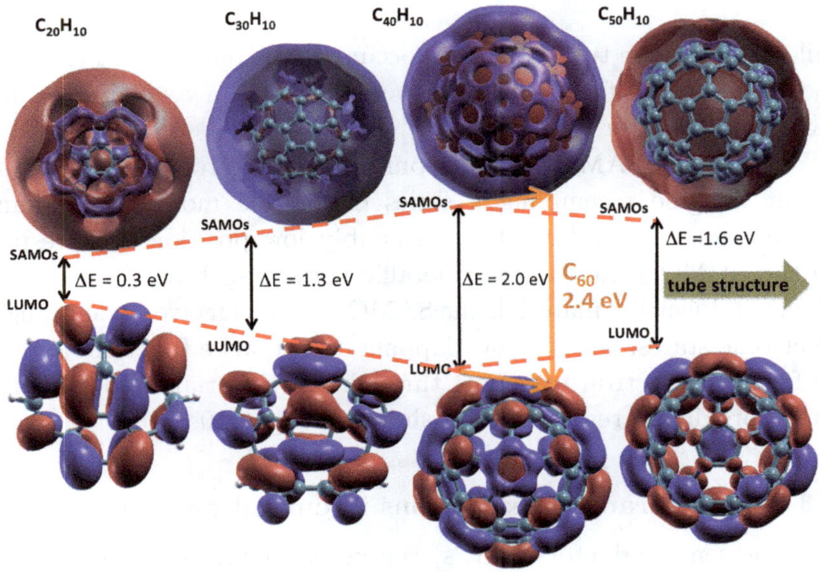

Figure 8.9. G_0W_0@LDA predicted trends across the series of buckybowls, $C_{20}H_{10}$, $C_{30}H_{10}$, $C_{40}H_{10}$, $C_{50}H_{10}$, C_{60}.

Figure 8.10. Single-point DFT calculation of $C_{20}H_{10}$ tiled onto a Cu(111) surface (a), illustrating the potential for inter-SAMO interaction in the molecular layer (b).

falloff toward the tube construct occurs at the onset of $C_{50}H_{10}$. The local curvature of C_{60} places it between $C_{40}H_{10}$ and $C_{50}H_{10}$, but closer to the latter.

To exploit SAMOs for the purpose of electron transport, one might imagine assembling a series of feasible molecular systems, which manifest SAMOs at a reasonably low position with respect to the LUMO states, onto a metallic surface such as, for example, Cu(111). In such a material, the SAMOs would survive as delocalized electronic states, with the overlapping diffuse wave functions serving to facilitate electron transport through the material, as depicted in the approximate representation shown in Fig. 8.10.[14a]

8.3 Charge transfer excitations in curved aromatics

Organic semiconductors (OSCs) composed of π-conjugated systems typically absorb/emit in the UV-visible region. Such devices support electronic excitations in the form of coulomb-bound electron–hole pairs (excitons), which eventually can recombine and either emit

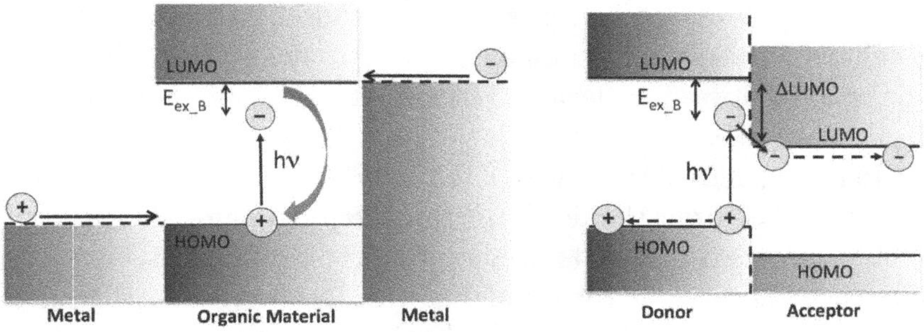

(a) (b)

Figure 8.11. Schematic representation of excitons and exciton binding energy (E_{ex_B}) in (a) OLED multilayer material, where electrons and holes injected from opposite sides recombine and emit light and (b) OPV device with an organic donor–acceptor (DA) interface that facilitates the charge-separation process of the photogenerated electrons and holes.

light (OLEDs) (Fig. 8.11(a)) or further separate and conduct current (OPVs) (Fig. 8.11(b)).

Focusing on the OPV device, in most cases, the large binding energy (on the order of eV)[32b] of the photoexcited electron–hole pair requires the presence of a donor–acceptor interface[51] to dissociate the bound exciton and produce the free carriers at the origin of the photocurrent. Although the mechanism underlying the associated charge separation is still under debate, it is believed that the fundamental intermediate states are charge-transfer excitations,[8,52] with the hole and the electron located on adjacent donor and acceptor units, respectively. For these systems, one can estimate an interfacial CT exciton binding energy of a few hundred meV,[51] which is one order of magnitude higher than the thermal energy at room temperature. How a photocurrent generation process can overcome such a high binding energy remains a critical issue but an unanswered question.

In this scenario, understanding structural and electronic modifications that drive transitions from photoinduced intramolecular (Frenkel) excitons to CT states where electron and hole are spatially separated, and vice versa, is an important goal for improving[32a] functionality of OSC devices. While such modifications can have profound influence on carrier generation and loss mechanism in devices, the underlying physical chemistry is still poorly understood. As such, insight from theory into fundamental electronic processes driving the conversion between molecular photoexcitations and free charges is highly desirable.

In an organic solar cell with a DA interface (Fig. 8.11(b)), in order to initiate CT from the donor to the acceptor material, the energy levels of the two materials need to be properly aligned so that the difference between the corresponding LUMOs overcomes the exciton binding energy.[4c] In fact, only when this alignment occurs is the electron transfer from the donor to the acceptor material a downhill energy process. The interface morphology of the photoactive DA complex also has a strong impact on the detailed mechanisms of charge-carrier generation, recombination, and transport of bound or separated charges.[4c] In this respect, organic bulk heterojunction solar cells (BHJs), where donor and acceptor components interpenetrate each other through the whole extent of the active layer, are considered the most effective for performance of the device (Fig. 8.12).[33]

Understanding the performance and mechanistic aspects of these devices relies heavily on a balanced interplay between experiment and theory. On each side, there are many components that

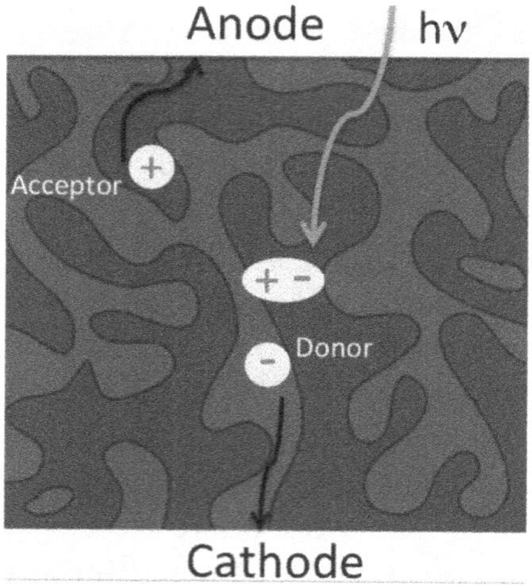

Figure 8.12. Schematic representation of a BHJ solar cell with distributed donor–acceptor interface over the extent of the whole active layer.

are otherwise unattainable without such a joint investigation, and together they enable a better understanding of device performance and, importantly, the ability to optimize performance of the constructed device. In the present example, a joint experimental and theoretical investigation was undertaken to consider the use of curved aromatics as acceptor components in the construction of a BHJ solar cell. In this particular construction, the BHJ solar cell consisted of a poly(3-n-hexylthiophene), P3HT, as the donor material, and a $C_{20}H_{10}$(Cor) derivatized with strong electron-withdrawing substituents as the acceptor components.[14b] Although P3HT is no longer considered a high-efficiency OPV material, it is the system of choice for such investigations on the fundamental photophysics of OPV devices.[33] For the acceptor component, fullerene-derived cages have been considered the most successful type of acceptors in OPV devices due to their high electron affinity and high electron mobility. The curved morphology has been found to facilitate the coupling between the acceptor molecules and multidimensional charge transport properties.[33] It was therefore also of interest to consider the set of curved bowl structures as acceptor materials. In particular, corannulene has strong accepting ability, which normalized per C atom goes beyond that of C_{60}. Together with the ability for tuning the electronic character via chemical functionalization, this property strongly motivates its use as an acceptor material in BHJ solar cells.

Two related $C_{20}H_{10}$-based acceptor materials were investigated, n-hexylphthalimide-corannulene (Cor-PI) and n-hexylnaphthalimide-corannulene (Cor-NI), each derivatized with substituents that enhance the electron-accepting abilities of the corannulene core (Fig. 8.13).[14b] On the experimental side, the *N*-hexyl derivative was a convenient choice to ensure sufficient solubility and film-forming properties of the compounds. The computational results show a quite large dihedral angle between the corannulene core and imide moieties of both Cor-PI and Cor-NI derivatives, due to the steric hindrance of the aromatic protons of the substituent and the corannulene core. However, this in fact can lead to improved miscibility with the P3HT donor material (Fig. 8.13).

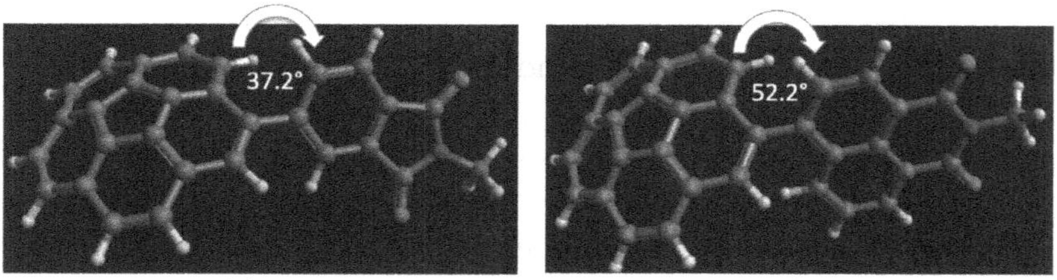

Figure 8.13. B97D/Def2-TVPP[45] optimized geometries of (a) Cor-PI and (b) Cor-Ni.

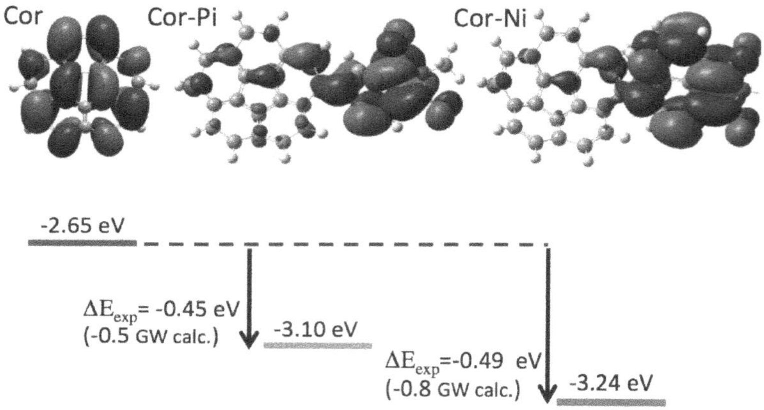

Figure 8.14. Depiction of calculated LUMO orbitals and orbital energy alignment results for Cor-PI and Cor-NI, with respect to Cor.

The electron-withdrawing substituents attached to the corannulene core serve to significantly lower the LUMO levels of these compounds compared to that of $C_{20}H_{10}$. This in turn leads to superior energy level alignment with the donor material in the ultimate device. As previously discussed, accurate and consistent predictions of energy level alignment values, such as IP and EA, necessitate the use of MBPT- and GW-based electronic structure approaches. Results as detected in experiments and predicted by theoretical calculations at the G_0W_0 level are shown together with LUMO depictions in Fig. 8.14.[14b] The computational analysis shows the LUMOs to be primarily localized on the imide moieties.

To achieve satisfactory device performance, the acceptor molecules in BHJ solar cells need to have appropriate energy level alignments at the DA interface and effective film forming properties. In the present case, the ΔLUMO at the Cor-NI(Cor-PI)/P3HT interface is overcoming the typical exciton binding energy in the P3HT donor material,[53] thereby fulfilling the energetic condition to initiate the separation process of electrons and holes at the DA interface. When blended with P3HT donors, the devices provided power conversion efficiencies (PCEs) of 0.19% for pristine corannulene. Introduction of electron-withdrawing groups to the corannulene core results in an increase in PCE to 0.32% for Cor-PI and to 1.03% for Cor-NI.[14b] The significant difference in device performance of the two systems is ascribed to variations in the crystallization properties. The crystal domain area in the Cor-NI/P3HT device construction is in fact much larger than in the Cor-PI/P3HT, indicative of a superior crystallization of the donor and acceptor materials in the composite and supportive of more efficient electron transport in the device.[14b]

These proposed corannulene derivatives are presented as the first applications of bowl-shaped fragments of fullerene as electron acceptors in BHJ solar cells. With PCE values over 1%, and the significant difference observed with simple changes in substituent, it is of great interest to further optimize the substituents for improved performance of the organic semiconductor. In this respect, one interesting modification of the corannulene motif involves attaching a symmetric distribution of withdrawing groups to the central core. Preliminary results of the effect of such a symmetric substitution on the electronic character have been investigated with G_0W_0 theory for the specific case of two n-hexylnaphthalimide substituents groups as shown in Fig. 8.15. From the orbital depiction, one observes that the charge density is primarily localized on the substituent-withdrawing groups with a small contribution on the central core. Importantly, one finds a further lowering of the LUMO energy, which is 0.2 eV (1.0 eV) lower than the monosubstituted molecule (clean molecule).

Further modifications of the corannulene core and the device fabrication are in now in progress.

(a) (b)

Figure 8.15. Depiction of LUMO orbitals for (a) Cor-NI and (b) Cor-di-NI.

8.4 Work function modification at the organic–metal interface

Understanding details of electronic transport across an interface between an active organic layer and a metallic surface is of considerable interest in the field of nanoscale electronics to enable better control in devices.[22,43] One important parameter governing electron transport at the metal–organic interface in organic device performance is the work function (WF).[54] In a bare metal surface, the inherent WF is a consequence of the intrinsic dipole caused by the spilling-out of electrons at the surface (Fig. 8.16). For weakly bound adsorbates (e.g., most organic molecules), an additional dipole opposite to that of the metal surface is induced, termed the interface dipole.[55]

Characterization of molecules on surfaces represents a major challenge for standard DFT approaches. This is in particular due to the inability of current XC functionals to adequately describe long-range interactions, which constitute an important component of the absorption process.[56] This inherent failing is even more severe when considering aromatic closed shell systems as adsorbates, where it becomes quite challenging to properly describe the large surface area of delocalized and polarizable electron densities of the π-system and possible charge rearrangement processes at the molecule–surface interface. In computational modeling of such a complex system, there is a particular sensitivity to methodology typically requiring advancements beyond standard DFT-based methods.

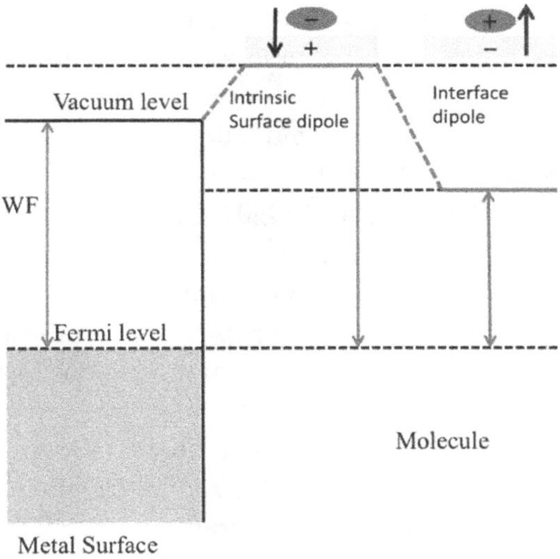

Figure 8.16. Schematic of surface WF (difference between vacuum level and Fermi level), dipole of a clean metal surface (intrinsic surface dipole), and induced dipole upon molecular adsorption (interface dipole).

Recent progress to include van der Waals interactions within standard DFT includes semi-empirical dispersion corrections (DFT-D)[45] and fully first-principles van der Waals density functionals (vdW-DF).[57] Among the most newly developed vdW-inclusive methods, further improvements have been achieved with approaches accounting for the nonlocal screening within the bulk and further refined vdW-DF.[56] Overall, those schemes have been successful in demonstrating that the additional attractive interaction is crucial for determination of adsorption geometries, energetics, and associated forces responsible for binding of molecules to a surface.[56,58]

To improve organic-device performance, low WF interfaces, appropriate for electron transport at the metal–organic interface, must be tailored. Recently, deposition of $C_{20}H_{10}$ and functionalized $C_{20}H_{10}$ onto copper surfaces has served to demonstrate the possibility for tuning the molecule–organic interface WF.[18] For example, experiments establish that the deposition of $C_{20}H_{10}$ and $C_{20}(CH_3)_5$ on a Cu(111) surface results in a significant decrease in surface WF of up to 1.4 eV, with corresponding interface dipole of 8–9 D, without inducing appreciable CT.[18] To understand the details of the adsorption process, electronic structure analysis at the metal–molecule interface was considered. The modeling of the adsorption process of $C_{20}H_{10}$ on the Cu(111) surface was carried out using a six-layer Cu(111) slab (~400 atoms) and $C_{20}H_{10}$ approaching the surface, as shown in Figs. 8.17(a) and 8.17(b).[59] In principle, corannulene can adsorb onto Cu(111) at several high symmetry sites of the copper surface, approaching the surface either in parallel, interacting with the five-membered ring, or in a tilted fashion, interacting with one six-membered ring. However, the molecule was observed by both experiment and theory to approach in a tilted bowl-up configuration, with one hexagonal ring sitting on top of a *fcc* hollow site.[18] A dispersion-enabled localized orbital DFT strategy (SIESTA),[60] DFT-D, was used to investigate the electronic structure and mechanistic details.[61]

Importantly, standard DFT without correction for dispersive interactions actually results in the corannulene detaching from the surface. The inclusion of van der Waals corrections has been shown to be a necessary component of the binding of the molecule onto the surface.[61] Concerning the adsorption process, it is interesting to look first at the component parts of the organic–metal interface construct.

(a) (b)

Figure 8.17. (a) $C_{20}H_{10}$ approaching a Cu(111) six-layer slab in a tilted bowl-up configuration; (b) $C_{20}H_{10}$ adsorption geometry with one hexagonal ring on top of an fcc hollow-site.

The organic component, $C_{20}H_{10}$, bowl shape has the majority of the electron density concentrated in the base of the bowl (manifesting its substantial dipole of 2.1 D), which is coming down onto the metal slab. The electronic structure of the dense copper surface is such that it spills out over the top of the slab of the metal atoms at the surface. As such, for the low coverage case of an isolated molecule coming onto the Cu(111) surface, this electron density spilling out of the surface is repelled down and sideways by the pushback of the electron density at the base of the $C_{20}H_{10}$ bowl approaching from above. This pushback effect in the substrate, together with a notable depletion of charge in the molecular framework, induces substantial charge separation and a remarkable interface dipole (calcd. 5.8 D; exptl. 6.4–8.8 D). The consequence is a significant decrease in the surface WF (calcd. $\Delta W = 1.37\,eV$; exptl. $1.4\,eV$).[18,61] Pauli repulsion hinders overlap of the electronic wave function of the molecule with that of the metal, giving rise to the characteristic deformation of the surface electronic charge density, also known as the "cushion effect"[55] as illustrated in Fig. 8.18.

Considering a more dense molecular packing requires accounting for depolarization effects from enhanced molecule–molecule dipolar interactions. This additional interaction affects the electronic

Figure 8.18. Depiction of charge density rearrangement of $C_{20}H_{10}$ approaching a Cu(111) surface. Blue: depletion of charge; red: accumulation of charge.

structure at the interface.[62] Increasing surface coverage from 0.25 monolayers (ML) up to 1 ML shows a rapid linear increase of WF (in absolute value) initially with increasing coverage, followed by saturation (Fig. 8.19), in agreement with experiment.[18] The total dipole value of the complex decreases with increasing coverage, ranging from 5.8 D at ~0.25 ML to 3.4 D at 1 ML (Fig. 8.19).

In these studies, at both low and full monolayer coverage, appropriate and carefully applied theory was able to clearly distinguish a physisorption phenomenon between the organic layer and metallic surface. The inclusion of van der Waals corrections in the methodology has been shown to be essential for representation of accurate binding of the molecule onto the surface.

Further studies have been carried out with functionalized corannulene derivatives as organic components in the organic–metal interface, in order to compare WF modification and CT, as well as issues of tiling onto a metallic surface. The systems investigated all revealed significant charge rearrangement at the interface with significant interface dipole and commensurate decrease in WF, without any CT (e.g., physisorbed). These investigations support and extend experimental efforts to shed light on mechanistic aspects of the adsorption process, and offer valuable new ideas toward interface design for electrodes in electronic devices based on organic compounds.

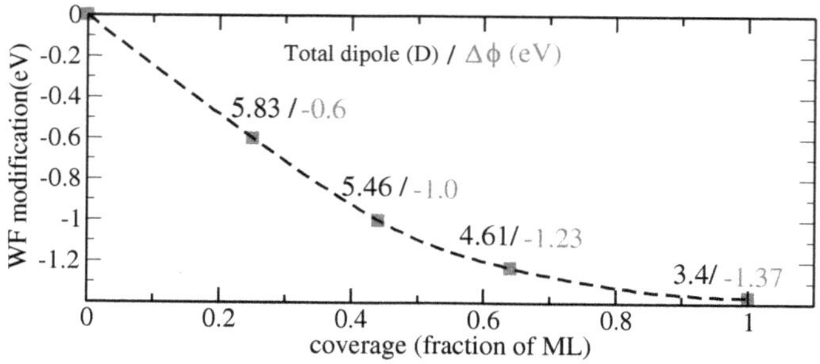

Figure 8.19. Work function decrease, dipole values vs. increasing coverage (ML) for $C_{20}H_{10}$ adsorbed on Cu(111).

8.5 Electronic transport through a molecular nanojunction

Fundamental device technology based on properties inherent in single organic molecules offers, in principle, unlimited possibilities for technological development, provided that the component molecule can take on diverse electronic functions with tuning through chemical design and synthesis.[16b] In this respect, a single molecule connected to macroscopic electrodes, i.e., molecular junctions, provide ideal systems to investigate charge transport on the nanoscale. Electronic transport phenomenon through single molecule junctions is a subject of intense current interest for both practical applications and for achievement of fundamental understanding of key physical phenomena that takes place in molecular scale charge transport.[10a,63]

Recently, it has been possible to manufacture a spatial gap in a single-walled carbon nanotube (SWCNT), and assemble an organic molecule component therein.[64] With this procedure, a single organic molecule can be covalently attached to SWCNT electrodes through stable and well-controlled amide linkages.[65] While such a system represents a significant improvement in the achievement of well-defined electronic contacts, there still remain factors that cannot be controlled in the experiments, causing the conductance value for any particular molecule to show considerable variance. In such a complex scenario, a detailed theoretical investigation of single molecules connected to macroscopic electrodes provides a powerful template for improving our knowledge of electronic transport phenomena in molecular junctions, which, in turn, should lead to advancements in experimental measurement capabilities as well as in the fundamental theoretical understanding of their construction.[66]

To facilitate this direction, a first-principles methodology based on DFT and electron transport techniques has been directed toward the investigation of charge transport phenomenon in molecular junctions. The ability to work with the tunable range of functional components based on the corannulene core enables the design of suitable junction motifs with potential for actual device technology. For this purpose, a set of junctions have been investigated that are designed with key functionalized corannulene units, f-$C_{20}H_{10}$, as the main active molecular elements. The functionalized corannulene is assembled between two CNT components that serve as the leads through an atomistic linking group (spacer) (Fig. 8.20(a)).

Various design factors can have dramatic impact on the electron transport properties through a molecular junction, i.e., optimal alignment of the orbital levels throughout the junction to enable the highest possible electron transport through the device.

For example, for the CNT−[spacer]−f-$C_{20}H_{10}$−[spacer]−CNT junction, the primary factors affecting the transport can be categorized in terms of: (i) the chemical composition of the linking-spacer; (ii) the chemical functionalization of the active element; and (iii) the possibility for chemical reactions occurring at the surface of the active

(a)

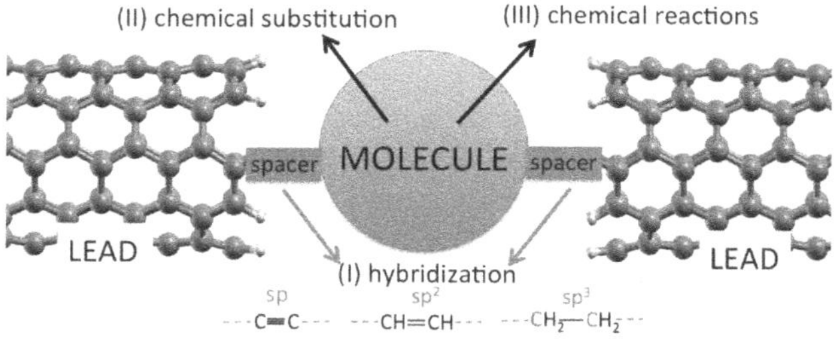

(b)

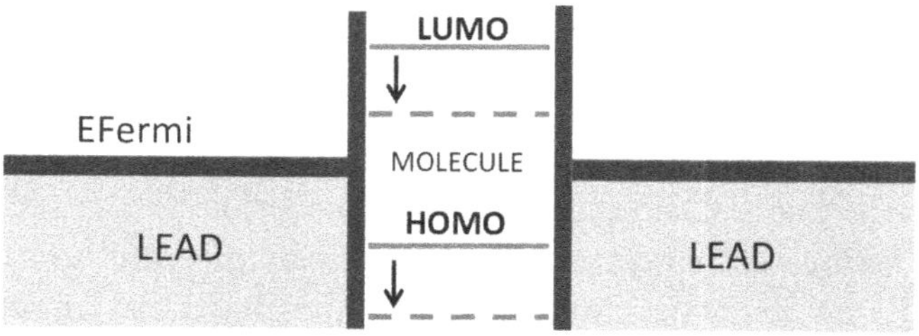

Figure 8.20. (a) Factors affecting transport properties through a molecular junction: (I) chemical composition of the linking-spacer, (II) chemical functionalization of the active element, (III) chemical reactions occurring at the surface of the active element. (b) Implications of the junction features on the alignment of the active element's HOMO and LUMO levels with respect to the Fermi level of the leads.

element (Fig. 8.20). The focus of this discussion will be on the chemical composition of the active molecular element, to illustrate the potential of molecular engineering for modulating electron transport through molecular junctions.

To investigate electron transport through a molecular junction, a standard single-particle Green's function method[67] is an optimal method of choice. This method combines the Landauer approach[68] with *ab initio* density functional theory. Within this approach, the quantum conductance, G, of the junction is related to the associated electronic scattering properties through the expression,

$$G = \frac{2e^2}{h} T(E_F) \tag{8.1}$$

where $T(E_F)$ is the transmittance at the Fermi energy E_F. The quantum transmittance can be simply expressed as the trace of a matrix product, using the Fisher–Lee expression,[69]

$$T = Tr(\Gamma_L G_C^r \Gamma_R G_C^a) \tag{8.2}$$

where G_C^r and G_C^a are the retarded and advanced lattice Green's functions of the conductor region, respectively, and $\Gamma_{L,R}$ represent the coupling between the conductor and the leads.[70] In the zero-bias and zero-temperature regime, the transmittance spectra is evaluated using the WanT software,[71] initialized from the ground-state electronic structure as computed by the Quantum ESPRESSO suite.[47]

For illustration of a prototypical CNT-[spacer]-molecule-[spacer]-CNT junction, the case of CNT-[1-ethynyl]-$C_{20}H_{10}$-[6-ethynyl]-CNT is illustrated here. Structurally, this composite structure is described as having a $C_{20}H_{10}$ active element linked to a (5,5)-CNT through carbon–carbon triple-bond linkers (Fig. 8.21).

To gauge the degree of interaction of the molecular orbitals of the active element inside the junction, a comparison of the 1,6-diethynylcorannulene projected density of states (PDOS) in the CNT-[1-ethynyl]-$C_{20}H_{10}$-[6-ethynyl]-CNT junction, with the density of states (DOS) of the isolated active element, 1,6-diethynylcorannulene, is considered (Fig. 8.22(a)). Here, the PDOS

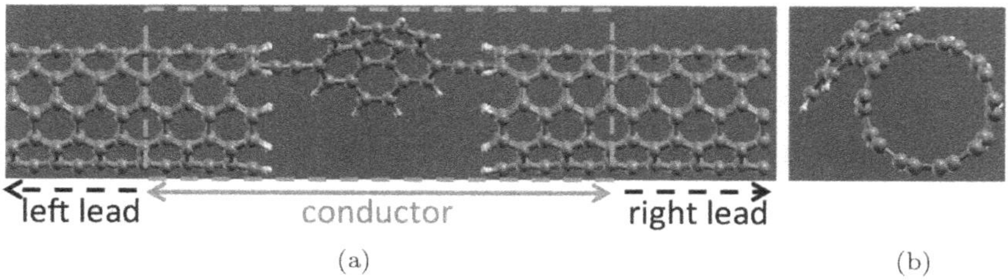

Figure 8.21. (a) Prototype junction geometry consisting of a 1,6-diethynylcorannulene linked to (5,5) CNT leads in the relaxed configuration: the conductor region includes the molecule and a portion of the leads, the lead regions are highlighted. (b) Side view of the relaxed configuration.

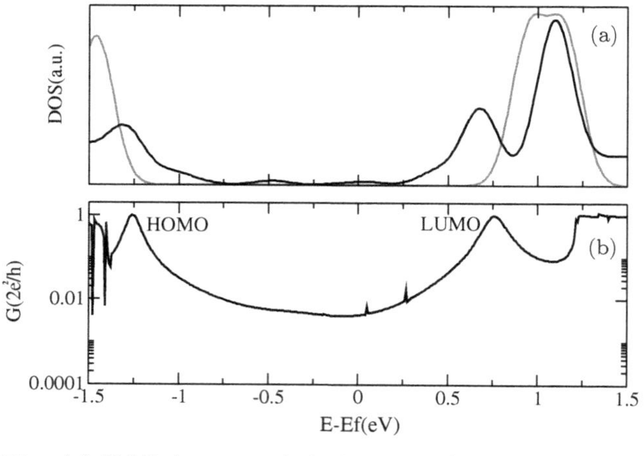

Figure 8.22. (a) DOS (curve with higher peaks) of the gas-phase molecule and PDOS of the molecule inside the junction (b) Quantum transmittance in the junction (in units of G0, the quantum of conductance).

of the active element in the junction (black) shows broader structures for HOMO and LUMO with respect to the sharp molecular states found for the isolated molecule (curve with higher peaks), with the molecular LUMO displaying a multipeak structure. The left-most peak of the LUMO structure in the PDOS is closer to the Fermi energy by ~0.4 eV than was the pristine isolated molecular LUMO. The relative position of the frontier molecular orbitals and the Fermi level indicates that the transport would be primarily mediated through the LUMO orbital in this junction.

Having a set of functional molecular units with tunable HOMO and LUMO levels facilitates the tailoring of the alignment of the orbital levels with the Fermi energy of the leads. In this way, the set of functionalized corannulene active elements is highly suitable for the purpose of developing "chemically tuned" molecular junctions. The corresponding tuning of the electronic conductance should be consistent and dependent on the magnitude and direction of the HOMO/LUMO shifts with respect to the Fermi energy.[72] In particular, the series of functionalized $C_{20}H_{10}$ systems illustrated in Fig. 8.23 represented an interesting set of active elements to demonstrate a relatively large and consistent lowering of HOMO (LUMO) energies, dependent on the number and type of substituents on the corannulene core. The single molecule

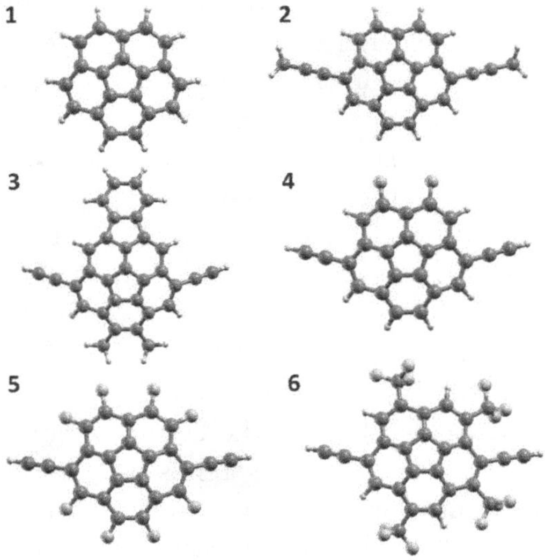

Figure 8.23. Molecules linked within a (5,5)-SWCNT in the molecular nanojunction systems: (**1**) corannulene; (**2**) 1,6-bispropynylcorannulene; (**3**) 2,7-diethynyl-4,5-dimethylindenocorannulene; (**4**) 5,10-diethynyl-2,3- difluorocorannulene; (**5**) 5,10-diethynyl-1,2,3,4,6,7,8,9-octafluorocorannulene; and (**6**) 5,10-diethynyl-2,4,6,8-tetrakis(trifluoromethyl)-corannulene.

Table 8.3. PBEPBE/Def2-TZVPP (PBE plane-wave) calculated molecular dipole moment (D), HOMO/LUMO energy (eV), and bowl-depth (Å).

Molecule	Dipole (Debye)	HOMO (eV)	LUMO (eV)	Bowl depth (Å)
1	2.09	−5.61 (−5.64)	−2.58 (−2.61)	0.90
2	2.63	−5.16 (−5.18)	−2.75 (−2.77)	0.89
3	2.93	−5.13 (−5.26)	−2.89 (−3.00)	1.08
4	2.02	−5.70 (−5.75)	−3.25 (−3.30)	0.88
5	0.39	−6.08 (−6.18)	−3.72 (−3.83)	0.80
6	2.10	−6.26 (−6.40)	−3.95 (−4.00)	0.76

PBEPBE/Def2-TZVPP (PBE plane-wave) calculated molecular dipole moment (D), HOMO/LUMO energy (eV), and bowl-depth (Å) results for this set are shown in Table 8.3.

As the actual alignment of the frontier orbitals of the molecule when assembled in a molecular junction strongly affects the charge transport properties through the complex, it is of interest to investigate the effect of the chemical modification on the molecular level alignment with respect to the Fermi energy of the complex in the molecular junction systems. Placement of the functionalized corannulene active elements into the nanojunction, using identical sp carbon–carbon triple-bond spacer linkers to the (5,5)-CNT leads, establishes the nanojunction systems. Four key nanojunction systems are considered for illustration which, for simplicity of notation, are labeled according to the main active molecule element introduced above: (**3**), (**4**), (**5**), and (**6**). To gauge how the tunability of the energy offset between HOMO (LUMO) and the Fermi energy due to chemical functionalization is reflected in the trends in quantum conductance through the junctions, a comparison of quantum conductance plots across the series with that of the pristine junction (**2**) (Fig. 8.24) is made.

In all four molecular nanojunctions considered, the two main peaks in the quantum conductance, one above and one below the Fermi level, can be ascribed to the HOMO and LUMO molecular

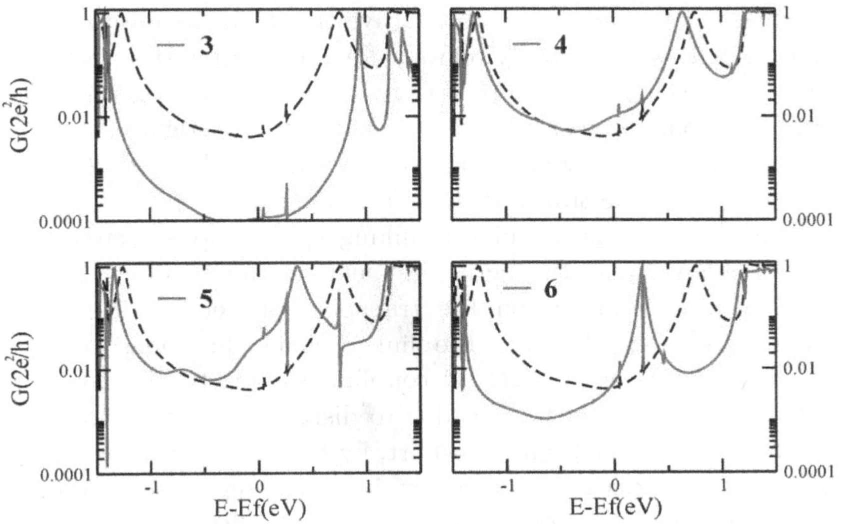

Figure 8.24. Quantum conductance for the functionalized molecular units in the assembled molecular nanojunctions.

levels, in agreement with the PDOS analysis (not shown here). Most importantly, for nanojunction systems (**5**) and (**6**), the shift of the LUMO-related peak relative to the parent system (**2**) corresponds to a significant modulation in the quantum conductance (∼ two orders of magnitude), at 0.36 eV and 0.26 eV, respectively. Such tunability is of significant importance for the design of molecular junctions because it shows the potential of molecular engineering as an efficient tool for adjusting interfacial electronic structure to modulate the intrinsic transport properties.

The presented results stress that the actual positions of the frontier orbitals of the molecule in the junction are the ultimate determinants of the conductance, which in principle differs from the corresponding frontier orbitals of isolated molecules in the gas phase. Theoretical predictions of HOMO/LUMO levels of the isolated gas-phase molecules and associated PDOS of the molecules in the junctions presented here are computed within a DFT-PBE framework. While it is well known that DFT-based methodology using standard local or semi-local density functionals may not provide quantitative prediction of IP and EA,[73] the relative trends in the data are more reliable. This is particularly relevant for the investigations presented here, where a common LEAD (armchair CNT) is used together with a set of functionalized corannulene constructs to investigate the impact of the chemical functionalization of the active molecular element in the junction on the transport phenomenon.

The chemical nature of the linking spacer ($sp/sp^2/sp^3$) bridging the LEAD and the active element of the junction also has an important influence on the transport properties through the junction complex. In fact, the linker choice functions to either (a) inhibit transport, largely decoupling the electronic interaction between molecule and LEADs due to disruption of π-electron delocalization; or (b) facilitate transport, by facilitating π delocalization through the junction. In particular, in the present case, a dramatic reduction in conductance is observed with modifications from sp/sp^2 to sp^3 hybridization in the linker spacers due to loss of conjugation with the completely saturated linker (sp^3).[16b] In this way, the linker design can greatly impact the quantum conductance, to the extent of enabling an on/off capability.

Several theoretical investigations examining prototype CNT–molecule–CNT systems have been reported, where the junction is further modified in order to account for specific experimental conditions.[64] These studies reveal that, in certain cases, in addition to rigid, high-transmission junctions, a multitude of chemical factors (binding chemistries, geometries, LEAD details) can have a strong impact on the transport properties. For example, in the model CNT–amide–(1,4-diisocyanatobenzene)–amide–CNT junction, transport studies have shown that CNT–molecule interface chemistry, interface geometry, CNT chirality, and molecular conformation are all key areas of focus.[74] However, modifications of such geometrical parameters, while changing resonance widths and coupling strengths, are still expected to preserve the molecular origin of the main conductance peaks. In fact, the transmission spectra still relate to the features of the isolated molecules, independent of geometrical details, preserving trends and transport properties.

In situations where more quantitative results are important in order to gain insight into a particular transport process, it is highly desirable to have accurate information about the molecular-level alignment with the electrode Fermi energy in the molecular junction. Although it is still rather challenging to show such a fundamental effect in a systematic manner, recent developments combining DFT-based schemes with GW-derived model corrections have been established (see Sec. 8.6.3).

8.6 Practical considerations

There has been considerable effort in the past decade toward the achievement of computationally affordable theoretical methods for accurate prediction of structure and properties of materials. Theoretical predictions of solids began decades ago, but only recently have solid-state quantum techniques become sufficiently reliable to be considered routine for investigation of solids. Reliable and computationally efficient *ab initio* predictive theories for solids that are able to provide atomic-scale insights into properties of bulk materials, interfaces, and nanostructures, are highly sought after. Unfortunately, no single theory exists that provides reliable predictive capabilities across a wide range of materials complexity, from single molecule to molecular crystals, to molecular adsorbates on metal surfaces, and nanojunction constructs. Instead, interdisciplinary efforts are being brought to bear to create a richly textured and substantive portfolio of methods, which promise quantitative predictions of materials and device properties as well as associated performance analysis.

As discussed in this chapter and as observed in the literature, even with methodology available, application of such techniques toward predictive analysis is not an overnight process. For many of these applications, considerable time, effort, and care are required to provide results that are at a level sufficient to be predictive and useful. Practically speaking, it is better not to apply any theory at all when adequate theory and associated resources are not available, as inadequate theory applied to such complicated problems is seen to be far from predictive and even misleading.

In many of the materials types considered here, accurate prediction of the properties associated with the supramolecular ordering of building blocks requires consideration of dispersive effects, electronic correlation, and quantum confinement. As such, first-principles approaches are called for, and, due to the complexity of effects and typical building blocks involved, this typically implies methods based on DFT. Given a carefully chosen and richly textured toolset of appropriate methodology, similar to the rich expanse of experimental techniques and instrumentation, the result is a suite of tools that provide multiple windows on phenomena that ultimately provide a more complete analysis in the end.

In this chapter, results of theoretical calculations, taken together with the comparison with experimental findings, have been applied to quite complicated materials device construction involving a rich class of curved aromatics. Despite the relatively large efforts expended on theoretical developments, there still remain many approximations that limit their impact in one way or another. Clearly, much room exists for improvements in computational methodology in terms of breakthroughs to reduce the extent of approximations made, as well as in algorithmic efficiency to bring out the capabilities of some of the methodology that are prohibitively out of reach of application. In this last section, we highlight the most recent methodological developments toward the calculation of excitation properties, electron-transport properties, and molecular adsorption processes on metal surfaces, as utilized throughout the examples described in this chapter, with a special emphasis here directed at practical considerations.

8.6.1 *Excitation properties prediction*

Within the GW-BSE framework, the heaviest part of the calculation is the evaluation of the self-energy operator, which describes the many-body effects and the polarizability of the system. In general, the efficiency of a GW calculation is strictly dependent on the number of unoccupied states that needs to be explicitly included in the polarizability and the electronic self-energy evaluation. These operations may become prohibitively expensive from the computational point of view, limiting the applicability of the method to low-dimensional systems, such as molecules, CNTs, graphene, and molecule–surface interfaces. It is therefore highly desirable to develop algorithms that are scalable to large systems without resorting to approximations. More recently, speed-up algorithms that are a function of the number of empty states have been developed and implemented in open-source (plane waves) computer codes, applicable toward the electronic structure and excitation properties of materials.[75] It is strongly recommended to take advantage of such algorithms, which enable the investigation of the excitation properties in systems of realistic size, such as organic crystals and donor–acceptor interfaces, while keeping the computational workload at an acceptable level. When considering only the case of single molecules in the gas phase, it makes more sense to use localized basis sets codes, which people are starting to now adapt to including the GW-BSE theory.[76] These codes are much less computationally expensive than are plane-wave–based codes.

8.6.2 *Electron-transport properties*

As has been shown,[77] the primary source of errors in the theoretical prediction of quantum transport in nanojunctions originates from the use of local or semi-local XC potentials within the DFT formalism. Most of the failures of the theory in this respect can be traced back to the self-interaction (SI) problem,[78] which causes the eigenstates of the molecule in the junction to be pushed to higher energies and gives rise to incorrect peak positions in the conductance spectra. A significant improvement in this direction would be to describe the electronic structure of the system according to the GW approach, and combining such a description with a Landauer-like approach for the electronic transport. However, schemes going beyond a DFT description can be highly computationally demanding for simulations involving realistic junction geometries, and are absolutely not recommended. Instead, an alternative framework is recommended that consists of a hybrid scheme. In such an approach, a hybrid functional (i.e., optimally tuned range-separated hybrid functional [OT-RSH][79]) would be employed for proper description of the molecular outer valence states,[79] together with a model GW-based self-energy correction (DFT+Σ^{80}). Within such a hybrid scheme, a GW-based self-energy correction method is applied for the determination of the eigenvalues of the combined lead–molecule–lead junction system, which takes into account the (i) inaccuracy of gas-phase frontier orbital energies derived from LDA as well as the (ii) nonlocal static correlations coming from surface polarization. Such a hybrid scheme has been shown to be quite accurate for a broad array of molecules with different linkers and, at the same time, represents a computationally less expensive method than a real GW scheme.[80]

8.6.3 *Molecular adsorption on metal surfaces*

Electronic structure calculations generally depend on the quality of the initial geometry of the molecule–metal substrate and the quality of the DFT functional. As illustrated in Sec. 8.1.4, schemes have recently been developed that include van der Waals corrections into DFT approaches. Such approaches have been shown to be quite reliable for description of adsorption geometries and energetics, providing results in good agreement with experiments[56] and with a reasonable computational cost. Having said that, a key quantity at the organic–metal interface is the energy level alignment of molecular electronic states with the metallic Fermi level. However, it is well known that standard DFT completely fails for the description of molecular energy level renormalization upon adsorption on a metal substrate,[43] even though prediction of substrate work functions and adsorbate-induced dipole effects can be described fairly accurately (Sec. 8.4). However, calculations of the combined molecule–metal interface system remain scarce within present-day GW due to the high computational workload. As such, a common approach is the customization of computational DFT-based methodologies with IP models to account for polarization effects.[43,81] These models usually describe a classic picture, where an electron on the surface of a polarizable material is attracted to the surface due to the interaction with an oppositely charged polarization distribution; the net result is equivalent to a classic Coulomb interaction between the electron and a fictitious positive image charge located below the surface (hence, the name of the IP state).[43,81] A recent refinement of this approach,[22] recommended for description of molecules physisorbed on a metal surface, involves the use of a DFT+Σ scheme, with an OT-RSH description of the molecule and a nonclassic image-charge model for the metal. This restores the asymptotically correct XC potential crucial for accurate energy level alignment prediction.[22] Importantly, the computational cost of this approach is small compared to the underlying DFT calculation of the complete molecule–metal interface, highlighting this scheme as a reliable and efficient theoretical toolset for calculating energy level alignment.

References

1. (a) Uoyama, H., Goushi, K., Shizu, K., Nomura, H. and Adachi, C. (2012). Highly efficient organic light-emitting diodes from delayed fluorescence, *Nature*, 492, pp. 234–240; (b) Congreve, D. l. N., Lee, J., N. J. Thompson, Hontz, E., Yost, S. R., Reusswig, P. D., Bahlke, M. E., Reineke, S., Voorhis, T. V. and Baldo, M. A. (2013). External quantum efficiency above 100% in a singlet-exciton-fission–based organic photovoltaic cell, *Science*, 340, pp. 334–337.
2. Forrest, S. R., (2004). The path to ubiquitous and low-cost organic electronic appliances on plastic, *Nature*, 428, pp. 911–918.
3. (a) Burroughes, J. H., Bradley, D. D. C., Brown, A. R., Marks, R. N., Mackey, K., Friend, R. H., Burns, P. L. and Holmes, A. B. (1990). Light-emitting diodes based on conjugated polymers, *Nature*, 347, pp. 539–541; (b) Zoppi, L., Martin-Samos L. and Baldridge, K. K. (2011). Effect of molecular packing on corannulene-based materials electroluminescence, *J. Am. Chem. Soc.*, 133, pp. 14002–14009; (c) Friend, R. H., Gymer, R. W., Holmes, A. B., Burroughes, J. H., Marks, R. N., Taliani, C., Bradley, D. D. C., Santos, D. A. D., Brédas J. L. and Lögdlund, M. (1999). Electroluminescence in conjugated polymers, *Nature*, 397, pp. 121–128.
4. (a) Günes, S., Neugebauer, H. and Sariciftci, N. S. (2007). Conjugated polymer-based organic solar cells, *Chem. Rev.*, 107, pp. 1324–1338; (b) Scharber, M. C., Wuhlbacher, D., Koppe, M., Denk, P., Waldauf, C., Heeger, A. J. and Brabec, C. L. (2006). Design rules for donors in bulk-heterojunction solar cells - towards 10% energy-conversion efficiency, *Adv. Mater.*, 18, pp. 789–794; (c) Deibel, C. and Dyakonov, V. (2010). Polymer–fullerene bulk heterojunction solar cells, *Rep. Prog. Phys.*, 73, pp. 096401-1–096401-139.

5. Sirringhaus, H. (2014). 25th anniversary article: Organic field-effect transistors: The path beyond amorphous silicon, *Adv. Mater.*, 26, pp. 1319–1335.
6. Filatov, A. S., Scott, L. T. and Petrukhina, M. A. (2010). π–π interactions and solid state packing trends of polycyclic aromatic bowls in the indenocorannulene family: Predicting potentially useful bulk properties, *Cryst. Grow. Des.*, 10, pp. 4607–4621.
7. Zoppi, L., Siegel, J. S. and Baldridge, K. K. (2013). Electron transport and optical properties of curved aromatics, *WIREs Comput. Mol. Sci.*, 3, pp. 1–12.
8. Scholes, G. D. and Rumbles, G. (2006). Excitons in nanoscale systems, *Nat. Mater.*, pp. 683–696.
9. Halls, J. J. M., Walsh, C. A., Greenham, N. C., Marseglia, C. A., Friend, R. H., Moratti, S. C. and Holmes, A. B. (1995). Efficient photodiodes from interpenetrating polymer networks, *Nature*, 376, pp. 498–500.
10. (a) Song, H., Reed, M. A. and Lee., T. (2011). Single molecule electronic devices, *Adv. Mater.*, 23, pp. 1583–1608; (b) Sharifzadeh, S., A. Biller, Kronik, L. and Neaton, J. B. (2012). Quasiparticle and optical spectroscopy of the organic semiconductors pentacene and PTCDA from first principles, *Phys. Rev. B*, 85, pp. 125307-1–125307-11; (c) Sharifzadeh, S., Darancet, P., Kronik, L. and Neaton, J. B. (2013). Low-energy charge-transfer excitons in organic solids from first-principles: The case of pentacene, *J. Phys. Chem. Lett.*, 4, pp. 2197–2201.
11. Aradhya, S. V. and Venkataraman, L. (2013). Single-molecule junctions beyond electronic transport, *Nat. Nanotech.*, 8, pp. 399–410.
12. Chaudhuri, D., Wettach, H., Schooten, K. J. v., Liu, S., Sigmund, E., Höger, S. and Lupton, J. M. (2010). Tuning the singlet–triplet gap in metal-free phosphorescent p-conjugated polymers, *Angew. Chem. Int. Ed.*, 49, pp. 7714–7717.
13. Borchardt, l., Fuchicello, A., Kilway, K. V., Baldridge, K. K. and Siegel, J. S. (1992). Synthesis and dynamics of the corannulene nucleus, *J. Am. Chem. Soc.*, 114, pp. 1921–1923.
14. (a) Zoppi, L., Martin-Samos, L. and Baldridge, K. K. (2014). Structure–property relationships of curved aromatic materials from first principles, *Acc. Chem. Res.*, 47, pp. 3310–3320; (b) Lu, R.-Q., Zheng, Y.-Q., Zhou, Y.-N., Yan, X.-Y., Lei, T., Shi, K., Zhou, Y., Pei, J., Zoppi, L., Baldridge, K. K., Siegel, J. S. and Cao, X.-Y. (2014). Corannulene derivatives as non-fullerene acceptors in solution-processed bulk heterojunction solar cells, *J. Mater. Chem. A*, 2, pp. 20515–20519; (c) Bauert, T., Zoppi, L., Koller, G., Siegel, J. S., Baldridge, K. K. and Ernst, K.-H. (2013). Quadruple anionic buckybowls by solid-state chemistry of corannulene and cesium, *J. Am. Chem. Soc.*, 135, pp. 12857–12860.
15. Lovas, F. J., McMahon, R. J., Grabow, J.-U., Schnell, M., Mack, J., Scott, L. T. and Kuczkowski, R. L. (2005). Interstellar chemistry: A strategy for detecting polycyclic aromatic hydrocarbons in space, *J. Am. Chem. Soc.*, 127, pp. 4345–4349.

16. (a) Wu, Y.-T., Bandera, D., Maag, R., Linden, A., Baldridge, K. K. and Siegel, J. S. (2008). Multi-ethynyl corannulenes: Synthesis, structure and properties, *J. Am. Chem. Soc.*, 130, pp. 10729–10739; (b) Zoppi, L., Ferretti, A. and Baldridge, K. K. (2015). Tuning electron transport through functionalized C20H10 molecular junctions, *J. Chem. Theory Comput.*, 11, pp. 4900–4910.
17. Mack, J., Vogel, P., Jones, D., Kaval, N. and Sutton, A. (2007). The development of corannulene-based blue emitters, *Org. Biomol. Chem.*, 5, pp. 2448–2452.
18. Bauert, T., Zoppi, L., Koller, G., Garcia, A., Baldridge, K. K. and Ernst, K.-H. (2011). Large interface dipole moments without charge transfer: Buckybowls on metal surfaces, *J. Phys. Chem. Lett.*, 2, pp. 2805–2809.
19. Zoppi, L., Ferretti, A. and Baldridge, K. K. (2013). Static and field-oriented properties of bowl-shaped polynuclear aromatic hydrocarbon fragments, *J. Chem. Theory Comput.*, 9, pp. 4797–4804.
20. Zoppi, L., Martin-Samos, L. and Baldridge, K. K. (2105). Buckybowl superatom states: A unique route for electron transport? *Phys. Chem. Chem. Phys.*, 17, pp. 6114–6121.
21. Lu, R.-Q., Zhou, Y.-N., Yan, X.-Y., Shi, K., Zheng, Y.-Q., Luo, M., Wang, X.-C., Pei, J., Xia, H., Zoppi, L., Baldridge, K. K., Siegeld, J. S. and Cao, X.-Y. (2015). Thiophene-fused bowl-shaped polycyclic aromatics with a dibenzo[a,g]corannulene core for organic field-effect transistors, *Chem. Commun.*, 51, pp. 1681–1684.
22. Egger, D. A., Liu, Z.-F., Neaton, J. B. and Kronik, L. (2015). Reliable energy level alignment at physisorbed molecule-metal interfaces from density functional theory, *Nano Lett.*, 15, pp. 2448–2455.
23. Kümmel, S. and Kronik, L. (2008). Orbital-dependent density functionals: Theory and applications, *Rev. Mod. Phys.*, 80, pp. 3–60.
24. Kaduk, B., Kowalczyk, T. and Voorhis, T. V. (2012). Constrained density functional theory, *Chem. Rev.*, 112, pp. 321–370.
25. (a) Difley, S. and Voorhis, T. V. (2011). Exciton/charge-transfer electronic couplings in organic semiconductors, *J. Chem. Theory Comput.*, 7, pp. 594–601; (b) Yost, S. R., Wang, L. P. and Voorhis, T. V. (2011). Molecular insight into the energy levels at the organic donor/acceptor interface: A quantum mechanics/molecular mechanics study, *J. Phys. Chem. C*, 115, pp. 14431–14436.
26. (a) Onida, G., Reining, L. and Rubio, A. (2002). Electronic excitations: Density-functional versus many-body Green's-function approaches, *Rev. Mod. Phys.*, 74, pp. 601–659; (b) Marques, M. A. L., Ulrich, C. A., Nogueira, F., Rubio, A., Burke, K. and Gross, E. K. U. (2006). *Time Dependent Density Functional Theory.* Lecture Notes in Physics (Berlin: Springer 2006); (c) Casida, M. E. (2009). Time-dependent density-functional theory for molecules and molecular solids, *J. Mol. Struct. THEOCHEM*, 914, pp. 3–18.

27. Albrecht, S., Reining, L., Sole, R. D. and Onida, G. (1998). Ab initio calculation of excitonic effects in the optical spectra of semiconductors, *Phys. Rev. Lett.*, 80, pp. 4510–4515.
28. Hummer, K. and Ambrosch-Draxl, C. (2005). Oligoacene exciton binding energies: Their dependence on molecular size, *Phys. Rev. B*, 71, pp. 081202(R)-1–081202(R)-4.
29. Bruno, M., Palummo, M., Marini, A., Sole, R. D., Olevano, V., Kholod, A. N. and Ossicini, S. (2005). Excitons in germanium nanowires: Quantum confinement, orientation, and anisotropy effects within a first-principles approach, *Phys. Rev. B,* 72, 153310-1–153310-4.
30. Fetter, A. L. and Walecka, J. D., *Quantum Theory of Many-Particle Systems* (McGraw-Hill, New York, 1971).
31. (a) Rohlfing, M. and Louie, S. G. (1999). Optical excitations in conjugated polymers, *Phys. Rev. Lett.*, 82, pp. 1959–1962; (b) Tiago, L., Chelikowsky, J. R. (2005). First-principles GW–BSE excitations in organic molecules, *Sol. State Commun.*, 136, pp. 333–337; (c) Ferretti, A., Mallia, G., Martin-Samos, L., Bussi, G. i., Ruini, A., Montanari, B. and Harrison, N. M. (2012). Ab initio complex band structure of conjugated polymers: Effects of hybrid density functional theory and GW schemes, *Phys. Rev. B*, 85, pp. 235105-1–235105-15.
32. (a) Duchemin, I. and Blase, X. (2013). Resonant hot charge-transfer excitations in fullerene-porphyrin complexes: Many-body Bethe-Salpeter study, *Phys. Rev. B*, 87, pp. 245412-1–245412-10; (b) Baumeier, B., Andrienko, D. and Rohlfing, M. (2012). Frenkel and charge-transfer excitations in donor–acceptor complexes from many-body Green's functions theory, *J. Chem. Theory Comput.*, 8, pp. 2790–2795.
33. Nardes, A. M., Ferguson, A. J., Wolfer, P., Gui, K., Burn, P. L., Meredith, P. and Kopidakis, N. (2014). Free carrier generation in organic photovoltaic bulk heterojunctions of conjugated polymers with molecular acceptors: Planar versus spherical acceptors, *ChemPhysChem*, 15, pp. 1539–1549.
34. Feng, M., Zhao, J. and Petek, H. (2008). Atomlike, hollow-core-bound molecular orbitals of C60, *Science*, 320, pp. 359–362.
35. Feng, M., Zhao, J., Huang, T., Zhu, X. and Petek, H. (2011). The electronic properties of superatom states of hollow molecules, *Acc. Chem. Res.*, 44, pp. 360–368.
36. Silkin, V. M., Zhao, J., Guinea, F., Chulkov, E. V., Echenique, P. M. and Petek, H. (2009). Image potential states in graphene, *Phys. Rev. B*, 80, pp. 121408-1–121408-4.
37. Hohenberg, P. and Kohn, W. (1964). Inhomogeneous electron gas, *Phys. Rev.*, 136, pp. B864–B871.
38. Kohn, W. and Sham, L. (1965). Self-consistent equations including exchange and correlation effects, *Phys. Rev.*, 140, pp. A1133–A1138.
39. (a) Hedin, L. (1965). New method for calculating the one-particle Green's function with application to the electron-gas problem, *Phys. Rev.*, 139, pp. A796–A823; (b) Hedin, L. and Lundqvist, S. (1969). Effects of

electron-electron and electron-phonon interactions on the one-electron states in solids. In *Solid State Physics: Advances in Research and Applications*, eds. H. Ehrenreich, F. Seitz and D. Turnbull, Vol. 23 (Academic Press, New York).
40. Marom, N., F. Caruso, Ren, X., Hofmann, O. T., Korzdorfer, T., Chelikowsky, J. R., Rubio, A., Scheffler, M. and Rinke P. (2012). Benchmark of GW methods for azabenzenes, *Phys. Rev. B.*, 86, pp. 1–16.
41. Körzdörfer, T. and Marom, N. (2012). Strategy for finding a reliable starting point for G0W0 demonstrated for molecules, *Phys. Rev. B*, 86, pp. 041110-1–041110-5.
42. Koval, P., Foerster, D. and Sanchez-Portal, D. (2014). Fully self-consistent GW and quasiparticle self-consistent GW for molecules, *Phys. Rev. B*, 89, pp. 155417-1–155417-19.
43. Neaton, J. B., Hybertsen, M. S. and Louie, S. G. (2006). Renormalization of molecular electronic levels at metal-molecule interfaces, *Phys. Rev. Lett.*, 97, pp. 216405-1–216405-4.
44. Kaasbjerg, K. and Thygesen, K. S. (2010). Benchmarking GW against exact diagonalization for semiempirical models, *Phys. Rev. B*, 81, pp. 085102-1–085102-8.
45. Grimme, S. J. (2006). Semiempirical GGA-type density functional constructed with a long-range dispersion correction, *J. Comput. Chem.*, 27, pp. 1787–1799.
46. Schmidt, M. W., Baldridge, K. K., Boatz, J. A., Elbert, S. T., Gordon, M. S., Jensen, J. H., Koseki, S., Matsunaga, M., Nguyen, K. A., Su, S., Windus, T. L. and Elbert, S. T. (1993). General atomic and molecular electronic structure system, *J. Comp. Chem.*, 14, pp. 1347–1363.
47. Giannozzi, P., Baroni, S., Bonini, N., Calandra, M., Car, R., Cavazzoni, C., Ceresoli, D., Chiarotti, G. L., Cococcioni, M., Dabo, I., Corso, A. D., Fabris, S., Fratesi, G., Gironcoli, S. D., Gebauer, R., Gerstmann, U., Gougoussis, C., Kokalj, A., Lazzeri, M., Martin-Samos, L., Marzari, N., Mauri, F., Mazzarello, R., Paolini, S., Pasquarello, A., Paulatto, L., Sbraccia, C., Scandolo, S., Sclauzero, G., Seitsonen, A. P., Smogunov, A., Umari, P. and Wentzcovitch, R. M. (2009). QUANTUM ESPRESSO: A modular and open-source software project for quantum simulations of materials, *J. Phys. Condens. Matter*, 21, pp. 395502-1–395502-19.
48. Martin-Samos, L. and Bussi, G. (2009). SaX: An open source package for electronic-structure and optical-properties calculations in the GW approximation, *Comp. Phys. Comm.*, 180, pp. 1416–1425.
49. Ernzerhof, M. and Scuseria, G. E. (1999). Assessment of the Perdew–Burke–Ernzerhof exchange-correlation functional, *J. Chem. Phys.*, 110, pp. 5029–5035.
50. Perdew, J. P. and Zunger, A. (1981). Self-interaction correction to density functional approximations for many-electron systems, *Phys. Rev. B*, 23, pp. 5048–5079.
51. Yang, Q., Muntwiler, M. and Zhu, X.-Y. (2009). Charge transfer excitons and image potential states on organic semiconductor surfaces, *Phys. Rev. B*, 80, pp. 115214-1–115214-8.

52. Barford, W. (2004). Theory of singlet exciton yield in light-emitting polymers, *Phys. Rev. B*, 70, pp. 205204-1–205204-8.
53. Deibel, C., Mack, D., Gorenflot, J., Scholl, A., Krause, S., Reinert, F., Rauh, D. and Dyakonov, V. (2010). Energetics of excited states in the conjugated polymer poly(3-hexylthiophene), *Phys. Rev. B*, 81, pp. 085202-1–085202-5.
54. Cahen, D. and Kahn, A. (2003). Electron energetics at surfaces and interfaces: Concepts and experiments, *Adv. Mater.*, 15, pp. 271–277.
55. Witte, G., Lukas, S., Bagus, P. S. and Wöll, C. (2005). Vacuum level alignment at organic/metal junctions: 'Cushion' effect and the interface dipole, *Appl. Phys. Lett.*, 87, pp. 263502-1–263502-3.
56. Carrasco, J., Liu, W., Michaelides, A. and Tkatchenko, A. (2014). Insight into the description of van der Waals forces for benzene adsorption on transition metal (111) surfaces, *J. Chem. Phys.*, 140, pp. 084704-1–084704-10.
57. Dion, M., Rydberg, H., Schröder, E., Langreth, D. C. and Lundqvist, B. I. (2004). Van der Waals density functional for general geometries, *Phys. Rev. Lett.*, 92, pp. 246401-1–246401-4.
58. (a) McNellis, E. E., Meyer, J. and Reuter, K. (2009). Azobenzene at coinage metal surfaces: Role of dispersive van der Waals interactions, *Phys. Rev. B*, 80, pp. 205414-1–205414-10; (b) Atodiresei, N., Caciuc, V., Lazić, P. and Blügel, S. (2009). Chemical versus van der Waals Interaction: The Role of the heteroatom in the flat absorption of aromatic molecules C6H6, C5NH5, and C4N2H4 on the Cu(110) surface, *Phys. Rev. Lett.*, 102, pp. 136809-1–136809-4; (c) Li, G., Tamblyn, I., Cooper, V. R., Gao, H.-J. and Neaton, J. B. (2012). Molecular adsorption on metal surfaces with van der Waals density functionals, *Phys. Rev. B*, 85, pp. 121409(R)-1–121409(R)-4; (d) Shi, X.-Q., Li, Y., Hove, M. A. V. and Zhang, R.-Q. (2012). Interactions between organics and metal surfaces in the intermediate regime between physisorption and chemisorption, *J. Phys. Chem. C*, 116, pp. 23603–23607.
59. Merz, L., Parschau, M., Zoppi, L., Baldridge, K. K., Siegel, J. S. and Ernst, K. H. (2009). Reversible phase transitions in a buckybowl monolayer, *Angew. Chem. Int. Ed.*, 48, pp. 1966–1969.
60. Soler, J. M., Artacho, E., Gale, D. J., Garcia, A., Junquera, J., Ordejon, P. and Sanchez-Portal, D. (2002). The SIESTA method for ab initio order-N materials simulation, *J. Phys. Condens. Mat.*, 14, pp. 2745–2779.
61. Zoppi, L., Garcia, A. and Baldridge, K. K. (2010). Development of methods for computational analysis of the binding of molecules on metallic surfaces: Application to corannulene on copper surface, *J. Phys. Chem. A*, 114, pp. 8864–8872.
62. Romaner, L., Heimel, G., Ambrosch-Draxl, C. and Zojer, E. (2008). The dielectric constant of self-assembled monolayers, *Adv. Funct. Mater.*, 18, pp. 3999–4006.
63. Reed, M. A., Zhou, C., Muller, C. J., Burgin, T. P. and Tour, J. M. (1997). Conductance of a molecular junction, *Science*, 278, pp. 252–254.

64. Guo, X., Small, J. P., Klare, J. E., Wang, Y., Purewal, M. S., Tam, I. W., Hong, B. H., Caldwell, R., Huang, L., Brien, S. O., Kim, P. and Nuckolls, S. (2006). Covalently bridging gaps in single-walled carbon nanotubes with conducting molecules, *Science*, 311, pp. 356–358.
65. Feldman, A. K., Steigerwald, M. L., Guo, X. and Nuckolis, C. (2008). Molecular electronic devices based on single-walled carbon nanotube electrodes, *Acc. Chem. Res.*, 41, pp. 1731–1741.
66. Jiaa, C. and Guo, X. (2013). Molecule–electrode interfaces in molecular electronic devices, *Chem. Soc. Rev.*, 42, pp. 5642–5660.
67. Meir, Y. and Wingreen, N. S. (1992). Landauer formula for the current through an interacting electron region, *Phys. Rev. Lett.*, 68, pp. 2512–2515.
68. (a) Agapito, L. A., Ferretti, A., Calzolari, A., Curtarolo, S. and Nardelli, M. B. (2013). Effective and accurate representation of extended Bloch states on finite Hilbert spaces, *Phys. Rev. B*, 88, pp. 165127-1–165127-7; (b) Landauer, R. (1970). Electrical resistance of disordered one-dimensional lattices, *Philos. Mag.*, 21, pp. 863–867.
69. Fisher, D. S. and Lee, P. A. (1981). Relation between conductivity and transmission matrix, *Phys. Rev. B*, 23, pp. 6851–6854.
70. Nardelli, M. B. (1999). Electronic transport in extended systems: Application to carbon nanotubes, *Phys. Rev. B*, 60, pp. 7828–7833.
71. Calzolari, A., Marzari, N., Souza, I. and Nardelli, M. B. (2004). Ab initio transport properties of nanostructures from maximally localized Wannier functions, *Phys. Rev. B*, 69, pp. 035108-1–035108-10.
72. Jia, C., Wang, J., Yao, C., Cao, Y., Zhong, Y., Liu, Z., Liu, Z. and Guo, X. (2013). Conductance switching and mechanisms in single-molecule junctions, *Angew. Chem. Int. Ed.*, 52, pp. 8660–8670.
73. (a) Borghi, G., Ferretti, A., Nguyen, N. L., Dabo, I. and Marzari, N. (2014). Koopmans-compliant functionals and their performance against reference molecular data, *Phys. Rev. B*, 90, pp. 075135-1–075135-16; (b) Nguyen N. L., Borghi G., Ferretti A., Dabo I. and Marzari N. (2015). First-principles photoemission spectroscopy and orbital tomography in molecules from Koopmans-compliant functionals, *Phys. Rev. Lett.*, 114, pp. 166405-1–166405-6.
74. (a) Valle, M. D., Gutiérez, R., Tejedor, C. and Cuniberti, G. (2007). Tuning the conductance of a molecular switch, *Nat. Nanotech.*, 2, pp. 176–179; (b) Ashraf, M. K., Bruque, N. A., Tan, J. L., Beran, G. J. O. and Lake, R. K. (2011). Conductance switching in diarylethenes bridging carbon nanotubes, 134, pp. 024524-1–024524-9; (c) Ke, S.-H., Baranger, H. U. and Yang, W. (2007). Contact transparency of nanotube-molecule-nanotube junctions, *Phys. Rev. Lett.*, 99, pp. 146802-1–146802-4; (d) Qian, Z., Hou, S., Ning, J., Li, R., Shen, Z., Zhao, X. and Xue, Z. (2007). First-principles calculation on the conductance of a single 1,4-diisocyanatobenzene molecule with single-walled carbon nanotubes as the electrodes, *J. Chem. Phys.*, 126, pp. 084705-1–084705-5; (e) Ren, W., Reimers, J. R., Hush, N. S., Zhu, Y., Wang, J. and

Guo, H. (2007). Models for the structure and electronic transmission of carbon nanotubes covalently linked by a molecular bridge via amide couplings, *J. Phys. Chem. C*, 111, pp. 3700–3704; (f) Bruque, N. A., Ashraf, M. K., Beran, G. J. O., Helander, T. R. and Lake, R. K. (2009). Conductance of a conjugated molecule with carbon nanotube contacts, *Phys. Rev. B*, 80, pp. 155455-1–155455-13.

75. (a) Deslippe, J., Samsonidze, G., Jain, M., Cohen, M. L. and Louie, S. G. (2013). Coulomb-hole summations and energies for GW calculations with limited number of empty orbitals: A modified static remainder approach, *Phys. Rev. B*, 87, pp. 165124-1–165124-6; (b) Deslippe, J., Samsonidze, G., Strubbe, D. A., Jain, M., Cohen, M. L. and Louie, S. G. (2012). Berkeley GW: A massively parallel computer package for the calculation of the quasiparticle and optical properties of materials and nanostructures, *Comp. Phys. Comm.*, 183, pp. 1269–1289.
76. (a) Bruneval, F. and Marques, M. A. L. (2012). Benchmarking the starting points of the GW approximation for molecules, *J. Chem. Theory Comput.*, 9, pp. 324–329; (b) Bruneval, F., Hamed, S. M. and Neaton, J. B. (2015). A systematic benchmark of the ab initio Bethe-Salpeter equation approach for lowlying optical excitations of small organic molecules, *J. Chem. Phys.*, 142, pp. 244101-1–244101-10.
77. Darancet, P., Ferretti, A., Mayou, D. and Olevano, V. (2007). Ab initio GW electron-electron interaction effects in quantum transport, *Phys. Rev. B*, 75, pp. 075102-1–075102-4.
78. Toher, C. and S. Sanvito, S. (2008). Effects of self-interaction corrections on the transport properties of phenyl-based molecular junctions, *Phys. Rev. B*, 77, pp. 155402-1–155402-12.
79. Egger, D. A., Weissman, S., Refaely-Abramson, S., Sharifzadeh, S., Dauth, M., Baer, R. I., Kümmel, S., Neaton, J. B., Zojer, E. and Kronik, L. (2014). Outer-valence electron spectra of prototypical aromatic heterocycles from an optimally tuned range-separated hybrid functional, *J. Chem. Theory Comput.*, 10, pp. 1934–1952.
80. Egger, D. A., Weissman, S., Refaely-Abramson, S., Sharifzadeh, S., Dauth, M., Baer, R., Kümmel, S., Neaton, J. B., Zojer, E. and Kronik, L. (2014). Control of single-molecule junction conductance of porphyrins via a transition-metal center, *J. Chem. Theory Comput.*, 11, pp. 1934–1952.
81. (a) Garcia-Lastra, J. M., Rostgaard, C., Rubio, A. and Thygesen, K. S. (2009). Polarization-induced renormalization of molecular levels at metallic and semiconducting surfaces, *Phys. Rev. B*, 80, pp. 245427-1–245427-7; (b) Thygesen, J. M. G.-L. a. K. S. (2011). Renormalization of optical excitations in molecules near a metal surface, *Phys. Rev. Lett.*, 106, 187402-1–187402-4.

Chapter 9

Modern Treatments of Aromaticity

Judy I-Chia Wu

Department of Chemistry, University of Houston,
Houston, TX 77204, USA

9.1 Overview

Aromaticity is unarguably one of the most intriguing time-dependent chemical concepts. This ever-developing branch of chemistry, originally conceived to characterize planar benzenoid-like hydrocarbons, is now inclusive of many rings and cages that are not "benzene-like" at all. Due to its elusive (i.e., "nonmeasurable") and multifaceted nature (i.e., characterized indirectly by energetic, magnetic, or structural properties) a universal quantification of aromaticity remains controversial, while much of its interpretations are continually disputed. The highly restricted textbook definition for aromaticity, i.e., molecules that are: (1) planar, (2) cyclic, (3) π-conjugated, and (4) follows the Hückel $4n + 2$ π-electron rule also warrants revision. This chapter highlights current progress and challenges encountered in textbook models of aromaticity and as well as energetic and magnetic quantifications of aromaticity.

Although benzene prevails in modern organic textbooks as the hallmark of aromaticity (perhaps due more to its aesthetic appeal than historical precedence[1]), it is not representative of the majority of aromatic molecules. None of the polycyclic aromatic hydrocarbons have equal bond lengths or undergo electrophilic substitution reactions exclusively. Fullerenes lack external hydrogens, can only undergo addition reactions, and follow the Hirsch $2(n+1)^2$ rule for spherical aromaticity.[2] The simplest σ-aromatic species, H_3^+, as well as other σ- and δ-electron delocalized aromatics, also do not resemble benzene. Conversely, many nonaromatic molecules can exhibit aromatic features. Quinoidal acenes display significantly downfield shifted ^{1}H NMR signals but are nonaromatic.[3] Both borazine and $(LiF)_3$ (D_{3h}) are planar, cyclic, have six π-electrons, and exhibit equalized ring bond lengths, but are essentially nonaromatic due to the large electronegativity difference between the B/N's and Li/F's. There is no "magic" in having six π-electrons!

With the advent of quantum chemical calculations, computational chemistry has become the key approach to understanding and quantifying aromaticity. This is largely due to the growing sophistication and versatility of theoretical tools, which can be improved more effectively than experimental approaches to isolate structural and electronic features related exclusively to aromaticity. The hierarchy of nucleus-independent chemical shifts (NICS)–based indices, by eliminating magnetic responses not related to aromaticity, is an example of such an advantage.[4–6] Energetic evaluations of aromaticity also benefit from the availability of theoretical reference models. These methodological improvements have resolved many key issues related to aromaticity.

9.2 Limitations to textbook models of aromaticity

Many textbook models for aromaticity, like the "Hückel $4n+2\,\pi$-electron" rule[7,8] and "Clar sextet" rule,[9,10] have interpretive merits but in practice have limitations. Hückel, in 1930, first demonstrated that rings of carbons with $4n+2$ π-electrons led to filled π-molecular orbital shells (e.g., the six π-electrons in benzene) equivalent to the filled atomic orbital shells of noble gases (e.g., the $2(n+1)^2$ valence electrons for noble gas atoms), thus resulting in unusual thermochemical stability.[7,8] With the exception of Möbius rings[11–13] and triplet state[14–16] species (which follow a reversed Hückel rule), this electron-counting rule is applicable to many closed-shell monocyclic species. Antiaromatic rings have $4n$ π-electrons and are destabilized.[17] Larger π-conjugated cycles become more polyene-like regardless of their π-electron counts. But polycyclic species (e.g., acenes, polybenzenoids, and porphyrinoids) with multiple π-conjugated circuits can exhibit contrasting (anti)aromatic features.

For example, dihydrodiazapentacenes have two *local* six π-electron circuits (i.e., two "Clar rings,"[9,10] see below) embedded in a *macrocyclic* 24 π-electron perimeter and thus are aromatically stabilized (do not easily oxidize to their 22 π-electron counterparts with one Clar ring, see Fig. 9.1) but display the expected upfield ^{1}H NMR shifts of antiaromatic compounds.[18] Many porphyrinoids with *macrocyclic* $4n$ π-electron perimeters but multiple *local* six π-electron circuits, e.g., amethyrin (24 π-electrons, four Clar sextets, see Fig. 9.1), orangarin (20 π-electrons, three Clar sextets),[19] and porphycene (20 π-electrons, four Clar sextets), are experimentally viable and have large HOMO-LUMO gaps, but exhibit antiaromatic spectroscopic features.[19–21]

The "Clar ring,"[9,10] often depicted by a circle notation in chemistry, refers to a "group of six electrons that resist disruption." Historically, the terms "six centric affinity" (Armstrong,[22] pre-electron period), "six aromatic electrons" (Crocker[23]), and "aromatic sextet" (Armit and Robinson[24]), also have been evoked to describe this association of six π-electrons and its chemical significance. Clar[9,10] proposed the selective placement of such "circles" and noted that,

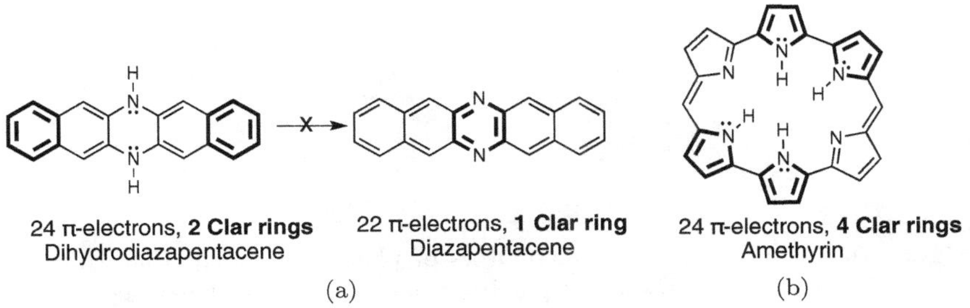

Figure 9.1. Polycyclic systems like (a) dihydrodiazapentacene (two Clar rings, in bold) and (b) amethyrin (four Clar rings, in bold) display contrasting (anti)aromatic features.

for polybenzenoids, resonance structures with the maximum number of Clar rings correspond most closely to their chemical properties. Thus, phenanthrene could be depicted as having two outer Clar rings and a central cyclohexene moiety or one central Clar ring and two cyclohexadiene units (Fig. 9.2(a)). But the former emphasizes the higher reactivity of the C9 and C10 positions toward cycloaddition and is the more critical resonance form.

Although often seen in textbooks and in current chemical literature, depictions of polybenzenoids with one "circle" inscribed in each of the six-membered rings are inaccurate and not encouraged. The Clar ("circle") notation has chemical significance and should be used properly. Isomers of polybenzenoids with greater numbers of Clar rings, e.g., phenanthrene (two) vs. anthracene (one) are more aromatically stabilized (Fig. 9.2(b)). Likewise, nanographene sheets with armchair ("phenanthrene-like") edges can accommodate more Clar rings, are more aromatic, and thus are less reactive compared to their zigzag ("anthracene-like") analogs of similar size (Fig. 9.2(c)).[25]

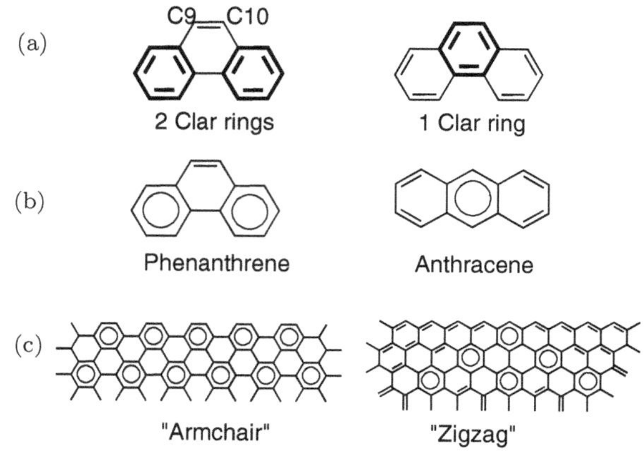

Figure 9.2. (a) Resonance contributors of phenanthrene (b) Clar ring(s) in phenanthrene and anthracene, (c) polybenzenoids with "armchair" and "zigzag" edges.

9.3 Energetic criteria

Aromaticity may be defined quantitatively as the energy lowering due to a continuous delocalization of electrons in "rings" and "cages." Like many other energetic quantities that relate the structures and energies of molecules, such as "π-conjugation," "group enthalpy increments," and "strain," "aromaticity" is not measurable directly experimentally but must be derived from other data and depends critically on the conventions employed to deduce the quantity of interest. The choice of needed reference molecules can be decisive (also for interpretation). Even literature estimates of the stabilization energy of benzene range over 30 kcal/mol![26]

Following Kekulé's[27] leading depiction of the fully symmetrical benzene (D_{6h}) (a single molecule with rapidly interchanging formal single and double bonds), Pauling, Sherman,[28] and Kistiakowsky[29] realized, independently, the first experimental resonance energy (RE) estimates for benzene. Pauling and Sherman[28] deduced a 37.4 kcal/mol RE value based on C–C–C, C–C=C, and C–H bond energies derived from conjugated polyenes. Kistiakowsky[29] proposed a +36.0 kcal/mol value by comparing the experimental heats of hydrogenation of benzene to three times that of cyclohexene (Fig. 9.3). Based on these early estimates, an approximately 30 kcal/mol "stabilization energy of benzene" continues to be widely

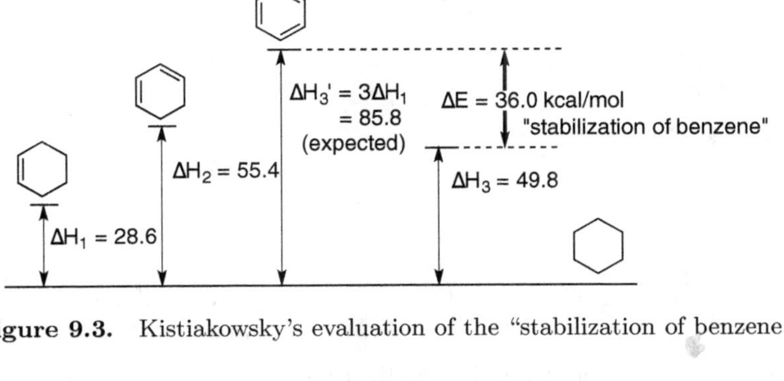

Figure 9.3. Kistiakowsky's evaluation of the "stabilization of benzene."

taught in many organic textbooks. However, both approaches neglect hyperconjugation stabilization in the –C=C groups and cyclohexene references employed, and suffer from conceptual imperfections. When corrected for hyperconjugation, the RE of benzene approaches 60 kcal/mol[26] (*cf.* Fig. 9.4, Eq. (1)).

Balanced chemical equations, which isolate specific energetic effects in molecules by recourse to selected references, also can be used to quantify aromaticity. Such procedures rely on the cancellation of energetic effects on both sides of the equation. Ideally,

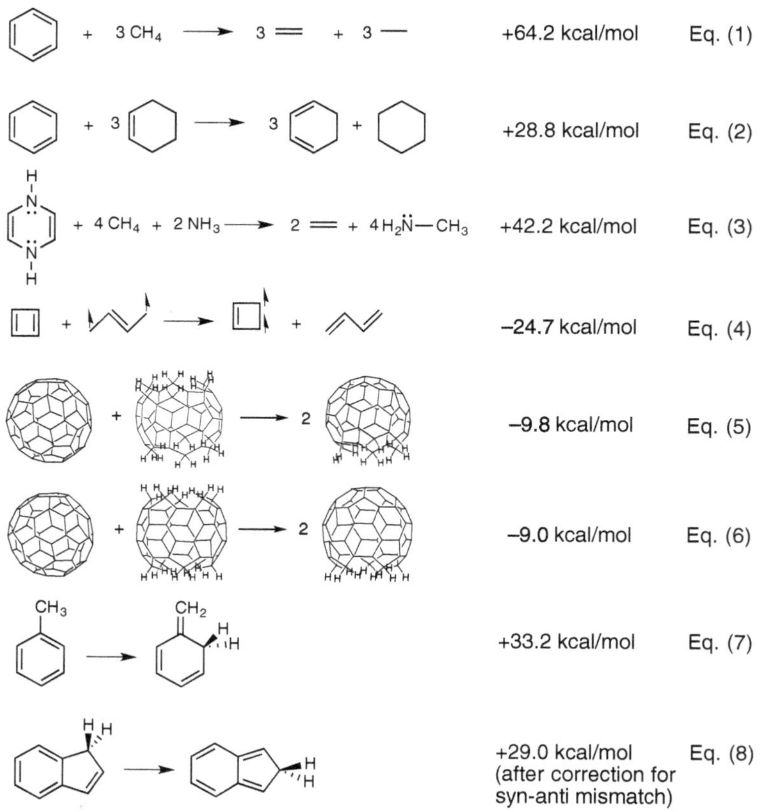

Figure 9.4. Equations for evaluating the REs and ASEs of (anti)aromatic compounds.

proper reference standards must be nonaromatic yet largely resemble the geometric and electronic features of the considered molecule. However, differences among historically inherited definitions and the use of various reference energies often raise problems. In particular, resonance energies (RE) and aromatic stabilization energies (ASE, also called the Dewar resonance energy) have different meanings and must be distinguished from one another.

REs are evaluated relative to unconjugated reference compounds and measure the net stabilization in a cyclic or acyclic polyene due to π-conjugation. For example, the RE of benzene, compared to three ethanes and three ethylenes is +64.2 kcal/mol (Fig. 9.4, Eq. (1)). Based on this definition, antiaromatic molecules also can have stabilizing REs due to π-conjugation. Dihydropyrazine is formally 8π-electron antiaromatic, but net stabilized by four enamine conjugations (RE = +42.2 kcal/mol, Fig. 9.4, Eq. (3)).[30]

ASEs measure the extra stabilization energy associated with the cyclic conjugation of an aromatic molecule and may be evaluated by comparing the molecule being considered to acyclic but π-conjugated polyene references with the same number and type of π-conjugations. For example, the ASE of benzene compared to three cyclohexadienes is +28.8 kcal/mol (Fig. 9.4, Eq. (2)); this is roughly half the RE of benzene, since ASEs reflect only the "extra" stabilization associated with aromaticity. Antiaromatic compounds exhibit extra cyclic destabilization and have negative ASEs.

Highly strained molecules, e.g., cyclobutadiene and the buckminsterfullerene (C60), pose additional challenges, as the reference compounds selected for estimating their ASEs must closely resemble the skeletal features of the molecule being examined. Apart from being antiaromatic, the high energy of cyclobutadiene arises mainly from four acute 90° CCC angles and π-π repulsion between the closely positioned parallel π bonds.[31] Equation (4) in Fig. 9.4, gives an upper bound estimate for the antiaromatic destabilization energy for singlet cyclobutadiene (−24.7 kcal/mol, note negative ASE value), since the reference triplet cyclobutadiene possesses similar structural features but is aromatic.[31] In C60, all of the carbons are pyramidalized. Thus, the reference compound selected for evaluating the ASE of C60 also must possess such strain. Cyranski *et al.*'s[32] strain-balanced equation for evaluating the ASE of C60, based on C60 with a six- or five-membered ring omitted, gives a negative ASE (*ca.* −9 kcal/mol, Fig. 9.4, Eqs. (5) and (6)), and is by far the best estimate. Surprisingly, C60 is not "superaromatic" or even aromatic but is modestly antiaromatically destabilized.

Alternatively, ASEs can be evaluated by the isomerization stabilization energy (ISE) of methylated (anti)aromatics and their quinoidal nonaromatic analogs.[33] This approach assumes that methylation has negligible energetic effects on aromaticity and has many advantages: (1) only one reference compound is needed; (2) ISEs can be corrected unambiguously for syn-anti mismatches and bond type differences; and (3) ISEs are applicable to triplet species, strained, and polycyclic aromatics/antiaromatics.[33–35] The closely related indene-isoindene isomerization (ISEII) approach, a modification of the ISE method by employing fused five-membered ring appendages, also gives very good ASE estimates.[35] As shown in Fig. 9.4, both the ISE of toluene (33.2 kcal/mol, Eq. (7)) and the ISEII approach (29.0 kcal/mol, Eq. (8)) for evaluating the ASE of benzene are in close agreement with that of Eq. (2) (28.8 kcal/mol). Based on the ISEII approach, Wannere *et al.*[35] reported insignificant antiaromatic destabilization energies for a series of [4n] annulenes ($n = 2 - 6$).

Energy-decomposition–based procedures (see Chapter 7), such as the block-localized wavefunction (BLW) method[36,37] offer practical means for evaluating REs and ASEs computationally. Theoretical reference standards are constructed by applying artificial electronic constraints to measure the effects of electron delocalization. REs are evaluated based on the energy difference between the fully delocalized wavefunction of the molecule considered and that of its most stable Lewis structure. The latter is described by a localized wavefunction, in which selected σ- or π-interactions (i.e., electron delocalization) can be mathematically "disabled" to quantify the corresponding energetic effect. ASEs can be derived indirectly based on comparisons of the computed RE of the molecule considered to that of selected reference compounds.

For the specific use of BLW applications, Mo and Schleyer[38] coined the term "extra cyclic resonance energy" (ECRE, equivalent to the ASE), in which the BLW-REs of (anti)aromatic rings may be compared to those of linear polyenes that have either: (1) the same number of π-electrons (ECRE 1) or (2) the same number of π-conjugations (ECRE 2). The latter follows the Dewar definition for ASEs and provides a more definitive energetic measurement of aromaticity. Positive/negative ECRE values indicate aromaticity/antiaromaticity. Nonaromatics have ECRE values close to zero. For example, the ECRE of benzene, by comparing the BLW-RE of benzene to three syn-butadienes, is 29.9 kcal/mol.[38] This is in excellent agreement with best ASE estimate (28.8 kcal/mol, see Fig. 9.4, Eq. (2)). Despite its usual association as the antiaromaticity paradigm, the ECRE of cyclobutadiene (−15.9 kcal/mol), by comparing its BLW-RE to two syn-butadienes, is very modest.[38]

Although aromatic/antiaromatic compounds are expected to be thermochemically more/less stable than they ought to be, other structural and electronic factors can also have considerable effects on the energies of molecules. These features must be considered when selecting references for evaluating REs and ASEs. For example, 1,2-, 1,3-, and 1,4-diazine are all equally aromatic but 1,2-diazine is approximately 20 kcal/mol higher in energy than the nearly thermoneutral 1,3- and 1,4- diazines due to in-plane N lone pair repulsion (Fig. 9.5(a)).[39] Both 2- and 3-hydroxyfuran are aromatic but exist in solution in their nonaromatic keto forms;[40] the keto tautomers

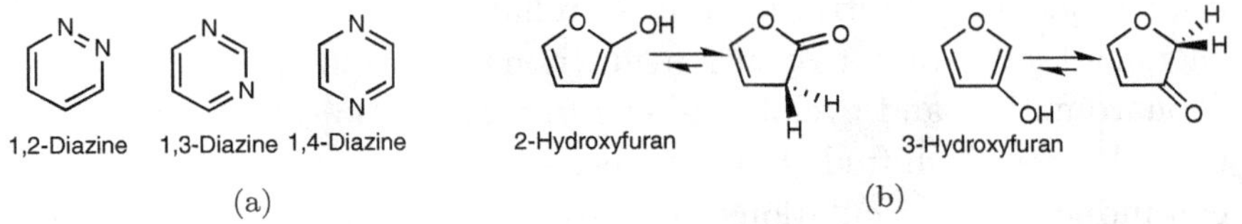

Figure 9.5. (a) 1,2-, 1,3-, and 1,4-diazines, and (b) the tautomeric forms of 2- and 3-hydroxyfuran.

benefit from reduced five-membered ring strain due to a sp3 CH2 group in the ring (Fig. 9.5(b)). The high energy of cyclobutadiene arises from π-π repulsion and acute angle strain rather than its supposed considerable antiaromatic destabilization.[31]

Discrepancies among various energetic evaluations of aromaticity often can be reconciled by conceptual and computational method differences. But RE and ASE evaluations are not always practical for more exotic aromatic molecules (e.g., inorganic, strained, nonplanar, or charged species) due to the lack of thermochemical and structural regularity as well as the absence of experimental standards for comparison. For these families of compounds, magnetic evaluations of aromaticity may be more decisive.

9.4 Magnetic criteria

Pauling's[41] ring current theory first envisioned that the p_z electrons of cyclic π-conjugated aromatic ring carbon atoms might circulate freely, moving from one carbon to an adjacent carbon under the presence of an external applied magnetic field. This assumption along with the advent of NMR spectroscopy, anticipated Pople's ring current model,[42] which explained the relationship between ^{1}H NMR chemical shifts and the induced ring current effects of (anti)aromatic molecules (Fig. 9.6). Aromatic molecules are diamagnetic and have deshielded outer H's that resonate downfield (7–9 ppm) relative to the typical olefinic H's (*ca.* 5.5 ppm); the inner H's of bridged or large aromatic annulenes are shielded and exhibit upfield ^{1}H chemical shifts. Antiaromatic molecules are paramagnetic and behave in the opposite manner (see Fig. 9.6).

Nevertheless, the 1H NMR chemical shifts of ring H's do not always identify (anti)aromaticity reliably. For example, the 1H chemical shifts of cyclopentadiene (nonaromatic), quinoidal acenes (nonaromatic), and cyclobutadiene (antiaromatic) are all unexpectedly downfield shifted. Since external applied magnetic fields also can induce local circulations from σ-bonds, lone pairs, and core electrons, a major challenge is to quantify magnetic responses associated exclusively to an induced aromatic ring current.

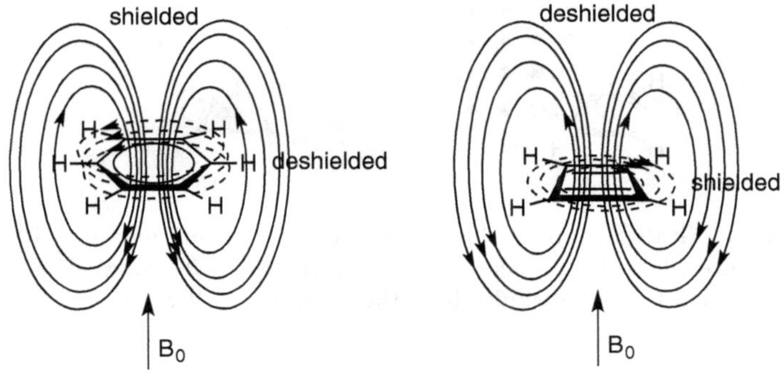

(a) (b)

Figure 9.6. Pople's ring current model. Induced ring current (dotted lines) and magnetic response (solid lines) in (a) aromatic and (b) antiaromatic compounds, with an external applied magnetic field (B0).

The NICS method[4–6] offers practical advantages for such purposes. NICS computations do not rely on recourse to reference compounds, are not size-dependent, can be computed at any point in the vicinity of the molecule considered (see Fig. 9.7), and can be systematically refined to isolate magnetic effects related to aromaticity (see Fig. 9.8). Aromatic molecules have negative NICS values in the ring (indicated by white "dots" in Fig. 9.7(a)) and positive NICS values outside of the ring (indicated by black "dots" in Fig. 9.7(a)), as a result of their diatropic ring currents. Antiaromatic molecules have positive NICS values in the ring (see black "dots," Fig. 9.7(b)) and negative NICS values outside of the ring (see white "dots," Fig. 9.7(b)), due to their paratropic ring currents. The signs of the computed shielding are reversed to conform to the experimental convention for upfield (negative) and downfield (positive) NMR chemical shifts.

Isotropic NICS(0) ("0" denotes NICS points computed at the heavy atom ring or cage centers of molecules) include blends of magnetic responses coming from the local circulations of electrons in bonds, lone pairs, and atoms cores, but can be systematically refined to give better results (Fig. 9.8(a)).[5] NICS(1), computed at

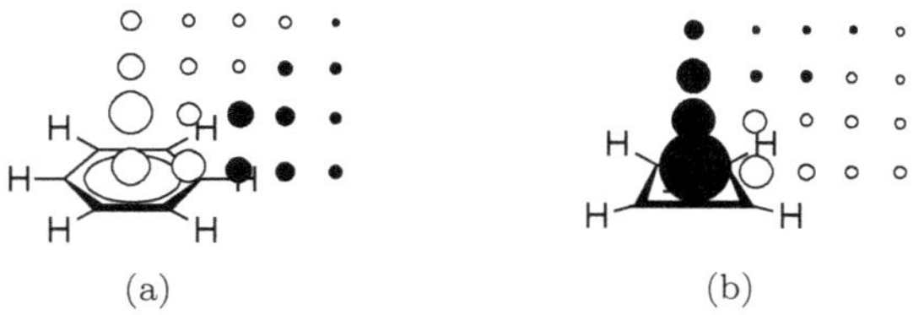

Figure 9.7. Grids of computed isotropic NICS for (a) benzene and (b) cyclobutadiene. White/black dots indicate negative/positive NICS values. The size of the "dots" correspond to the magnitude of the computed NICS values.

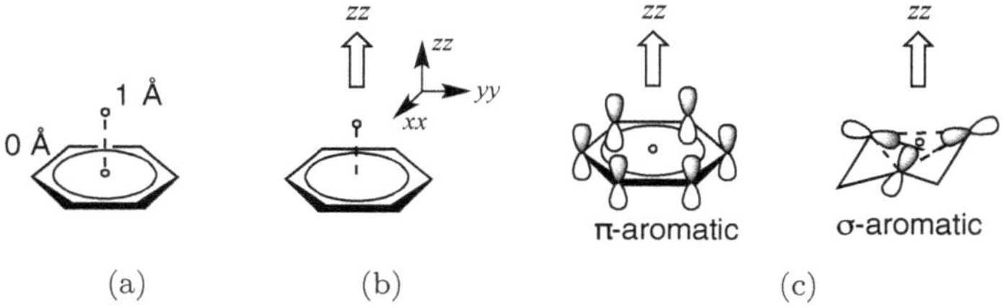

Figure 9.8. Hierarchy of refined NICS indices: (a) isotropic NICS(0) and NICS(1), (b) NICS(1)zz, and (c) NICS(0)MOzz for quantifying π- and σ-aromaticity.

1 Å above the ring center, reduce such contaminations but only to some extent (Fig. 9.8(a)). NICS(1)zz corrects for this problem by including only the out-of-plane (zz) tensor component, but is not optimal either, especially for metal clusters which tend to have considerable local magnetic shieldings from the metal atoms themselves (Fig. 9.8(b)). NICSMOzz extracts the out-of-plane (zz) tensor component of the isotropic NICS, includes only contributions from the molecular orbitals (MOs) relevant to aromaticity, and is the most sophisticated NICS index (Fig. 9.8(c)). Such refinements are pertinent, especially for small three- and four-membered rings, which have nonnegligible σ-MO and in-plane (xx and yy) tensor component contributions even at 1 Å above the ring centers.

Other computational procedures that quantify and visualize the induced ring currents of (anti)aromatic compounds, e.g., the aromatic ring current shielding (ARCS) method,[43] anisotropy of the induced current density (ACID),[44] through-space NMR shieldings

(TSNMRS) that can be visualized as iso-chemical shielding surfaces (ICSS),[45] and current density map plots[46] are also available and provide complementary chemical insights.

9.5 Practical advice

Modern treatments of aromaticity largely rely on the use of theoretical tools, but controversies remain especially when various geometric, electronic, energetic, or magnetic criteria disagree. The following guidelines may help minimize ambiguities when different standards and computational procedures are employed to understand the (anti)aromaticities of molecules:

(1) Polycyclic systems can have "multifaceted" (anti)aromatic character; magnetic evaluations are useful for predicting spectroscopic features, but not indicative of thermochemical stability or reactivity.
(2) Aromatic/antiaromatic compounds are typically thermochemically *more/less stable than they ought to be*, but other stabilizing or destabilizing structural features also can play prominent roles and must be considered.
(3) Reactivity is *not* a valid criterion for aromaticity, as it depends on the relative energy of the molecule being considered and the reaction transition state.

Nearly 200 years since Faraday's 1825 isolation of benzene,[47] research in the field of aromaticity has evolved enormously from solving the structure of benzene to explosive developments in the definition and quantification of aromaticity. However, conceptual developments of aromaticity are important, not only for the classification of compounds but also for establishing general rules to predict the rates, mechanisms, and outcomes of chemical reactions as well as for designing functional organic molecules with unusual structures and properties. We are now equipped with the tools and understanding necessary to push the boundaries of aromaticity research toward such applications. As our understanding of aromaticity continues to evolve, so must the research questions we ask. New aspects of this fascinating branch of chemistry await discovery.

References

1. Naphthalene, originally named "naphthalin" was discovered in 1821 as a white solid with a pungent odor. Other polybenzenoid hydrocarbons, being crystalline, also were identified as coal tar constituents well before Faraday's 1825 discovery of benzene. Kidd, J. (1821). Observation on naphthalene, a peculiar substance resembling a concrete essential oil, which is produced during the decomposition of coal tar, by exposure to a red heat, *Philos. Transact.*, 111, pp. 209–221.
2. Hirsch, A., Chen, Z. and Jiao, H. (2000). Spherical aromaticity in I_h symmetrical fullerenes: The $2(n+1)^2$ rule, *Angew. Chem. Int. Ed.*, 39, pp. 3915–3917.
3. Wannere, C. S., Corminboeuf, C., Allen, W. D., Schaefer, H. F. III and Schleyer, P. V. R. (2005). Downfield proton chemical shifts are not reliable aromaticity indicators, *Org. Lett.*, 7, pp. 1457–1460.
4. Schleyer, P. V. R., Maerker, C., Dransfeld, A., Jiao, H. and Hommes, N. J. R. V. E. (1996). Nucleus-independent chemical shifts: A simple and effective aromaticity probe, *J. Am. Chem. Soc.*, 118, pp. 6317–6318.
5. Fallah-Bagher-Shaidaei H., Wannere, C. S., Corminboeuf, C., Puchta, R. and Schleyer, P. V. R. (2006). Which NICS aromaticity index for planar π rings is best? *Org. Lett.*, 8, pp. 863–866.
6. Chen, Z., Wannere, C. S., Corminboeuf, C., Puchta, R. and Schleyer, P. V. R. (2005). Nucleus-independent chemical shifts (NICS) as an aromaticity criterion, *Chem. Rev.*, 105, pp. 3842–3888.
7. Hückel, E. Z. (1930). Zur quantentheorie der doppelbindung, *Zeitschrift für Physik*, 60, pp. 423–456.
8. Hückel, E. Z. (1931). Quantentheoretische beiträge zum benzolproblem, *Zeitschrift für Physik*, 70, pp. 204–286.
9. Clar, E. (1964). *Polycyclic Hydrocarbons* (Academic Press, New York).
10. Clar, E. (1972). *The Aromatic Sextet* (Wiley, London).
11. Heilbronner, E. (1964). Hückel molecular orbitals of möbius-type conformations of annulenes, *Tetrahedron Lett.*, 29, pp. 1923–1928.
12. Rzepa, H. S. (2005). Möbius aromaticity and delocalization, *Chem. Rev.*, 105, pp. 3697–3715.
13. Craig, D. P. and Paddock, N. L. (1958). A novel type of aromaticity, *Nature*, 181, pp. 1052–1053.
14. Baird, N. C. (1972). Quantum organic photochemistry. II. Resonance and aromaticity in the lowest pi... pi state of cyclic hydrocarbons, *J. Am. Chem. Soc.*, 94, pp. 4941–4948.
15. Gogonea, V., Schleyer, P. V. R. and Schreiner, P. R. (1998). Consequence of triplet aromaticity in 4n π-electron annulenes: Calculation of magnetic shieldings for open-shell species, *Angew. Chem. Int. Ed.*, 37, pp. 1945–1948.

16. Ottosson, H. (2012). Organic photochemistry: Exciting excited-state aromaticity, *Nat. Chem.*, 4, pp. 969–971.
17. Breslow, R. (1973). Antiaromaticity, *Acc. Chem. Res.*, 6, pp. 393–398.
18. Wu, J. I., Wannere, C. S., Mo, Y., Schleyer, P. V. R. and Bunz, U. H. F. (2009). 4N π electrons but stable: N,N-dihydropetacenes, *J. Org. Chem.*, 74, pp. 4343–4349.
19. Sondheimer, F., Wolovsky, R. and Amiel, Y. (1962). Unsaturated macrocyclic compounds XXIII. The synthesis of the fully conjugated macrocyclic polyenes cyclooctadecanonaene ([18]annulene), cyclotetracosadodecaene ([24]annulene), and cyclotriacontapentadecaene ([30]annulene), *J. Am. Chem. Soc.*, 84, pp. 274–284.
20. Wu, J. I., Fernandez, I. and Schleyer, P. V. R. (2013). Description of aromaticity in porphyrinoids, *J. Am. Chem. Soc.*, 135, pp. 315–321.
21. Vogel, E. (1993). The porphyrins from the "annulene chemists" perspective, *Pure Appl. Chem.*, 65, pp. 143–152.
22. Armstrong, H. E. (1890). The structure of cycloid hydrocarbons, *Proc. Chem. Soc. London*, 6, pp. 101–105.
23. Crocker, E. C. (1922). Application of the octet theory to single-ring aromatic compounds, *J. Am. Chem. Soc.*, 44, pp. 1618–1630.
24. Armit, J. W. and Robinson, R. (1925). Polynuclear heterocyclic aromatic types. Part III. Some anhydronium bases, *J. Chem. Soc. Transact.*, 127, pp. 1604–1618.
25. Stein, S. E. and Brown, R. L. (1987). Pi-electron properties of large condensed polyaromatic hydrocarbons, *J. Am. Chem. Soc.*, 109, pp. 3721–3729.
26. Wodrich, M. D., Wannere, C. S., Mo, Y., Jarowski, P. D., Houk, K. N. and Schleyer, P. V. R. (2007). The concept of protobranching and its many paradigm shifting implications for energy evaluations, *Chem. Eur. J.*, 13, pp. 7731–7744.
27. Kekulé, A. (1866). *Lehrbuch der Organischen Chemie* (Enke, Erlangen, Germany).
28. Pauling, L. and Sherman, J. (1933). The nature of the chemical bond. VII. The calculation of resonance energy in conjugated systems, *J. Chem. Phys.*, 1, pp. 679–686.
29. Kistiakowsky, G. B., Ruhoff, J. R., Smith, H. A. and Vaughan, W. E. (1963). Heats of organic reactions. IV. Hydrogenation of some dienes and of benzene, *J. Am. Chem. Soc.*, 58, pp. 146–153.
30. Murray, J. S., Seminario, J. M. and Politzer, P. (1994). Does antiaromaticity imply net destabilization? *Int. J. Quantum Chem.*, 49, pp. 575–579.
31. Wu, J. I., Mo, Y., Evangelista, F. A. and Schleyer, P. V. R. (2012). Is cyclobutadiene really highly destabilized by antiaromaticity? *Chem. Commun.*, 48, pp. 8437–8439.
32. Cyrański, M. K, Howard, S. T. and Chodkiewicz, M. L. (2004). Bond energy, aromatic stabilization energy and strain in IPR fullerenes, *Chem. Commun.*, 21, pp. 2458–2459.

33. Schleyer, P. V. R. and Pühlhofer, F. (2002). Recommendations for the evaluation of aromatic stabilization energies, *Org. Lett.*, 4, pp. 2873–2876.
34. Zhu, J., An, K. and Schleyer, P. V. R. (2013). Evaluation of triplet aromaticity by the isomerization stabilization energy, *Org. Lett.*, 15, pp. 2442–2445.
35. Wannere, C. S., Moran, D., Allinger, N. L., Hess, B. A. Jr., Lawrence J. Schaad, P. V. R. and Schleyer (2003). On the stability of large [4n]annulenes, *Org. Lett.*, 5, pp. 2983–2986.
36. Mo Y., Gao, J. and Peyerimhoff, S. D. (2000). Energy decomposition analysis of intermolecular interactions using a block-localized wave function approach, *J. Chem. Phys.*, 112, pp. 5530–5538.
37. Y. Mo, Song, L. C. and Lin, Y. C. (2007). Block-localized wavefunction (BLW) at the density functional theory (DFT) level, *J. Phys. Chem. A*, 111, pp. 8291–8301.
38. Mo, Y. and Schleyer, P. V. R. (2006). An energetic measure of aromaticity and antiaromaticity based on the Pauling-Wheland resonance energies, *Chem. Eur. J.*, 12, pp. 2009–2020.
39. Wang, Y., Wu, J. I., Li, Q. S. and Schleyer, P. V. R. (2010). Aromaticity and relative stabilities of azines, *Org. Lett.*, 12, pp. 4824–4827.
40. Friedrichsen, W., Traulsen, T., Elguero, J. and Katritzky, A. R. (2000). Tautomerism of Heterocycles: Five-Membered Rings with One Heteroatom, *Adv. Heterocycl. Chem.*, 76, pp. 85–156.
41. Pauling, L. (1936). The diamagnetic entropy of aromatic molecules, *J. Chem. Phys.*, 4, pp. 673–677.
42. Pople, J. A. (1956). Proton magnetic resonance of hydrocarbons, *J. Chem. Phys.*, 24, pp. 1111–1112.
43. Jusélius, J. and Sundholm, D. (1999). Ab initio determination of the induced ring current in aromatic molecules, *Phys. Chem. Chem. Phys.*, 1, pp. 3429–3435.
44. Geuenich, D., Hess, K., Köhler, F. and Herges, R. (2005). Anisotropy of the Induced Current Density (ACID), a general method to quantify and visualize electronic delocalization, *Chem. Rev.*, 105, pp. 3758–3772.
45. Klod, S. and Kleinpeter, E. (2001). An initio calculation of the anisotropy effect of multiple bonds and the ring current effect of arenes-application in conformational and configurational analysis, *J. Chem. Soc., Perkin Transact.*, 2, pp. 1893–1898.
46. Steiner, E. and Fowler, P. W. (1996). Ring currents in aromatic hydrocarbons, *Int. J. Quantum Chem.*, 60, pp. 609–616.
47. Faraday, M. (1825). On new compounds of carbon and hydrogen and on certain other products obtained during the decomposition of oil by heat, *Philos. Transact. Royal Soc. London*, 115, pp. 440–446.

Chapter 10

Weak Intermolecular Interactions

Rajat Maji and Steven E. Wheeler

Department of Chemistry, Texas A&M University,
College Station, Texas 77842, USA

10.1 Introduction

Weak noncovalent interactions are prevalent in organic systems, and accurately capturing their impact is vital for the reliable description of myriad chemical phenomena. These interactions impact everything from molecular conformations and stability to the outcome of stereoselective organic reactions and the function of biological macromolecules. These noncovalent interactions have long posed a challenge to popular quantum chemical methods, hampering efforts to provide reliable computational predictions for many problems in organic chemistry.[1] However, recent years have witnessed tremendous advances in efficient computational methods suitable for the description of these noncovalent interactions, which has enabled reliable computational studies of many organic problems that would not have been feasible a decade ago.

There have been a number of excellent reviews in recent years covering noncovalent interactions relevant to organic systems. For instance, Salonen *et al.*[2] have provided a general review of noncovalent interactions involving aromatic rings, while recent reviews of Takahashi *et al.*[3] and Nishio[4] have focused on weak hydrogen bonds and CH/π interactions. Wagner and Schreiner[5] have recently reviewed dispersion interactions in a broad range of chemical contexts. There have also been more focused reviews on noncovalent interactions in the context of organic chemistry, including the reviews by Krenske and Houk[6] on noncovalent interactions as control elements in chemical reactions, Johnston and Cheong[7] on nonclassical CH $\cdots$ O interactions, Singh and Das[8] on lone-pair/π interactions, Zhao *et al.*[9] on anion–π interactions, and Kennedy *et al.*[10] on cation–π interactions in small molecule catalysis.

Herein, we proceed by first discussing the general classes of noncovalent interactions and their physical nature, followed by discussions of the many challenges and pitfalls associated with describing these weak noncovalent interactions computationally. This is followed by representative examples of noncovalent interactions drawn from across the spectrum of organic, biological, and supramolecular systems. Finally, we conclude with practical considerations with regard to applications to organic systems in which noncovalent interactions play important roles.

10.2 Nature of noncovalent interactions

Favorable noncovalent interactions can occur between diverse functional groups, and there is a plethora of "named" noncovalent interactions in the literature.[2] However, all of these interactions arise from combinations of the same fundamental physical interactions. Noncovalent interactions can be classified based on the relative importance of these different physical effects, including electrostatic interactions (i.e., Coulombic interactions between fixed partial charges), induction or polarization effects (i.e., interactions arising from the polarization of one molecule due to its proximity to another), dispersion interactions (i.e., interaction of an instantaneous dipole in one molecule with an induced dipole in another), and exchange repulsion or Pauli repulsion (interactions due to overlapping electron distributions). Among these, dispersion interactions and induction/polarization interactions are always stabilizing, while exchange repulsions are always unfavorable.

Electrostatic interactions can be either repulsive or attractive and are often discussed in terms of interactions involving charge–charge,

charge–dipole, dipole–dipole, charge–quadrupole, etc. Often, the first nonzero term of such electrostatic multipole expansions dominates, and the contribution of the higher-order contributions is generally small. For instance, the interaction of two neutral dipolar molecules can often be understood in terms of the leading dipole–dipole term based on the favorable orientation of the two molecular dipoles. However, for complexes of larger molecular systems, the use of such multipole expansions is often on shaky physical ground. The multipole expansion of an electrostatic interaction is always convergent for large intermolecular distances. That is, the interaction of two molecules at large separation (i.e., where the distance between molecules is much larger than the dimension of either molecule) can be written exactly as a sum of multipolar interaction terms. As molecules move closer together, this expansion becomes more protracted. Ultimately, when the distance between charge centers is smaller than the radius of either molecule, the multipole expansion of the electrostatic interaction diverges![11] While the leading term in such divergent multipole expansions can still be qualitatively correct, one must be cognizant of the fact that this is the first term in a divergent mathematical expression. Thus, while simple concepts such as the favorable alignment of molecular dipole moments can often serve as a qualitative guide to intermolecular interactions, such models become increasingly unreliable for larger molecules.

Alternatively, noncovalent interactions between molecules can be classified based on the identity of the interacting functional groups, and tremendous efforts have been expended in recent years to understand the origin of these different named interactions. Several key classes of noncovalent interactions are discussed below, with a particular emphasis on π-stacking interactions, ion–π interactions, and XH/π interactions, among others.[12–14]

π-Stacking interactions are generally defined as attractive interactions between aromatic rings. However, there has been some recent debate regarding the role of aromaticity in these interactions and even the name "π-stacking interaction" has recently come under fire.[15–17] Grimme[15] showed, that for aromatic systems smaller than anthracene, there does not appear to be anything special about π-stacking interactions involving aromatic systems. That is, saturated cyclic systems (e.g., cyclohexane) interact just as strongly as their comparably sized aromatic counterparts. Bloom and Wheeler[16] examined the impacts of aromaticity more directly, showing that the π-electron delocalization associated with aromaticity actually hinders π-stacking interactions! That is, planar nonaromatic cyclic conjugated species can engage in stronger π-stacking interactions than their aromatic counterparts. Finally, Martinez and Iverson[17] reviewed both experimental and computational literature on diverse π-stacking interactions. They came to the conclusion that the name itself is highly misleading, because the attractive nature of these interactions is unrelated to the π-electron systems. Following Grimme,[15] we elect to use the term π-stacking primarily as a geometric descriptor. That is, we consider two π-systems to be "stacked" if they are in a roughly parallel arrangement with significant overlap.

For the simplest system that exhibits π-stacking interactions, the benzene dimer, one generally considers three prototypical configurations: sandwich, parallel-displaced, and T-shaped (see Fig. 10.1). Among these, the sandwich dimer is a saddle point on the potential energy surface, and lies about 1 kcal/mol higher in energy than

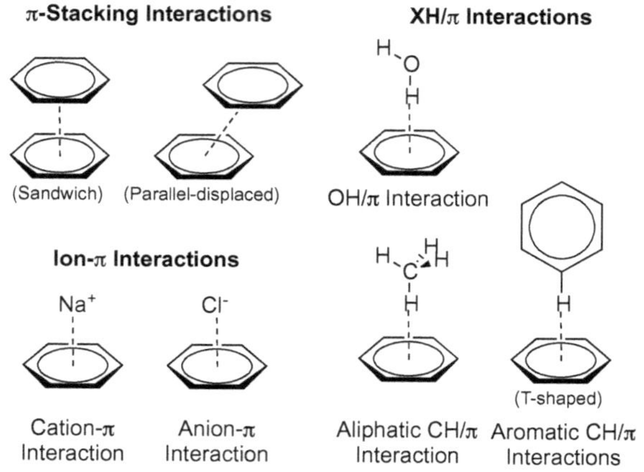

Figure 10.1. Prototypical noncovalent interactions involving aromatic rings.

the parallel-displaced and T-shaped configurations. Among these, we consider the sandwich and parallel-displaced configurations to be stacked; the T-shaped dimer is an example of an aromatic CH/π interaction (*vide infra*). This distinction between the sandwich and parallel-displaced interactions on the one hand and the T-shaped configuration on the other is justified not only on geometric grounds but also on physical grounds; whereas the first two configurations are driven primarily by dispersion interactions in the gas phase, the T-shaped interaction is largely electrostatic in origin.[18] The strength of π-stacking interactions can vary considerably across systems, depending on the presence of substituents and heteroatoms as well as the size of the interacting rings. Generally, the strength of π-stacking interactions increases with the increasing size of the interacting arenes. For instance, the π-stacking interaction between two stacked naphthalenes is much stronger than that between two stacked benzenes.[15]

There has been considerable effort aimed at understanding the impact of substituents on the strength of π-stacking interactions, which can be substantial.[13,14,19–34] Since the early 1990s, the prevailing model of substituent effects in π-stacking interactions was that championed by Hunter *et al.*[29–32] In this electrostatic model, π-stacking interactions are maximized when the two interacting arenes have complementary electrostatic character. That is, strong π-stacking interactions arise when an "electron-rich" ring interacts with an "electron-poor" ring. This view is based on the underlying assumption that substituents modulate the strength of π-stacking interactions by altering the π-electron density of the rings. However, computational work over the last decade, as well as experimental examples of strong π-stacking interactions between electron-poor rings, has upended this widely entrenched model.[19–25] Wheeler *et al.*[13,14,33,34] have introduced an alternative view, dubbed the local, direct interaction model of substituent effects in π-stacking interactions. In this model, the impact of substituents is primarily the result of direct, through-space electrostatic interactions between the substituents on one ring and the electric field of the other ring. The practical ramification is that the overall electrostatic character of the interacting arenes is unimportant; instead, the strength of π-stacking interactions depends on the presence of favorable or unfavorable local interactions around the periphery of the interactions rings.[13,14,33,34]

Ion–π interactions include cation–π and anion–π interactions, in which an atomic or polyatomic ion interacts with the face of an aromatic ring. Cation–π interactions have been known for decades, popularized in large part by Dougherty *et al.*[35–38] in the mid-1990s. These interactions are predominantly electrostatic in origin, although polarization of the arene by the cation also contributes to binding[39] Anion–π interactions were only recognized more recently and are also largely electrostatic in nature. However, dispersion and induction effects are more important to binding in these systems than they are in cation–π interactions. For instance, Geronimo *et al.*[40] demonstrated that favorable anion–π interactions can arise even in systems in which the electrostatic component of the interaction is slightly repulsive. In general, the addition of electron-withdrawing substituents to an arene enhances anion–π interactions and hinders cation–π interactions; the incorporation of nitrogens into the arene has a similar impact.

Both anion–π and cation–π interactions are often discussed in terms of charge–quadrupole interactions, which is the leading term in the multipolar expansion of the electrostatic interaction between an ion and a symmetric (nondipolar) arene. However, the interaction distances in these complexes are often smaller than the radius of the arene, rendering electrostatic multipole expansions invalid. Regardless, the strength of these interactions across similar arenes is generally correlated with the Q_{zz} component of the arene quadrupole moment; potential problems arise when considering particularly large arenes, in which case these charge–quadrupole models becomes notably worse.

Electrostatic potentials (ESPs), which implicitly account for all order of multipoles of the arene, provide a more reliable predictor of the strength of both cation–π and anion–π interactions.[41–43] For instance, Mecozzi *et al.*[41] demonstrated the predictive power of ESP plots in the context of cation–π interactions two decades ago.[41] More recently, Wheeler and Houk[42,43] reported very strong linear correlations between ESP values computed at the position of the ion and the interaction energy in model anion–π and cation–π interactions (see Fig. 10.2). This correlation can break down, however, when arenes with drastically different polarizabilities are considered. In such cases, variations induction effects spoil the correlation of total interaction energies with the electrostatic component captured by the ESPs.

XH/π interactions can describe any interaction between any X-H bond and the face of an arene. Common examples include CH/π, NH/π, and OH/π interactions, although Zhang *et al.*[44] recently reported the first examples of BH/π interactions. The nature of these interactions depends largely on the nature of the X-atom. For instance, Bloom *et al.*[45] showed that as one progresses from

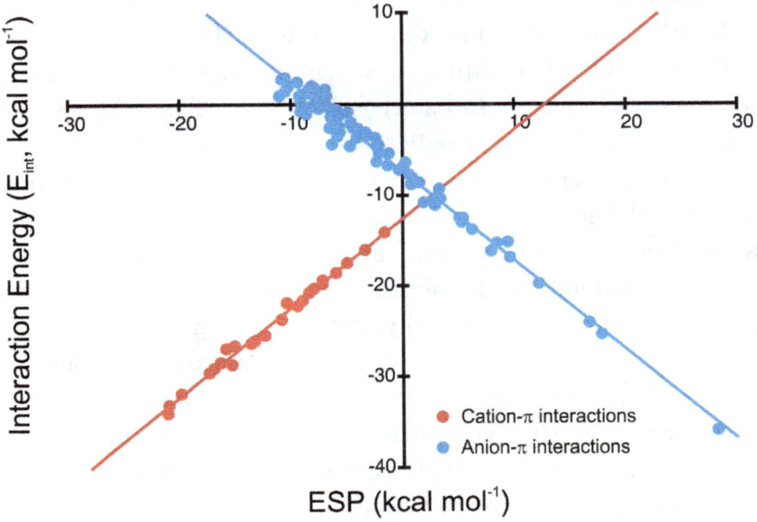

Figure 10.2. Interaction energies (kcal/mol) of model cation–π and anion–π interactions vs. the ESP at the position of ion above the center of substituted benzenes. Data are from Refs. [42, 43] and were computed at the M05-2X/6-31+G(d) and M06-2X/6-31+G(d) levels of theory for the cation–π and anion–π interactions, respectively.

BH/π to FH/π interactions, there is a gradual shift from largely dispersion-driven interactions to interaction dominated by electrostatics. Similarly, for a given type of XH/π interaction, there can be variation in the electrostatic component with changes in hybridization. For instance, whereas sp^3-hybridized CH/π interactions (e.g., $CH_4 \cdots$ benzene) are largely dispersion-driven, there is a significant contribution from electrostatic effects in sp-hybridized CH/π interactions (e.g., acetylene$\cdots$ benzene).[46,47]

One final, less well-appreciated noncovalent interaction that has emerged as a key factor in a surprising number of organic systems is the $CH \cdots O$ interaction. These nonclassical hydrogen bonds were recently reviewed by Johnston and Cheong[7] and have been shown to be key stereocontrolling elements in a wide range of organocatalyzed reactions.

The presence of a formal charge on one or more interacting species can significantly impact the strength and geometry of these noncovalent interactions. For instance, in 2002, Cannizzaro and Houk[48] reported remarkably strong $CH \cdots O$ interactions in $R_3N^+C-H \cdots O=C$ complexes, which are important in the context of molecular recognition and stereoselective catalysis. Moreover, this enhanced interaction was predicted to persist even in water. Subsequent work by Nepal and Scheiner[49] and Adhikari and Scheiner[50] has examined the impact of ionic charge more broadly, considering a number of model noncovalent interactions. In 2014, Nepal and Scheiner[49] examined the impact of ionic charge on CH/π interactions in model complexes of tetraalkylammonium cations with benzene (see Fig. 10.3). They found that such complexes are considerably more strongly bound than their neutral counterparts. Similarly, computations revealed that complexes of methylamines and thioethers with *N*-methyacetamide are strengthened considerably by an additional methyl group on the proton donor. These effects are tempered somewhat by polar solvents, but the ionic complexes were predicted to retain their favored status even in water.

In cases where both interacting molecules bear formal charges, one must be careful not to convolute the strength of a specific

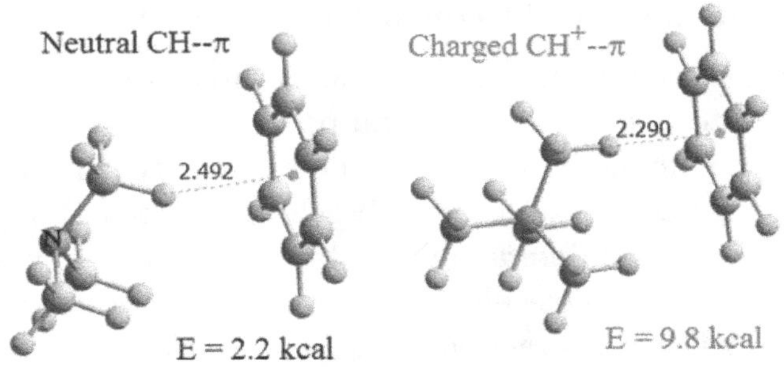

Figure 10.3. Model complexes studied by Nepal and Scheiner[49] to understand the impact of charge on CH/π interactions.
Source: Adapted with permission from *J. Phys. Chem. A*, 2014, 118, p. 9575. © 2014 American Chemical Society.

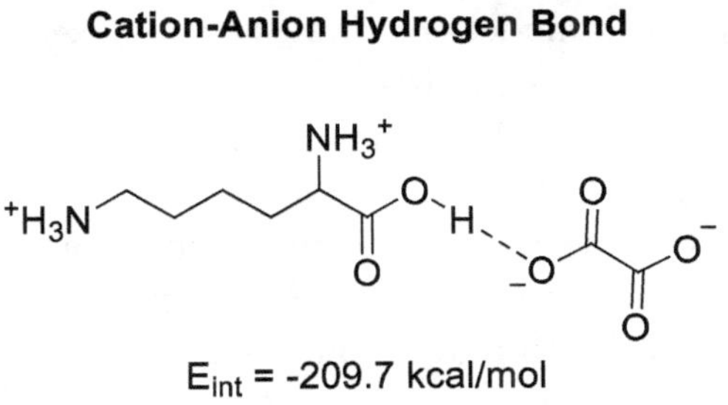

Figure 10.4. Example of a cation–anion hydrogen bond from D'Oria and Novoa.[51]

noncovalent interaction with the overall intermolecular Coulombic interaction between two charged species. For instance, in 2011, D'Oria and Novoa[51] introduced the concept of cation–anion hydrogen bonds (see Fig. 10.4), which exhibit binding energies sometimes exceeding 200 kcal/mol! However, these complexes are cases in which the interacting species bear complementary ionic charges, and the large reported gas-phase interaction energies are primarily the result of the Coulombic interactions between the two charged species. The hydrogen-bonding interactions themselves are likely not much stronger than the conventional hydrogen bonds.

10.3 Methods to study noncovalent interactions

Much of our understanding of noncovalent interactions involving aromatic rings stems from experimental probes of these interactions. In particular, a number of groups have devised molecular balances and supramolecular complexes that enable the experimental quantitation of these noncovalent interactions.[52–60] For instance, Zhao *et al.*[52] recently designed a molecular balance to probe the deuterium isotope effect on CH/π interactions, finding that this effect is either very small or nonexistent. Other torsional balance systems have provided unprecedented insight into the nature of noncovalent interactions, including the effects of solvents.[58–60] Much of the work in this area was reviewed by Mati and Cockroft in 2010.[61]

Complementary information has been gleaned from computational studies of both model noncovalent complexes and more realistic systems. However, the application of popular electronic structure methods to systems in which noncovalent interactions play key roles is rife with pitfalls.[1] For more than a decade, the venerable B3LYP functional was the workhorse of computational organic chemistry. By providing reliable structures, thermochemistry, and reaction barrier heights at a modest computational cost, B3LYP was the obvious choice for the vast majority of computational studies of medium-sized organic molecules. However, significant weaknesses in B3LYP became apparent as attention turned to larger molecular systems.

The major weakness stemmed from the inability of B3LYP and other conventional DFT functionals to describe dispersion effects. For instance, B3LYP and other conventional functionals predict purely repulsive interaction potentials for the benzene sandwich dimer, whereas reliable *ab initio* methods indicate a binding energy approaching 2 kcal/mol (see Fig. 10.5). Until about a decade ago, capturing dispersion-driven noncovalent interactions required the use of computationally demanding *ab initio* methods. In particular, coupled cluster theory (e.g., CCSD(T)) with large basis sets has been applied to many small model noncovalent complexes to provide benchmark-quality interaction potentials. Conventional second-order Møller–Plesset perturbation theory (MP2), which is considerably cheaper

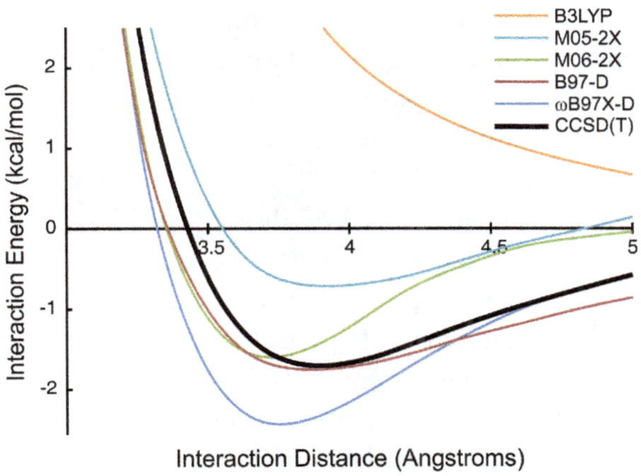

Figure 10.5. Interaction potentials for the benzene sandwich dimer computed using popular DFT functionals compared to benchmark CCSD(T) data from Ref. [63]. All DFT computations utilized the $6-311+G(d,p)$ basis set.

and can be applied to larger molecular systems, tends to overestimate dispersion interactions. However, the spin-component–scaled variant, SCS-MP2,[62] largely corrects this deficiency and has been used to provide reliable interaction energies for many noncovalent complexes.

Fortunately, there have been a number of advances in DFT methods in the last decade, and many functionals are now available for reliable studies of noncovalent interactions in organic systems. The most popular approaches employ the semi-empirical dispersion corrections of Grimme *et al.*[64,65] (so-called –D functionals, e.g., B97-D, ωB97X-D, etc.) and the M05 and M06 suites of functionals from Zhao *et al.*[66–71] (e.g., M05-2X and M06-2X). Such methods provide varying degrees of accuracy when applied to model stacked systems (see Fig. 10.5) and have proved reliable when applied to large organic systems in which noncovalent interactions play key roles. Consequently, these methods are now widely used in the computational organic chemistry community. However, it should be noted that the M06 family of functionals is particularly sensitive to the choice of integration grid, and the use of default grids in some popular electronic structure packages can lead to substantial errors in both predicted reaction energies and interaction energies.[72–75] Other computational tools available for quantifying individual noncovalent interactions in organic systems including Bader's quantum theory of atoms-in-molecules (QTAIM)[76–78] and the natural bond orbital (NBO) approach of Foster and Weinhold[79] and Reed *et al.*[80]

There are also a number of qualitative tools that are widely used to understand noncovalent interactions in organic systems. Chief among these are molecular ESP plots (see Fig. 10.6). Unfortunately, these ESP plots are often misinterpreted and misused. The primary problem arises from the connections between the electron density and ESP. Many organic chemists conflate ESP in a region with the local electron density. Common descriptions derived from ESP plots such as "electron-rich" and "electron-poor" only serve to exacerbate this problem. However, ESPs and electron densities are distinct, and, while they often track each other, there are countless examples where they do not. Most importantly, a change in the ESP in some region of space does not necessarily indicate a local change in the electron density. This can be seen most clearly for aromatic molecules: in 2009, Wheeler and Houk[81] showed that substituent effects on the ESPs of substituted arenes are dominated by the through-space effects of the substituents and not by any substantial changes in the π-electron density of the arene. For instance, the drastic differences among the ESPs over the centroids of the rings shown in Fig. 10.6 are due almost entirely to the through-space electrostatic effects of the substituents; any small differences in the π-electron densities of these rings have negligible impact on the ESPs. Similarly, Wheeler

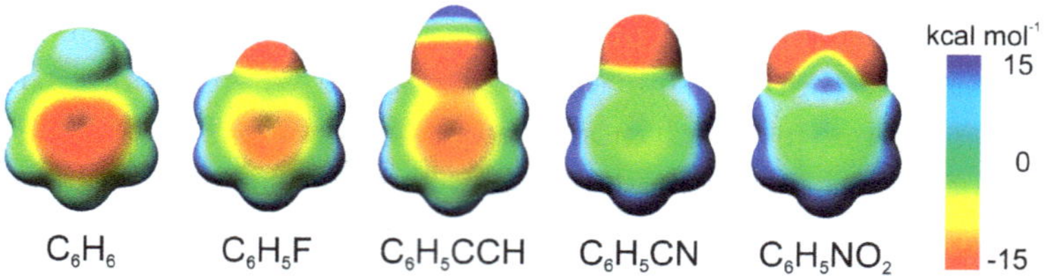

Figure 10.6. Molecular ESPs of several monosubstituted benzenes.

and Bloom[82] recently showed that changes in the ESPs above the centroids of many *N*-heterocycles are not due to changes in the π-electron density, as is commonly assumed.

Finally, Contreras-García *et al.*[83] and Johnson *et al.*[84] introduced the now widely used noncovalent interation (NCI) method, which provides a graphical representation of repulsive and attractive inter- and intra-molecular interactions based on an analysis of the electron density and its gradient. The resulting "NCI plots" provide a useful guide for comparing weak inter- and intra-molecular interactions among different organic systems.

10.4 Examples of noncovalent interactions in organic systems

Having established the broad range of computational tools that are now available to study noncovalent interactions in the context of organic systems, we next discuss representative examples in which insight into noncovalent interactions has been gleaned from careful computational studies.

10.4.1 *Noncovalent interactions as conformational controlling elements*

Intramolecular noncovalent interactions can have considerable impact on the conformations of organic molecules. For instance, Nishio[4] provided intriguing examples of the impact of noncovalent interactions on molecular conformations, even suggesting that some well-established phenomena like the alkyl ketone effect and anomeric effect are artifacts of stabilizing noncovalent interactions. Similarly, work by Takahashi *et al.*[3,85,86] also suggested that the relatively small energy difference between axial and equatorial conformers of halogenated cyclohexanes, as compared to alkyl cyclohexanes, may be ascribed to stabilizing 1,3-diaxial $X \cdots H$ noncovalent interactions.

Nishio *et al.*[87] have also shown that noncovalent interactions can play key roles in the conformations of larger molecules. For instance, levopimaric acid adopts an unusual folded conformation, as opposed

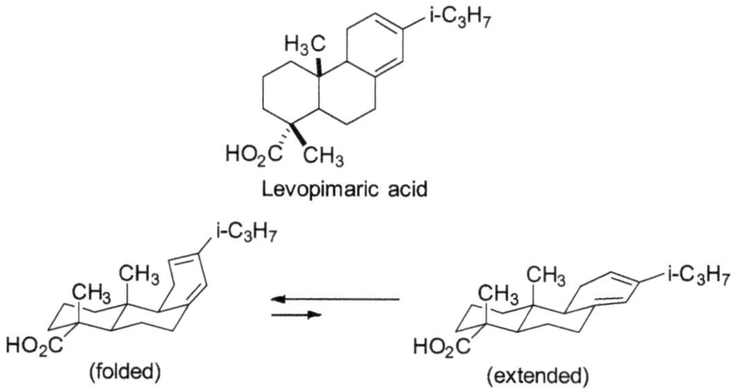

Figure 10.7. Folded and extended conformers of levopimaric acid studied by Nishio *et al.*, who showed computationally that CH/π interaction stabilizes the folded conformer.

to the more sterically relieved extended conformation one might expect (see Fig. 10.7). Takahashi *et al.*[88] showed computationally that for model compounds, the folded conformer benefits from stabilizing CH/π interactions between the conjugated diene ring and nearby methyl group.

Jones *et al.*[89] have also presented a compelling study in which they demonstrated the switching of conformational preference in flouroamides through noncovalent CH$\cdots$O interactions. α-Flouroamides are known to have a strong tendency to adopt *trans*-planar conformations in which the fluorine is *anti* to the carbonyl, minimizing electrostatic repulsions (see Fig. 10.8(a)).[90] However Jones *et al.*[89] envisioned that gradually increasing fluorine substitution, along with the incorporation of a suitable proton acceptor, could override this inherent conformational bias. Computed torsional potential energy scans for model fluoroamides confirmed that the conformer with a *trans*-planar OCCF dihedral angle is favored by 6 kcal/mol over the corresponding *cis*-planar conformation in the case of $CH_3NHCOCH_2F$; for $CH_3NHCOCHF_2$ this energy difference is reduced to 4 kcal/mol. A more elaborate system was then devised in which a carbamate group was installed that could interact with the CHF_2 group through CH$\cdots$O interactions (see Fig. 10.8(b)).

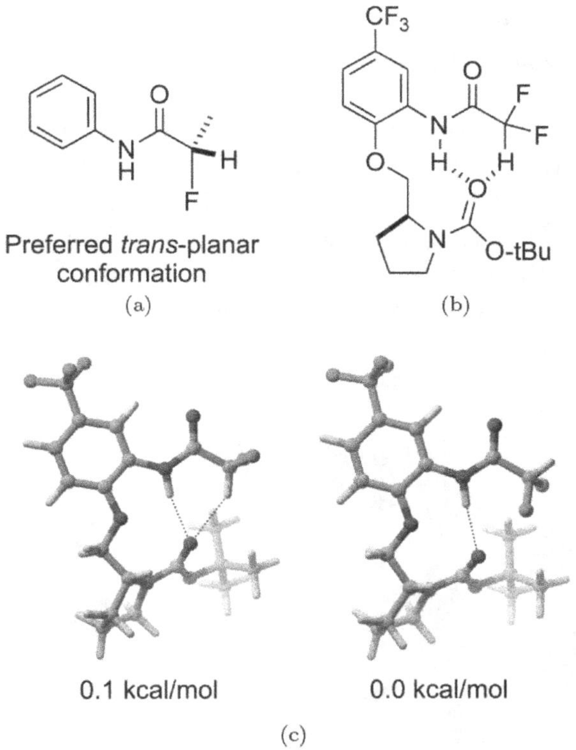

Figure 10.8. (a) Strongly preferred *trans*-planar conformer of α-fluoroamides; (b) modified system devised by Jones *et al.*[89] to prove the ability of CH $\cdots$ O interactions to impact conformations; (c) lowest-lying computed conformers, in which the *cis*-planar conformer is nearly isoenergetic with the *trans*-planar conformer due to favorable CH $\cdots$ O interactions in the former.

Ultimately, it was shown that this CH $\cdots$ O interaction is sufficient to overcome the inherent bias for *trans*-planar configurations of fluoroamides and render the two planar conformations roughly isoenergetic (see Fig. 10.8(c)).

10.4.2 *Noncovalent interactions in organic reactions*

Noncovalent interactions have also emerged as a powerful strategy for controlling the outcomes of organic reactions, with applications across both catalytic and noncatalytic transformations. A number of groups have reviewed this area[6–10]; here, we present several selected examples of organic reactions in which noncovalent interactions play vital roles.

Maity *et al.*[91] documented the importance of weak interactions in the enantioselectivity of a phosphoric acid catalyzed aromatic aza-Claisen rearrangement (Fig. 10.9). Through computations, they showed that the catalyst engages with the substrate via NH···O and CH···O interactions; the nine-anthracenyl group of the catalyst blocks the bottom *Si* face, forcing the reaction to take place

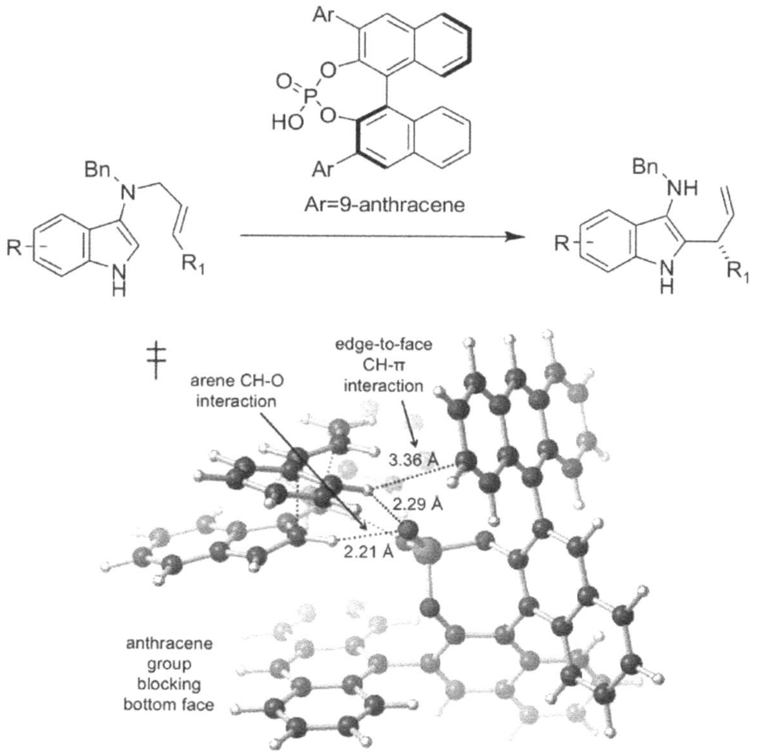

Figure 10.9. Phosphoric acid catalyzed aromatic Aza-Claisen of Maity *et al.*[91] along with their computed TS structure featuring a number of pivotal noncovalent interactions.
Source: Adapted with permission from *J. Am. Chem. Soc.*, 2013, 135, p. 16380. © 2013 American Chemical Society.

on the less crowded *Re* face. Moreover, it was shown that the higher selectivity in the case of aromatic substituents, compared to aliphatic substituents, can be attributed to additional stabilizing edge-to-face CH/π interactions in former cases.

Odagi *et al.*[92] examined the role of noncovalent interactions in the stereoselectivity of an oxidative kinetic resolution of a tetralone derived β-ketoester using a guanidine-bisurea organocatalyst. In addition to hydrogen-bonding interactions, CH/π and π-stacking interactions were shown to stabilize the TS leading to the favored isomer (see Fig. 10.10). Although Odagi *et al.*[92] did not quantify the impact of individual noncovalent interactions, their overall predicted free energy differences between diastereomeric TS structures were in perfect agreement with experimental stereoselectivities.

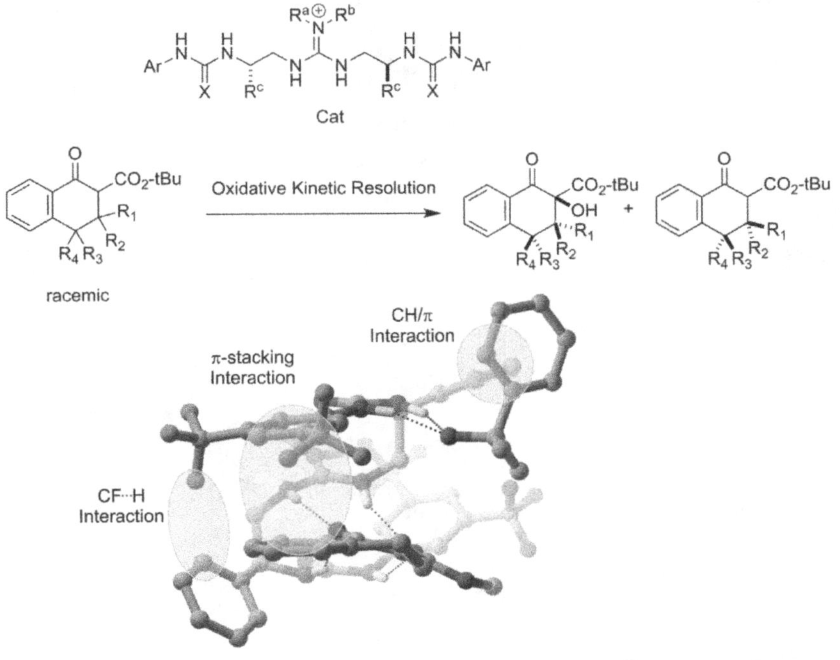

Figure 10.10. Oxidative kinetic resolution of Odagi *et al.*[92] along with a key TS structure featuring nonconventional CF $\cdots$ H, π-stacking, and CH/π interactions (nonpolar hydrogens omitted for clarity).

Finally, we note that a thorough understanding of noncovalent interactions can pave the way for the design of improved catalysts. Jang *et al.*[93] used their understanding of nonclassical hydrogen bonds (NCHBs) to design an imidazolium-derived *N*-heterocyclic carbene catalyst for asymmetric homoenolate additions to acyl phosphonates (see Fig. 10.11). They envisioned that formation of the major (*S*,*S*) and minor (*R*,*R*) enantiomeric products stem from the nucleophilic attack of the acyl phosphonate carbonyl by the homoenol, whereby differential stabilization of the phosphonyl oxygen by the aryl protons of the catalyst through NCHBs is mainly responsible for the observed stereoselectivity. A careful computational analysis of competing transition states revealed a number of important noncovalent contacts favoring formation of either the major or minor products (Fig. 10.11). This analysis suggested that methyl substitutions at the *meta* positions of the terminal phenyl ring would further destabilize the TS

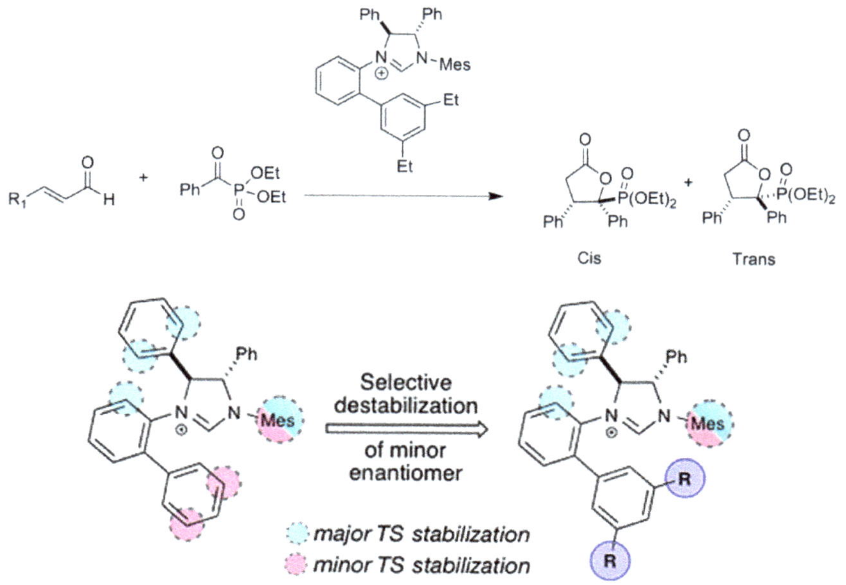

Figure 10.11. NHC-catalyzed asymmetric homoenolate additions to acyl phosphonates, along with the computationally derived strategy for enhancing stereoselectivity.

structure leading to minor enantiomer, which was then demonstrated experimentally. Replacing these methyl groups with even bulkier ethyl groups further increased the *er* to 94:6.

10.5 Noncovalent interactions in supramolecular systems

Noncovalent interactions also play vital roles in supramolecular chemistry, and noncovalent interactions involving aromatic rings have proved particular useful in sensing applications. For instance, Watt *et al.*[94] reported a tripodal urea-based receptor that shows excellent selectivity toward nitrates (Fig. 10.12). They observed that anion-binding by this receptor followed the general trend $NO_3^- > Cl^- > Br^- > I^-$. However, this substrate specificity is lost upon removal of the three fluorines from the central phenyl ring. This suggested that anion–π interactions between the bound anion and central phenyl ring, which would only be favorable in the case of the trifluorophenyl case, are important for both binding and selectivity.

Other functional supramolecular systems have been designed that rely on anion–π interactions, including the anion receptor cage based

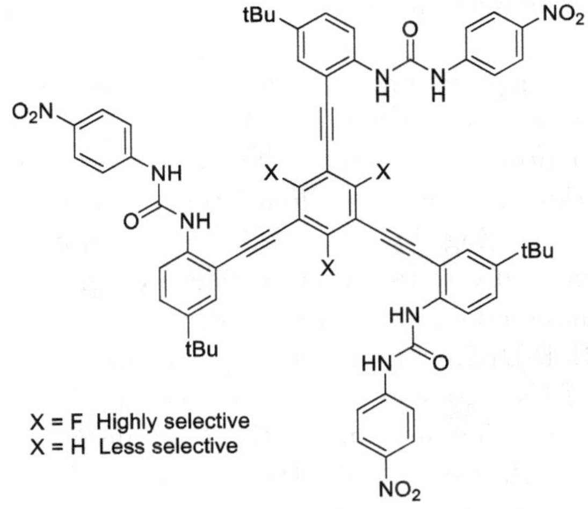

Figure 10.12. Tripodal urea based anion receptor of Watt *et al.*[94]

on triazine linked by trialkylamines pioneered by Mascal *et al.*,[95] the naphthalene diimide (NDI) based fluoride sensor by Guha and Saha,[96] the NDI-based prism and macrobicyclic cyclophane derivative of Schneebeli *et al.*[97] and Hafezi *et al.*,[98] and the NDI-based "anion–π slides" of Perez-Velasco *et al.*[99]

10.6 Noncovalent interactions in biological systems

Noncovalent interactions also play vital roles in myriad biological systems,[100] impacting everything from the structure and stability of DNA and proteins to the binding of ligands by proteins. As such, understanding these noncovalent interactions is important for understanding biological function. Understanding π-stacking interactions is particularly important within the context of drug design. For example, Steuber *et al.*[101] developed a potent inhibitor for aldol reductase featuring an *m*-nitrophenyl ring. The X-ray data and docking simulations revealed that the binding of this inhibitor is driven in large part by the stacking interaction of this *m*-nitrohphenyl ring with Trp111 chain in the binding pocket (see Fig. 10.13). Moreover, removal of the nitro group eroded the binding affinity by almost an order of magnitude, demonstrating the importance of substituent effects on π-stacking interactions and their subsequent impact on ligand binding.

The π-stacking interactions are also important in DNA-intercalation phenomena, which have been widely studied using computational quantum chemistry.[102–105] For example, Řeha *et al.*[103] studied π-stacking interactions of four intercalators used in antitumor chemotherapy, showing that the binding is driven by a combination of electrostatic and dispersion interactions. Hargis *et al.*[105] studied the stacking interactions of DNA-base pairs with benzo[a]pyrene diol epoxide, (+)BaP-DE2, which is the major carcinogenic metabolite of components of tobacco smoke and soot (see Fig. 10.14). DFT optimizations showed that in some of the most favorable noncovalent complexes of (+)BaP-DE2 with the GC base pair, the epoxide is perfectly poised for backside nucleophilic attack by the exocyclic

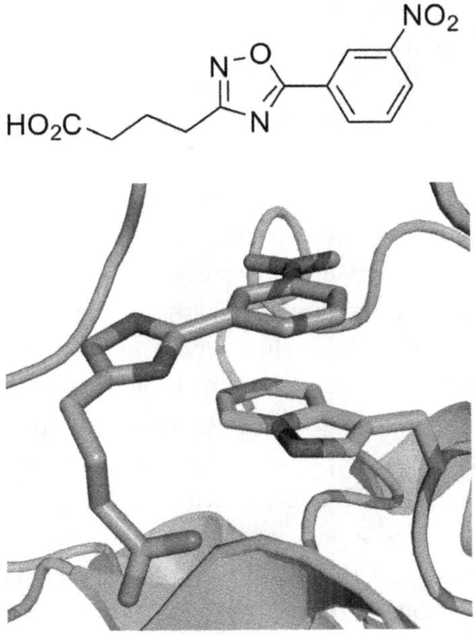

Figure 10.13. Aldol reductase inhibitor from Ref. [101] whose binding is driven in part by a π-stacking interaction between a nitrophenyl ring and Trp111 (PDB code: 2IKG).

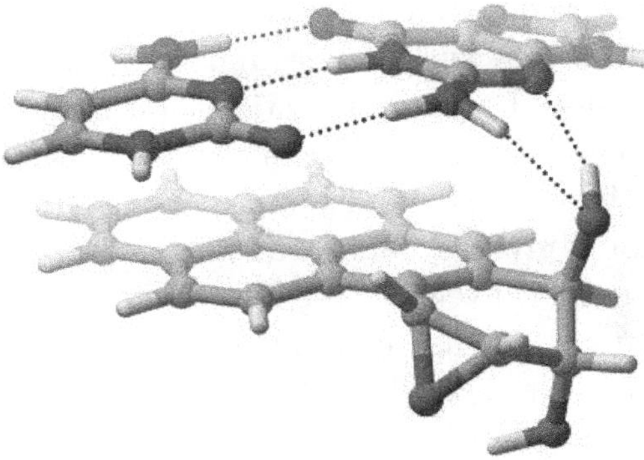

Figure 10.14. Stacked complex of benzo[a]pyrene diol epoxide with the GC base pair in which the exocyclic amino group of guanine is ideally positioned for backside nucleophilic attack of the epoxide, from Ref. [105].

amine of guanine. This provided a potential explanation for the strong tendency of this carcinogen to form covalent adducts with GC-rich regions of double-stranded DNA.

Other noncovalent interactions abound in biological systems. For instance, Hong and Tantillo[106] have probed the role of CH/π interactions as modulators of carbocation structure, with important implications in the cation-rearrangements operative in terpene biosynthesis. Finally, we note there is rapidly growing interest in anion–π interactions in biological systems, driven in part by computational analyses of the PDB that have revealed many close contacts between anions and aromatic rings in biological macromolecules. For instance, in 2013, Bauzá *et al.*[107] quantified the involvement of anion–π interactions in flavin-dependent oxidoreductases[107] X-ray crystal structures along with DFT computations indicated that the π-system of the flavin stabilizes a key anionic intermediate via an anion–π interaction. Similarly, the inhibition of ureate oxidase by cyanide has been explained based on attractive anion–π interactions between CN^- and the uric acid moiety.[108]

10.7 Practical considerations

Modern DFT methods have opened the door for reliable studies of a broad range of organic systems in which noncovalent interactions play key roles. However, such studies must be carried out with due caution, since many of these interactions are driven by dispersion interactions that not well described by many once-popular methods. Although the impact of dispersion interactions might fortuitously cancel for a given system, explaining the surprisingly good results derived from conventional DFT functionals in many cases, in general it is necessary use methods capable of capturing dispersion interactions when studying any organic systems larger than a few dozen atoms. Luckily, there are now many widely available DFT functionals that provide accurate descriptions of these interactions. In general, the second- and third-generation empirical dispersion corrections from Grimme *et al.*[64,65] (the so-called -D2 and -D3 methods) are the simplest to employ and can be paired with any well-behaved DFT functional. We have found that the B97-D functional,[64,109] when used with a triple-ζ basis set such as def2-TZVP,[110] provides reliable predictions across many different noncovalent interactions and organic systems. Moreover, when paired with density fitting techniques, B97-D computations are inexpensive and can be routinely applied to systems with 100s of atoms or to 1000s of systems with dozens of atoms. However, as always, one must be careful to reliably describe all properties of interest. In this regard, it should be noted that B97-D provides reaction barrier heights and overall reaction thermochemistry that are often in significant error relative to experiment or more robust computational methods. More sophisticated functionals, including ωB97X-D,[111] ameliorate many of these problems, providing accurate predictions of not only noncovalent interactions but also reaction kinetics and thermochemistry. Of course, this comes with some increase in computational cost.

Acknowledgments

This work was supported in part by the National Science Foundation (Grant CHE-1266822) and the Welch Foundation (Grant A-1775).

References

1. Müller-Dethlefs, K. and Hobza, P. (2000). Noncovalent interactions: A challenge for experiment and theory, *Chem. Rev.*, 100, pp. 143–168.
2. Salonen, L. M., Ellermann, M. and Diederich, F. (2011). Aromatic rings in chemical and biological recognition: Energetics and structures, *Angew. Chem. Int. Ed.*, 50, pp. 4808–4842.
3. Takahashi, O., Kohno, Y. and Nishio, M. (2010). Relevance of weak hydrogen bonds in the conformation of organic compounds and bioconjugates: Evidence from recent experimental data and high-level ab initio MO calculations, *Chem. Rev.*, 110, pp. 6049–6076.
4. Nishio, M. (2005). CH/π hydrogen bonds in organic reactions, *Tetrahedron*, 61, pp. 6923–6950.
5. Wagner, J. P. and Schreiner, P. R. (2015). London dispersion in molecular chemistry — Reconsidering steric effects, *Angew. Chem. Int. Ed.*, 54, pp. 12274–12296.
6. Krenske, E. H. and Houk, K. N. (2013). Aromatic interactions as control elements in stereoselective organic reactions, *Acc. Chem. Res.*, 46, pp. 979–989.

7. Johnston, R. C. and Cheong, P. H.-Y. (2013). C-H$\cdots$O non-classical hydrogen bonding in the stereomechanics of organic transformations: Theory and recognition, *Org. Biomol. Chem.*, 11, pp. 5057–5064.
8. Singh, S. K. and Das, A. (2015). The $n \longrightarrow \pi^*$ interaction: A rapidly emerging non-covalent interaction, *Phys. Chem. Chem. Phys.*, 17, pp. 9596–9612.
9. Zhao, Y., Cotelle, Y., Sakai, N. and Matile, S. (2016). Unorthodox interactions at work, *J. Am. Chem. Soc.*, 138, pp. 4270–4277.
10. Kennedy, C. R., Lin, S. and Jacobsen, E. N. (2016). The cation–π interaction in small-molecule catalysis, *Angew. Chem. Int. Ed.*, doi: 10.1002/anie.201600547.
11. Stone, A. J. (1996). *The Theory of Intermolecular Forces* (Oxford University Press, Oxford).
12. Raju, R. K., Bloom, J. W. G., An, Y. and Wheeler, S. E. (2011). Substituent effects in non-covalent interactions with aromatic rings: Insights from computational chemistry, *ChemPhysChem*, 12, pp. 3116–3130.
13. Wheeler, S. E. (2013). Understanding substituent effects in non-covalent interactions involving aromatic rings, *Acc. Chem. Res.*, 46, pp. 1029–1038.
14. Wheeler, S. E. and Bloom, J. W. G. (2014). Toward a more complete understanding of non-covalent interactions involving aromatic rings, *J. Phys. Chem. A*, 118, pp. 6133–6147.
15. Grimme, S. (2008). Do special noncovalent π-π stacking interactions really exist? *Angew. Chem. Int. Ed.*, 47, pp. 3430–3434.
16. Bloom, J. W. G. and Wheeler, S. E. (2011). Taking aromaticity out of aromatic interactions, *Angew. Chem. Int. Ed.*, 50, pp. 7847–7849.
17. Martinez, C. R. and Iverson, B. L. (2012). Rethinking the term "π-stacking," *Chem. Sci.*, 3, pp. 2191–2201.
18. Sinnokrot, M. O. and Sherrill, C. D. (2006). High-accuracy quantum mechanical studies of π-π interactions in benzene dimers, *J. Phys. Chem. A*, 110, pp. 10656–10668.
19. Watt, M., Hardebeck, L. K. E., Kirkpatrick, C. C. and Lewis, M. (2011). Face-to-face arene-arene binding energies: Dominated by dispersion but predicted by electrostatic and dispersion/polarizability substituent constants, *J. Am. Chem. Soc.*, 133, pp. 3854–3862.
20. Sinnokrot, M. O. and Sherrill, C. D. (2003). Unexpected substituent effects in face-to-face π-stacking interactions, *J. Phys. Chem. A*, 107, pp. 8377–8379.
21. Sinnokrot, M. O. and Sherrill, C. D. (2004). Substituent effects in π-π interactions: Sandwich and T-shaped configurations, *J. Am. Chem. Soc.*, 126, pp. 7690–7697.
22. Ringer, A. L., Sinnokrot, M. O., Lively, R. P. and Sherrill, C. D. (2006). The effect of multiple substituents on sandwich and T-shaped π-π interactions, *Chem. Eur. J.*, 12, pp. 3821–3828.

23. Arnstein, S. A. and Sherrill, C. D. (2008). Substituent effects in parallel-displaced π-π interactions, *Phys. Chem. Chem. Phys.*, 10, pp. 2646–2655.
24. Hohenstein, E. G., Duan, J. and Sherrill, C. D. (2011). Origin of the surprising enhancement of electrostatic energies by electron-donating substituents in substituted benzene sandwich dimers, *J. Am. Chem. Soc.*, 133, pp. 13244–13247.
25. Parrish, R. M. and Sherrill, C. D. (2014). Quantum-mechanical evaluation of π-π versus substituent-π interactions in π stacking: Direct evidence for the Wheeler-Houk picture, *J. Am. Chem. Soc.*, 136, pp. 17386–17389.
26. Cozzi, F., Cinquini M., Annunziata R., Dwyer T. and Siegel J. S. (1992). Polar/π interactions between stacked aryls in 1,8-diarylnaphthalenes, *J. Am. Chem. Soc.*, 114, pp. 5729–5733.
27. Cozzi, F., Cinquini, M., Annunziata, R. and Siegel, J. S. (1993). Dominance of polar/π over charge-transfer effects in stacked phenyl interactions, *J. Am. Chem. Soc.*, 115, pp. 5330–5331.
28. Cozzi, F., Annunziata, R., Benaglia, M., Cinquini, M., Raimondi, L., Baldridge, K. K. *et al.* (2003). Through-space interactions between face-to-face, center-to-edge oriented arenes: Importance of polar–π effects, *Org. Biomol. Chem.*, 1, pp. 157–162.
29. Hunter, C. A. and Sanders, J. K. M. (1990). The nature of π-π interactions, *J. Am. Chem. Soc.*, 112, pp. 5525–5534.
30. Hunter, C. A., Lawson, K. R., Perkins, J. and Urch, C. J. (2001). Aromatic interactions, *J. Chem. Soc., Perkin Trans.*, 2, pp. 651–669.
31. Cockroft, S. L., Hunter, C. A., Lawson, K. R., Perkins, J. and Urch, C. J. (2005). Electrostatic control of aromatic stacking interactions, *J. Am. Chem. Soc.*, 127, pp. 8594–8595.
32. Cockroft, S. L., Perkins, J., Zonta, C., Adams, H., Spey, S. E., Low, C. M. R. *et al.* (2007). Substituent effects on aromatic stacking interactions, *Org. Biomol. Chem.*, 5, pp. 1062–1080.
33. Wheeler, S. E. (2011). Local nature of substituent effects in stacking interactions, *J. Am. Chem. Soc.*, 133, pp. 10262–10274.
34. Raju, R. K., Bloom, J. W. G. and Wheeler, S. E. (2013). Broad transferability of substituent effects in π-stacking interactions provides new insights into their origin, *J. Chem. Theory Comput.*, 9, pp. 3479–3490.
35. Ma, J. C. and Dougherty, D. A. (1997). The cation-π interaction, *Chem. Rev.*, 97, pp. 1303–1324.
36. Dougherty, D. A. (2013). The cation-π interaction, *Acc. Chem. Res.*, 46, pp. 885–893.
37. Gallivan, J. P. and Dougherty , D. A. (1999). Cation-π interactions in structural biology, *Proc. Natl. Acad. Sci. USA*, 96, pp. 9459–9464.
38. Dougherty, D. A. (2008). Physical organic chemistry on the brain, *J. Org. Chem.*, 73, pp. 3667–3673.

39. Cubero, E., Luque, F. J. and Orozco, M. (1998). Is polarization important in cation-π interactions? *Proc. Natl. Acad. Sci. USA*, 95, pp. 5976–5980.
40. Geronimo, I., Singh, N. J. and Kim, K. S. (2011). Can electron-rich π systems bind anions? *J. Chem. Theory Comput.*, 7, pp. 825–829.
41. Mecozzi, S., West, A. P., Jr. and Dougherty, D. A. (1996). Cation-π interactions in simple aromatics: Electrostatics provide a predictive tool, *J. Am. Chem. Soc.*, 118, pp. 2307–2308.
42. Wheeler, S. E. and Houk, K. N. (2009). Substituent effects in cation/π interactions and electrostatic potentials above the center of substituted benzenes are due primarily to through-space effects of the substituents, *J. Am. Chem. Soc.*, 131, pp. 3126–3127.
43. Wheeler, S. E. and Houk, K. N. (2010). Are anion/π interactions actually a case of simple charge-dipole interactions? *J. Phys. Chem. A*, 114, pp. 8658–8664.
44. Zhang, X., Dai, H., Yan, H., Zou, W. and Cremer, D. (2016). B–H$\cdots\pi$ Interaction: A new type of nonclassical hydrogen bonding, *J. Am. Chem. Soc.*, 138, pp. 4334–4337.
45. Bloom, J. W. G., Raju, R. K. and Wheeler, S. E. (2012). Physical nature of substituent effects in xh/π interactions, *J. Chem. Theory Comput.*, 8, pp. 3167–3174.
46. Shibasaki, K., Fujii, A., Mikami, N. and Tsuzuki, S. (2006). Magnitude of the CH/π interaction in the gas phase: Experimental and theoretical determination of the accurate interaction energy in benzene-methane, *J. Phys. Chem. A*, 110, pp. 4397–4404.
47. Tsuzuki, S., Honda, K., Uchimaru, T., Mikami, M. and Tanabe, K. (2000). The magnitude of the ch/π interaction between benzene and some model hydrocarbons, *J. Am. Chem. Soc.*, 122, pp. 3746–3753.
48. Cannizzaro, C. E. and Houk, K. N. (2002). Magnitudes and chemical consequences of $R_3N^+{-}C{-}H\cdots O{=}C$ hydrogen bonding, *J. Am. Chem. Soc.*, 124, pp. 7163–7169.
49. Nepal, B. and Scheiner, S. (2014). Effect of ionic charge on the CH$\cdots\pi$ hydrogen bond, *J. Phys. Chem. A*, 118, pp. 9575–9587.
50. Adhikari, U. and Scheiner, S. (2013). Magnitude and mechanism of charge enhancement of CH$\cdots$O hydrogen bonds, *J. Phys. Chem. A*, 117, pp. 10551–10562.
51. D'Oria, E. and Novoa, J. J. (2011). Cation–anion hydrogen bonds: A new class of hydrogen bonds that extends their strength beyond the covalent limit. A theoretical characterization, *J. Phys. Chem. A*, 115, pp. 13114–13123.
52. Zhao, C., Parrish, R. M., Smith, M. D., Pellechia, P. J., Sherrill, C. D. and Shimizu, K. D. (2012). Do deuteriums form stronger CH-π interactions? *J. Am. Chem. Soc.*, 134, pp. 14306–14309.
53. Oki, M. (1990). 1,9-disubstituted triptycenes: An excellent probe for weak molecular interactions, *Acc. Chem. Res.*, 23, pp. 351–356.

54. Gung, B. W., Xue, X. and Reich, H. J. (2005). The strength of parallel-displaced arene–arene interactions in chloroform, *J. Org. Chem.*, 70, pp. 3641–3644.
55. Carroll, W. R., Pellechia, P. and Shimizu, K. D. (2008). A rigid molecular balance for measuring face-to-face arene–arene interactions, *Org. Lett.*, 10, pp. 3547–3550.
56. Yamada, S., Yamamoto, N. and Takamori, E. (2015). A molecular seesaw balance: Evaluation of solvent and counteranion effects on pyridinium-π interactions, *Org. Lett.*, 17, pp. 4862–4865.
57. Motherwell, W. B., Moïse, J., Aliev, A. E., Nič, M., Coles, S. J., Horton, P. N. *et al.* (2007). Noncovalent functional-group–arene interactions, *Angew. Chem. Int. Ed.*, 46, pp. 7823–7826.
58. Yang, L., Brazier, J. B., Hubbard, T. A., Rogers, D. M. and Cockroft, S. L. (2015). Can dispersion forces govern aromatic stacking in an organic solvent? *Angew. Chem. Int. Ed.*, 55, pp. 912–916.
59. Muchowska, K. B., Adam, C., Mati, I. K. and Cockroft, S. L. (2013). Electrostatic Modulation of Aromatic Rings via Explicit Solvation of Substituents, *J. Am. Chem. Soc.*, 135, pp. 9976–9979.
60. Yang, L., Adam, C. and Cockroft, S. L. (2015). Quantifying solvophobic effects in nonpolar cohesive interactions, *J. Am. Chem. Soc.*, 137, pp. 10084–10087.
61. Mati, I. K. and Cockroft, S. L. (2010). Molecular balances for quantifying non-covalent interactions, *Chem. Soc. Rev.*, 39, pp. 4195–4205.
62. Grimme, S. (2003). Improved second-order Møller–Plesset perturbation theory by separate scaling of parallel- and antiparallel-spin pair correlation energies, *J. Chem. Phys.*, 118, pp. 9095–9102.
63. Sinnokrot, M. O. and Sherrill, C. D. (2004). Highly accurate coupled cluster potential energy curves for the benzene dimer: Sandwich, T-shaped, and parallel-displaced configurations, *J. Phys. Chem. A*, 108, pp. 10200–10207.
64. Grimme, S. (2006). Semiempirical GGA-type density functional constructed with a long-range dispersion correction, *J. Comp. Chem.*, 27, pp. 1787–1799.
65. Grimme, S., Antony, J., Ehrlich, S. and Krieg, H. (2010). A Consistent and accurate *ab initio* parametrization of density functional dispersion correction (DFT-D) for the 94 elements H-Pu, *J. Chem. Phys.*, 132, p. 154104.
66. Zhao, Y., Schultz, N. E. and Truhlar, D. G. (2006). Design of density functionals by combining the method of constraint satisfaction with parametrization for thermochemistry, thermochemical kinetics, and noncovalent interactions, *J. Chem. Theory Comput.*, 2, pp. 364–382.
67. Zhao, Y. and Truhlar, D. G. (2006). Density functional for spectroscopy: No long-range self-interaction error, good performance for Rydberg and charge-transfer states, and better performance on average than B3LYP for ground states, *J. Phys. Chem. A*, 110, p. 13126.

68. Zhao, Y. and Truhlar, D. G. (2006). A New local density functional for main-group thermochemistry, transition metal bonding, thermochemical kinetics, and noncovalent interactions, *J. Chem. Phys.*, 125, p. 194101.
69. Zhao Y. and Truhlar, D. G. (2007). Density functionals for noncovalent interaction energies of biological importance, *J. Chem. Theory Comput.*, 3, pp. 289–300.
70. Zhao, Y. and Truhlar, D. G. (2008). The M06 suite of density functionals for main group thermochemistry, thermochemical kinetics, noncovalent interactions, excited states, and transition elements: Two new functionals and systematic testing of four M06 functionals and twelve other functionals, *Theo. Chem. Acc.*, 120, pp. 215–241.
71. Zhao, Y. and Truhlar, D. G. (2008). Density functionals with broad applicability in chemistry, *Acc. Chem. Res.*, 41, pp. 157–167.
72. Gräfenstein, J. and Cremer, D. (2007). Efficient density-functional theory integrations by locally augmented radial grids, *J. Chem. Phys.*, 127, p. 164113.
73. Gräfenstein, J., Izotov, D. and Cremer, D. (2007). Avoiding singularity problems associated with Meta-GGA (Generalized Gradient Approximation) exchange and correlation functionals containing the kinetic energy density, *J. Chem. Phys.*, 127, p. 214103.
74. Johnson, E. R., Becke, A. D., Sherrill, C. D. and DiLabio, G. A. (2009). Oscillations in meta-generalized-gradient approximation potential energy surfaces for dispersion-bound complexes, *J. Chem. Phys.*, 131, p. 034111.
75. Wheeler, S. E. and Houk, K. N. (2010). Integration grid errors for meta-GGA-predicted reaction energies: Origin of grid errors for the M06 suite of functionals, *J. Chem. Theory Comput.*, 6, pp. 395–404.
76. Tognetti, V. and Joubert, L. (2014). Density functional theory and Bader's Atoms-in-Molecules Theory: Towards a vivid dialogue, *Phys. Chem. Chem. Phys.*, 16, pp. 14539–14550.
77. Bader, R. F. W. (1991). A quantum theory of molecular structure and its applications, *Chem. Rev.*, 91, pp. 893–928.
78. Bader, R. F. W. (1985). Atoms in molecules, *Acc. Chem. Res.*, 18, pp. 9–15.
79. Foster, J. P. and Weinhold, F. (1980). Natural hybrid orbitals, *J. Am. Chem. Soc.*, 102, pp. 7211–7218.
80. Reed, A. E. and Curtiss, L. A. and Weinhold, F. (1988). Intermolecular interactions from a natural bond orbital, donor-acceptor viewpoint, *Chem. Rev.*, 88, pp. 899–926.
81. Wheeler, S. E. and Houk, K. N. (2009). Through-space effects of substituents dominate molecular electrostatic potentials of substituted arenes, *J. Chem. Theory Comput.*, 5, pp. 2301–2312.
82. Wheeler, S. E. and Bloom, J. W. G. (2014). Anion-π interactions and positive electrostatic potentials of N-heterocycles arise from the positions of the nuclei, not changes in the π-electron distribution, *Chem. Commun.*, 50, pp. 11118–11121.

83. Contreras-García, J., Johnson, E. R., Keinan, S., Chaudret, R., Piquemal, J.-P., Beratan, D. N. *et al.* (2011). NCIPLOT: A program for plotting noncovalent interaction regions, *J. Chem. Theory Comput.*, 7, pp. 625–632.
84. Johnson, E. R., Keinan, S., Mori-Sánchez, P., Contreras-García, J., Cohen, A. J. and Yang, W. (2010). Revealing noncovalent interactions, *J. Am. Chem. Soc.*, 132, pp. 6498–6506.
85. Takahashi, O., Yasunaga, K., Gondoh, Y., Kohno, Y., Saito, K. and Nishio, M. (2002). The Conformation of 2-phenylpropionaldehyde and alkyl 1-phenylethyl ketones as evidenced by ab initio calculations. Relevance of the CH/π; and CH/O interactions in stereochemistry, *Bull. Chem. Soc. Jpn.*, 75, pp. 1777–1783.
86. Takahashi, O., Yamasaki, K., Kohno, Y., Ueda, K., Suezawa, H. and Nishio, M. (2009). The origin of the relative stability of axial conformers of cyclohexane and cyclohexanone derivatives: Importance of the CH/n and CH/π; hydrogen bonds, *Bull. Chem. Soc. Jpn.*, 82, pp. 272–276.
87. Nishio, M., Umezawa, Y., Fantini, J., Weiss, M. S. and Chakrabarti, P. (2014). CH/π hydrogen bonds in biological macromolecules, *Phys. Chem. Chem. Phys.*, 16, pp. 12648–12683.
88. Takahashi, O., Yamasaki, K., Kohno, Y., Ueda, K., Suezawa, H. and Nishio, M. (2009). The conformation of levopimaric acid investigated by high-level *ab initio* MO calculations. Possibility of the CH/π hydrogen bond, *Tetrahedron*, 65, pp. 3525–3528.
89. Jones, C. R., Baruah, P. K., Thompson, A. L., Scheiner, S. and Smith, M. D. (2012). Can a C–H$\cdots$O interaction be a determinant of conformation? *J. Am. Chem. Soc.*, 134, pp. 12064–12071.
90. Banks, J. S., Batsanov, A., A. K. Howard, J., O'Hagan, D., Rzepa, H. and Martin-Santamaria, S. (1999). The preferred conformation of α-fluoroamides, *J. Chem. Soc., Perkin Trans.*, 2, pp. 2409–2411.
91. Maity, P., Pemberton, R. P., Tantillo, D. J. and Tambar, U. K. (2013). Brønsted acid catalyzed enantioselective indole Aza-Claisen rearrangement mediated by an arene CH–O interaction, *J. Am. Chem. Soc.*, 135, pp. 16380–16383.
92. Odagi, M., Furukori, K., Yamamoto, Y., Sato, M., Iida, K., Yamanaka, M. *et al.* (2015). Origin of stereocontrol in guanidine-bisurea bifunctional organocatalyst that promotes α-hydroxylation of tetralone-derived β-ketoesters: Asymmetric synthesis of β- and γ-substituted tetralone derivatives via organocatalytic oxidative kinetic resolution, *J. Am. Chem. Soc.*, 137, pp. 1909–1915.
93. Jang, K. P., Hutson, G. E., Johnston, R. C., McCusker, E. O., Cheong, P. H. Y. and Scheidt, K. A. (2014). Asymmetric homoenolate additions to acyl phosphonates through rational design of a tailored N-heterocyclic carbene catalyst, *J. Am. Chem. Soc.*, 136, pp. 76–79.
94. Watt, M. M., Zakharov, L. N., Haley, M. M. and Johnson, D. W. (2013). Selective nitrate binding in competitive hydrogen bonding solvents: Do

anion–π interactions facilitate nitrate selectivity? *Angew. Chem. Int. Ed.*, 52, pp. 10275–10280.
95. Mascal, M., Yakovlev, I., Nikitin, E. B. and Fettinger, J. C. (2007). Fluoride-selective host based on anion–π interactions, ion pairing, and hydrogen bonding: Synthesis and fluoride-ion sandwich complex, *Angew. Chem. Int. Ed.*, 46, pp. 8782–8784.
96. Guha, S. and Saha, S. (2010). Fluoride ion sensing by an anion-π interaction, *J. Am. Chem. Soc.*, 132, pp. 17674–17677.
97. Schneebeli, S. T., Frasconi, M., Liu, Z., Wu, Y., Gardner, D. M., Strutt, N. L. *et al.* (2013). Electron sharing and anion–π recognition in molecular triangular prisms, *Angew. Chem. Int. Ed.*, 52, pp. 13100–13104.
98. Hafezi, N., Holcroft, J. M., Hartlieb, K. J., Dale, E. J., Vermeulen, N. A., Stern, C. L. *et al.* (2015). Modulating the binding of polycyclic aromatic hydrocarbons inside a hexacationic cage by anion–π interactions, *Angew. Chem. Int. Ed.*, 54, pp. 456–461.
99. Perez-Velasco, A., Gorteau, V. and Matile, S. (2008). Rigid oligoperylenediimide rods: Anion–π slides with photosynthetic activity, *Angew. Chem. Int. Ed.*, 47, pp. 921–923.
100. Cerny J. and Hobza P. (2007). Non-covalent interactions in biomacromolecules, *Phys. Chem. Chem. Phys.*, 9, pp. 5291–5303.
101. Steuber, H., Heine, A. and Klebe, G. (2007). Structural and thermodynamic study on aldose reductase: Nitro-substituted inhibitors with strong enthalpic binding contribution, *J. Mol. Biol.*, 368, pp. 618–638.
102. Li, S., Cooper, V. R., Thonhauser, T., Lundqvist, B. I. and Langreth, D. C. (2009). Stacking interactions and DNA intercalation, *J. Phys. Chem. B*, 113, pp. 11166–11172.
103. Řeha, D., Kabeláč, M., Ryjáček, F., Šponer, J., Šponer, J. E., Elstner, M., *et al.* (2002). Intercalators. 1. Nature of stacking interactions between intercalators (ethidium, daunomycin, ellipticine, and 4′6-diaminide-2-phenylindole) and DNA base pairs. *Ab Initio* quantum chemical, density functional theory, and empirical potential study, *J. Am. Chem. Soc.*, 124, pp. 3366–3376.
104. Langner, K. M., Kedzierski, P., Sokalski, W. A. and Leszczynski, J. (2006). Physical nature of ethidium and proflavine interactions with nucleic acid bases in the intercalation plane, *J. Phys. Chem. B*, 110, pp. 9720–9727.
105. Hargis, J. C., Schaefer, H. F., Houk, K. N. and Wheeler, S. E. (2010). Non-covalent interactions of a benzo[a]pyrene diol epoxide with DNA base pairs: Insight into the formation of adducts of (+)-BaP DE-2 with DNA, *J. Phys. Chem. A*, 114, pp. 2038–2044.
106. Hong, Y. J. and Tantillo, D. J. (2013). C-H/π interactions as modulators of carbocation structure-implications for terpene biosynthesis, *Chem. Sci.*, 4, pp. 2512–2518.
107. Bauzá, A., Quiñonero, D., Deyà, P. M. and Frontera, A. (2013). On the importance of anion–π interactions in the mechanism of sulfide: Quinone oxidoreductase, *Chem. Asian J.*, 8, pp. 2708–2713.

108. Estarellas, C., Frontera, A., Quiñonero, D. and Deyà, P. M. (2011). Relevant anion–π interactions in biological systems: The case of urate oxidase, *Angew. Chem. Int. Ed.*, 50, pp. 415–418.
109. Becke, A. (1997). Density-functional thermochemistry. V. Systematic optimization of exchange-correlation functionals, *J. Chem. Phys.*, 107, pp. 8554–8560.
110. Weigend, F. and Ahlrichs, R. (2005). Balanced basis sets of split valence, triple zeta valence and quadruple zeta valence quality for H to Rn: Design and assessment of accuracy, *Phys. Chem. Chem. Phys.*, 7, pp. 3297–3305.
111. Chai, J.-D. and Head-Gordon, M. (2008). Systematic optimization of long-range corrected hybrid density functionals, *J. Chem. Phys.*, 128, p. 084106.

Chapter 11

Predicting Reaction Pathways from Reactants

Romain Ramozzi[*,†], *W. M. C. Sameera*[*] *and Keiji Morokuma*[*]

[*]*Fukui Institute for Fundamental Chemistry*
Kyoto University, Kyoto 606-8103, Japan
[†]*Université de Lorraine, CS 25233, F-54052 Nancy cedex, France*

11.1 Introduction

The determination of reaction pathways of organic reactions is fundamental to rationalize experimental observations, as well as to develop new synthetic methodologies. However, this task is particularly challenging for organic reactions. From experimental studies, the full characterization of a mechanism is extremely difficult. Computations provide information that are complementary to the experiment and become essential to elucidate the reaction mechanisms in detail. Development of accurate quantum chemical methods nowadays allows us to determine reaction pathways for complex reactions. Setting aside dynamic effects (see Chapter 13), a reaction is visualized qualitatively to follow the minimum energy pathway (MEP) from a reactant through a transition state to the product or to an intermediate. The MEP is also called the intrinsic reaction coordinate (IRC). Therefore, in order to clarify a reaction mechanism, a systematic determination of the structure and energies of all intermediates and transition states (TSs) through all the possible pathways is required. In this chapter, we discuss the various methods available to determine reaction pathways from the reactants. Different approaches will be discussed using selected examples. Section 11.2 focuses on the methods that require a chemical knowledge of the system to determine the pathway. Section 11.3 will develop some of the most recent automatic methods without requiring a guess.

11.2 Determining pathways from the reactants based on chemical guesses

Until very recently, no systematic method was practically available to determine all reaction pathways for a reaction system without a guess. With the lack of such an approach, pathways are usually proposed or assumed on the basis of our chemical knowledge. Furthermore, TSs and intermediates are theoretically calculated along these assumed or guessed pathways. Many methods have been developed to optimize the critical structures efficiently starting from a guess. In the first part of the chapter, we will discuss some common theoretical methods for this purpose.

11.2.1 *Determining a TS structure from a guess structure*

Many methods have been developed to locate a TS structure.[1–4] A first way is to start with a guess for the TS structure. The minimization method, using only the energy gradient, is not very efficient and the use of a Hessian (second derivative of energy) is usually preferred. A well-known method is the Berny algorithm.[5,6] Starting from a guess structure, an exact or approximate Hessian is calculated. Often, more than one negative eigenvalue of the Hessian is obtained, and the larger one is followed for walking in the uphill direction, while the energy gradient is lowered at the same time. Once a new structure is obtained, the Hessian matrix is updated from the previous one with the gradient information. This process is repeated until the exact TS structure, the structure at which the gradient is zero and Hessian has only one negative eigenvalue, is obtained. Sometimes, it may be necessary to recalculate a new Hessian after a few cycles for a more efficient optimization. Apart from the Berny algorithm, many different methods can be found in the literature, including eigenvector following,[7] the trust radius method,[8–10] or the rational function optimization.[11] While all these methods are relatively efficient and widely used, a good guess for the TS structure is always needed. Therefore, alternative methods were developed to determine the TS structure without a guess.

If one coordinate is involved in the reaction, it is possible to follow it by doing a relaxed scan.[12,13] If multiple coordinates are involved, the fast marching method[14] is more efficient. However, it is often difficult to know the reaction coordinates in advance. An alternative approach is to start from the reactant and product structures to determine the TS structure. These are called double-ended methods, and include the synchronous transit method,[15] the sphere contraction optimization method,[16] the saddle optimization method,[17] the self-penalty walk method,[18] the ridge method,[19] the nudged elastic band method,[20–23] the string method,[24] and the growing string method.[25] In following sub-sections, we will introduce few popular methods along this line.

11.2.1.1 *The quadratic synchronous transit method*

When the reaction coordinate is unknown, the first approach is to use the linear synchronous transit method (LST). In this approach, the TS structure is obtained starting from the initial and final structures (i.e., the reactant and product or intermediate(s)). A line is then interpolated between these structures. The highest energy structure along this line is then used as a guess to search the TS structure using any method, such as the Berny optimization mentioned above. However, a more efficient method is the Quadratic Synchronous Transit (QST) method.[15] In the QST method, a parabola (instead of a line) is used between the reactant and the product by adding the third point. This point can be determined with a preliminary LST calculation, which is then called the QST2 method, or with a guess structure of the TS, which is called the QST3 method. At the end, the minimization from the maximum of a parabola gives the TS structure. In some cases, the QST method fails to determine the TS structure. This is particularly true when the reaction coordinate drastically changes its nature during the reaction step.

11.2.1.2 *The nudged elastic band and the string methods*

The NEB method is a chain-of-states method.[20–23] To determine the MEP, a line is used between the initial and final structures, and states between them are considered. These states are geometric configurations of the system, and one can reach from the initial geometry to the final one. They are assumed to be linked by spring forces. The NEB pathway is then defined by all the states (including the initial and final structures) and aims to converge to the MEP. To do so, a force, called F_i^{NEB} in Fig. 11.1, is applied to all states i. This force can

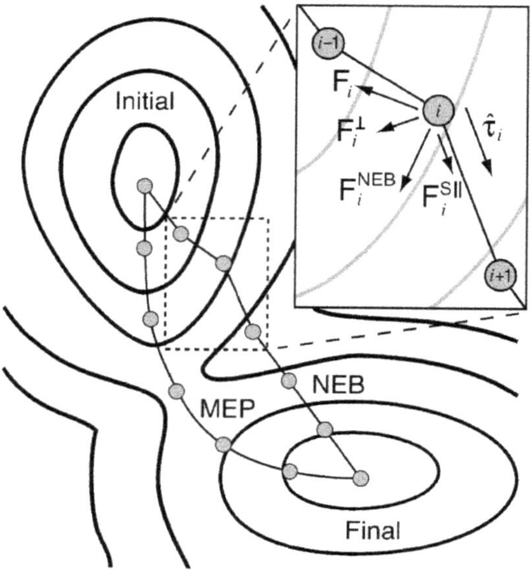

Figure 11.1. Decomposition of the NEB force F_i^{NEB} applied to the state i.
Source: Reproduced from Ref. [23] with permission from American Institute of Physics.

be decomposed into two independent forces: $F_i^{S\|}$ is the spring force along the tangent $\hat{\tau}_i$ and ensures that the states are not slipping into the initial and final geometries during the optimization; $F_i^{\perp}$ perpendicular to the band and is due to the potential. After optimization of the NEB pathway to the MEP, the highest state in energy is used as a guess for the TS structure optimization. In the NEB method, the springs between the states have no physical meaning and may add some numerical noise during the optimization. To avoid this, an alternative method, the string method, was proposed.[24] This method considers a string between the initial and final geometries with different states without any springs between them. To avoid the states from slipping down during the optimization, they are regularly redistributed along the string. This method was found to be more reliable than the NEB method.

11.3 Determining reaction pathways using experimental results

In this section, we will present different approaches to investigate the reaction mechanism using experimental results. Different tools, such as NMR or mass spectrometry, can be used to determine *in situ* the existence and the nature of an intermediate. Sometimes, experimentalists may also isolate some of the intermediates involved in the reaction.

11.3.1 *NMR and kinetic studies*

The 1,2-hydrogen shift from an atom to a metal is well known and usually occurs in unsaturated complexes and also observed for saturated complexes. Both experimental and theoretical chemists conducted investigations to understand the mechanism of this reaction.[26,27] A concerted experimental and theoretical study of this isomerization has been performed, involving a ruthenium complex **3** as shown in Fig. 11.2.

Addition of $TripSnH_3$ on complex **3** changes the solution color from brown to green. This color change implies the formation of a

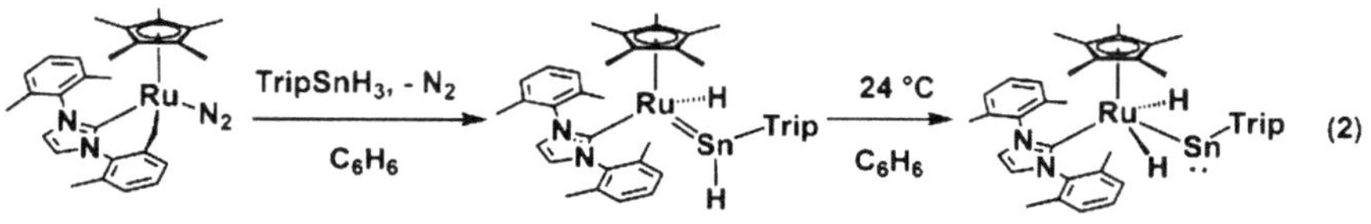

3 4 5

Figure 11.2. 1,2-hydrogen shift from **4** to **5**. Trip = 2,4,6-iPr$_3$C$_6$H$_2$.
Source: Reproduced from Ref. [26] with permission from the American Chemical Society.

new complex **4**, which is then transformed into a new red color complex **5**. In this case, the intermediate **4** is supposed to be a stannylene complex. Despite their efforts, experimentalists were unable to purify this intermediate. However, various experiments were conducted to characterize it. Among them, the ^{1}H NMR experiments of the reaction mixture showed a quantitative conversion of the complex **3**, and the SnH resonance led to the structure of intermediate **4**. NMR confirmed the conversion of **4** into a new complex **5** in a few hours. Complex **5** was characterized by X-ray analysis. As the metal is electronically saturated, the isomerization of **4** to **5** is unusual. To investigate this 1,2-hydrogen migration, various kinetics experiments were conducted. First, a first-order dependence is observed on the concentration of **4**. Second, the Kinetic Isotope Effect (KIE) k_H/k_D was calculated to be 1.3, suggesting that the Sn-H migration is involved in the rate-determining step. Finally, radicals are not involved in the reaction, as a radical inhibitor does not change the reactivity of the system. A unimolecular reaction was then supposed to occur and the activation parameters were measured to be $\Delta H^\ddagger = 20.43 \pm 0.38\,\text{kcal mol}^{-1}$, $\Delta S^\ddagger = -3.49 \pm 1.29\,\text{cal mol}^{-1}\,\text{K}^{-1}$. Starting from these observations, a DFT investigation was conducted.[26] The results are summarized in Fig. 11.3.

From the reactant **I**, two pathways were proposed for the formation of the intermediate **IV**. Among them, the green pathway is the MEP. After an oxidative addition of TripSnH$_3$ on the reactant, the intermediate **II** is formed. A reductive elimination occurs and forms the 16-electron complex **III**. These two intermediates, **II**

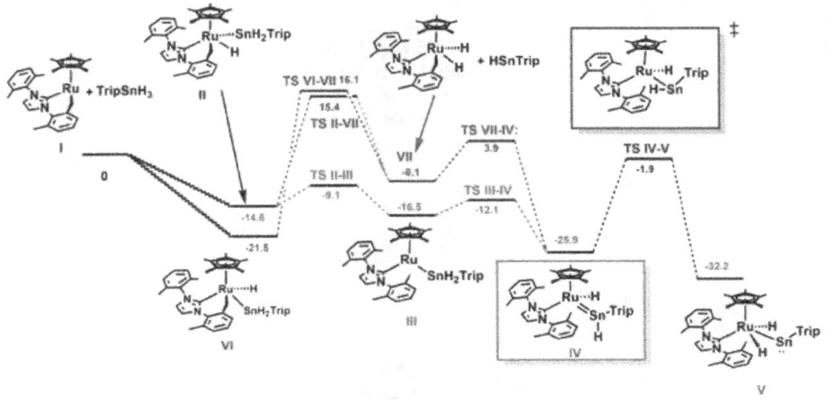

Figure 11.3. DFT results for the 1,2 hydrogen shift in a saturated ruthenium complex at the PBE0-D3BJ/Def2-TZVPP, (SMD=benzene)//PBE0-D3BJ/Def2-DZVP level.
Source: Reproduced from Ref. [26] with permission from the American Chemical Society.

and **III,** were not expected from the experiments. Then, the saturated complex **IV** is formed with a 25.9 kcal mol^{-1} stabilization relative to the reactant. From **IV**, the final product is obtained in one step and its formation is rate-determining with an activation energy of 24 kcal mol^{-1}. A KIE of 1.1 was predicted from calculations. The value of the barrier, the KIE prediction, and the strong stabilization of the intermediate **IV** are in all good agreement with the experiments.

11.3.2 *Trapping an intermediate*

11.3.2.1 *The C(sp^3)-H activation of the methylquinoline*

The first example discussed here is a Rh(III)-catalyzed C-H bond activation.[28,29] Starting from the methylquinoline, the addition of a Rh(III) complex in the presence of copper acetate leads to the C(sp^3)-H activation of the methyl group followed by the insertion of an alkyne (Fig. 11.4).[28]

In the original experimental paper, one of the intermediates, namely **A** in Fig. 11.4, was isolated by using sodium acetate as a base.

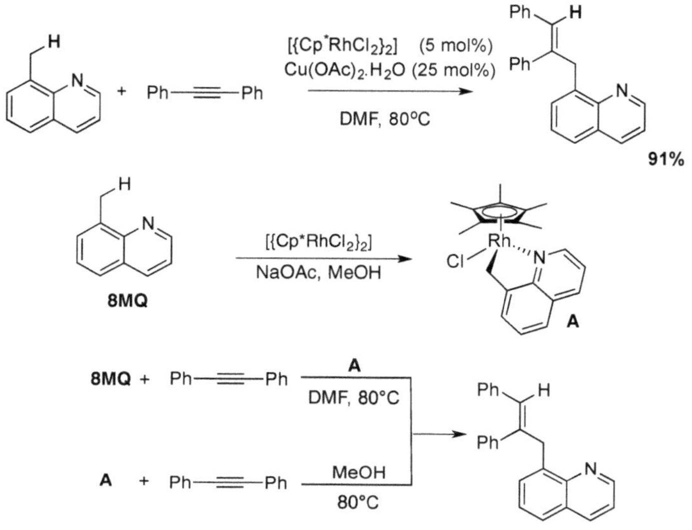

Figure 11.4. Rh(III)-catalyzed C(sp^3)-H activation of the methylquinoline.[28]

NMR, mass spectrometry, and X-ray analysis confirmed the structure of complex **A**. This intermediate was then used as a catalyst and as a reactant to continue the reaction (see Fig. 11.4), which led to a proposed mechanism involving this intermediate. The intermediate **A** was considered to form after exchanging one of the chlorine ligands of the initial rhodium monomer complex with an acetate anion to activate the C–H bond via an intramolecular concerted metalation–deprotonation (CMD) mechanism. The insertion of the alkyne then occurs via a neutral pathway from this intermediate.

Afterward, a theoretical investigation of the mechanism was performed to rationalize the full mechanism using the QST3 method to determine the TS structures.[29] As expected from the experiments, **A** is proved to be an active intermediate. However, the formation of **A** does not occur via an intramolecular CMD mechanism. Indeed, under the experimental conditions, no acetate anion is available to replace the chlorine ligand. Here, the copper acetate acts as an external base to deprotonate the methyl to the metallation of the reactant. Moreover, this intermediate was found to be stabilized around by ~20 kcal mol^{-1} relative to the reactant, which explains why this intermediate can be isolated experimentally. Finally, a cationic mechanism occurs for the 1,2 insertion of the alkyne, which is in contrast to the original mechanism. Indeed, an exchange between the nitrogen of the quinoline and the alkyne occurs via a cationic rhodium(III) complex **B**, which permits the insertion of the alkyne and gives rise to the final product (see Fig. 11.5). All steps were found to be reversible, except the 1,2-insertion step, which insures the efficiency of the process.

The C–H bond activation barrier was estimated to be 26.4 kcal/mol. This barrier is 5 kcal/mol higher than that for the

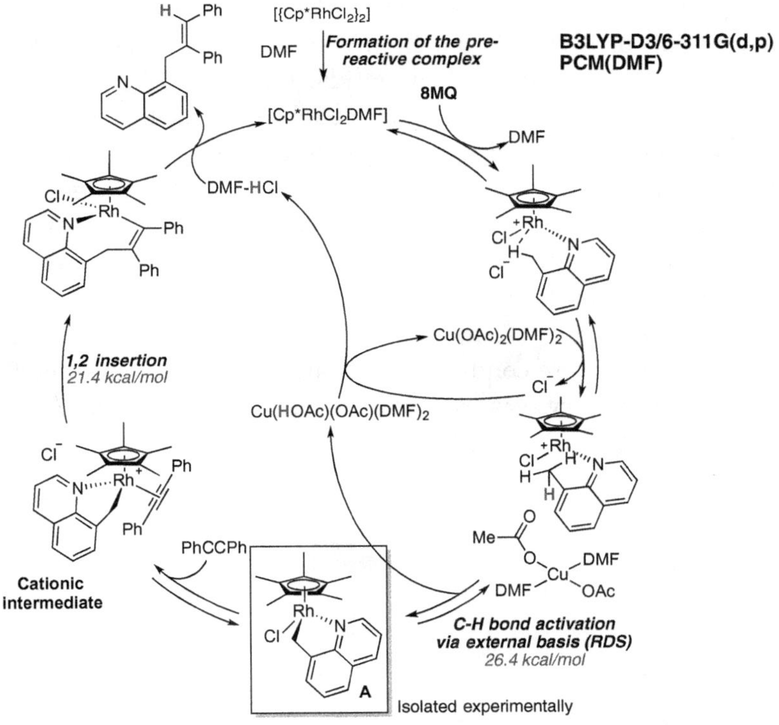

Figure 11.5. Revised mechanism for the Rh(III)-catalyzed C(sp^3)-H activation of the methylquinoline.

Source: Adapted from Ref. [29] with permission from John Wiley & Sons.

1,2-insertion of the alkyne. Therefore, C–H bond activation step is the rate-determining step, which is in good agreement with the KIE experiments. Indeed, when a deuterated methyl group is used for the methylquinoline, the k_H/k_D ratio is measured to be 4.0. When the free acetate anion is used as a base, theoretical calculations also showed a very stable intermediate after the C–H activation, resulting a very high barrier for the 1,2 insertion. This result is also in a good agreement with the experiments, as the reaction cannot occur with acetate anion as a base, despite its basicity higher than that of copper acetate.

11.3.2.2 *Ugi-type reactions*

The Ugi reaction was developed in 1960s and is a multicomponent reaction involving an aldehyde, an amine, an isocyanide, and a carboxylic acid to form a dipeptidic structure. A variation has been proposed using an activated phenol as the acidic partner.[30,31] This reaction, the so-called Ugi–Smiles reaction, leads to the formation of an aryl-amide (see Fig. 11.6). The mechanisms of these two couplings were recently investigated theoretically.[32] For a typical reaction involving two reactants, it may be easy to determine all intermediates and TS structures. However, the situation is much more complicated when four reactants react together. Therefore, various experiments were conducted to characterize some intermediates and the string theory was used to determine the TS structures.

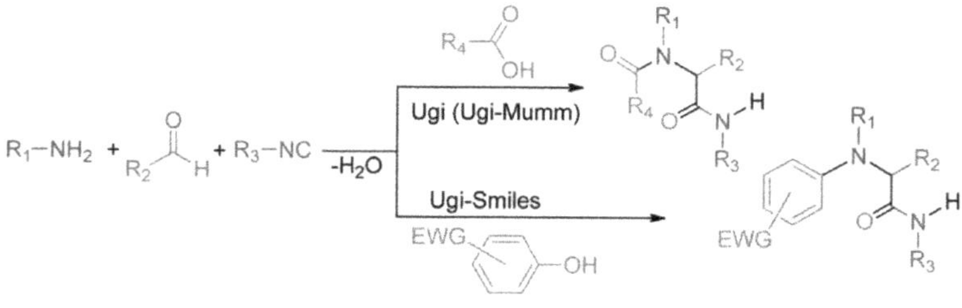

Figure 11.6. The Ugi-type reaction.

Source: Reproduced from Ref. [32] with permission from the American Chemical Society.

First, the imidate intermediate was isolated, and the final Ugi product was obtained starting from this intermediate. In the case of the Ugi reaction, a Mumm rearrangement, an acyl transfer step, occurs to form the final product. In the case of the Ugi–Smiles reaction, a Smiles rearrangement, an intramolecular aromatic nucleophilic substitution, takes place. These observations led to a conclusion that the imidate intermediate is involved in the reaction. The remaining question is: how can this intermediate be formed from all reactants? Twelve possible pathways are possible to form the imidate starting from the four components. However, only two of them are possible from experiments. For instance, experiments showed that the acidic partner (the carboxylic acid or the activated phenol) and the isocyanide could not react together under the reaction conditions. Therefore, all pathways involving this combination do not need to be considered. The remaining pathways, namely path A and B, are depicted in Fig. 11.7. In these two pathways, the amine and aldehyde react together to give the electrophilic iminium species after deprotonation of the acidic partner. From this intermediate, two nucleophiles are able to react: the isocyanide or the conjugated base of the acidic partner (an acetate or a phenolate). In path A,

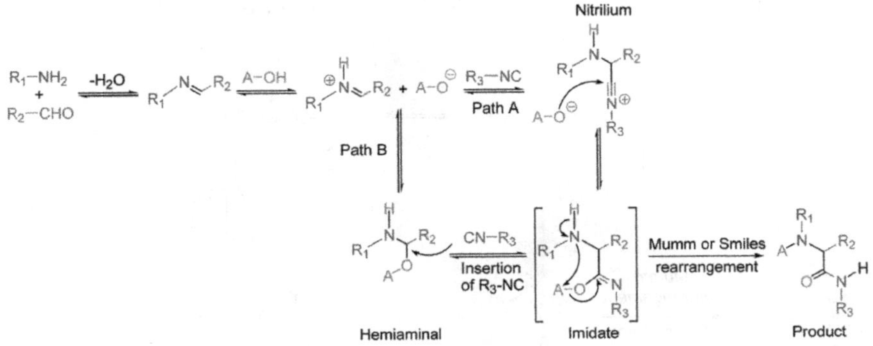

Figure 11.7. Initially proposed mechanisms for the Ugi-type reactions. A–OH is a carboxylic acid in the Ugi reaction or a activated phenol in the Ugi–Smiles reaction.

Source: Reproduced from Ref. [32] with permission from the American Chemical Society.

the isocyanide addition gives the nitrilium intermediate, which then reacts with the acetate to form the imidate. In path B, the addition of the conjugated base on the hemiaminal permits the insertion of the isocyanide in the C–O bond to form the imidate. Whatever the path followed by the system, the imidate is formed and the final Ugi product is obtained after the rearrangement.

Theoretical calculations showed that in both reactions only the path A can take place and the nitrilium formation was rate-determining. While the TS structure of the insertion of the isocyanide in the C–O bond was found (path B), and the associated minimum energy path showed that the iminium and the base were the reactants (and not the hemiaminal). Moreover, the energy of this TS structure was found to be much higher than one of the nitrilium formation (path A). Finally, the imidate formation was found to be irreversible, which suggested that the imidate could be trapped experimentally. The final Ugi mechanism is depicted in Fig. 11.8. The mechanism for the Ugi–Smiles reaction is rather similar to the one shown in Fig. 11.8, except that an activated phenol replaces the carboxylic acid. However, the final rearrangement, the Smiles step, was found to be rate-determining for the reaction. Recently, two mass spectrometry analyses characterized all intermediates involved in path A, including the nitrilium species.[33,34] The hemiaminal, which would

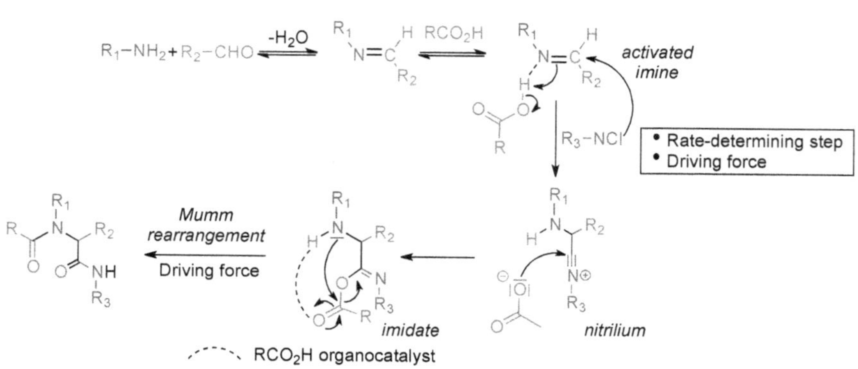

Figure 11.8. Mechanism of the Ugi reaction. A second carboxylic molecule is acting as an organocatalyst to promote the Mumm rearrangement.

be expected from path B, was not observed, confirming that this pathway is not favored.

11.4 Determining pathways without preconception

Locating all intermediates and all pathways for a system is very time-consuming, and nothing insures that the search based on intuition will be good. During the past few years, different approaches have been developed to investigate reaction pathways by reducing human effort and performing the search in a systematic way. Such an approach is particularly valuable in discovering unknown or unexpected reaction paths as well as in determining anticipated reaction paths. Systematic approaches can be separated into two main categories. The first one is searching TS structure from predetermined reactive coordinates. Among these methods, metadynamics[35] and chemical flooding[36] methods are often too expensive to explore the entire PES systematically. The second category is to use approximate reaction path methods, such as the heuristic approaches and the artificial force induced reaction (AFIR) method.

11.4.1 *Heuristic approaches*

Heuristic approaches aim to propose a chemical reaction network from the reactants, which can be determined in an efficient way. Over the past few years, various approaches were proposed. Rappoport *et al.*[37] used heuristic rules based on the formal bond orders and Hammond's postulate to study prebiotic chemistry. Bergeler *et al.*[38] used rules derived from conceptual electronic-structure theory to study an organometallic system.[38]

In this subsection, we discuss an approach developed by Zimmerman.[39,40] To determine all possible intermediates from a reactant structure, a few rules were proposed to generate them automatically by checking the atom connections of the system and counting the number of neighbors of an atom. This number allows determining the connectivity changes of the structure. As only elementary steps are searched, a limited number of bond-breaking/making are possible. The breaking and making connections are set up to two for each step. Moreover, at least one connection must change during the action. Using this connectivity rule, computational cost can be reduced in determining all intermediates systematically from a given structure. The generated structures are optimized at a low level (e.g., molecular mechanics) before optimized at a higher level (e.g., DFT). From the last calculations, energies of all intermediates are then known. Therefore, a cutoff can be chosen to remove high-energy intermediates that could not contribute to the reaction. To conclude the search of MEP, TS structures have to be determined. In this case, the growing string method is used to determine the TS structure.[25] The overall procedure is summarized in Fig. 11.9.

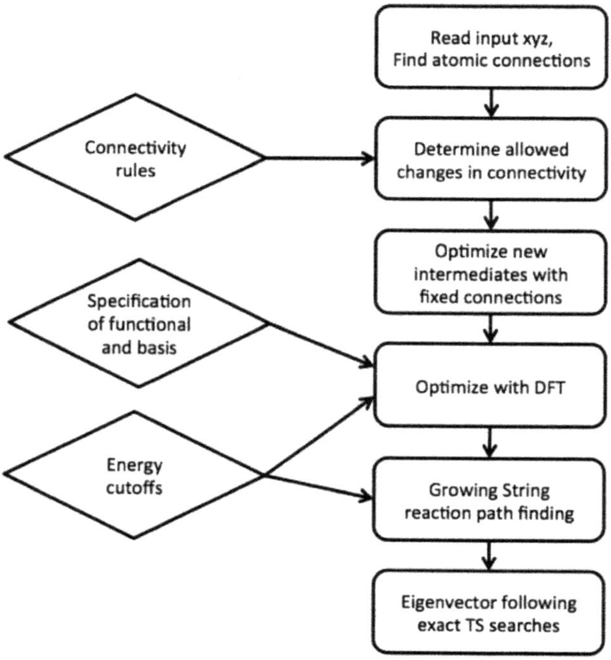

Figure 11.9. Overall procedure proposed by Zimmerman to systematically determine intermediates and TS structure from an initial structure.

Source: Reproduced from Ref. [39] with permission from John Wiley & Sons.

This method was used for various simple reactions, such as nucleophilic additions or the Diels–Alder reaction.[40] It was also applied to more complex reactions, such as the enantioselective synthesis of piperidines using a phosphoric acid as a chiral catalyst (see Fig. 11.10).[41]

The reaction was expected to occur via an ionic species **11** or via a concerted mechanism through the TS structure **10** (see Fig. 11.11). However, from the DFT calculations at the ωB97X-D/6-311+G(d,p)//B3LYP/6-31G(d,p) level, these pathways were significantly higher in energy (by around 30 kcal mol^{-1}) to reach **10**. However, an alternative mechanism involving the intermediate (**12**) was found. The TS structure leading to this intermediate was located at 17 kcal mol^{-1}, which is much lower in energy than that the concerted TS structure. From **12**, a barrier of 12 kcal mol^{-1} is necessary to give the final product.

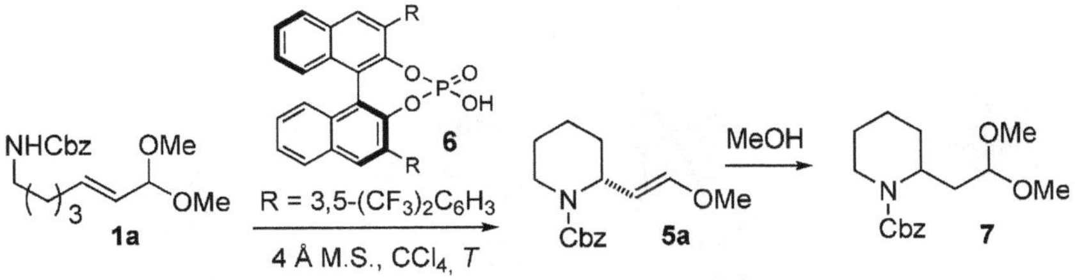

Figure 11.10. Enantioselective synthesis of piperidines.
Source: Reproduced from Ref. [41] with permission from John Wiley & Sons.

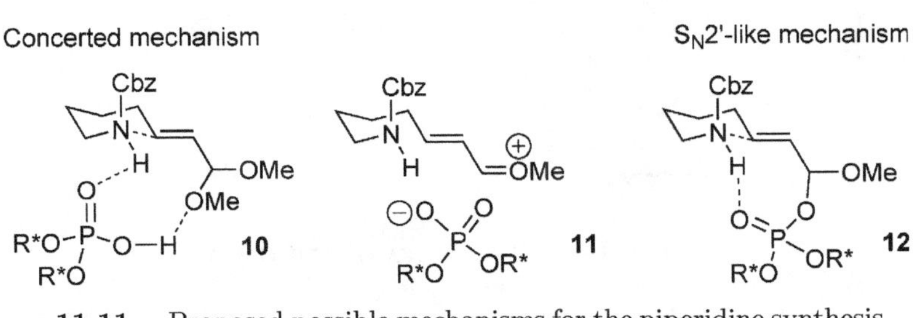

Figure 11.11. Proposed possible mechanisms for the piperidine synthesis.
Source: Reproduced from Ref. [41] with permission from John Wiley & Sons.

11.4.2 *The artificial force induced reaction method*

11.4.2.1 *Description of the method*

The AFIR method, as implemented in the Global reaction route mapping (GRRM) strategy,[42,43] is a systematic approach to explore all low-energy reaction paths of chemical reactions. This approach determines approximate local minima (LMs) and TSs of reaction paths automatically, which is used as starting points to locate true LMs and TSs (see Fig. 11.12). The method has been discussed in detail in several review articles.[43,44]

In the AFIR method, an artificial force is applied between the reactants. For instance, the potential energy surface $E(r_{\mathrm{AB}})$ of a reaction between reactant A and reactant B as a function of the distance between A and B (r_{AB}) is shown in Fig. 11.13. A barrier exists on the potential energy surface, $E(r_{\mathrm{AB}})$. In the presence of an artificial force, a linear function of r_{AB} with an appropriate constant (α), the barrier disappears on the AFIR function $F(r_{\mathrm{AB}})$ and one can follow

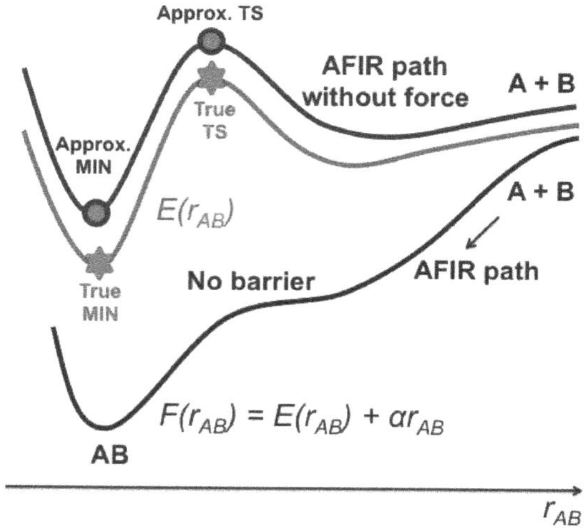

Figure 11.12. The potential surface ($E(r_{\mathrm{AB}})$) and the AFIR energy function ($F(r_{\mathrm{AB}})$) along the reaction coordinate (r_{AB}).

Source: Reproduced from Ref. [44] with permission from the American Chemical Society.

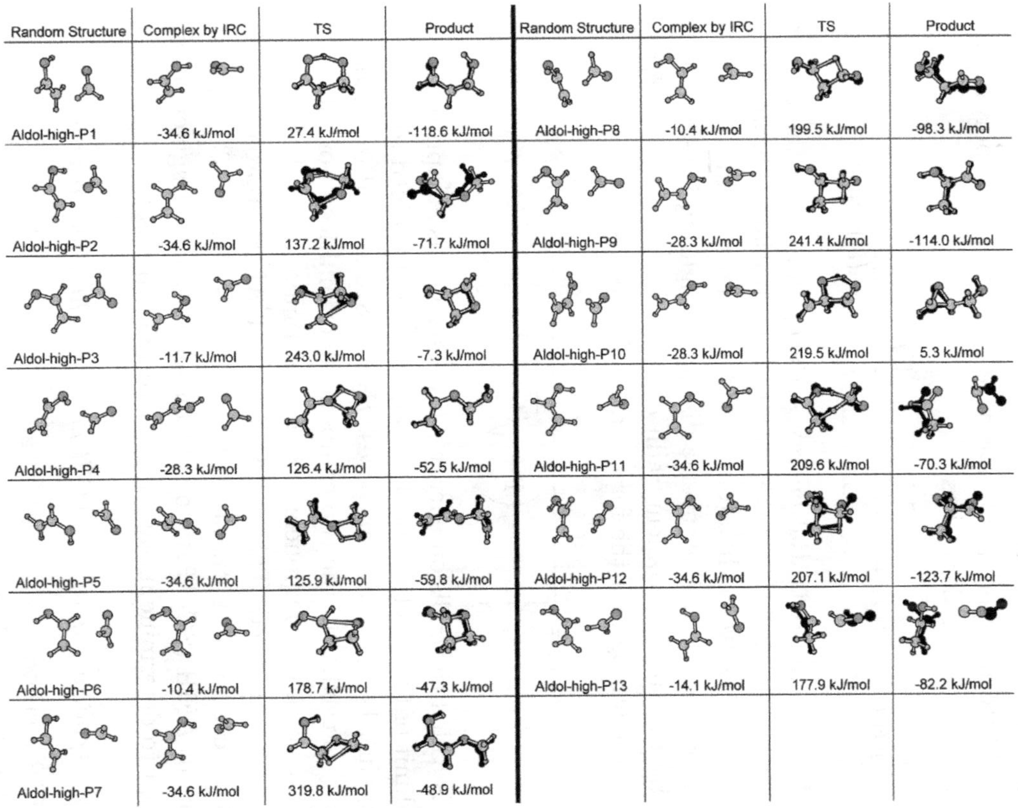

Figure 11.13. The reaction paths between vinyl alcohol and formaldehyde obtained from AFIR search. *Source*: Reproduced from Ref. [42] with permission from the American Chemical Society.

it (called the AFIR path) to reach a minimum. The minimum and maximum of $E(r_{\mathrm{AB}})$ on the AFIR path can be taken as approximate LMs and TSs, respectively. Starting with these structures, the true LMs and TSs are optimized without the artificial force.

The AFIR function $F(Q)$ for a reaction between polyatomic molecules A and B is defined as

$$F(Q) = E(Q) + \alpha \frac{\sum_{i\in A}\sum_{j\in B}[(R_i + R_j)/r_{ij}]^p r_{ij}}{\sum_{i\in A}\sum_{j\in B}[(R_i + R_j)/r_{ij}]^p}$$

where $E(Q)$ is the potential energy surface, and Q are the atomic coordinates. The second term represents the artificial force, where r_{ij} is the distance between the atoms i and j of the reactant molecules A and B, respectively. The inverse distance weighting is used and is scaled by the sum of covalent radii of atoms i and j, $(R_i + R_j)$.[42] The α is the strength of the force, and is defined as

$$\alpha = \frac{\gamma}{\left[2^{-\frac{1}{6}} - \left(1 + \sqrt{1 + \frac{\gamma}{\sigma}}\right)^{-\frac{1}{6}}\right] R_0}$$

The γ is called the collision energy parameter and is an approximate upper limit of the reaction barrier that will be determined (discussed later again). For practical purposes, $R_0 = 3.8164$ Å and $\sigma = 1.0061$ kJ mol^{-1} is used.[45] In the two components or multicomponent reactions, two or more components are placed randomly, and the AFIR function is minimized. This multicomponent AFIR approach (MC-AFIR) is continued until no new product is found. The single-component AFIR (SC-AFIR) method is an automatic fragmentation approach to determine intramolecular reaction paths.[46]

11.4.2.2 *Applications of the AFIR method: The Aldol reaction*

The AFIR method was applied for a systematic mechanistic study of the Aldol reaction.[43] In this case study, the aldol reaction between vinyl alcohol and formaldehyde (two-component process) was investigated. The B3LYP functional and the 6-31G basis sets were used for the AFIR search with the artificial force (γ) of 100 kJ mol^{-1} between vinyl alcohol and formaldehyde. This γ value suggests that TSs with energy below $\sim$100 kJ mol^{-1} will be found. The AFIR search and optimization gave only one path (P1, Fig. 11.13) that represented the aldol product formation. In order to ascertain that any other path has not been missed, the AFIR search with $\gamma = 1000$ kJ mol^{-1} was performed. This gave 13 AFIR paths (Fig. 11.13), one path (P1) for the aldol product formation found above plus 12 new pathways for hydrogen transfer and biradical reactions all higher in energy than 100 kJ mol^{-1}. In Fig. 11.13, two structures are overlaid for TSs and products; the structures in back are approximate TS and product structures obtained by the AFIR search, and those in color are true TS and product structures optimized from the approximate structures without artificial force. One can see that approximate AFIR structures are very similar to the true optimized structures.

As the next step, an H_2O molecule was introduced to the vinyl alcohol and formaldehyde system, and AFIR search was performed with γ of 100 kJ mol^{-1} among the three complements. Finally, eight different paths were found, and the aldol product was obtained from two reaction pathways with different conformations (W1-P2 and W1-P5). In addition to the main aldol reaction paths, two new types of paths were also appeared for methylene-diol formation with the enol-keto tautomerization (W1-P6 and W1-P8) and without the enol-keto tautomerization (W1-P3, W1-P4, and W1-P7) (Fig. 11.14). This systematic mechanistic study on the aldol reaction suggests that the AFIR method is a very useful method for the study of not only the lowest reaction pathway but also many unexpected and higher energy pathways in multicomponent organic reactions.

The microiteration AFIR method was also applied for the study of aldol reaction in $(H_2O)_{299}$ cluster.[47] In this study, the reaction center consisting of the reactant molecules of H_2CO, $H_2C{=}C(H)OH$ and a H_2O molecule was calculated with the B3LYP/6-31G method, while $(H_2O)_{299}$ cluster was treated with the AMBER force field in the ONIOM fashion. Starting from 100 random conformations, AFIR microiteration search with $\gamma = 100$ kJ mol^{-1} and optimization automatically determined the four independent aldol formation pathways

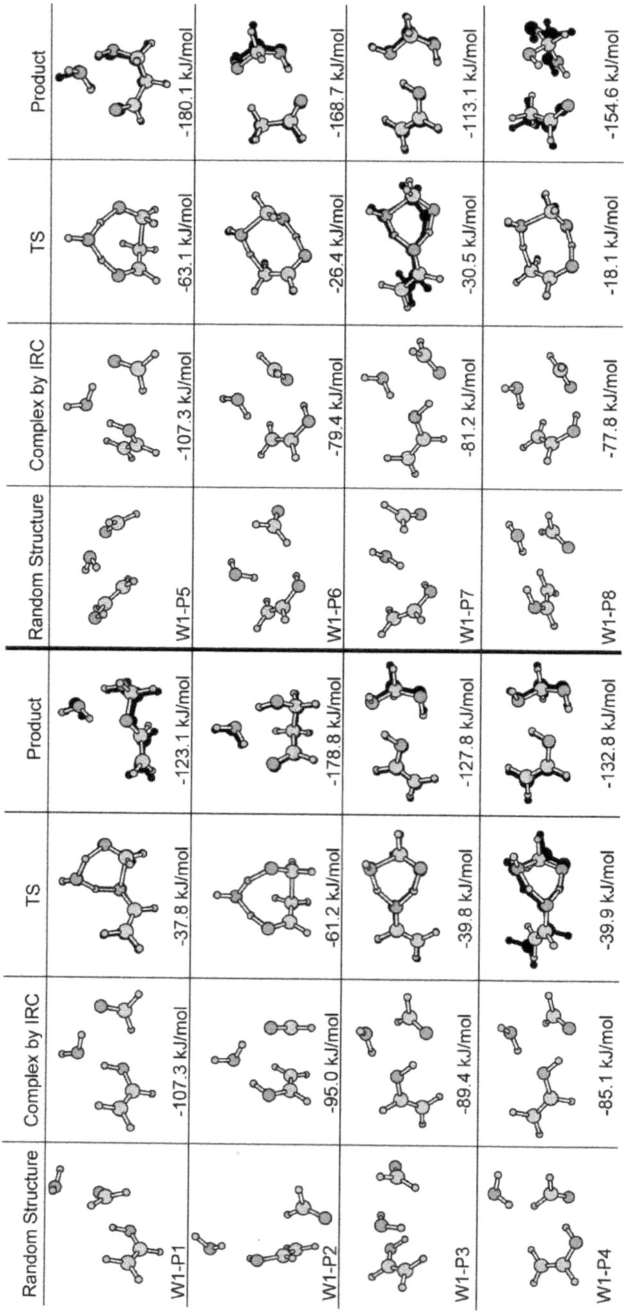

Figure 11.14. The reaction paths between vinyl alcohol, formaldehyde, and a H_2O molecule from AFIR search. *Source:* Reproduced from Ref. [42] with permission from the American Chemical Society.

that are similar to those obtained in the corresponding gas-phase calculations.

11.4.2.3 *Applications of the AFIR method: The Passerini reaction*

Passerini reaction is a multicomponent reaction between a carboxylic acid, an aldehyde (or ketone), and an isocyanide, giving rise to an α-acyloxycarboxamide.[48–50] Organic textbooks have listed the original mechanism of three components for nearly 100 years. Recently this mechanism has been systematically reexamined using the AFIR method.[47] In this study, HCOOH, HCHO, and CH_3NC were used as the reactants, and all possible reaction paths were determined. M06/6-31+G(d,p) method in gas phase was used for the AFIR search, and the artificial force of 150 kJ mol^{-1} was applied between HCOOH and HCHO, between HCOOH and CH_3NC, or between HCHO and CH_3NC. All located reaction paths to reach the final product for the three component reactions were, however, showed high-energy barriers. The best path led to intermediate **21** (Fig. 11.15), but could not proceed further to reach the desired product **6**. When one extra HCOOH molecule was introduced to the intermediate **21**, the AFIR search provided a path leading to the intermediate **27** via TS **49**, and the final product **6** was obtained through TS **53**. At TS **49**, the fourth component (HCOOH) transfers a proton from the bottom to the top of the structure, and therefore HCOOH molecule acts as a catalyst. Similarly, HCOOH is a proton transfer catalyst at TS **53**. Thus, the extra HCOOH molecule participates in proton exchange twice, and the overall reaction proceeds through low-energy barriers (Fig. 11.15), and is consistent with the experimental results. Therefore, the Passerini reaction is not a three-component reaction but a four-component organocatalytic reaction with the second HCOOH molecule working as a catalyst.

Solvent effects on the Passerini reaction were systematically investigated using the AFIR method[50] M06-2X/6-31+G(d,p) method, including polarizable continuum model (PCM) for solvation treatments, was used for this AFIR search. Then, approximate

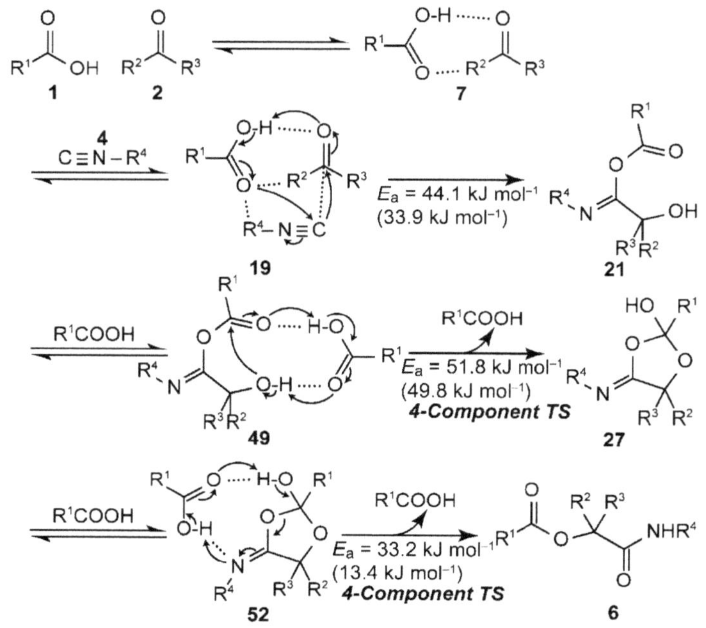

Figure 11.15. The mechanism of the Passerini reaction from the AFIR search. The values in parenthesis represent the solvation energies (single point on gas-phase geometries) from the polarizable continuum model (PCM).

Source: Reproduced from Ref. [49] with permission from John Wiley & Sons.

stationary points were fully optimized without the artificial force with mPW2PLYP-D/6-311++G(d,p) method including PCM. This study suggests that the nitrilium intermediate formation takes place in solution and is the rate-determining step of the Passerini reaction. An AFIR search shows that an extra carboxylic acid molecule assists both the nitrilium intermediate formation and the subsequent Mumm rearrangement (Fig. 11.16). A protic solvent, such as methanol, increases the barrier for the rate-determining step due to the extra stabilization of intermediate through the hydrogen bonds network. On the other hand, aprotic solvents, such as dichloromethane, do not form such a hydrogen bonds network with the solvent. This is consistent with the experimental results: the Passerini reaction is more efficient in aprotic solvents such as dichloromethane.

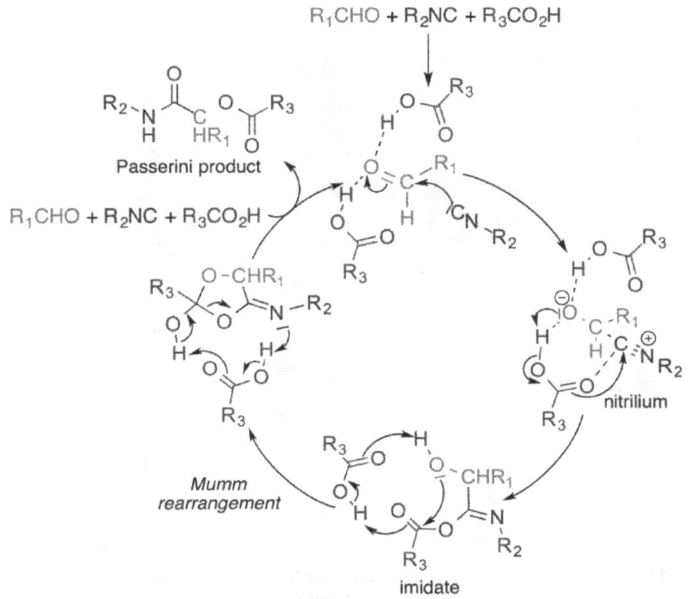

Figure 11.16. Revised mechanism for the Passerini reaction based on the AFIR study.

Source: Reproduced from Ref. [50] with permission from the American Chemical Society.

11.4.2.4 *Applications of the AFIR method: The hydroformylation of alkenes*

Hydroformylation is a well-known carbon–carbon doublebond functionalization reaction and is catalyzed by transition metal complexes.[51] The Heck–Breslow mechanism has been accepted for cobalt-catalyzed hydroformylation.[52] Many reaction pathways covering the entire catalytic cycle have been systematically studied for the first time using the AFIR method.[53] In this study, $HCo(CO)_3$ **1** was used as the active catalyst species, while CO, H_2, and C_2H_4 were used as the reactants (see Fig. 11.17). Starting from many random relative orientations and approach directions between a reactant molecule and the catalyst, the AFIR search was performed with B3LYP/6-31G method with $\gamma = 100\,\text{kJ mol}^{-1}$, and structures were fully optimized

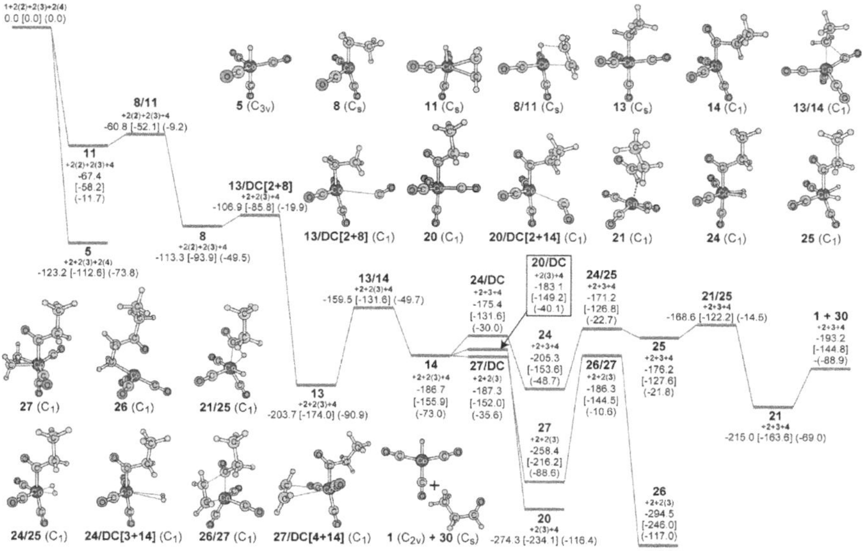

Figure 11.17. Energy profiles for the $HCo(CO)_3$-catalyzed hydroformylation and competing paths obtained from the AFIR search. Energies in square brackets and parentheses show the relative energies including zero-point energy and free-energy corrections at 403.15 K and 200 atm, respectively.

Source: Reproduced from Ref. [42] with permission from the American Chemical Society.

with the M06/6-311++G(d,p) method. The AFIR search showed that the reaction of $HCo(CO)_3$ **1** with ethylene via an ethylene complex $[CH_2CH_2]HCo(CO)_3$ **11** to give intermediate $CH_3CH_2Co(CO)_3$ **8** is energetically the most favorable. Starting from **8**, the most favorable reaction is the addition of CO to give $CH_3CH_2Co(CO)_3$ **14**. Starting from **14**, the most favorable path is the addition of H_2 to give the intermediate $[CH_3CH_2(CO)H]CoH(CO)_3$ **25**, which completes the catalytic cycle by releasing the hydroformylation product $CH_3CH_2(CO)H$ **30**, and regenerates the active catalyst **1**. The calculated mechanism confirmed the Heck–Breslow mechanism. This is the first example in which AFIR was used for a systematic semi-automatic exploration of a full catalytic cycle of transition metal catalysis.

A most impressive technical fact from this study is that the entire calculations locating and optimizing about 100 intermediate and TS structures took only 5.8 days on two CPUs (Intel Xeon, X5670, 2.93 GHz, 6 core).[51] Recently the first step of the hydroformylation between $HCo(CO)_3$ and $CH_3CH{=}CH_2$ was also studied with the SC-AFIR method.[52]

11.5 Practical consideration and perspective

This chapter has shown that automatic and systematic approaches are particularly efficient for the determination of complex reaction pathways of complex reaction systems. Indeed, these methods go beyond the traditional methods, and explore unknown and unexpected reaction pathways as well as known pathways. Therefore, the results are less biased by expectations that may or may not be correct. Since these methods explore many possible pathways, computational cost is still substantial. A practical technique in the AFIR method is to perform systematic wide-range search of reaction pathways to locate many approximate TSs and intermediates at a lower level of theory, such as the semiempirical methods, ONIOM(QM:MM) or ONIOM(QM:QM) methods,[54] or relatively inexpensive DFT approach with a small basis set. Full optimization, without artificial force, may be performed from these approximate structures at a higher level of theory with a large basis set.

To conclude the chapter, we emphasize that automatic search methods find many pathways in a systematic way. Capability of finding all possible pathways provides an excellent opportunity to "predict" the outcome of a reaction from the reactant. Indeed, automatic search methods find not only the most favorable pathway but also many competing or unfavorable pathways. With the knowledge of different possible pathways, one may be able to "control" the barrier height of the favorable pathway of a reaction, thus affording "theoretical design" of new reaction systems. Of course, these "predictions" and "designs" are based on theoretical simulations, and experimental confirmation is always required. Thus, collaboration between theory and experiment is now reaching a new area for further development of organic reactions.

References

1. Schlegel, H. B. (2003). Exploring potential energy surfaces for chemical reactions: An overview of some practical methods, *J. Comput. Chem.*, 24, pp. 1514–1527.
2. Jensen, F. (2007). *Introduction to Computational Chemistry*, 2nd edn. (Wiley, Chichester).
3. Schlegel, H. B. (2011). Geometry optimization, *Wiley Interdiscip. Rev.: Comput. Mol. Sci.*, 1, pp. 790–809.
4. Cheeseman, J. B. and Frisch, A. E. (2015). *Exploring Chemistry with Electronic Structure Methods*, 3rd edn. (Gaussian, Wallingford).
5. Schlegel, H. B. (1982). Optimization of equilibrium geometries and transition structures, *J. Comput. Chem.*, 3, pp. 214–218.
6. Farkas, Ö and Schlegel, H. B. (1999). Methods for optimizing large molecules. II Quadratic Search, *J. Chem. Phys.*, 111, pp. 10806–10814.
7. Cerjanand, C. J. and Miller, W. H. (1981). On finding transition states, *J. Chem. Phys.*, 75, pp. 2800–2806.
8. Golab, J. T., Yeager, D. L. and Jørgensen, P. (1983). Proper characterization of MC SCF stationary points, *Chem. Phys.*, 78, pp. 175–199.
9. Helgaker, T. (1991). Transition-state optimizations by trust-region image minimization, *Chem. Phys. Lett.*, 182, pp. 503–510.
10. Culot, P., Dive, G., Nguyen, V. H. and Ghuysen, J. M. (1992). A quasi-Newton algorithm for first-order saddle-point location, *Theor. Chim. Acta*, 82, pp. 189–205.
11. Banerjee, A., Adams, N., Simons, J. and Shepard, R. (1985). Search for stationary points on surfaces, *J. Phys. Chem.*, 89, pp. 52–57.
12. Hayes, D. M. and Morokuma, K. (1972). Theoretical studies of carbonyl photochemistry. I. Ab initio potential energy surfaces for the photodissociation $H_2CO^* \rightarrow H + HCO$, *Chem. Phys. Lett.*, 12, pp. 539–543.
13. Jaffe, R. L., Hayes, D. M. and Morokuma, K. (1974). Photodissociation of formaldehyde: Potential energy surfaces for $H_2CO \rightarrow H_2 + CO$, *J. Chem. Phys.*, 60, pp. 5108–5109.
14. Burger, S. K. and Ayers, P. W. (2010). Dual grid methods for finding the reaction path on reduced potential energy surfaces, *J. Chem. Theory Comput.*, 6, pp. 1490–1497.
15. Halgren, T. A. and Lipscomb, W. N. (1977). The synchronous-transit method for determining reaction pathways and locating molecular transition states, *Chem. Phys. Lett.*, 49, pp. 225–232.
16. Ishida, K., Morokuma, K. and Komornicki, A. (1977). The intrinsic reaction coordinate. An *ab initio* calculation for $HNC \rightarrow HCN$ and $H^- + CH_4 \rightarrow CH_4 + H^-$, *J. Chem. Phys.*, 66, p. 2153.

17. Dewar, M. J. S., Healy, E. F. and Stewart, J. J. P. (1984). Location of transition states in reaction mechanisms, *J. Chem. Soc., Faraday Trans. 2*, 80, pp. 227–233.
18. Elber, R. and Karplus, M. (1987). A method for determining reaction paths in large molecules: Application to myoglobin, *Chem. Phys. Lett.*, 139, pp. 375–380.
19. Ionova, I. V. and Carter, E. A. (1993). Ridge method for finding saddle points on potential energy surfaces, *J. Chem. Phys.*, 98, p. 6377.
20. Müller K. (1980). Reaction paths on multidimensional energy hypersurfaces, *Angew. Chem. Int. Ed.*, 19, pp. 1–13.
21. Jonsson, H., Mills, G. and Jacobsen, K. W. (1998). Nudged elastic band for finding minimum energy paths of transitions. In *Classical and Quantum Dynamics in Condensed Phase Simulations*, B. J. Berne, (ed.) Chapter 13 (World Scientific, Singapore), pp. 285–404.
22. Henkelman, G., Uberuaga, B. P. and Jónsson, H. (2000). A climbing image nudged elastic band method for finding saddle points and minimum energy paths, *J. Chem. Phys.*, 113, p. 9901.
23. Sheppard, D., Terrell, R. and Henkelman, G. (2008). Optimization methods for finding minimum energy paths. *J. Chem. Phys.*, 128, p. 134106.
24. Weinen, E., Ren, W. and Vanden-Eijnden, E. (2002). String method for the study of rare events, *Phys. Rev. B*, 66, p. 052301.
25. Peters, B., Heyden, A., Bell, A. T. and Chakraborty, A. (2004). A growing string method for locating transition states: Comparison to NEB and string methods, *J. Chem. Phys.*, 120, p. 7877.
26. Liu, H.-J., Guihaume, J., Davin, T., Raynaud, C., Eisenstein, O. and Tilley, T. D. (2014). 1,2-hydrogen migration to a saturated ruthenium complex via reversal of electronic properties for tin in a stannylene-to- metallostannylene conversion, *J. Am. Chem. Soc.*, 36, pp. 13991–13994.
27. Hayes, P. G., Gribble, C. W., Waterman, R. and Tilley, T. D. (2009). A hydrogen-substituted osmium stannylene complex: Isomerization to a metallostannylene complex via an unusual α-hydrogen migration from tin to osmium, *J. Am. Chem. Soc.*, 131, pp. 4606–4607.
28. Liu, B., Zhou, T., Li, B., Xu, S., Song, H. and Wang, B. (2014). Rhodium(III)-catalyzed alkenylation reactions of 8-methylquinolines with alkynes by C(sp3)-H activation, *Angew. Chem. Int. Ed.*, 53, pp. 4191–4195.
29. Jiang, J., Ramozzi, R. and Morokuma, K. (2015). Rh^{III}-catalyzed $C(sp^3)$-H bond activation by an external base metalation/deprotonation mechanism: A theoretical study, *Chem. Eur. J.*, 21, pp. 11158–11164.
30. Ugi, I., Meyr, R., Fetzer, U. and Steinbrückner, C. (1959). Versuche mit Isonitrilen, *Angew. Chem.*, 71, pp. 373–388.
31. El Kaïm, L., Grimaud, L. and Oble, J. (2005). Phenol Ugi–Smiles systems: Strategies for the multicomponent N-arylation of primary amines with isocyanides, aldehydes, and phenols, *Angew. Chem. Int. Ed.*, 44, pp. 7961–7964.

32. Chéron, N., Ramozzi, R., El Kaïm, L., Grimaud, L. and Fleurat-Lessard, P. (2012). Challenging 50 years of established views on Ugi reaction: A theoretical approach, *J. Org. Chem.*, 77, pp. 1361–1366.
33. Medeiros, G. A., da Silva, W. A., Bataglion, G. A., Ferreira, D. A. C., de Oliveira, H. C. B., Eberlin, M. N. and Neto, B. A. D. (2014). Probing the mechanism of the Ugi four-component reaction with charge-tagged reagents by ESI-MS(/MS), *Chem. Commun.*, 50, pp. 338–340.
34. Iacobucci, C., Reale, S., Gal, J.-F. and De Angelis, F. (2014). Insight into the mechanisms of the multicomponent Ugi and Ugi–Smiles reactions by ESI-MS(/MS), *Eur. J. Org. Chem.*, pp. 7087–7090.
35. Laio, M. and Parrinello, M. (2002). Escaping free-energy minima, *Proc. Natl. Acad. Sci. USA*, 99, pp. 12562–12566.
36. Mueller, E. M., de Meijere, A. and Grubmueller, H. (2002). Predicting unimolecular chemical reactions: Chemical flooding, *J. Chem. Phys.*, 116, pp. 897–905.
37. Rappoport, D., Galvin, C. J., Zubarev, D. Y. and Aspuru-Guzik, A. (2014). complex chemical reaction networks from heuristics-aided quantum chemistry, *J. Chem. Theory Comput.*, 10, pp. 897–907.
38. Bergeler, M., Simm, G. N., Proppe, J. and Reiher, M. (2015). Heuristics-guided exploration of reaction mechanisms, *J. Chem. Theory Comput.*, 11, pp. 5712–5722.
39. Zimmerman, P. M. (2013). Automated discovery of chemically reasonable elementary reaction steps, *J. Comput. Chem.*, 34, pp. 1385–1392.
40. Zimmerman, P. M. (2015). Navigating molecular space for reaction mechanisms: An efficient, automated procedure, *Mol. Simul.*, 41, pp. 43–54.
41. Sun, Z., Winschel, G. A., Zimmerman, P. M. and Nagorny, P. (2014). Enantioselective synthesis of piperidines through the formation of chiral mixed phosphoric acid acetals: Experimental and theoretical studies, *Angew. Chem. Int. Ed.*, 53, pp. 11194–11198.
42. Maeda, S. and Morokuma, K. (2011). Finding reaction pathways of type $A + B \rightarrow X$: Toward systematic prediction of reaction mechanisms, *J. Chem. Theory Comput.*, 7, pp. 2335–2345.
43. Maeda, S., Ohno K. and Morokuma, K. (2013). Systematic exploration of the mechanism of chemical reactions: The global reaction route mapping (GRRM) strategy by the ADDF and AFIR methods, *Phys. Chem. Chem. Phys.*, 15, pp. 3683–3701.
44. Sameera, W. M. C., Maeda, S. and Morokuma, K. (2016). Computational catalysis using the artificial force induced reaction (AFIR) method, *Acc. Chem. Res.*, 49(4), pp. 763–773.
45. Collins, M.A. (2002). Molecular potentialenergy surfaces for chemical reaction dynamics, *Theor. Chem. Acc.*, 108, pp. 313–324.
46. Maeda, S., Taketsugu, T. and Morokuma, K. (2014). Exploring transition state structures for intramolecular pathways by the artificial force induced reaction method, *J. Comput. Chem.*, 35, pp. 166–173.

47. Maeda, S., Abe, E., Hatanaka, M., Taketsugu, T. and Morokuma, K. (2012). Exploring potential energy surfaces of large systems with artificial force induced reaction method in combination with ONIOM and microiteration, *J. Chem. Theory Comput.*, 8, pp. 5058–5063.
48. Passerini, M. and Simone, L. (1921). Sopra gli isonitrili (I). Composto del p-isonitril-azobenzolo con acetone ed acido acetico, *Gazz. Chim. Ital.*, 51, pp. 126–129.
49. Maeda, S., Komagawa, S., Uchiyama, M. and Morokuma, K. (2011). Finding reaction pathways for multicomponent reactions: The Passerini reaction is a four-component reaction, *Angew. Chem. Int. Ed.*, 50, pp. 644–649.
50. Ramozzi, R. and Morokuma, K. (2015). Revisiting the Passerini reaction mechanism: Existence of the nitrilium, organocatalysis of its formation, and solvent effect, *J. Org. Chem.*, 80, pp. 5652–5657.
51. Cornils, B., Herrmann, W. A. and Rasch, M. (1994). Otto Roelen, pioneer in industrial homogeneous catalysis, *Angew. Chem. Int. Ed. Engl.*, 33, pp. 2144–2163.
52. Heck, R. F. and Breslow, D. S. (1961). The reaction of cobalt hydrotetracarbonyl with olefins, *J. Am. Chem. Soc.*, 83, pp. 4023–4027.
53. Maeda, S. and Morokuma, K. (2012). Toward predicting full catalytic cycle using automatic reaction path search method: A case study on HCo(CO)3-catalyzed hydroformylation, *J. Chem. Theory Comput.*, 8, pp. 380–385.
54. Chung, L. W., Sameera, W. M. C., Ramozzi, R., Page, A. J., Hatanaka, M., Petrova, G. P., Harris, T. V., Li, X., Ke, Z., Liu, F., Li, H.-B., Ding, L. and Morokuma, K. (2015). The ONIOM method and its applications, *Chem. Rev.*, 115, pp. 5678–5796.

Chapter 12

Unusual Potential Energy Surfaces and Nonstatistical Dynamic Effects

Charles Doubleday

Department of Chemistry, Columbia University, NY 10027, New York, USA

12.1 Introduction

In reaction mechanisms, nonstatistical dynamics refers to the competition between chemical reaction and vibrational relaxation, which includes intramolecular vibrational energy redistribution (IVR)[1] and intermolecular energy transfer through collisions. Statistical theories like transition state theory (TST) and Rice–Ramsperger–Kassel–Marcus (RRKM) theory rely on fast vibrational relaxation to guarantee a Boltzmann distribution at a given temperature or a microcanonical distribution at a given energy. A reactive intermediate, however, might decompose at a rate that is competitive with IVR, which would lead to nonstatistical dynamics of the intermediate. Competition between chemical reaction and IVR may occur either because the chemical process is fast or because IVR is slowed by a phase space bottleneck (slow energy transfer between subsets of vibrational modes). We will consider examples of both.

Excellent recent reviews of nonstatistical dynamics in organic reactions are available.[2] In this chapter, we discuss characteristics of the potential energy surface (PES) that are associated with nonstatistical dynamics, and the information that has been obtained from dynamical studies.

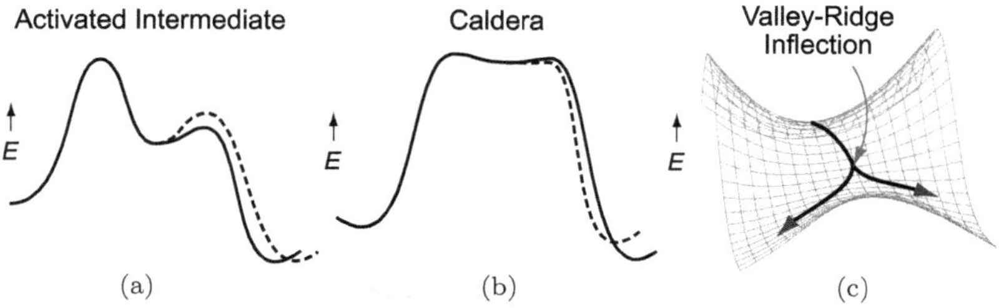

Figure 12.1. Three PES archetypes: (a) an intermediate activated by a large exothermicity from the initial TS; (b) a caldera, with a shallow and flat local minimum; (c) post-TS bifurcation due to a VRI in the minimum energy path.

Figure 12.1 shows three PES types in which an initial transition state (TS) is followed by a branching region that leads to two products. All of these can lead to nonstatistical dynamics. In the first case, an intermediate is activated by converting the potential energy of the TS into kinetic energy of the intermediate, which exceeds the barriers for product formation. The question is how fast the energized molecules move through the local minimum and form products, compared to the time scale for IVR. With the large energy release from the TS, the potential for nonstatistical dynamics clearly exists. The caldera is typical of singlet biradical intermediates. In contrast to the previous case, the only activation this intermediate possesses is the kinetic energy of the reaction coordinate that propelled it onto the caldera. However, the flat PES offers little resistance to motion across it, lifetimes are short, and the dynamics is typically nonstatistical. In a post-TS valley-ridge inflection (VRI), the reaction path valley becomes a ridge at the VRI (i.e., one of the transverse modes has negative curvature). Trajectories that pass through the initial TS then bifurcate toward two products in the region of the VRI. With perhaps 100 fs between passage through the TS and bifurcation toward products, the time available for IVR is minimal.

12.2 Gas-phase S_N2 reactions and multiple recrossing

In a series of papers, Cho *et al.*,[3a] Hase,[3b] and Manikandan *et al.*[3c] have critically examined the dynamics of gas-phase S_N2 reactions

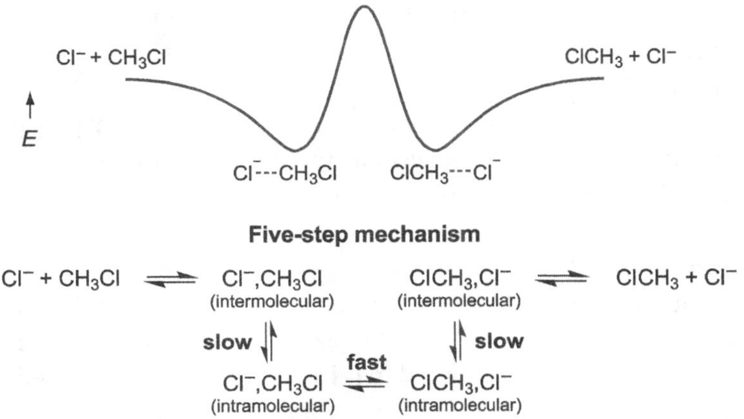

Figure 12.2. Upper: PES for gas-phase S_N2 reaction. Lower: mechanism for the reaction requires five steps. Once the complex is formed, the rate-determining step is vibrational energy transfer between intermolecular and intramolecular modes.

(Fig. 12.2). Here we discuss a result that is broadly applicable to the current topic. In a study of $Cl^- + CH_3Cl$,[3a] the researchers examined the lifetime of the ion–dipole complex with respect to dissociation. They initialized trajectories at the central barrier at 300 K and compared the dynamics to an RRKM calculation of the lifetime of the complex at the mean energy of the trajectory study. The RRKM rate constant for dissociation of the complex was 0.25 ps^{-1}, but *not one trajectory dissociated*, even with propagation times up to 40 ps. Instead of dissociating, trajectories carried out multiple recrossings of the central barrier.

RRKM theory assumes that the full phase space is statistically populated, but the long trajectory-based lifetime indicates that a phase space bottleneck prevents vibrational energy from flowing among all 12 modes. The S_N2 model proposed by Hase is shown at the bottom of Fig. 12.2. When the complex is formed from reactants, the three intermolecular modes are excited (Cl^-–C stretch and two Cl^-–CH_3Cl bends). These low-frequency modes are coupled only weakly to the high-frequency CH_3Cl modes ("intramolecular" modes), which must be excited to cross the barrier. Energy transfer between inter- and intramolecular modes is slow and rate-determining. Once energy is transferred to the intramolecular modes, multiple recrossings occur prior to dissociation.

This striking example shows the large effect that a phase space bottleneck can have on reaction dynamics. Another example is the vinylidene–acetylene rearrangement. In classical trajectories for $D_2C{=}C\colon \rightarrow DC{\equiv}CD$, Hayes *et al.*[4] reported multiple recrossings that repeatedly regenerated vinylidene, even though the reverse barrier is 48 kcal/mol. In acetylene, this large exothermicity is localized in the bending modes, which couple weakly with stretching, allowing repeated regeneration of vinylidene. In agreement with Carter's classical results, Schork and Köppel[5] reported multiple recurrences of unit probability for vinylidene formation in their five-dimensional quantum dynamics calculations of $H_2C{=}C\colon \rightarrow HC{\equiv}CH$.

In both the S_N2 and vinylidene–acetylene cases, it was the lowest frequency modes that retained their energy long enough to allow repeated chemical bonding changes. The large differences in frequency between these modes and the rest contributed to the slow vibrational energy transfer, as did the small number of atoms.

12.3 Activated intermediates

12.3.1 *Hydroboration regioselectivity*

In a combined experimental and computational study, Oyola and Singleton[6] presented credible evidence that the regiochemistry of hydroboration (Fig. 12.3) is not fully explained by TST. Using CCSD(T) with extrapolation to an infinite basis set, they found a converged value of $\Delta\Delta G^{\ddagger} = 2.5 \pm 0.1$ kcal/mol, implying regioselectivity of 98:2 to 99:1 according to TST. Their experimental ratio was 90:10, equivalent to $\Delta\Delta G^{\ddagger} = 1.3$ if the ratio is interpreted by TST. When Oyola and Singleton initialized trajectories at the variational TS for approach of reactants, 10–13% Markovnikov product was formed, in good agreement with experiment. Trajectories initialized at the optimized complex gave 1% Markovnikov product as expected for a complex at thermal equilibrium.

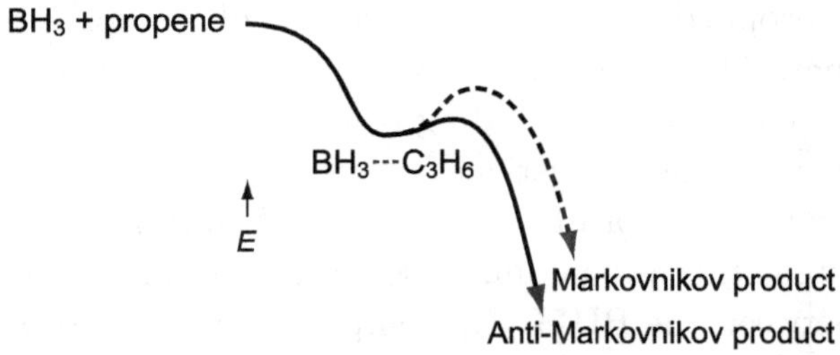

Figure 12.3. Qualitative PES for hydroboration studied by Oyola and Singleton.[6]

The time dependence of the nonstatistical product ratio is informative. Trajectories that formed products early, in 0.8 ps or less, gave the largest deviation from statistical, 21% Markovnikov product. Trajectories with lifetimes of 0.8–5 ps gave 7.5% Markovnikov, and the rest were assigned the statistical fraction. They also found experimentally that the regioselectivity is relatively insensitive to temperature, varying from 10.0% Markovnikov at 21°C to 11.2% at 70°C. Minimal sensitivity of the product ratio to temperature has been seen as characteristic of nonstatistical dynamics.[7]

Soon after this work appeared, Zheng *et al.*[8] and Glowacki *et al.*[9] used alternative theoretical methods to compute regioselectivities that agreed with experiment and did not require trajectory calculations. Glowacki *et al.* applied an energy grained master equation (ME) to model competitive cooling and product formation in the vibrationally excited borane–propene complex. ME rate coefficients are computed by RRKM theory. Their results fit the experimental regioselectivity with two adjustable parameters: collision frequencies in the range of $1–3 \times 10^{13}$ s^{-1} and energy loss per collision of 2.6–3.1 kcal/mol. The experimental temperature dependence was reproduced, although the range of collision frequencies and energy per collision implies an uncertainty about as large as the temperature dependence.

In Truhlar's model, the rate constant is a sum of statistical (TST) and nonstatistical parts. The nonstatistical contribution is based on phase space theory (PST),[10] in which the probability of forming a product is proportional to the volume of phase space available to that product. This is modified by nonstatistical factors applied to the reaction coordinate, according to the prescription of nonstatistical PST.[11,12] The nonstatistical factors alter the statistical product branching ratio in favor of the *product with higher energy.*

This interesting prediction arises from the importance of strong coupling between the $BH_3^{\bullet}C_3H_6$ complex states and product states. The interaction energy between these states implies a characteristic frequency of energy transfer, and a corresponding minimum time required for strong coupling. Highly energetic molecules move too swiftly; they presumably escape the complex. The time available for strong coupling to product P is proportional to $E_P^{-1/2}$ ($\propto$ velocity^{-1}), where E_P is the energy above product P. At a given total energy, E_P is less for the higher energy product (Markovnikov addition), and its nonstatistical contribution is greater.

The three methods — trajectories, master equation, and nonstatistical PST — are remarkably different, not only in formalism but also in their physical models. This fortunate application of several methods to the same problem was an exciting development, and one hopes to see more discussion of different approaches for understanding nonstatistical dynamics.

12.3.2 *Tetrazine + cyclopropene Diels–Alder reaction*

Tetrazine reacts with alkenes in a two-step sequence of forward and reverse Diels–Alder reactions to give diazine products with expulsion of N_2. Törk *et al.*[13] recently investigated the PES and dynamics of reactions with several alkenes, including cyclopropene (Fig. 12.4). Adduct **4** decomposes by loss of syn or anti N_2, with the syn barrier 9 kcal/mol higher than anti. The situation is similar to that in Fig. 12.3 except that **4** is a much more energetic intermediate. RRKM calculation of the anti/syn ratio gave anti:syn = 96:4 at an energy equal to 2.3 kcal/mol above the zero-point level of **3**. Trajectories initialized at **3** with quasi-classical microcanonical normal mode barrier sampling gave anti:syn = 8:1, which indicates a 3:1 dynamical preference for the higher-energy syn route. The half-life of **4** is under

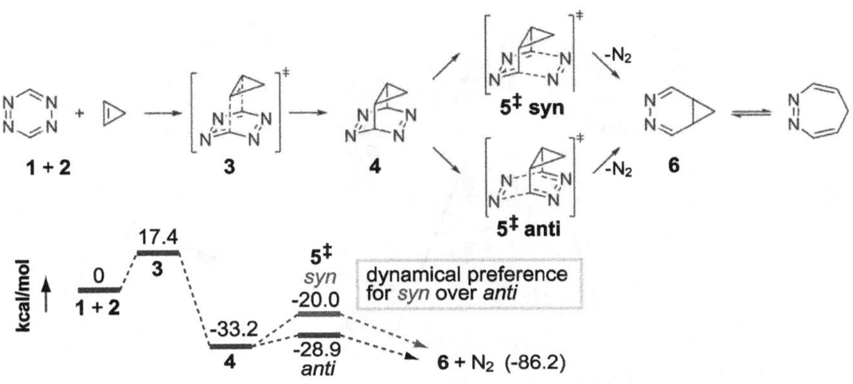

Figure 12.4. Tetrazine + cyclopropene PES (ΔG°, kcal/mol).

100 fs; the survival probability decays as a step function, gated by a 830 cm^{-1} mode involving cyclopropyl rocking + CC stretching.

In contrast to the BH_3 + propene case, the energy grained ME with a helium bath gave only the RRKM result at collision frequencies up to 10^{17} s^{-1}. This ME failure is clearly due to the very short lifetime of **4**. The comparison with BH_3 + propene may be useful in gauging the applicability of ME methods to other activated intermediates.

12.3.3 *Acetone radical cation*

Peterson and Carpenter[14] reported AM1-SRP trajectories for the isomerization of acetone enol radical cation to acetone radical cation **A**, which rapidly loses a methyl radical (Fig. 12.5). Experiments had shown that the newly formed methyl is lost preferentially, a nonstatistical effect. Carpenter initialized quasi-classical trajectories microcanonically at the **A** minimum and at **TS**, at an energy 10 kcal/mol above the zero-point energy of **TS**. Trajectories initialized at **A** gave statistical results: equal amounts of the two methyls, and a survival probability (fraction of trajectories not decomposed at a given time) that is cleanly exponential with a half-life of 409 fs. For trajectories initialized at the **TS**, the integrated product ratio was 1.13 in favor of the newly formed methyl, in agreement with experimental values

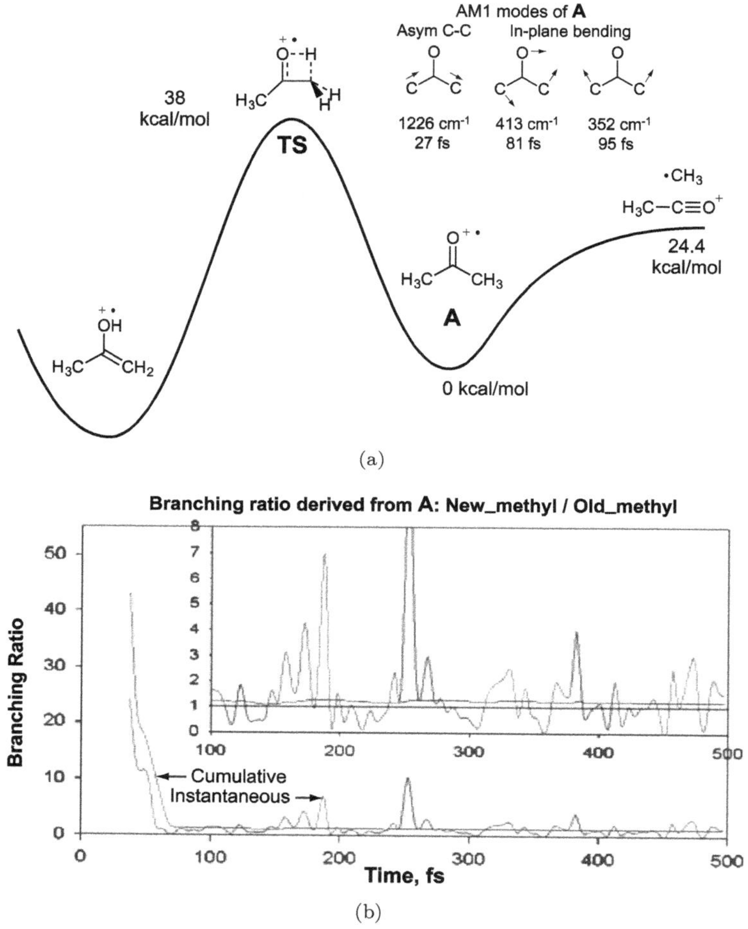

Figure 12.5. (a) PES for acetone enol radical cation computed by AM1-SRP, and AM1 frequencies for three modes (SRP parameters were not reported). (b) Branching ratio vs. time (ratio of newly formed methyl product vs. original methyl product).

of 1–1.4. But the *time development* of the product ratio (bottom of Fig. 12.5) is particularly informative.

In the first 60 fs, the new methyl dissociates almost exclusively. Carpenter points out that because the **TS** is bent, potential energy release excites the asymmetric CCO bend (top of Fig. 12.5), which brings **A** toward a geometry that favors dissociation (CCO angles of

90° and 180°). For dissociation to occur, the asymmetric CC stretch must also be excited, and this may be facilitated by the 3:1 resonance of the CC stretch with the asymmetric CCO bend (1,226 and 413 cm^{-1} in Fig. 12.5).

After the initial 100 fs, strong oscillations in the branching ratio occur for the next 400 fs. The peak-to-trough period of these oscillations, in which maximal loss of the new methyl is followed by a minimal loss of the new methyl (or maximal loss of old methyl), has a period in the range of 7–15 fs. The AM1 asymmetric CC stretching frequency of **A** has a period of 27 fs. One-half this period would be the switching time from one C–CH_3 bond to the other. Irregular spacing of the oscillations suggests that energy flow into the asymmetric CC stretch is modulated by coupling to the modes that are activated by potential energy release from the **TS**, including CH stretching and CCH, HCH, and CCO bending.

In addition, the survival probability of **A** oscillates (not shown here). The amplitude is small, but the decay has a step-function component with a period of 95 fs, which matches the AM1 period for the symmetric CCO bend in Fig. 12.5. That is, the probability of methyl loss (either one) is partially suppressed every 95 fs by the symmetric bend. A possible reason is that the symmetric bend distorts **A** away from the favored geometry for dissociation (CCO angles of 90° and 180°). Carpenter also showed the oscillations of the average CCO angles vs. time for each methyl, with periods of 85 and 93 fs, comparable to the AM1 bending periods for **A**.

A remarkable result reported in this paper is the very rapid decomposition of **A** (75% decay in 500 fs), even though it lies in a 24 kcal/mol energy well — as if the barrier to methyl loss were of little consequence. On the other hand, the strong oscillatory control of the product ratio implies an efficient flow of energy into the asymmetric CCO bend and CC stretch. That is, energy released from the **TS** becomes sequestered in a few modes coupled to methyl loss, and IVR competes poorly. The situation is similar to the gas phase S_N2 reaction and vinylidene rearrangement described above, in which even a 48 kcal/mol barrier is irrelevant if vibrational energy is localized in reactive modes.

12.4 Dynamics on a caldera: Newton's first law

Dynamic matching[15,16] has proven to be a potent idea for understanding nonstatistical product ratios. If more than one product is dynamically accessible from a given TS, dynamic matching provides a way of estimating the product ratio by applying Newton's first law, or conservation of momentum. This idea underlies the Polanyi rules[17] for predicting vibrational vs. translational energy partitioning in atom–diatom reactions, and Carpenter showed how conservation of momentum can be applied in certain organic reaction mechanisms.

Dynamic matching refers to the requirement that trajectory momenta must be directed toward a particular product in order to form that product directly. Otherwise, trajectories form another product, get reflected back to reactants, or become trapped to form an intermediate. Peterson and Carpenter[15] described a vector model based on two assumptions: (1) that trajectories move in straight lines and (2) that the transition vector (imaginary frequency eigenvector at the TS) is parallel to the average momentum of trajectories passing through the TS. The probability of forming a given product is then proportional to the dot product of the transition vector with a vector pointing from the TS to the product (i.e., cosine of the angle between the vectors). The term "dynamic matching" was originally[16] applied to the relative timing of internal rotations in a biradical, which could form a specific product if their angular velocities matched. This usage and the dot product both use Newton's first law, in the form of conservation of linear or angular momentum.

The calculation discussed here[18] is from a collaboration with the Houk group on the degenerate rearrangement of bicyclo[3.1.0]hex-2-ene. It was inspired by the elegant experimental work of Baldwin and Keliher,[19] and the dynamics is based on the PES of Suhrada and Houk.[20] In their PES, shown in Fig. 12.6, the biradical lies in a shallow and flat caldera — basically a pool table with four wide corner pockets. This presents a nearly ideal case for applying the Carpenter dot product.

Figure 12.6(a) shows the directions taken by trajectories that form each product (colored curved arrows). The green arrow

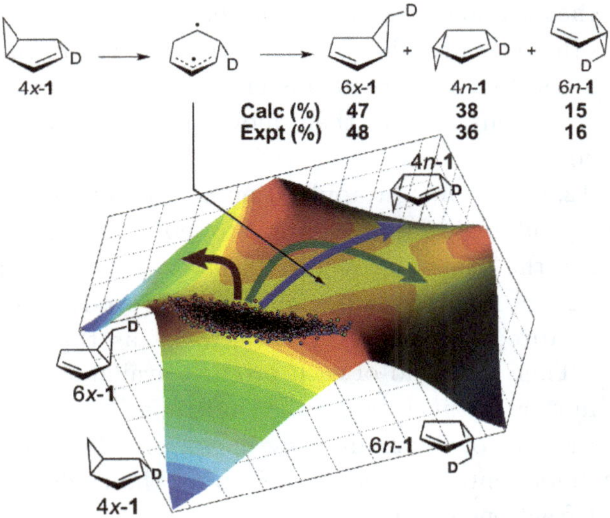

(a)

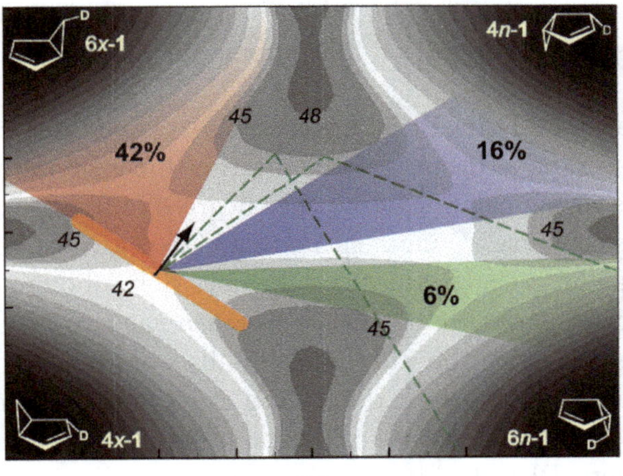

(b)

Figure 12.6. (a) Trajectory-derived vs. experimental product ratios; colored curved arrows show the route taken by trajectories that form each product. (b) Contour diagram of PES (kcal/mol); colored sections show product fractions computed for direct motion across the caldera by a cos θ distribution within 45 kcal/mol contours. Dotted lines indicate bank-shot trajectories that rebound into the $6n - \mathbf{1}$ minimum.

illustrates that the lower right product $6n - \mathbf{1}$ is formed mainly by trajectories that execute a bank shot off the upper hill. The initial TS distribution is shown as a broad cluster of points. Trajectories swarm over the surface, but most of them move in straight lines except for the bank shot.

Figure 12.6(b) shows an application of the dot product method to estimate product ratios. Defining θ $(-90° \leq \theta \leq 90°)$ to be the angle between the transition vector (black arrow) and a trajectory's momentum at the TS, we assume a cos θ (dot product) distribution of the initial directions taken by trajectories as they emerge onto the caldera. Then the yield of each product available to trajectories that move in straight lines across the caldera is $(1/2)\int \cos\theta \, d\theta$, integrated over ranges of θ delimited by lines from the saddle point to the 45 kcal/mol contours, about $3RT$ above the caldera. This gives the product fractions shown in the three colored sections. This only accounts for 64% of trajectories; the rest move toward hills and their outcome is uncertain. The dotted lines show how a contribution from trajectories that rebound off the upper hill into $6n - \mathbf{1}$ might be estimated.

The flatness of the PES suggests, and trajectories confirm, the essentially complete decoupling of the two-dimensional (2D) biradical motions in the caldera from other degrees of freedom. Therefore, trajectories conserve 2D momentum to a high degree — both the ones that go straight to the product and the ones that rebound off the wall. There is no fundamental difference between the two types of trajectories. They all move the way billiard balls would move on a similar surface, and the entrance and exit TSs are dynamically matched.

In a recent study of classical and quantum dynamics on an analytical caldera PES, Collins *et al.*[21] also found bank-shot trajectories, which they call "generalized dynamical matching" because they rebound off a wall into a product well. Importantly, quantum dynamics on this surface exhibits a quantum analog of dynamic matching. The expectation value of the momentum of the initial wavepacket as it enters the caldera determines the product channel through which most of the probability density passes.

12.5 Reaction path bifurcation and recrossing

Wang *et al.*[22] studied the Diels–Alder reaction of acrolein with butenone experimentally and computationally (Fig. 12.7). Reaction of **1** + **2** led to **3** + **4** concurrently, and the **3**/**4** ratio varied over time. Prolonged heating of **3** + **4** gave 98% **3**. The activation energy for conversion of **1** + **2** to **3** + **4** is 20.8 ± 1.2 kcal/mol, and the relative rate of formation of **3** vs. **4** was 2.5 ± 0.4.

Computationally, the lowest energy TS for reaction of **1** + **2** is **TS1**, whose intrinsic reaction coordinate (IRC) leads to **3**. **TS2** mediates the Cope rearrangement between **3** and **4**. The TS

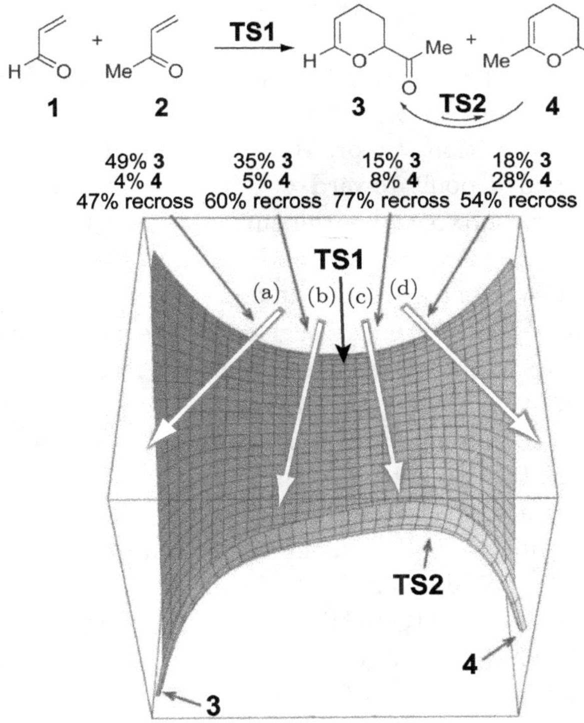

Figure 12.7. Qualitative PES of Wang *et al.*[22] for the Diels–Alder reaction of **1** + **2**. As trajectories pass through **TS1** (upper saddle point), they encounter a bifurcation region and can form either product. The bifurcation is due to the proximity of **TS2** which interconverts products **3** and **4**.

for **1** + **2** → **4** is 4.9 kcal/mol higher than **TS1**. If this is the lowest direct route to **4**, then TST does not explain the **3**:**4** relative rate of 2.5. However, **TS1** and **TS2** are situated on the PES in a way that creates a bifurcation region, such that trajectories that pass through **TS1** can form either **3** or **4**.

The authors initialized 296 trajectories at **TS1** with quasi-classical normal mode sampling at 80°C. The trajectories produced **3** and **4** in a ratio of 2.7, fortuitously close to the experimental value 2.5. Only 41% of trajectories actually formed **3** or **4**. The remaining 59% moved past **TS1** into the region of **TS2**, where they were reflected back to **1** + **2**. This might seem irrelevant because no product was formed. However, closer inspection reveals that recrossing influences the product ratio.

Figure 12.7 divides the angular space of trajectory momenta at **TS1** into four regions *a, b, c,* and *d* according to their direction relative to the transition vector. Regions *a* and *b* point toward **3** and regions *c* and *d* point toward **4**. The outcomes of trajectories in regions *a*–*d* are shown above the surface. Trajectories in regions *a* and *b* favor **3** by ratios of 12:1 and 7:1, and the sum of regions *c* and *d* favor **4** by a much smaller amount. There are two apparent reasons for this. First, the entire surface is tilted in favor of **3**, the thermodynamic product. Second, the recrossing fraction is higher for *c*, *d* than for *a*, *b*. Recrossing performs a masking operation, or the inverse of dynamic matching: trajectories that hit the reflecting wall in the vicinity of **TS2** do not form a product, and their absence from the product mixture affects the product ratio.

This description is similar to the caldera case described earlier. In Figs. 12.6 and 12.7, dynamic matching approximates the product ratio reasonably well, and there is a contribution to the product ratio from trajectories that hit a wall and change direction. On the caldera it was an angular reflection that formed a product; here it is a straight reversal back to reactants. In the caldera case, we argued that the direct and reflected trajectories are not different but are the expected results of conservation of momentum, justified by the decoupling of 2D biradical motion from the other degrees of freedom. The current case is similar, but the decoupling of 2D motion in Fig. 12.7 is less extensive than in Fig. 12.6. For example, it is not obvious from the PES of Fig. 12.7 how recrossing should vary among *a*–*d*. Nevertheless, the similarity to the caldera case is substantial.

12.6 Bifurcation cascades

Hong and Tantillo[23] have given strong evidence that terpene carbocation rearrangement cascades, long known for their complexity, are complex for reasons that had not been anticipated. They studied the reaction shown in Fig. 12.8, in which copalyl diphosphate, present in certain plants, is enzymatically converted to miltiradiene. This involves loss of diphosphate, followed by a carbocation rearrangement cascade. In their Density Functional Theory examination of the carbocation PES, they found that the rearrangements are characterized not by a series of stable carbocation minima connected by IRCs via saddle points, but by a cascade of bifurcations. This is a qualitatively different type of PES: a set of interconnected saddle points linked by IRC paths that terminate not at a local minimum but at another saddle point. From the highest energy TS, the PES allows trajectories to move down in energy through multiple bifurcations to

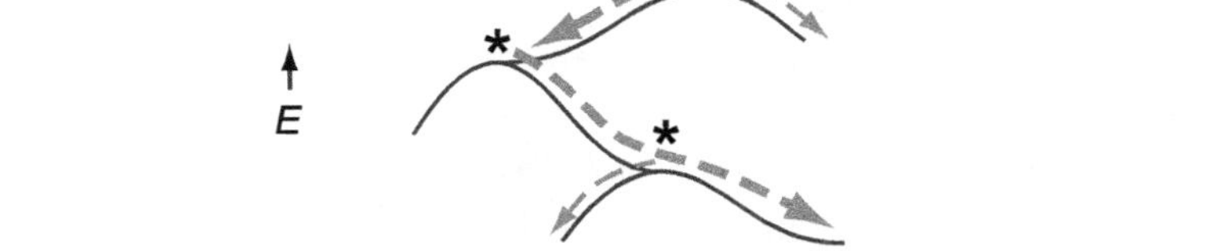

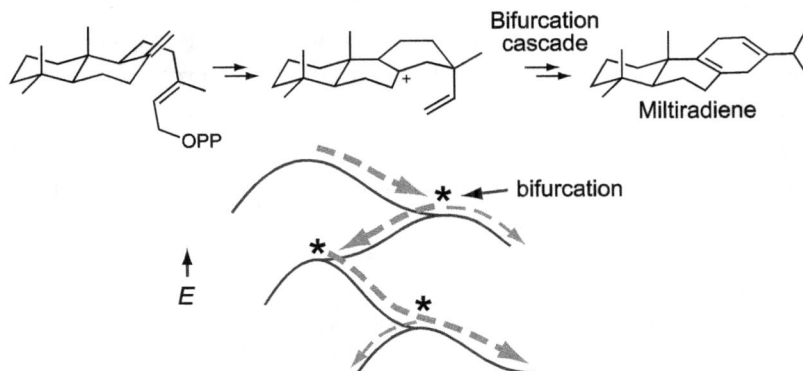

Figure 12.8. Qualitative PES with a bifurcation cascade based on Ref. [23]. Dotted lines are possible trajectory paths.

18 low-energy carbocation minima with only one intervening high-energy minimum.

In a system this complex, the simplest way to predict the rearrangement products is to run trajectories. When trajectories were initialized at the highest TS, 11 low-energy cations were formed. However, two of these accounted for 91% of the products, and the one formed in the largest amount is the direct precursor to miltiradiene. This is a surprising amount of dynamical selectivity on a PES in which 18 product ions can be formed by competing routes of monotonically decreasing energy.

This work raises a number of interesting questions and a caveat. How common are bifurcation cascades in carbocation rearrangements? Do they occur in other mechanisms? Is there a generalization of dynamic matching that could identify the most probable products? How does the enzyme alter the PES and dynamics to achieve a high yield of miltiradiene? The caveat is that these calculations were carried out in the gas phase. In any case, the dynamical question raised by these calculations is an important one, i.e., how to understand dynamically determined selectivity on a highly complex PES.

12.7 Practical considerations: Running trajectory calculations

A good starting point for investigating dynamics is to compute a relaxed PES scan (constrained optimizations with two coordinates fixed) to generate a 2D-PES of the region of interest. One usually needs this simply in order to think about nonstatistical dynamics.

Direct dynamics trajectories, in which atomic forces are computed at each step from electronic structure code, can require a large amount of computer time.

Most electronic structure packages have routines to integrate the classical equations of motion. In general application they are all adequate for this purpose. To have flexibility in choosing initial coordinates and momenta, a dynamics code is essential. Two codes with which the author is familiar are the Venus package coupled with NWChem in Venus-NWChem[24] and ProgDyn.[25]

In a typical application, trajectories are initialized at the highest-energy TS that leads to the region of interest. Initial coordinates and momenta are chosen by quasi-classical normal mode sampling at a given temperature (a standard option in the codes mentioned). Quasi-classical means that harmonic zero-point energy is included, with additional energy added in discrete quanta to simulate a quantum Boltzmann distribution of vibrational levels. To identify reactive trajectories, we run two trajectories from a TS, one in each direction from the same sampled point, which are stopped when reactants or products have been formed according to stopping criteria specified by the user (based on geometric parameters and/or time).

After computing trajectories, one has a set of coordinates and momenta for each trajectory step. In typical codes, a reactant and product have been identified and the lifetimes of trajectories have been reported. This might be all that one needed, but often one requires more. To answer the interesting questions that arise, one writes programs.

References

1. Gruebele, M. and Wolynes, P. G. (2004). Vibrational energy flow and chemical reactions, *Acc. Chem. Res.*, 37, pp. 261–267.
2. (a) Lourderaj, U., Park, K. and Hase, W. L. (2008). Classical trajectory simulations of post-transition state dynamics, *Int. Rev. Phys. Chem.*, 27, pp. 361–403; (b) Ess, D. H., Wheeler, S. E., Iafe, R. G., Xu, L., Çelebi-Ölçüm, N. and Houk, K. N. (2008). Bifurcations on potential energy surfaces of organic reactions, *Angew. Chem. Int. Ed.*, 47, pp. 7592–7601; (c) Lourderaj, U. and Hase, W. L. (2009). Theoretical and computational studies of non-RRKM unimolecular dynamics, *J. Phys. Chem. A*, 113, pp. 2236–2253; (d) Yamataka, H. (2010). Molecular dynamics simulations and mechanism of organic reactions: Non-TST behaviors, *Adv. Phys. Org. Chem.*, 44, pp. 173–222; (e) Rehbein, J. and Carpenter, B. K. (2011). Do we fully understand what controls chemical selectivity? *Phys. Chem. Chem. Phys.*, 13, pp. 20906–20922; (f) Carpenter, B. K. (2013). Energy disposition in reactive intermediates, *Chem. Rev.*, 113, pp. 7265–7286; (g) Rehbein, J. (2015). Chemistry in motion — off the MEP, *Tet. Lett.*, 56, pp. 6931–6943.
3. (a) Cho, Y. J., Vande Linde, S. R., Zhu, L. and Hase, W. L. (1992). Trajectory studies of S_N2 nucleophilic substitution. II. Nonstatistical central barrier recrossing in the Cl^- + CH_3Cl system, *J. Chem. Phys.*, 96, pp. 8275–8287; (b) Hase, W. L. (1994). Simulations of gas-phase chemical

reactions: Applications to S_N2 nucleophilic substitution, *Science*, 226, pp. 998–1002; (c) Manikandan, P., Zhang, J. and Hase, W. L. (2012). Chemical dynamics simulations of $X^- + CH_3Y \rightarrow XCH_3 + Y^-$ gas-phase S_N2 nucleophilic substitution reactions. Nonstatistical dynamics and nontraditional reaction mechanisms, *J. Phys. Chem. A*, 116, pp. 3061–3080.
4. Hayes, R. L., Fattal, E., Govind, N. and Carter, E. A. (2001). Long live vinylidene! A new view of the $H_2C{=}C: \rightarrow HC \vdots CH$ rearrangement from *ab initio* molecular dynamics, *J. Am. Chem. Soc.*, 123, pp. 641–657.
5. Schork, R. and Koppel, H. (2001). Barrier recrossing in the vinylidene–acetylene isomerization reaction: A five-dimensional *ab initio* quantum dynamical investigation, *J. Chem. Phys.*, 115, pp. 7907–7923.
6. Oyola, Y. and Singleton, D. A. (2009). Dynamics and the failure of transition state theory in alkene hydroboration, *J. Am. Chem. Soc.*, 131, pp. 3130–3131.
7. Carpenter, B. K. (1992). Intramolecular dynamics for the organic chemist, *Acc. Chem. Res.*, 25, pp. 520–528.
8. Zheng, J., Papajak, E. and Truhlar, D. G. (2009). Phase space prediction of product branching ratios: Canonical competitive nonstatistical model, *J. Am. Chem. Soc.*, 131, pp. 15754–15760.
9. Glowacki, D. R., Liang, C. H., Marsden, S. P., Harvey, J. N. and Pilling, M. J. (2010). Alkene hydroboration: Hot intermediates that react while they are cooling, *J. Am. Chem. Soc.*, 132, pp. 13621–13623.
10. Pechukas, P. and Light, J. C. (1965). On detailed balancing and statistical theories of chemical kinetics, *J. Chem. Phys.*, 42, pp. 3281–3291.
11. Serauskas, R. V. and Schlag, E. W. (1966). Nonstatistical energy-transfer theory, *J. Chem. Phys.*, 45, pp. 3706–3711.
12. Truhlar, D. G. (1975). Intermediate coupling probability matrix approach to chemical reactions. Dependence of the reaction cross section for $K + HCl \rightarrow KCl + H$ on initial translational and vibrational energy, *J. Am. Chem. Soc.*, 97, pp. 6310–6317.
13. Törk, L., Jimeńez-Osesś, G., Doubleday, C., Liu, F. and Houk, K. N. (2015). Molecular dynamics of the Diels–Alder reactions of tetrazines with alkenes and N_2 extrusions from adducts, *J. Am. Chem. Soc.*, 137, pp. 4749–4758.
14. Nummela, J. A. and Carpenter, B. K. (2002). Nonstatistical dynamics in deep potential wells: A quasiclassical trajectory study of methyl loss from the acetone radical cation, *J. Am. Chem. Soc.*, 124, pp. 8512–8513.
15. Peterson, T. H. and Carpenter, B. K. (1992). Estimation of dynamic effects on product ratios by vectorial decomposition of a reaction coordinate. Application to thermal nitrogen loss from bicyclic azo compounds, *J. Am. Chem. Soc.*, 114, pp. 767–769.
16. Carpenter, B. K. (1995). Dynamic matching: The cause of inversion of configuration in the [1,3] sigmatropic migration? *J. Am. Chem. Soc.*, 117, pp. 6336–6344.
17. Polanyi, J. C. (1972). Some concepts in reaction dynamics, *Acc. Chem. Res.*, 5, pp. 161–168.

18. Doubleday, C., Suhrada, C. P. and Houk, K. N. (2006). Dynamics of the degenerate rearrangement of bicyclo[3.1.0]hex-2-ene, *J. Am. Chem. Soc.*, 128, pp. 90–94.
19. Baldwin, J. E. and Keliher, E. J. (2002). Activation parameters for three reactions interconverting isomeric 4- and 6-deuteriobicyclo[3.1.0]hex-2-enes, *J. Am. Chem. Soc.*, 124, pp. 380–381.
20. Suhrada, C. P. and Houk, K. N. (2002). Potential surface for the quadruply degenerate rearrangement of bicyclo[3.1.0]hex-2-ene, *J. Am. Chem. Soc.*, 124, pp. 8796–8797.
21. Collins, P., Kramer, Z. C., Carpenter, B. K., Ezra, G. S. and Wiggins, S. (2014). Nonstatistical dynamics on the caldera, *J. Chem. Phys.*, 141, p. 034111.
22. Wang, Z., Hirschi, J. S. and Singleton, D. A. (2009). Recrossing and dynamic matching effects on selectivity in a Diels–Alder reaction, *Angew. Chem. Int. Ed.*, 48, pp. 9156–9159.
23. Hong, Y. H. and Tantillo, D. J. (2014). Biosynthetic consequences of multiple sequential post-transition-state bifurcations, *Nat. Chem.*, 6, pp. 104–111.
24. Lourderaj, U., Sunb, R., Kohaleb, S. C., Barnesc, G. L., de Jongd, W. A., Windus, T. L. and Hase, W. L. (2014). The VENUS/NWChem software package. Tight coupling between chemical dynamics simulations and electronic structure theory, *Comput. Phys. Commun.*, 185, pp. 1074–1080.
25. Current ProgDyn version is in the Supporting Info of the following: Biswas, B. and Singleton, D. A. (2015). Controlling selectivity by controlling the path of trajectories, *J. Am. Chem. Soc.*, 137, pp. 14244–14247.

Chapter 13

The Distortion/Interaction Model for Analysis of Activation Energies of Organic Reactions

K. N. Houk, Fang Liu, Yun-Fang Yang and Xin Hong

Department of Chemistry and Biochemistry and Department of Chemical and Biomolecular Engineering, University of California, Los Angeles, California 90095-1569, USA

13.1 Introduction

Before quantum mechanical calculations and orbital models became common, organic chemists often relied on valence bond and resonance theory ideas to rationalize reactivities and regioselectivities. Resonance structures or qualitative ideas about inductive effects were used to estimate partial charges. The 1940s to 1960s were the era of Hammett and Taft linear free energy relationships.[1] Substituent effects were obtained from experiment in terms of electron-donation or withdrawal by substituents that were thought to alter charges through inductive and resonance effects. In radical reactions or those involving radical intermediates, radical character was sometimes invoked to describe reactivities and regioselectivities.

This situation changed with Fukui's development of frontier molecular orbital (FMO) theory,[2] which describes reactivity and regioselectivity in terms of the energies and shapes of FMOs. Soon thereafter, the Woodward–Hoffmann discussion of what became known as the Conservation of Orbital Symmetry, or Woodward–Hoffmann Rules, spread these ideas into mainstream organic chemistry.[3] The highest occupied molecular orbital (HOMO) is the most important controller of reactivity for nucleophilic species, and the lowest unoccupied molecular orbital (LUMO) is critical for electrophilic species. The singly occupied molecular orbital (SOMO) is the FMO of radical species. A good example of the application of these ideas is our (the only one of us who was alive at the time) application to understanding the relative reactivities of the cyano-substituted alkenes, acrylonitrile through TCNE (tetracyanoethylene), in the Diels–Alder reactions with cyclopentadiene and 9,10-dimethylanthracene,[4] reactions that had been studied experimentally by Sauer *et al.*[5] The reactivities in these cases are clearly related to the energy gap between the diene HOMO and the dienophile (here the cyanoalkenes) LUMO (Fig. 13.1).[4]

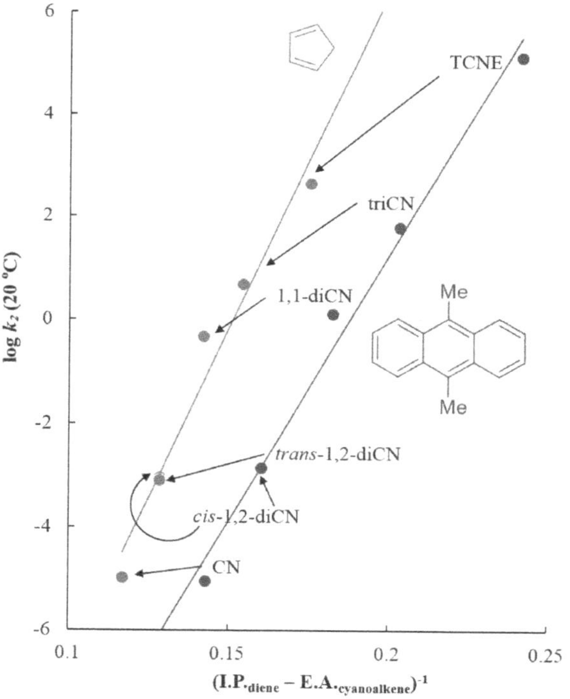

Figure 13.1. Plot of rates of cycloadditions of dienes with cyanoalkenes vs. the inverse of the diene HOMO–dienophile LUMO energy gap.

Such applications became the standard way to understand, and even predict, the relative reactivities of alkenes in various reactions. Considerations of the coefficient of the frontier orbitals led to the understanding of regioselectivities of cycloadditions, not in terms of charge but in terms of different overlap between the FMOs of two reactants in the different regioselective modes (Fig. 13.2). These overlaps are influenced by the nature of the atom and the sizes of the coefficients at the interacting centers.[6]

An elaboration of FMO theory was developed after Density Functional Theory (DFT) became the norm for calculations on organic systems. Although DFT commonly uses Kohn–Sham orbitals to determine electron densities, the Kohn–Sham theorem teaches us that all properties, such as reactivity, can be understood in terms of electron densities, without consideration of orbitals. Conceptual DFT considers reactivity in terms of electron densities and the changes in electron densities upon ionization or electron addition, to model electrophilic or nucleophilic reactions, respectively.[7]

All of these methods concentrate on characteristics of reactants to understand and predict reactivity. These theories are necessarily limited by the assumption that reactivity can be understood from reactant properties. Molecules change shape and are deformed along the pathway from reactants to products. One relatively early recognition that distortions should be considered to assess interactions between two molecules was in Morokuma's energy decomposition analysis, where he defined deformation energy as the energy

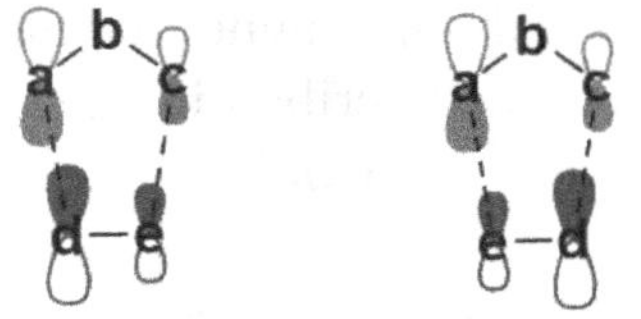

(a) (b)

Figure 13.2. FMO interactions in regioisomeric 1,3-dipolar cycloaddition transition states. The LUMO of an unsymmetrical 1,3-dipole overlaps more with the HOMO of an unsymmetrical dipolarophile in transition state (a) than in transition state (b).

to distort reactants in order to facilitate interaction.[8] Starting in the 1970s, Morokuma and Winick,[8a] Morokuma,[8b] and Umeyama and Morokuma[8c] published a series of papers on the study of hydrogen bonding using energy decomposition analysis. An examination of interaction energy components — electrostatic, polarization, exchange repulsion, and charge transfer — and their coupling was carried out for complexes such as $(H_2O)_2$, $(HF)_2$, and (H_3N-HF). Such analysis provided insights as to the origins of hydrogen bonding and the geometrical structures of hydrogen-bonded complexes.

The Morokuma energy decomposition analysis was then applied to studies of reactions such as the S_N2 reaction $H^- + CH_4 \rightarrow CH_4 + H^-$,[9] and the unimolecular dissociation reaction $H_2CCH_2F \rightarrow H_2C{=}CHF + H$,[10] Morokuma defined the deformation energy as the energy required to deform the olefin to its geometry in the transition state (TS). In this particular case, deformation energy was found to be the main source of the activation barrier for the H atom addition to fluoroethylene, the reverse of the dissociation reaction.

Various researchers, including our group, also recognized that these deformations, or distortions as we have called them, could be used to explain anomalies, such as the greater reactivities of alkynes than the correspondingly substituted alkenes.[11] We showed that the alkyne is easier to distort, and is more reactive with nucleophiles in spite of having a higher LUMO energy. However, the emergence of a general model to analyze the role of reactant distortion on activation barriers did not occur until this century. The Distortion/Interaction (D/I) Model, as we called it, or Activation Strain Model, as Matthias Bickelhaupt named it, has now been applied to many types of reactions, as described in this chapter.

13.2 Distortion/Interaction model

Aware of the significance of distortion in determining reactivities, our group developed the D/I model to analyze the relative importance of distortion and interaction energies on reactivities, and applied the model first to 1,3-dipolar cycloadditions.[12] Figure 13.3 illustrates

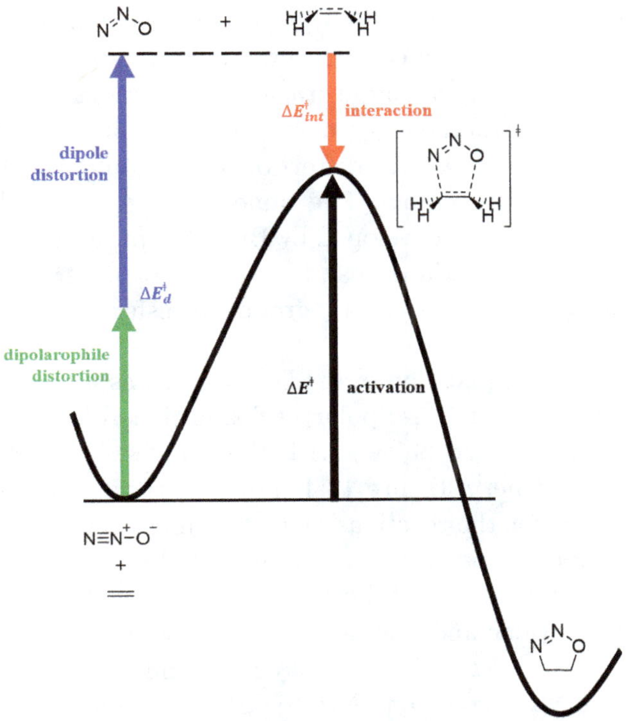

Figure 13.3. The D/I model. $\Delta E^{\ddagger}$ is the activation potential energy of the reaction, and it is divided into $\Delta E^{\ddagger}_{d}$ and $\Delta E^{\ddagger}_{int}$. The distortion energy of the individual components is represented in blue for the dipole and green for the dipolarophile.

how the activation energy of a reaction can be analyzed as a sum of distortion energy and interaction energy. The example given involves the reaction of nitrous oxide, a simple 1,3-dipole, with ethylene. The curved black line represents the energy vs. reaction coordinate diagram for the reaction of interest. The key step in D/I analysis is to compute the transition structure of the reaction, and from that to determine the energy required to distort the ground state dipole and dipolarophile into the geometries they take in transition state structure without interacting, as shown by the blue and green arrows in Fig. 13.3. The difference between the activation energy and distortion energy is the interaction energy, which is often a negative number (red arrow in Fig. 13.3).[12] The interaction energy can be further dissected into its components — electrostatic, polarization, exchange repulsion, and charge transfer interactions. We usually do not carry out further energy decompositions of interaction energies, since our focus is the distortion energy and the interaction between distorted reactants. The same analysis, called the Activation Strain Model, was developed by Bickelhaupt, and has been used to study oxidative additions and a variety of other organometallic reactions, S_N2 reactions, group transfers, and pericyclic reactions.[13]

The original application of the D/I model was to understand the reactivity patterns in 1,3-dipolar cycloadditions of three classes of dipoles **1**–**9** (as shown below) with the simplest dienophiles, ethylene, and acetylene.[12] Figure 13.4 shows overlays of the B3LYP/6-311G(d,p) TSs for the reactions of **1–9** with ethylene and acetylene. Each overlay contains six TS structures of three betaines from the same class in reactions with both ethylene and acetylene.

Clearly ethylene and acetylene TSs have very similar geometries! Within each class of 1,3-dipoles, there is a trend from late to early TS as the termini vary from O, to NH, to CH_2. This change in geometries is accompanied by a drop in activation barriers for the corresponding

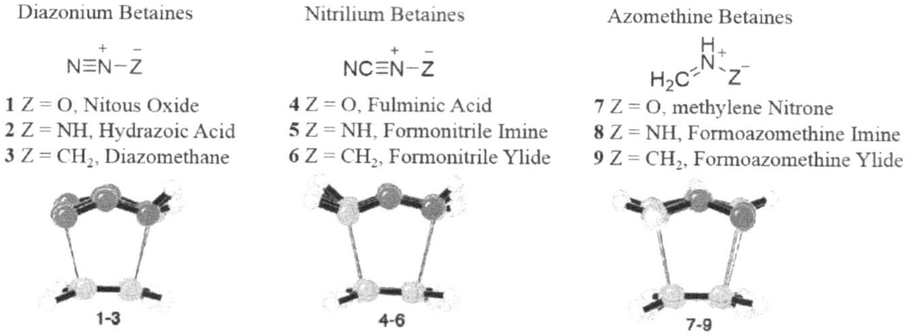

Figure 13.4. 1,3-Dipoles **1**–**9** and overlay of TS structures.

Source: Reprinted with permission from Ref. [12b]. © 2008, American Chemical Society.

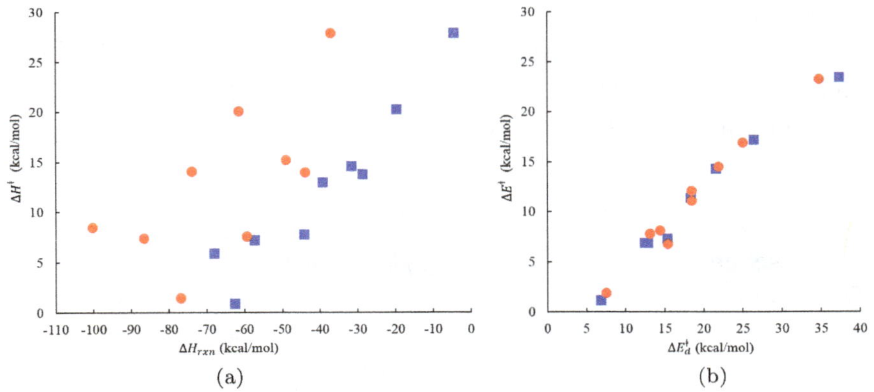

Figure 13.5. (a) Plot of activation enthalpy $\Delta H^{\ddagger}$ vs. reaction enthalpy ΔH_{rxn}. (b) Plot of activation energy $\Delta E^{\ddagger}$ vs. distortion energy $\Delta E_d^{\ddagger}$ (red dots for reactions with acetylene and blue squares for reactions with ethylene).

reactions. However, there is no general correlation between reactivity and reaction thermodynamics from the comparison of activation enthalpies to reaction enthalpies (Fig. 13.5(a)).

Using the D/I model discussed earlier, we found a linear correlation between activation energies and distortion energies (Fig. 13.5(b)) for the 18 reactions studied, indicating that differences in the reaction barriers result from differences in the distortion energies at the TSs. The differences in interaction energies are small compared to the changes in distortion energies and result from variations in charge transfer and electrostatic effects. For the 18 1,3-dipolar cycloadditions studied, the energies to distort the 1,3-dipoles and dipolarophiles to transition-state geometries control the reactivity patterns. Conceptually, and complementary to the Hammond Postulate, when distortion energies are low, favorable interaction energies overcome unfavorable distortion energies quickly as reactants approach each other, and the TSs are "early." When distortion energies are large, the interaction energies required to overcome this destabilization are large, the activation energies are high, and the TSs are late. There is a large range of distortion energies (7–40 kcal/mol) for different 1,3-dipoles, and consequently, a large range of reactivities. We have discussed how these distortion energies are related to the HOMO–LUMO gaps in the 1,3-dipoles. For example, the highly reactive ozone has a very small HOMO–LUMO gap and is easily distorted, while azides have much larger HOMO–LUMO gaps and are relatively unreactive.[12]

13.3 Diels–Alder reactions

In subsequent publications, the D/I model was applied to many Diels–Alder reactions. We first explored an extensive series of 1,4-cycloadditions of ethylene to a series of polycyclic aromatic hydrocarbons **10–17**.[14] This system had been studied earlier by Zhong *et al.*,[15] who showed a connection between activation energy and the energy of reaction. We explored whether the D/I analysis would also work for this series of reactions. A total of 29 possible 1,4-addition modes was explored, as shown in Fig. 13.6(a). The structures of hydrocarbons **10–17** and their possible Diels–Alder cycloaddition modes are named according to the system used by Zhong *et al.* Note that there can be more than one mode under the same class. For instance the cycloadditions of ethylene with anthracene at both the (5,10) and (1,4) positions are considered [1,1] mode (Fig. 13.6(b)).

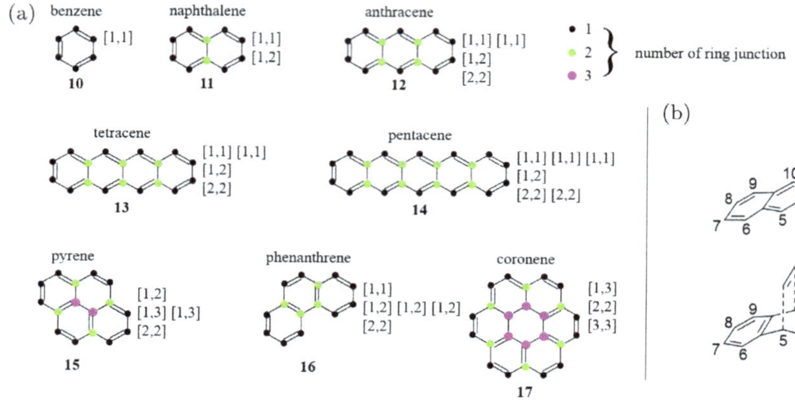

Figure 13.6. (a) Eight polycyclic aromatic hydrocarbons with 29 possible 1,4-addition modes. (b) Two [1,1] reactions of anthracene with ethylene.

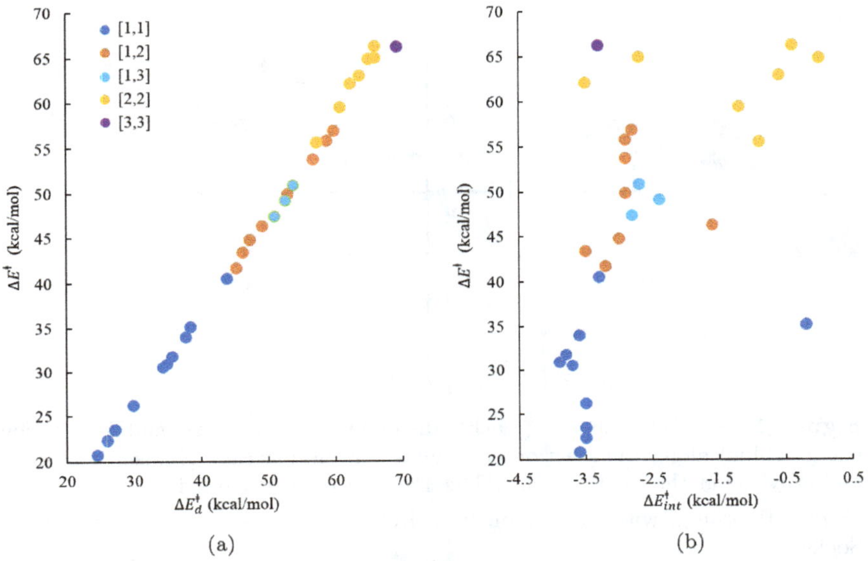

Figure 13.7. (a) Plot of activation energy $\Delta E^{\ddagger}$ vs. total distortion energy $\Delta E^{\ddagger}_{d}$. (b) Plot of activation energy $\Delta E^{\ddagger}$ vs. interaction energy $\Delta E^{\ddagger}_{int}$. Data points are color coded according to the cycloaddition modes.

For each reaction studied, the TS was analyzed by D/I model, and a linear correlation was observed between distortion energy and activation energy, as shown in Fig. 13.7(a). The interaction energy is more stabilizing in reactions with lower activation energy for these cases, but there is no tight correlation between the interaction and activation energies (Fig. 13.7(b)).

To give deeper insights into how distortion and interaction affect reactivities, the distortion and interaction analyses were carried out along the intrinsic reaction coordinate (IRC). This technique has been advocated by van Zeist and Bickelhaupt for other reaction types.[13] Using the reaction between pentacene and ethylene at the (6,13) atoms as an example, Fig. 13.8 shows how total energy, distortion energy, and interaction energy evolve along the reaction coordinate. Before the TS, there is no major stabilizing interaction between the reactants. The substantial increase in distortion energy in the vicinity of the TS arises in order to achieve better alignment of the

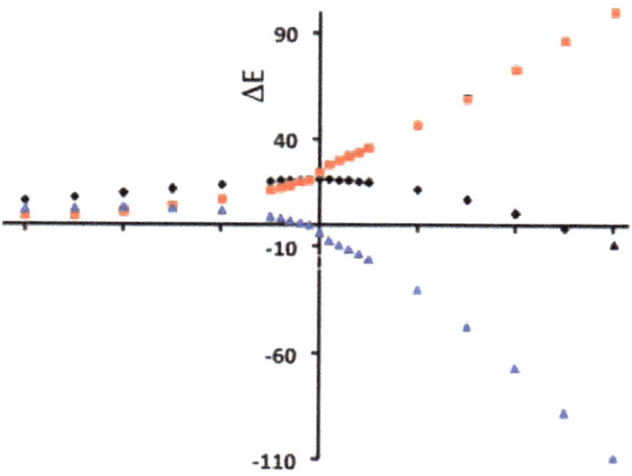

Figure 13.8. Total energy (black), distortion energy (red), and interaction energy (blue) along the reaction coordinate for the reaction between pentacene and ethylene at the (6, 13) atoms. The TS is on the vertical axis.

Source: Reprinted with permission from Ref. [14]. © 2009 American Chemical Society.

two reactants; this alignment and distortion produces better orbital overlap and results in a more favorable interaction energy. After the TS, the interaction energy becomes the major factor and overtakes the increase in distortion energy. The activation energy is mostly determined by the distortion energy, and the TS is the point where the slope of the change in interaction energy is equal in magnitude, and opposite in sign, to the slope of the change in distortion energy.

Our recent studies of cycloalkenones and cycloalkenes showed that their reactivities in Diels–Alder reactions are distortion controlled.[16] Figure 13.9 shows examples. Here we show the distortion energies of dienophile (green arrows), distortion energies of diene (blue arrows), interaction energies (red arrows), and activation energies (black arrows) of the cycloaddition reactions between cyclopropenone, cyclobutenone, cyclopentenone, cyclohexenone, and (*E*)-pent-3-en-2-one and their alkene counterparts (cyclopropene, cyclobutene, cyclopentene, cyclohexene, and *cis*-2-butene).

Little variation in interaction energies was observed for each class, 14.1–15.9 kcal/mol for alkenones and 10.5–11.0 kcal/mol for alkenes,

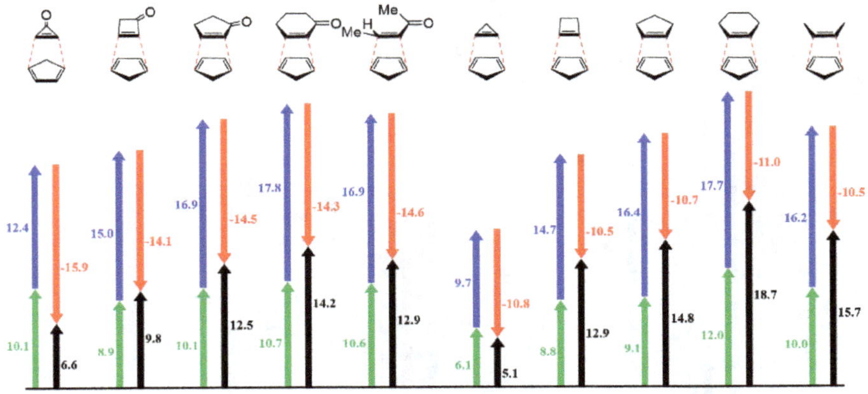

Figure 13.9. D/I diagram for Diels–Alder reactions of cyclopentadiene with cycloalkenones and cycloalkenes and acyclic analog.

while the activation energies range from 6.6 to 14.2 kcal/mol and 5.1 to 18.7 kcal/mol, respectively. The diene is constant in each series and so the difference in diene distortion energies results from the position of the TS; an early TS gives low diene distortion and vice versa. The difference in dienophile distortion energies arises from the ease of out-of-plane distortion of each molecule, which is the prominent distortion in the TS. Our calculations show that there is an increase in out-of-plane bending force constants as the size of the dienophile change from three-membered ring to six-membered ring. This can be understood qualitatively by the fact that more angle-strained cycloalkenes are already distorted toward an sp^3 transition-state–like geometry.

There is also an obvious difference in reactivities in the two series. The larger interaction energies for cycloalkenones make these uniformly more reactive than cycloalkenes. This is the type of substituent effect that is explained well by FMO theory.

In the cycloadditions of tetrazines, variation in both distortion and interaction contribute to the different activation barriers of the Diels–Alder reactions with inverse-electron-demand dienes **18**–**24**.[17a] Figure 13.10 shows the decomposition of activation energies for the reactions of **18**–**24** with *trans*-2-butene. There is a clear trend from less favorable to more favorable interaction energy as

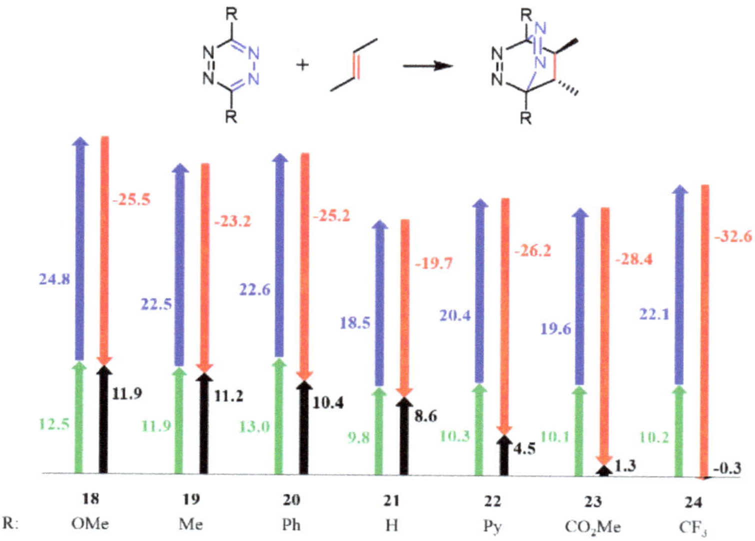

Figure 13.10. D/I diagram of cycloadditions of tetrazines **18**–**24** with *trans*-2-butene.

the substituents on tetrazines change from electron-donating groups to electron-withdrawing groups. This trend can be explained by the FMO theory. Electron-withdrawing groups lower the LUMO of tetrazines and achieve better interactions with the HOMO of dienophiles in inverse-electron-demand Diels–Alder reactions. We also discovered that the substituents on tetrazines also have a big effect on diene distortion energies (Fig. 13.10). By scanning the out-of-plane distortion, we showed that electron-donating groups stabilize the tetrazines ground states and cause them to be less easy distorted.[17a]

Recently we applied the D/I model to the study of the cycloadditions of benzene **25** and 10 azabenzenes **26**–**35** with ethylene.[17b] Multiple cycloaddition modes were investigated and illustrated in Fig. 13.11(a). The cycloaddition sites are indicated with dashed lines and are color-coded according to the types of bonds that are formed. All 21 reactions were analyzed with D/I model, and Figs. 13.11(b) and 13.11(c) show the correlations between distortion (interaction)

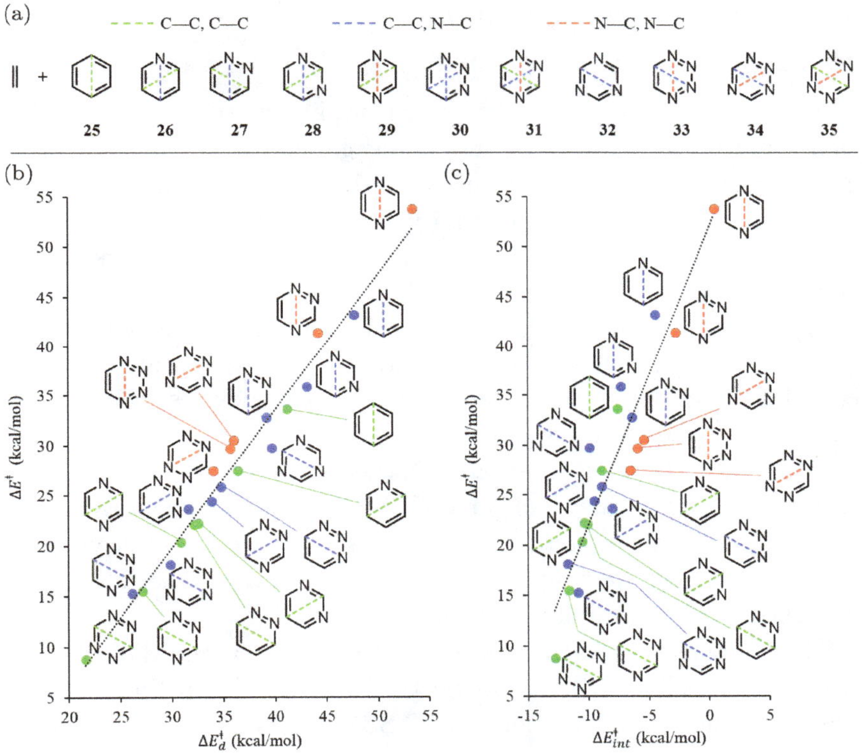

Figure 13.11. (a) Diene **25–35** and different cycloaddition modes. (b) Plot of activation energy $\Delta E^{\ddagger}$ vs. total distortion energy $\Delta E^{\ddagger}_{d}$. (c) Plot of activation energy $\Delta E^{\ddagger}$ vs. interaction energy $\Delta E^{\ddagger}_{int}$.

energy and activation energy. The graph shows a general trend from less reactive to more reactive as the number of hetero atoms in the azabenzene increases. The introduction of heteroatoms causes a decrease in distortion energies and also more favorable orbital interactions with the dienophile. We noted the fact that formation of C–N bond is much less favorable than the formation of C–C bond, in agreement with our previous studies.[18] The unfavorable formation of C–N bond is due to the less efficient orbital overlap of the interacting π-orbital on the more electronegative nitrogen atom.

13.4 Organometallic reactions

13.4.1 *C–X bond activation*

While the D/I model was developed originally to study cycloadditions, it can be applied to the whole gamut of organic and inorganic reactions. In this section, we describe how we have used this model to analyze various reactivity and regioselectivity patterns in organometallic reactions. In the transition metal-catalyzed cross-coupling reactions involving aryl halides and pseudohalides, it is generally assumed that the reactivity of a C–X bond is determined by the C–X bond strength. However, in the Pd/PR_3-catalyzed cross-coupling reactions with polyhalogenated heterocycles, the regioselectivity in the C–X bond activation does not correlate with the bond strength.[19] With substrate **36**, the weaker C–X bond is cleaved (Fig. 13.12), while the stronger C–X bond is cleaved in substrate **39** (Fig. 13.13). To understand the origins of the regioselectivities in this reaction, we have applied the D/I analysis on the TSs of the competing C–X bond activations.[20]

For substrate **36**, the distortion, particularly the distortion of heterocycle, controls the regioselectivity. As shown in Fig. 13.12, in the two competing insertions via **TS37** and **TS38**, the distortion

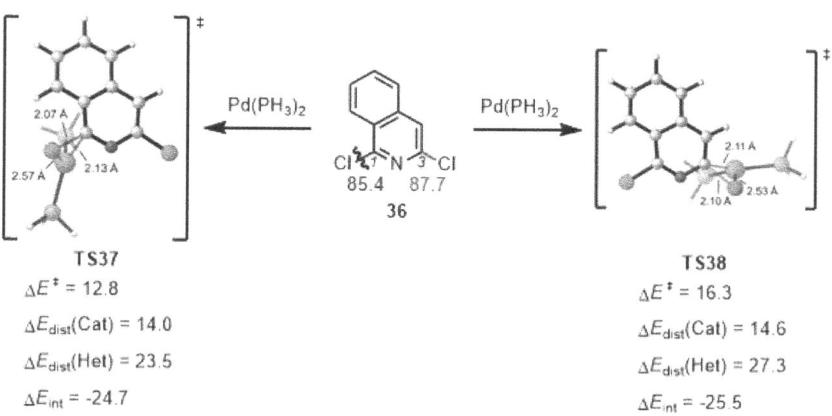

Figure 13.12. Energies and D/I analysis of TSs of [$Pd(PH_3)_2$]-mediated C–Cl oxidative addition of aryl chloride **36**. Energies are in kcal/mol.

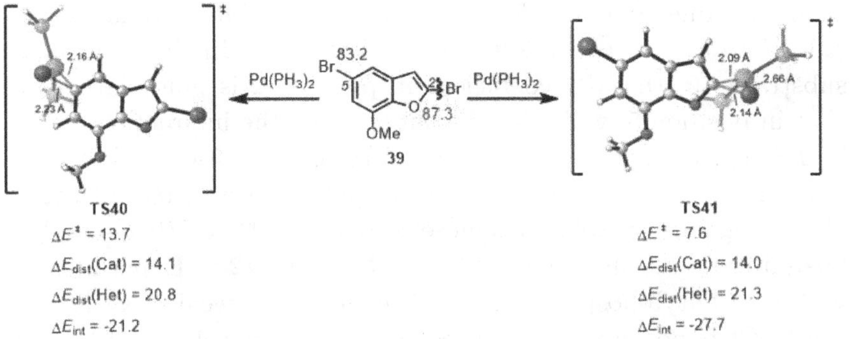

Figure 13.13. Energies and D/I analysis of TSs of $[Pd(PH_3)_2]$-mediated C–Cl oxidative addition of aryl bromide **39**. Energies are in kcal/mol.

of heterocycle is 23.5 kcal/mol for position 1 through **TS37** and 27.3 kcal/mol for position 3 through **TS38**. This 3.8 kcal/mol energy difference is the major contribution to the 3.5 kcal/mol difference in the electronic barrier, leading to the exclusive insertion in position 1. Although a number of geometric changes within heterocycle occur during the C–X bond activation, the distortion of the heterocycle is mainly due to the stretching of the C–X bond. Therefore, the homolytic bond dissociation energy (BDE), which reflects the intrinsic easiness of stretching C–X bond, is related to the activation barrier. The homolytic BDE of C–Cl bond in position 1 is 85.4 kcal/mol, and that of C–Cl bond in position 3 is 87.7 kcal/mol. The weaker C–Cl bond in position 1 leads to the smaller distortion of heterocycle in the oxidative addition TS, and eventually the regioselectivity of substrate **36**.

By contrast, with substrate **39**, the interaction between the heterocycle and palladium catalyst determines the regioselectivity (shown in Fig. 13.13). Because C–Br bonds are much weaker than C–Cl bonds, the C–Br bond TSs are earlier in the reaction coordinate, and the distortion of heterocycles in the competing C–Br activations, via **TS40** and **TS41**, are very similar (20.8 kcal/mol for **TS40** and 21.3 kcal/mol for **TS41**). Instead, the interaction energies favor position 2 significantly (−21.2 kcal/mol for **TS40** and −27.7 kcal/mol for **TS41**). The change in interaction energies is mainly due to the different d_{xy}-π^* orbital interactions between palladium and heterocycle in the two reactions. In the π^* orbital of substrate **39**, the MO coefficient in position 2 is much larger than that in position 5, which significantly favors the interaction energies in **TS41**, leading to the selective insertion of position 2.

Similar D/I analyses have been applied to understand the origins of ligand-controlled chemoselectivity in the Pd/PR_3-catalyzed Suzuki cross-coupling with chloroaryl triflate **42**.[21] Experimentally, with *tert*-butylphosphine ($PtBu_3$) ligand, the weaker C–X bond, the C–Cl bond, is functionalized, while the tricyclohexylphosphine (PCy_3) ligand leads to a reversed chemoselectivity, and the C–OTf bond is cleaved selectively.[22]

With the monoligated palladium catalyst ($PdPMe_3$), the distortion of aryl substrate controls the chemoselectivity (Fig. 13.14). Because the distortion of substrate is mainly due to the C–X bond stretching during the oxidative addition, the weaker C–Cl bond requires much less distortion in the process compared to the C–OTf bond. This makes the distortion energy of substrate 32.1 kcal/mol for the cleavage of C–Cl bond through **TS44** and 52.6 kcal/mol for the cleavage of C–OTf bond through **TS43**. The 20 kcal/mol energy difference in substrate distortion is the determining factor that makes the oxidative addition barrier of C–Cl bond about 5 kcal/mol lower than that of C–OTf bond.

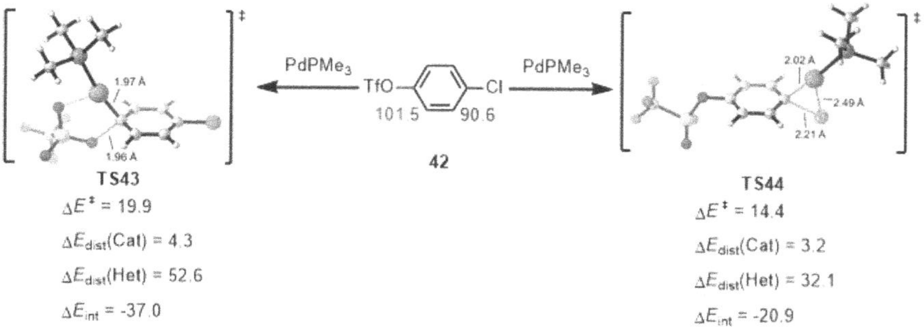

Figure 13.14. Energies and D/I analysis of TSs of Pd(PMe_3)-mediated C–Cl and C–OTf oxidative addition of substrate **42**. Energies are in kcal/mol.

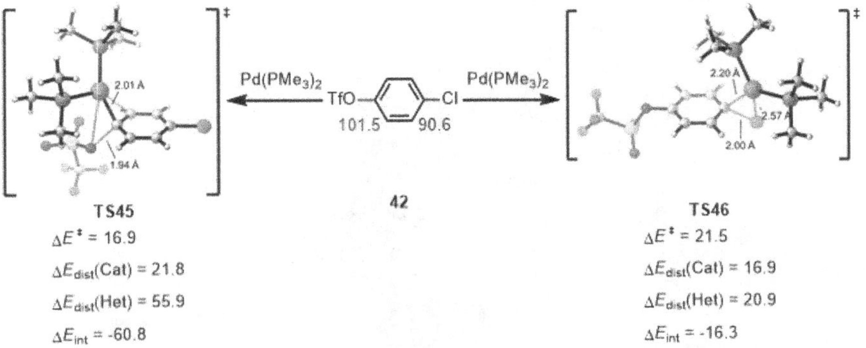

Figure 13.15. Energies and D/I analysis of TSs of [$Pd(PMe_3)_2$]-mediated C–Cl and C–OTf oxidative addition of substrate **42**. Energies are in kcal/mol.

With the bis-ligated palladium catalyst, $Pd(PMe_3)_2$, the interaction energy overrides the effect of substrate distortion and reverses the chemoselectivity (Fig. 13.15). The secondary PMe_3 coordination makes the palladium more nucleophilic and significantly strengthens the orbital interaction between palladium and aryl substrate. The C–OTf bond, which is the site of lowest LUMO energy, has much stronger interaction with the palladium as compared to C–Cl bond. Therefore, although the distortion of substrate still favors the C–Cl bond activation, the interaction energy between the palladium catalyst and distorted aryl substrate is over 40 kcal/mol larger in C–OTf bond activation than that in the C–Cl bond activation. This stabilizes the TS for C–OTf bond activation and switches the chemoselectivity.

13.4.2 *C–H bond activation*

In the computational study of the meta-selectivity of palladium-catalyzed C–H bond activation with a nitrile-containing template, the D/I analysis gave good insight into the origins of the site-selectivity. An example of the experimental substrate from the Jin-Quan Yu group[23] is given in Fig. 13.16. We discovered that a novel Pd-Ag heterodimeric C–H activation pathway is operating in this reaction, and the calculated site-selectivity agrees well with the

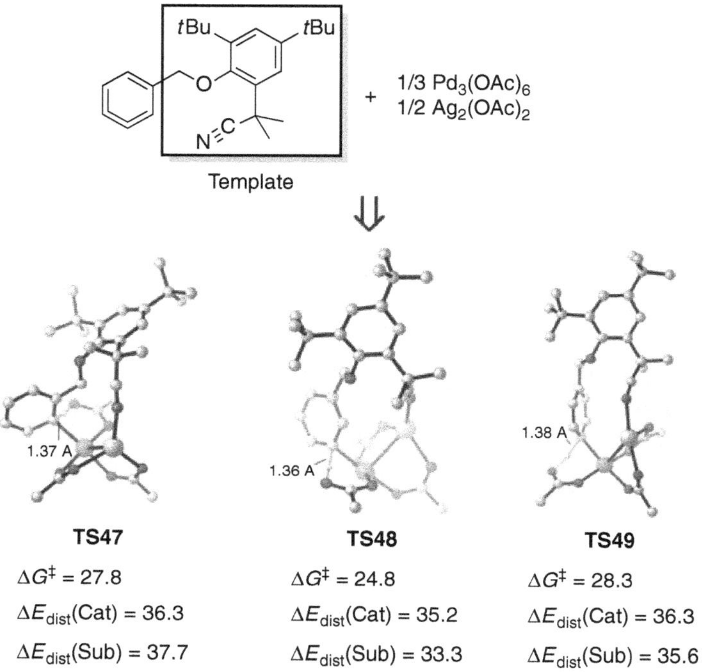

Figure 13.16. Energies and D/I analysis of regioisomeric TSs of Pd-Ag dimer-mediated C–H activation. Energies are in kcal/mol.

experimental product ratio.[23] The meta-C–H activation TS, **TS48**, is 3.0 kcal/mol more stable than the ortho-C–H activation TS, **TS47**, and 3.5 kcal/mol more stable than the para-C–H activation TS, **TS49**.

The distortion of the substrate controls the C–H activation barriers. As shown in Fig. 13.16, the substrate distortion of the meta-TS requires 33.3 kcal/mol energy, while the same distortion is 37.7 kcal/mol in ortho-TS, and 35.6 kcal/mol in the para-TS. However, unlike the cases in the palladium-catalyzed C–X bond activations, the change of substrate distortion is mainly due to distortion of template, instead of the stretching of C–H bond. In the three competing C–H activation TSs, the C–H bond distances are almost identical. The template distorts differently in the three TSs to allow the palladium to approach the C–H bond in a similar fashion. Therefore, our computation shows that the designed template is indeed the determining factor for the observed meta-selectivity, and TS with the least distortion is favored.

13.4.3 *C–O bond activation*

In the computational study of Ni/PR_3-catalyzed C–O activation of aryl esters, a D/I analysis elucidated the controlling factors in the change of chemoselectivity, depending on whether the ligand was mono- or bidentate (Fig. 13.17).[24] With the bidentate dcype ligand, the nickel catalyst cleaves the C–O bond of aryl ester **50** in a similar three-centered fashion. Because of the similar approach, the interactions between the distorted catalyst and substrate are similar in the competing TSs, **TS51** and **TS52**. The distortion of the substrate, especially the stretching of the C–O bond, determines the barrier for the C–O bond activation. Therefore, the weaker C–O bond, C(acyl)–O bond, is cleaved with the [Ni(dcype)] catalyst, as in many other transition metal-catalyzed C–X bond activations.

With the monodentate PCy3 ligand, the interaction energy controls the activation barrier and thus the chemoselectivity (Fig. 13.18). This is because the monodentate bulky phosphine ligand allows the nickel to have an additional coordination site, and a five-centered

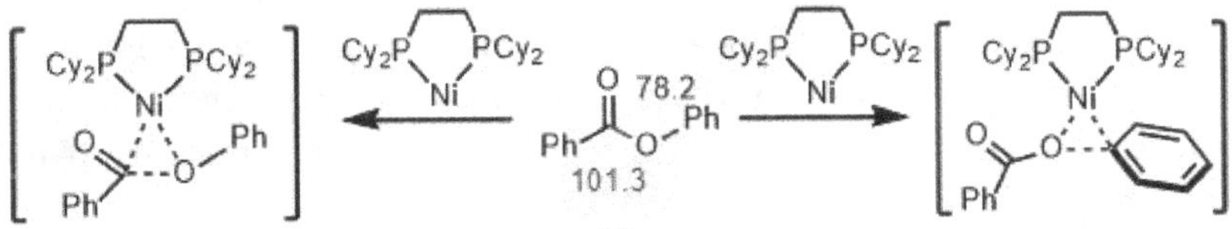

TS51
$\Delta\Delta G^{\ddagger} = 0.0$
$\Delta E = -21.4$
$\Delta E_{dist}(cat) = 5.2$
$\Delta E_{dist}(sub) = 35.4$
$\Delta E_{int} = -62.0$

50

TS52
$\Delta\Delta G^{\ddagger} = +12.6$
$\Delta E = -6.8$
$\Delta E_{dist}(cat) = 6.1$
$\Delta E_{dist}(sub) = 51.3$
$\Delta E_{int} = -64.2$

Figure 13.17. Energies and D/I analysis of TSs of Ni(dcype)-mediated C–O oxidative addition of substrate **50**. Energies are in kcal/mol.

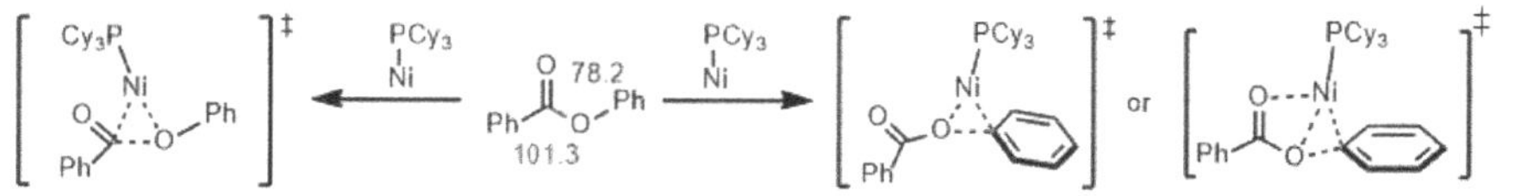

Figure 13.18. Energies and D/I analysis of TSs of [Ni(PCy_3)]-mediated C–O oxidative addition of substrate **50**. Energies are in kcal/mol.

C(aryl)–O activation TS (**TS55**) is possible in addition to the three-membered TSs (**TS53** and **TS54**). The five-membered ring TS has an additional strong nickel–oxygen interaction, which makes the TS **TS55** significantly more stable than the other competing C–O activation TSs, switching the chemoselectivity.

13.5 Nucleophilic additions to arynes

13.5.1 *Aryne and indolyne distortions and nucleophilic regioselectivities*

Over the last decade, Neil Garg and his group have pioneered the synthetic applications of arynes, reactive intermediates known since the work of Wittig[25] and established definitively by Roberts *et al.*[26] in 1953. Beginning with their work on indolynes, shown to be useful in the synthesis of indole alkaloid derivatives,[27,28] the Garg group showed that nucleophilic additions to these reactive intermediates occur with high regioselectivities in several cases. Our group collaborated with the Garg group to understand those systems, and this has flowered into a general theoretical model to explain and predict the regioselectivities of nucleophilic additions to unsymmetrical arynes, and more recently, strained aliphatic acetylenes. Paul Cheong, then a postdoc in our group, undertook DFT calculations and came up with an explanation of the origins of regioselectivities for 4,5-indolynes, and eventually for other substituted arynes, as well as 5,6- and 6,7-indolynes (Fig. 13.19).[27,28] This was in the early days

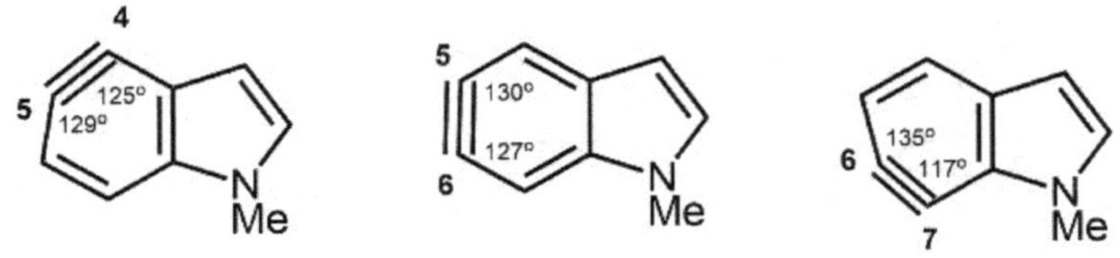

Figure 13.19. Structures of 4,5-indolyne, 5,6-indolyne, and 6,7-indolyne.

of the applications of the D/I model and opened up the potential for this model to explain some rather subtle issues in organic chemistry. Paul's pioneering work on arynes was followed up by great contributions by many in our group, notably Peng Liu, Robert Paton, Joel Mackey, Gonzalo Jiménez-Osés, and Yun-Fang Yang.

The internal angles computed for these indolynes (Fig. 13.20) show that they are distorted. Nucleophilic attack at the flatter site (larger internal angle) is favored because it requires the minimum geometrical and energetic change in going from the indolynes to TS geometry. These are reactions in which regioselectivity is controlled by distortion.

The internal angles of 4,5-indolyne show that the aryne is distorted, particularly at C-3a ($\theta_{CCC} = 110°$ vs. $\theta_{HCC} = 126°$ in pyrrole). Figure 13.20 shows the B3LYP/6-31G(d)-optimized structures of 4,5-indolyne and the TSs for aniline addition. The computed distortion energies for **TS56** and **TS57** are 3.5 kcal/mol and 4.9 kcal/mol, respectively. The favored **TS56** has the lower distortion energy. Nucleophilic addition at C-5 releases some strain at C-3a (with θ_{CCC} opening from 110° to 118°) in **TS56**, but attack at C-4 adds to the unfavorable distortion at C-3a (with θ_{CCC} closing from 110° to 108°) in **TS57**.

Based on the studies of 4,5-indolyne, nucleophilic attacks on 5,6- and 6,7-indolynes were studied computationally to predict the regioselectivities. The D/I model successfully predicts the regioselectivity that was observed experimentally.[29,30] The terminus of the aryne that is linear correlates with the experimentally favored site of nucleophilic

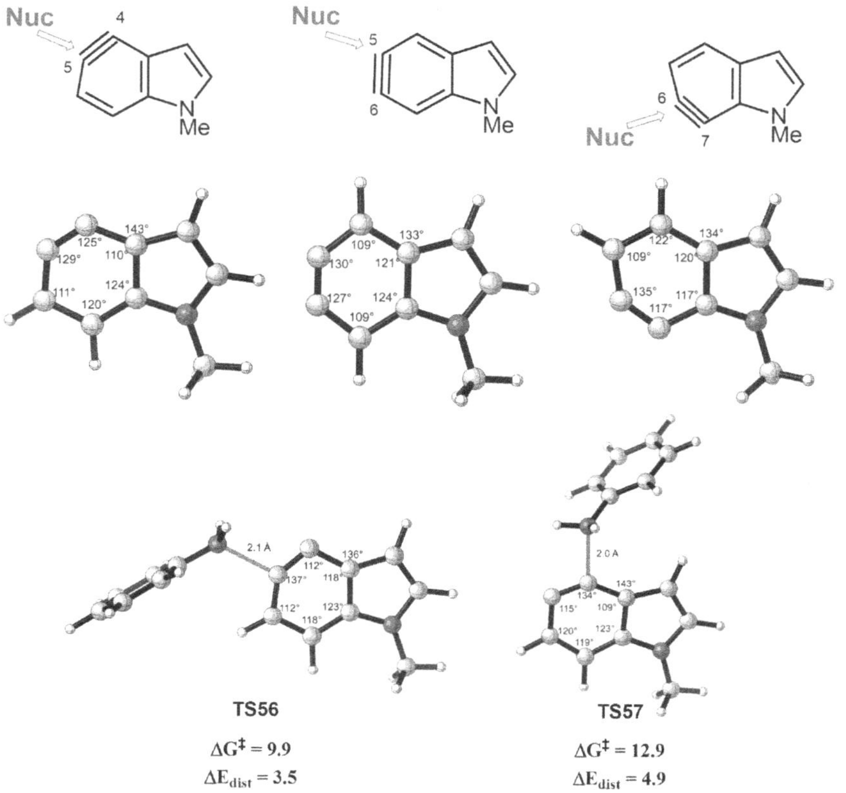

Figure 13.20. B3LYP/6-31G(d)-optimized structures of 4,5-indolyne, 5,6-indolyne, 6,7-indolyne, and the transition structures for aniline addition to 4,5-indolyne.

attack. The larger difference in internal angles correlates to the higher degree of regioselectivity.

13.5.2 *Cyclopentynes and cyclohexynes: Distortions and nucleophilic regioselectivities*

Cyclohexynes are strained alkynes, and Garg's group has begun to use these for the construction of bicyclic heterocycles. Figure 13.21 shows the M06-2X/6-311+G(2d,p)-optimized structures of cyclohexyne and cyclopentyne.[31] The calculations reveal that

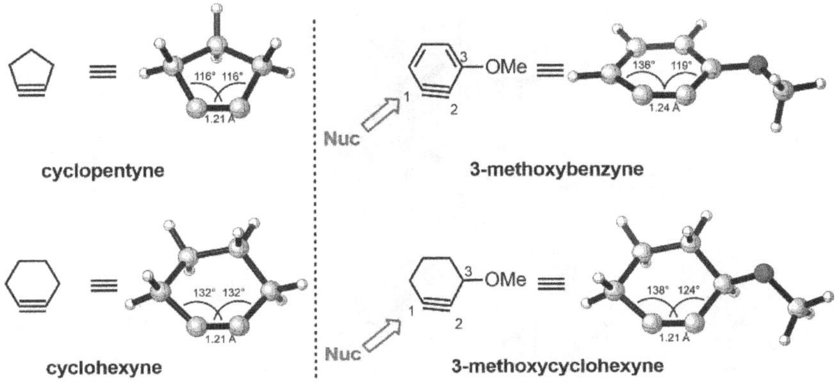

Figure 13.21. Optimized structures of cyclopentyne, cyclohexyne, 3-methoxybenzyne, and 3-methoxycyclohexyne obtained using PCM(THF)/M06-2X/6-311+G(2d,p).

cyclopentyne has a puckered geometry with an internal angle of 116°. There is significant angle-strain of this structure compared to linear alkynes, and such a large strain distorts the in-plane π-bond in a way that cyclopentyne has ~10% calculated diradical character. This kind of strain is released in cyclohexyne, with more relaxed internal angle in cyclohexyne (132°). Both cyclopentyne and cyclohexyne are symmetric structures, with C_s and C_2 symmetry, respectively. Introducing the inductively withdrawing methoxy group at C3 of 3-methoxybenzyne distorts the aryne significantly. Similarly, such substitution can also distort 3-methoxycyclohexyne, with internal angles at C1 and C2 as 138° and 124°, respectively. According to Bent's rule,[32] the inductively withdrawing nature of the C3 methoxy group causes rehybridization of C2. The more linear alkyne terminus C1 of 3-alkoxycycloalkynes is preferred for the nucleophilic addition.

13.5.3 *Piperidynes and pyridynes: Distortions and nucleophilic regioselectivities*

We collaborated again with Garg on the reactions of 3,4-pyridyne and piperidyne,[33] and Garg demonstrated the potential of the latter

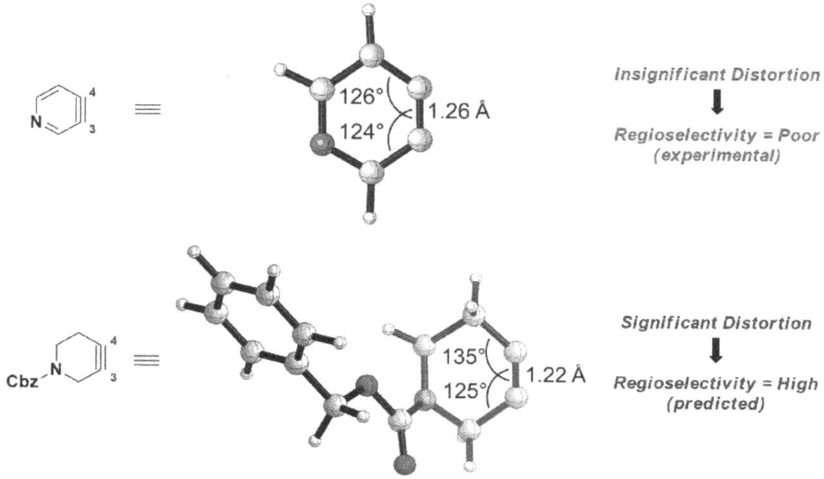

Figure 13.22. Optimized structures of 3,4-pyridyne and piperidyne obtained at the B3LYP/6-31G(d) level.

in natural product synthesis. Geometry optimization using DFT calculations at the B3LYP/6-31G(d) revealed that 3,4-piperidyne is significantly distorted with approximately 10° difference in internal angles at C4 and C3 (Fig. 13.22). Based on our previous studies that the degree of distortion present in the ground state of arynes and strained alkynes is correlated with the regioselectivity, we predicted that nucleophilic addition should occur preferentially at the more linear terminus C4. As a comparison, the 3,4-pyridyne has very similar internal angles at C4 and C3. It is predicted to react with poor regioselectivity.

The detailed D/I model analysis was used to study the competing TSs for the nucleophilic addition of morpholine to both 3,4-piperidyne and 3,4-pyridyne (Fig. 13.23), and it is consistent with the experimental regioselectivity.

13.6 Practical considerations

We end this chapter with a brief discussion about some practical aspects of performing D/I analysis and also comment on the

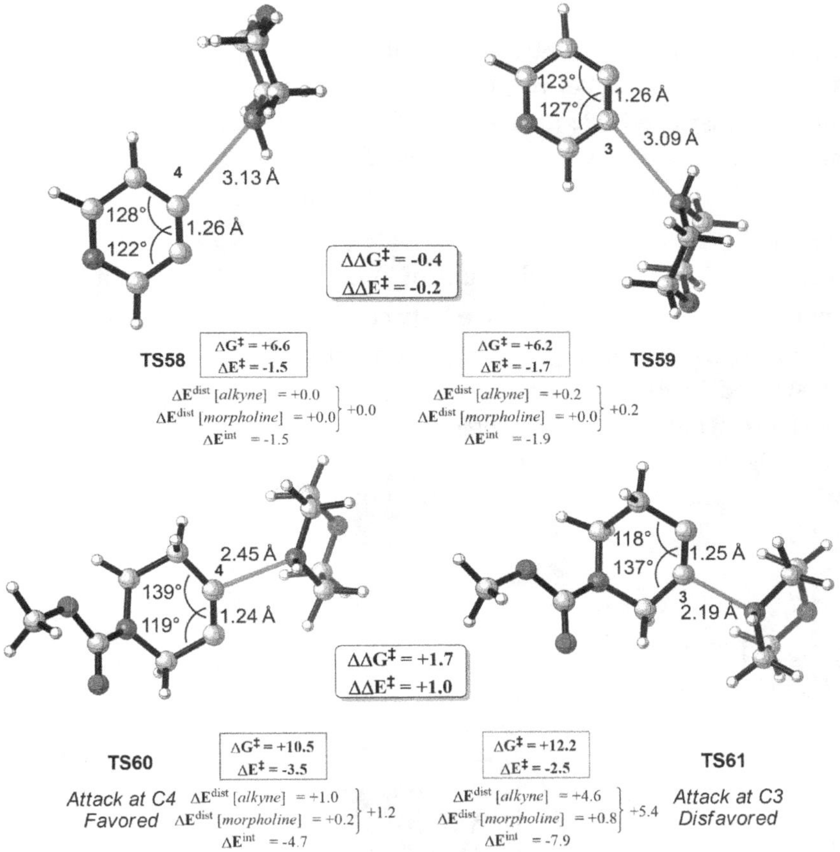

Figure 13.23. Optimized TSs for nucleophilic addition of morpholine to 3,4-pyridyne and piperidyne at the B3LYP/6-31G(d) level. Single-point energies were calculated at the B3LYP-D3/6-311+G(d,p) level with the CPCM solvent model for MeCN. Energies are provided in kcal/mol.

relationship of this model to Marcus Theory, another well-known theory of reactivity.

13.6.1 *Performing a D/I analysis*

We demonstrate the standard procedure in D/I analysis using the Diels–Alder reaction between dimethyltetrazine and ethylene as an example. The first step in a D/I treatment is to locate the lowest energy geometries of reactants (R), products (P), and the transition structure. The potential energies of these are noted E_{4e}, E_{2e}, E_P, and E_{TS} in Fig. 13.24, respectively. Single-point energy calculations on both distorted fragments ($R^‡$), freezing the geometries that they have in the TS structure give $E^‡_{4e}$ and $E^‡_{2e}$. The difference between $E^‡_{4e}$ and E_{4e} is the distortion of diene, $\Delta E^‡_{d\_4e}$; and the difference between $E^‡_{2e}$ and E_{2e} is the distortion of dienophile, $\Delta E^‡_{d\_2e}$. The total distortion energy, $\Delta E^‡_d$, is the sum of $\Delta E^‡_{d\_4e}$ and $\Delta E^‡_{d\_2e}$. Interaction energy, $\Delta E^‡_{int}$, is the difference between TS energy E_{TS} and separate distorted fragments $E^‡_{4e} + E^‡_{2e}$. The relationships between them are illustrated in Fig. 13.24.

Note that we use electronic potential energies in the D/I analysis, rather than free energies. Because the distorted geometries of the two reactants are not stationary points on the

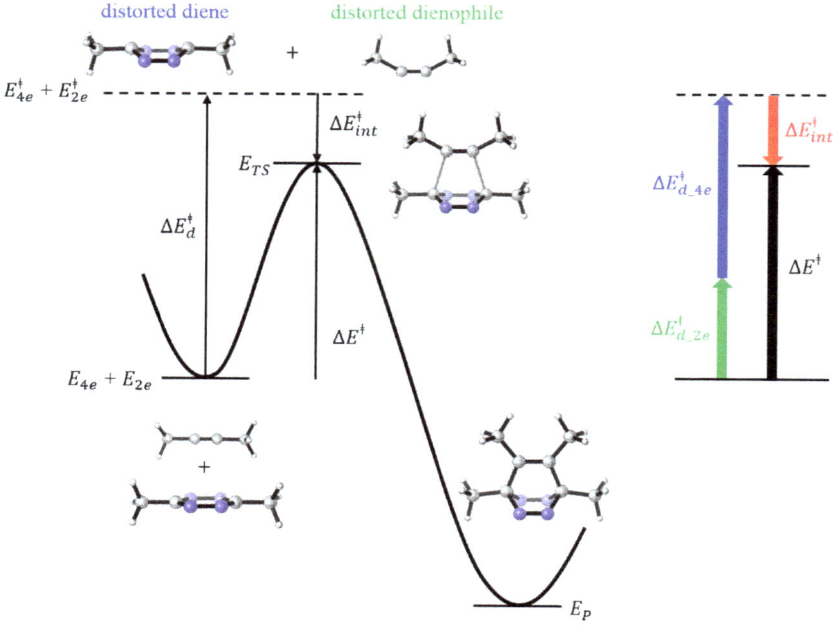

Figure 13.24. D/I model.

potential energy surface, the normal harmonic frequency analysis cannot be used for computations of *zero point energy* (*ZPE*), thermal energies, and entropies. Usually, the differences in potential energies, E, and free energies (which includes $E + ZPE +$ thermal corrections and $-T\Delta S$) are not greatly different.

13.6.2 *Relationship of distortion energies to Marcus reorganization energies*

The question is often raised about how distortion energy differs from Marcus's reorganization energy, λ. In the Marcus theory of electron transfer, the activation energy, and thus the rate constant, is related to the thermodynamic parameters of the system. The potential energy for distortion of the system along a vibrational reaction coordinate is represented by two simple parabolas (before and after electron transfer). Reorganization energy λ is defined as the energy to distort the nuclear configuration of the reactants into that of the products without allowing electron transfer. This is represented in Fig. 13.25(a).

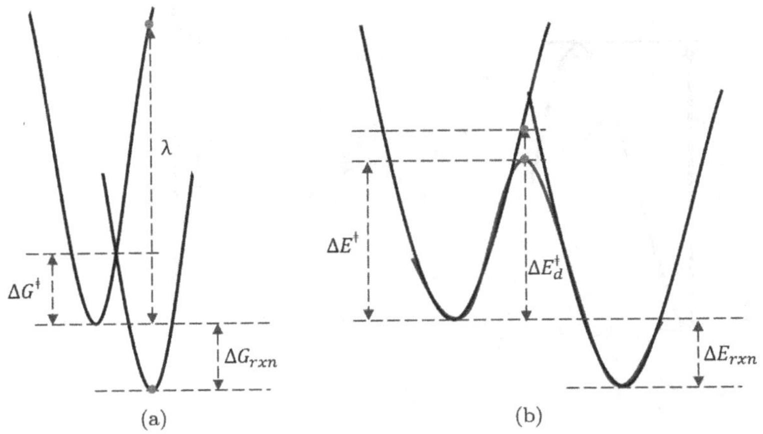

Figure 13.25. Energy quantities involved in Marcus theory (a) in comparison to D/I model (b).

By contrast, distortion energy in our D/I model defines the energy to distort the reactants into their geometries in TSs. This is shown in Fig. 13.25(b). We also show how the normal shape of a potential curve for a concerted cycloadditions can be roughly approximated by two parabolas as used in Marcus theory.

Although these two modes are quantitatively very different, they both reflect the fact that reactivity is determined by a combination of thermodynamics (reflected in ΔG_{rxn} and ΔE_{rxn}, which influences the position of the TS) and the energies required to distort reactants toward the product geometry.

13.6.3 *D/I analysis of unimolecular reactions*

Although we have described only bimolecular reactions here, the D/I model (a.k.a. Activation Strain Model) has been applied to intramolecular reactions with a consistent fragmentation scheme illustrated in Fig. 13.26. The idea is to compute the distortion and interaction energies of the reacting components, and separately

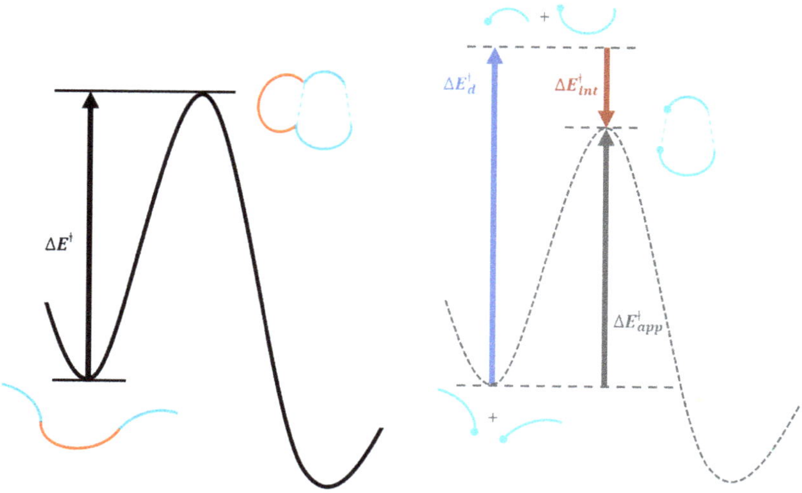

Figure 13.26. *Left*: Energy diagram for intramolecular reaction. *Right*: D/I model for intramolecular reaction with tether removed.

the distortion energy of the tether that holds the two fragments together.

In Fig. 13.26, the black solid curve on the left represents the potential surface of an intramolecular cycloaddition reaction. The substrate contains two types of regions: reactive components (colored in blue) and tether(s) (colored in orange). The two reactive components in ground state structures are separated, and hydrogen atoms are added to the atoms in which covalent bonds have been broken (Fig. 13.26, bottom right). Similarly, single-point energies are calculated for the interacting reactive components in the TS (Fig. 13.26, middle right), and the separated reactive components that maintains the geometries in TS (Fig. 13.26, top right). The sum of total distortion, $\Delta E^{\ddagger}_{d}$, and interaction energy, $\Delta E^{\ddagger}_{\text{int}}$, gives the apparent activation energy, $\Delta E^{\ddagger}_{\text{app}}$, as illustrated in Fig. 13.26. The difference between $\Delta E^{\ddagger}$ and $\Delta E^{\ddagger}_{\text{app}}$ is taken to be the distortion energy of tether, $\Delta E^{\ddagger}_{d\_\text{tether}}$. This assumes, of course, that the distortion of the tether and distortion of reacting parts are additive. With this fragmentation, intramolecular reactions can be treated with the D/I model.[34]

13.7 Concluding remarks

The D/I model has proven to be a useful way to analyze and understand the origins of reactivity phenomena in organic and organometallic chemistry. Combined with FMO theory and more complete energy decomposition analyses, the D/I model reveals all the factors controlling reactivities and selectivities.

References

1. Anslyn, E. V. and Dougherty, D. A. (2006). *Physical Organic Chemistry* (University Science Books), www.uscibooks.com.
2. (a) Fukui, K., Yonezawa, T. and Shingu, H. (1952). A molecular orbital theory of reactivity in aromatic hydrocarbons, *J. Chem. Phys.*, 20(4), pp. 722–725; (b) Fleming, I. (1978). *Frontier Orbitals and Organic Chemical Reactions* (Wiley, London).
3. Woodward, R. B. and Hoffmann, R. (1969). The conservation of orbital symmetry, *Angew. Chem. Int. Ed.*, 8(11), pp. 781–853.

4. Houk, K. N. (1975). The FMO theory of cycloaddition reactions, *Acc. Chem. Res.*, 8, pp. 361–369.
5. Sauer, J., Wiest, H. and Mielert, A. (1964). Die reaktivität von dienophilen gegenüber cyclopentadien und 9,10-dimethyl-anthracen, *Chem. Ber.*, 97, pp. 3183–3207.
6. Houk, K. N., Sims, J., Duke, Jr. R. E., Strozier, R. W. and George, J. K. (1973). Frontier molecular orbitals of 1,3-dipoles and dipolarophiles, *J. Am. Chem. Soc.*, 95, pp. 7287–7301.
7. Ess, D. H., Jones, G. O. and Houk, K. N. (2006). Conceptual, qualitative, and quantitative theories of 1,3-dipolar and Diels-Alder cycloaddition used in synthesis, *Adv. Synth. Catal.*, 348, pp. 2337–2361.
8. (a) Morokuma, K. and Winick, J. R. (1970). Molecular orbital studies of hydrogen bonds: Dimeric H_2O with the Slater minimal basis set, *J. Chem. Phys.*, 52, pp. 1301–1306; (b) Morokuma, K. (1971). Molecular orbital studies of hydrogen bonds. III. $C{=}O \cdots H-O$ hydrogen bond in $H_2CO \cdots H_2O$ and $H_2CO \cdots 2H_2O$, *J. Chem. Phys.*, 55, pp. 1236–1244; (c) Umeyama, H. and Morokuma, K. (1977). The origin of hydrogen bonding. An energy decomposition study, *J. Am. Chem. Soc.*, 99, pp. 1316–1332.
9. Joshi, B. D. and Morokuma, K. (1977). Force decomposition analysis along reaction coordinate, *J. Chem. Phys.*, 67, pp. 4880–4883.
10. Kato, S. and Morokuma, K. (1980). Potential energy characteristics and energy partitioning in chemical reactions: *Ab initio* MO study of $H_2CCH_2F \rightarrow H_2CCHF + H$ reaction, *J. Chem. Phys.*, 72, pp. 206–217.
11. Strozier, R. W., Caramella, P. and Houk, K. N. (1979). Influence of molecular distortions upon reactivity and stereochemistry in nucleophilic additions to acetylenes, *J. Am. Chem. Soc.*, 101, pp. 1340–1343.
12. (a) Ess, D. H. and Houk, K. N. (2007). Distortion/interaction energy control of 1,3-dipolar cycloaddition reactivity, *J. Am. Chem. Soc.*, 129, pp. 10646–10647; (b) Ess, D. H. and Houk, K. N. (2008). Theory of 1,3-dipolar cycloadditions: Distortion/interaction and frontier molecular orbital models, *J. Am. Chem. Soc.*, 130, pp. 10187–10198.
13. (a) van Zeist, W. J. and Bickelhaupt, F. M. (2010). The activation strain model of chemical reactivity, *Org. Biomol. Chem.*, 8, pp. 3118–3127; (b) Fernandez, I. and Bickelhaupt, F. M. (2014). The activation strain model and molecular orbital theory: Understanding and designing chemical reactions, *Chem. Soc. Rev.*, 43, pp. 4953–4967; (c) Wolters, L. P. and Bickelhaupt, F. M. (2015). The activation strain model & molecular orbital theory, *WIREs Comput. Mol. Sci.*, 5, pp. 324–343.
14. Hayden, A. E. and Houk, K. N. (2009). Transition state distortion energies correlate with activation energies of 1,4-dihydrogenations and Diels–Alder cycloadditions of aromatic molecules, *J. Am. Chem. Soc.*, 131, pp. 4084–4089.
15. Zhong, G., Chan, B. and Radom, L. J. (2007). Hydrogenation of simple aromatic molecules: A computational study of the mechanism, *J. Am. Chem. Soc.*, 129, pp. 924–933.

16. (a) Paton, R. S., Kim, S., Ross, A. G., Danishefsky, S. J. and Houk, K. N. (2011). Experimental Diels–Alder reactivities of cycloalkenones and cyclic dienes explained through transition-state distortion energies, *Angew. Chem. Int. Ed.*, 50, pp. 10366–10368; (b) Liu, F., Paton, R. S., Kim, S., Liang, Y. and Houk, K. N. (2013). Diels–Alder reactivities of strained and unstrained cycloalkenes with normal and inverse-electron-demand dienes: Activation barriers and distortion/interaction analysis, *J. Am. Chem. Soc.*, 135, pp. 15642–15649.
17. (a) Liu, F., Liang, Y. and Houk, K. N. (2014). Theoretical elucidation of the origins of substituent and strain effects on the rates of Diels–Alder reactions of 1,2,4,5-tetrazines, *J. Am. Chem. Soc.*, 136, pp. 11483–11493; (b) Yang, Y.-F., Liang, Y., Liu, F. and Houk, K. N. (2016). Diels–Alder reactivities of benzene, pyridine, and di-, tri-, and tetrazines: The roles of geometrical distortions and orbital interactions, *J. Am. Chem. Soc.*, 138, pp. 1660–1667.
18. (a) Houk, K. N. (1972). Regioselectivity and reactivity in the 1,3-dipolar cycloadditions of diazonium betaines (diazoalkanes, azides, and nitrous oxide), *J. Am. Chem. Soc.*, 94, pp. 8953–8955; (b) Houk, K. N., Sims, J., Duke, R. E., Strozier, R. W. and George, J. K. (1973). Frontier molecular orbitals of 1,3-dipoles and dipolarophiles, *J. Am. Chem. Soc.*, 95, pp. 7287–7301; (c) Houk, K. N., Sims, J., Watts, C. R. and Luskus, L. J. (1973). Origins of reactivity, regioselectivity, and periselectivity in 1,3-dipolar cycloadditions, *J. Am. Chem. Soc.*, 95, 7301–7315.
19. Fairlamb, I. J. S. (2007). Regioselective (site-selective) functionalisation of unsaturated halogenated nitrogen, oxygen and sulfur heterocycles by Pd-catalysed cross-couplings and direct arylation processes, *Chem. Soc. Rev.*, 36, pp. 1036–1045 and references therein.
20. Legault, C. Y., Garcia, Y., Merlic, C. A. and Houk, K. N. (2007). Origin of regioselectivity in palladium-catalyzed cross-coupling reactions of polyhalogenated heterocycles, *J. Am. Chem. Soc.*, 129, pp. 12664–12665.
21. Schoenebeck, F. and Houk, K. N. (2010). Ligand-controlled regioselectivity in palladium-catalyzed cross coupling reactions, *J. Am. Chem. Soc.*, 132, pp. 2496–2497.
22. Littke, A. F., Dai, C. Y. and Fu, G. C. (2000). Versatile catalysts for the Suzuki cross-coupling of arylboronic acids with aryl and vinyl halides and triflates under mild conditions, *J. Am. Chem. Soc.*, 122, pp. 4020–4028.
23. Yang, Y.-F., Cheng, G.-J., Liu, P., Leow, D., Sun, T.-Y., Chen, P., Zhang, X., Yu, J.-Q., Wu, Y.-D. and Houk, K. N. (2014). Palladium-catalyzed meta-selective C–H bond activation with a nitrile-containing template: Computational study on mechanism and origins of selectivity, *J. Am. Chem. Soc.*, 136, pp. 344–355.
24. For the computational study, see: Hong, X., Liang, Y. and Houk, K. N. (2014). Mechanisms and origins of switchable chemoselectivity of Ni-catalyzed C(aryl)–O and C(acyl)–O activation of aryl esters with phosphine ligands, *J. Am. Chem. Soc.*, 136, pp. 2017–2025. For the original experimental works, see (a) Muto, K., Yamaguchi, J. and Itami, K. (2012). Nickel-catalyzed

C–H/C–O coupling of azoles with phenol derivatives, *J. Am. Chem. Soc.*, 134, pp. 169–172; (b) Amaike, K., Muto, K., Yamaguchi, J. and Itami, K. (2012). Decarbonylative C–H coupling of azoles and aryl esters: Unprecedented nickel catalysis and application to the synthesis of muscoride A, *J. Am. Chem. Soc.*, 134, pp. 13573–13576.

25. Wittig, G. (1954). Fortschritte auf dem Gebiet der organischen Aniono-Chemie, *Angew. Chem.*, 66, pp. 10–17.
26. Roberts, J. D., Simmons, H. E., Carlsmith, L. A. and Vaughan, C. W. (1953). Rearrangement in the reaction of chlorobenzene-1-C14 with potassium amide, *J. Am. Chem. Soc.*, 75(13), pp. 3290–3291.
27. Im, G. Y. J., Bronner, S. M., Goetz, A. E., Paton, R. S., Cheong, P. H. Y., Houk, K. N. and Garg, N. K. (2010). Indolyne experimental and computational studies: Synthetic applications and origins of selectivities of nucleophilic additions, *J. Am. Chem. Soc.*, 132(50), pp. 17933–17944.
28. Cheong, P. H. Y., Paton, R. S., Bronner, S. M., Im, G. Y. J., Garg, N. K. and Houk, K. N. (2010). Indolyne and aryne distortions and nucleophilic regioselectivites, *J. Am. Chem. Soc.*, 132(4), pp. 1267–1269.
29. Goetz, A. E. and Garg, N. K. (2013). Regioselective reactions of 3,4-pyridynes enabled by the aryne distortion model, *Nat. Chem.*, 5(1), pp. 54–60.
30. Medina, J. M., Mackey, J. L., Garg, N. K. and Houk, K. N. (2014). The role of aryne distortions, steric effects, and charges in regioselectivities of aryne reactions, *J. Am. Chem. Soc.*, 136(44), pp. 15798–15805.
31. Medina, J. M., McMahon, T. C., Jiménez-Osés, G., Houk, K. N. and Garg, N. K. (2014). Cycloadditions of cyclohexynes and cyclopentyne, *J. Am. Chem. Soc.*, 136(42), pp. 14706–14709.
32. Bent, H. A. (1961). An appraisal of valence-bond structures and hybridization in compounds of the first-row elements, *Chem. Rev.*, 61(3), pp. 275–311.
33. McMahon, T. C., Medina, J. M., Yang, Y.-F., Simmons, B. J., Houk, K. N. and Garg, N. K. (2015). Generation and regioselective trapping of a 3,4-piperidyne for the synthesis of functionalized heterocycles, *J. Am. Chem. Soc.*, 137(12), pp. 4082–4085.
34. (a) Krenske, E. H., Houk, K. N., Holmes, A. B. and Thompson, J. (2011). Entropy versus tether strain effects on rates of intramolecular 1,3-dipolar cycloadditions of N-alkenylnitrones, *Tetrahedron Lett.*, 52, pp. 2181–2184; (b) He, C. Q., Chen, T. Q., Patel, A., Karabiyikoglu, S., Merlic, C. A. and Houk, K. N. (2015). Distortion, tether, and entropy effects on transannular Diels–Alder cycloaddition reactions of 10-18-membered rings, *J. Org. Chem.*, 80, pp. 11039–11047.

Chapter 14

Spreadsheet-Based Computational Predictions of Isotope Effects

O. Maduka Ogba, John D. Thoburn† and Daniel J. O'Leary**

**Department of Chemistry, Pomona College, USA*
†Department of Chemistry, Randolph-Macon College, Ashland, VA, 23005, USA

14.1 Historical perspective and motivation

As early as 1919, calculations were used to guide Lindemann and Aston's[1,2] efforts to separate isotopes, a feat realized a decade or so later in 1931 with Keesom and Van Dijk's[3] partial separation of the neon isotopes, and Urey *et al.*'s[4] landmark enrichment and detection of deuterium. Shortly thereafter, Urey *et al.*[5–7] showed that it was possible to calculate equilibrium constants for isotope exchange reactions from spectroscopic data. These equilibrium calculations were simplified by Bigeleisen and Mayer's[8] (B-M) introduction of the reduced partition function formalism, which was developed for isotopic separation efforts in the Manhattan Project and later published in 1947. Two years later, Bigeleisen[9] described how the B-M formalism could be joined with rate theory, including Wigner's[10] tunneling correction, to predict kinetic isotope effects (KIEs) on reaction rates.

By 1980, Hout *et al.*[11] showed it was possible to estimate "with reasonable accuracy" equilibrium isotope effects (EIEs) by coupling the B-M formalism with harmonic vibrational frequencies computed using *ab initio* molecular orbital theory.[11] The mid-1980s saw similar approaches begin to manifest in organic chemistry, with reports from Williams,[12] Anet and Kopelevich,[13] and Kopelevich[14] regarding the conformational preference of deuterium in cyclohexane and related heterocycles. Beginning in 1989, Saunders *et al.*[15] made *ab initio*–based B-M calculations widely possible with their distribution of the QUIVER program. By 1995, Olson *et al.*[16] used the program, augmented with the Bell tunneling correction,[17] to analyze conformational KIEs in amides. QUIVER-based applications are numerous: at the time of this writing, the paper (and presumably, the program) has been cited some 120 times, primarily by the Singleton (34), Houk (21), and Berti (10) groups. Originally a FORTRAN program, it is now available as a Python-based application.[18] Other widely distributed B-M packages include BEBOVIB[19] and ISOEFF.[20]

While integrated programs such as these have proven indispensable for many applications, we developed an alternative spreadsheet approach for calculating KIEs within the B-M formalism and coupled with variational transition state (TS) theory with multidimensional tunneling.[21] Other aspects of our work have required decomposing IEs into their enthalpic and entropic (*H*-*S*) components; these calculations use a different set of partition functions to yield the same overall answer and are entirely analogous with B-M calculations because of their reliance upon vibrational frequencies and moments of inertia.[22–26]

The focus of this chapter is a tutorial-style implementation of a dual (*H*-*S* and B-M) spreadsheet approach to calculating IEs. After introducing the requisite equations, we outline their incorporation into a spreadsheet that analyzes an easily computed example from the literature. By working through this example, the reader will be enabled to predict IEs using this dualistic approach or, at the very least, gain a better appreciation for results produced by fully integrated programs. IE calculations are now fairly routine and accurately predict experimental observations in the vast majority of cases. To make this point, the chapter concludes with a description of several studies that have featured KIE or EIE estimations, so that the utility of these calculations can be fully appreciated.

14.2 General steps for calculating isotope effects

The spreadsheet approach to IE calculations, which we implement with the Gaussian computational package[27] in a Linux environment, requires four key steps: (1) geometry optimization and frequency calculations for the relevant ground state (GS) structures for EIE calculations and, additionally, the TS structures for KIE calculations; (2) generation of Gaussian freqchk outputs for isotopically substituted structures; (3) extraction and spreadsheet insertion of freqchk-generated vibrational frequencies and moments of inertia or rotational temperatures for B-M or *H-S* IE calculations; and (4) evaluation of K_H/K_D for EIEs or k_H/k_D for KIEs (Fig. 14.1). Each of these general steps will be described in detail for a small molecule with a known KIE, highlighting the utility of the spreadsheet technique and validating its results with comparison to literature values.

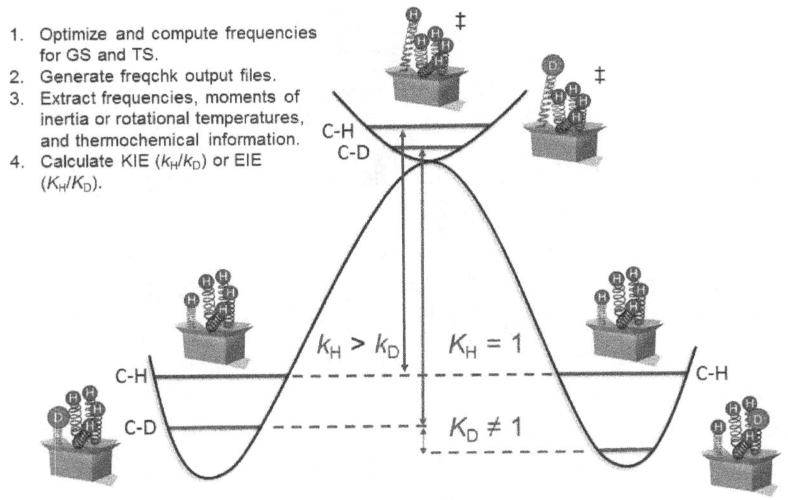
1. Optimize and compute frequencies for GS and TS.
2. Generate freqchk output files.
3. Extract frequencies, moments of inertia or rotational temperatures, and thermochemical information.
4. Calculate KIE (k_H/k_D) or EIE (K_H/K_D).
‡
‡
C-H
C-D
$k_H > k_D$
$K_H = 1$
C-H
C-D
$K_D \neq 1$
C-H

Figure 14.1. General steps for computing IEs and conceptualization of GS and TS structures as collections of springs attached to different masses (H or D). The resulting vibrational frequency differences define the energy for each unique molecular system and produce rate differences (KIEs) or shifts in equilibria (EIEs).

14.2.1 *The B-M equation*

The B-M equation for a KIE ($k_{\mathrm{H}}/k_{\mathrm{D}}$) can be written in the form of Eqs. (14.1)–(14.5) for a reaction involving unlabeled (H) and labeled (D) species. The reactant and TS terms are identified with a superscripted R and ‡, respectively:

$$\frac{k_{\mathrm{H}}}{k_{\mathrm{D}}} = \mathrm{SYM} \times \mathrm{MMI} \times \mathrm{EXC} \times \mathrm{ZPE} \tag{14.1}$$

$$\mathrm{SYM} = \frac{\sigma_{\mathrm{H}}^{R}/\sigma_{\mathrm{D}}^{R}}{\sigma_{\mathrm{H}}^{\ddagger}\sigma_{\mathrm{D}}^{\ddagger}} \tag{14.2}$$

$$\mathrm{MMI} = \left(\frac{M_{\mathrm{H}}^{\ddagger}M_{\mathrm{D}}^{R}}{M_{\mathrm{D}}^{\ddagger}M_{\mathrm{H}}^{R}}\right)^{\frac{3}{2}} \left(\frac{I_{\mathrm{H1}}^{\ddagger}I_{\mathrm{H2}}^{\ddagger}I_{\mathrm{H3}}^{\ddagger}I_{\mathrm{D1}}^{R}I_{\mathrm{D2}}^{R}I_{\mathrm{D3}}^{R}}{I_{\mathrm{D1}}^{\ddagger}I_{\mathrm{D2}}^{\ddagger}I_{\mathrm{D3}}^{\ddagger}I_{\mathrm{H1}}^{R}I_{\mathrm{H2}}^{R}I_{\mathrm{H3}}^{R}}\right)^{\frac{1}{2}} \tag{14.3}$$

$$\mathrm{EXC} = \prod_{n}^{3n^{\ddagger}-7} \frac{1/(1-e^{-u_n(\mathrm{H})^{\ddagger}})}{1/(1-e^{-u_n(\mathrm{D})^{\ddagger}})} \Bigg/ \prod_{n}^{3n-6} \frac{1/(1-e^{-u_n(\mathrm{H})^{R}})}{1/(1-e^{-u_n(\mathrm{D})^{R}})} \tag{14.4}$$

$$\mathrm{ZPE} = \frac{e^{-\frac{1}{2}\sum_{n}^{3n^{\ddagger}-7} u_n(\mathrm{H})^{\ddagger}}}{e^{-\frac{1}{2}\sum_{n}^{3n-6} u_n(\mathrm{H})^{R}}} \Bigg/ \frac{e^{-\frac{1}{2}\sum_{n}^{3n^{\ddagger}-7} u_n(\mathrm{H})^{\ddagger}}}{e^{-\frac{1}{2}\sum_{n}^{3n-6} u_n(\mathrm{H})^{R}}} \tag{14.5}$$

where $u_n = hv_n/kT$, and v_n are the vibrational frequencies (cm^{-1}) provided in Gaussian output files.

The SYM term consists of a ratio of symmetry numbers for reactant and TS isotopomers. Symmetry numbers are used in entropy calculations and are related to the number of indistinguishable but nonidentical positions into which a molecule can be turned by rigid rotation. Values of σ are derived from the point group order.[28] Isotopic substitution can affect symmetry numbers, especially in small and/or high-symmetry molecules. For example, H_2O (point group C_{2v}, $\sigma = 2$) and HOD (point group C_s, $\sigma = 1$) have different symmetry numbers. In most cases, the SYM term is unity, but exceptions can occur. It is also worth noting that Gaussian assigns correct rotational symmetry numbers to structures bearing default isotopes but does not redetermine σ for isotopically substituted systems.[24]

The mass-dependent part of the mass and moment of inertia (MMI) term in Eq. (14.3) is equal to one in most cases. The principal moments of inertia in the second term of Eq. (14.3) are provided in Gaussian output files. Using the Redlich–Teller product rule, it is also possible to express the inertial MMI term as a product of all vibrational frequencies using Eq. (14.6):

$$\left(\frac{I^{\ddagger}_{\mathrm{H1}}I^{\ddagger}_{\mathrm{H2}}I^{\ddagger}_{\mathrm{H3}}I^{R}_{\mathrm{D1}}I^{R}_{\mathrm{D2}}I^{R}_{\mathrm{D3}}}{I^{\ddagger}_{\mathrm{D1}}I^{\ddagger}_{\mathrm{D2}}I^{\ddagger}_{\mathrm{D3}}I^{R}_{\mathrm{H1}}I^{R}_{\mathrm{H2}}I^{R}_{\mathrm{H3}}}\right)^{\frac{1}{2}} = \left(\frac{v^{\ddagger}_{i(\mathrm{H})}}{v^{\ddagger}_{i(\mathrm{D})}}\right)\prod_{n}^{3n-6}\frac{v_{n(\mathrm{D})^{R}}}{v_{n(\mathrm{H})^{R}}}\Bigg/ \times \prod_{n}^{3n^{\ddagger}-7}\frac{v_{n(\mathrm{D})^{\ddagger}}}{v_{n(\mathrm{H})^{\ddagger}}} = \mathrm{VP} \qquad (14.6)$$

where the right-hand side is referred to as the vibrational product (VP) MMI term.[29] This approach has been shown to minimize computational errors in the MMI term.[30] It has been our experience that Gaussian outputs for large systems sometimes return only a partial list of inertial moments. In these cases, Eq. (14.6) can provide a workaround for calculating the MMI term. Another option is to use the Gaussian-printed rotational constants, which are inversely related to the inertial moments. While there are multiple ways to compute the MMI term, it should be noted that it is often close to unity and rarely governs the overall IE magnitude for unimolecular transformations. This is especially true for medium to large molecular systems, where inertial differences due to isotopic substitution are expected to be small. It can, however, have a large contribution in bimolecular transformations.[29]

The EXC and ZPE terms (Eqs. (14.4) and (14.5)) often play pivotal roles in determining the overall IE and are calculated using only real vibrational frequencies, hence the indexing difference for GS and TS normal modes (TSs are identified by having one imaginary frequency, flagged as a negative frequency in the output). When constructing the spreadsheet for a KIE calculation, therefore, it is important to not include imaginary (negative) TS frequencies in the calculation of these terms.

The simplified B-M equation, excluding the SYM term (which has the same form as Eq. (14.2)), for an EIE ($K_{\mathrm{H}}/K_{\mathrm{D}}$) can be written in a similar manner for a system containing protium (H) and deuterium (D), albeit with indexing modifications to account for the absence of imaginary frequencies. The reactant and product terms in Eqs. (14.7)–(14.11) are identified with R and P, respectively.

$$\frac{K_{\mathrm{H}}}{K_{\mathrm{D}}} = \mathrm{MMI} \times \mathrm{EXC} \times \mathrm{ZPE} \tag{14.7}$$

$$\mathrm{MMI} = \left(\frac{M_{\mathrm{H}}^{P} M_{\mathrm{D}}^{R}}{M_{\mathrm{D}}^{P} M_{\mathrm{H}}^{R}}\right)^{\frac{3}{2}} \left(\frac{I_{\mathrm{H1}}^{P} I_{\mathrm{H2}}^{P} I_{\mathrm{H3}}^{P} I_{\mathrm{D1}}^{R} I_{\mathrm{D2}}^{R} I_{\mathrm{D3}}^{R}}{I_{\mathrm{D1}}^{P} I_{\mathrm{D2}}^{P} I_{\mathrm{D3}}^{P} I_{\mathrm{H1}}^{R} I_{\mathrm{H2}}^{R} I_{\mathrm{H3}}^{R}}\right)^{\frac{1}{2}} \tag{14.8}$$

$$\mathrm{EXC} = \prod_{n}^{3n-6} \frac{1/(1 - e^{-u_n(\mathrm{H})^P})}{1/(1 - e^{-u_n(\mathrm{D})^P})} \Bigg/ \prod_{n}^{3n-6} \frac{1/(1 - e^{-u_n(\mathrm{H})^R})}{1/(1 - e^{-u_n(\mathrm{D})^R})} \tag{14.9}$$

$$\mathrm{ZPE} = \frac{e^{-\frac{1}{2}\sum_{n}^{3n-6} u_n(\mathrm{H})^P}}{e^{-\frac{1}{2}\sum_{n}^{3n-6} u_n(\mathrm{H})^R}} \Bigg/ \frac{e^{-\frac{1}{2}\sum_{n}^{3n-6} u_n(\mathrm{D})^P}}{e^{-\frac{1}{2}\sum_{n}^{3n-6} u_n(\mathrm{D})^R}} \tag{14.10}$$

$$\left(\frac{I_{\mathrm{H1}}^{P} I_{\mathrm{H2}}^{P} I_{\mathrm{H3}}^{P} I_{\mathrm{D1}}^{R} I_{\mathrm{D2}}^{R} I_{\mathrm{D3}}^{R}}{I_{\mathrm{D1}}^{P} I_{\mathrm{D2}}^{P} I_{\mathrm{D3}}^{P} I_{\mathrm{H1}}^{R} I_{\mathrm{H2}}^{R} I_{\mathrm{H3}}^{R}}\right)^{\frac{1}{2}} = \prod_{n}^{3n-6} \frac{v_{n(\mathrm{D})^R}}{v_{n(\mathrm{H})^R}} \Bigg/ \prod_{n}^{3n-6} \frac{v_{n(\mathrm{D})^P}}{v_{n(\mathrm{H})^P}} = \mathrm{VP} \tag{14.11}$$

14.2.2 *Computing enthalpy–entropy contributions to isotope effects*

The H-S thermodynamic values factor into KIE or EIE expressions according to Eqs. (14.12)–(14.18), using the reactant (R), transition state (‡, KIE) or product (P, EIE) label convention and defining the universal gas constant as R_g:

$$\left(\frac{k_{\mathrm{H}}}{k_{\mathrm{D}}}\right)_{\mathrm{enthalpy}} \mathit{or} \left(\frac{K_{\mathrm{H}}}{K_{\mathrm{D}}}\right)_{\mathrm{enthalpy}} = e^{\Delta\Delta H/R_g T} \tag{14.12}$$

$$\Delta\Delta H = (H_{\mathrm{D}}^{\ddagger/P} - H_{\mathrm{D}}^{R}) - (H_{\mathrm{H}}^{\ddagger/P} - H_{\mathrm{H}}^{R}) \tag{14.13}$$

$$\left(\frac{k_{\text{H}}}{k_{\text{D}}}\right)_{\text{entropy}} \textit{ or } \left(\frac{K_{\text{H}}}{K_{\text{D}}}\right)_{\text{entropy}} = e^{\Delta\Delta S/R_g} \tag{14.14}$$

$$\Delta\Delta S = \left(S_{\text{H}}^{\ddagger/P} - S_{\text{H}}^{R}\right) - \left(S_{\text{D}}^{\ddagger/P} - S_{\text{D}}^{R}\right) \tag{14.15}$$

The enthalpy and entropy contributions combine as their products to determine the overall IE:

$$KIE = \left(\frac{k_{\text{H}}}{k_{\text{D}}}\right)_{\text{Gibbs}} = e^{\Delta\Delta H^{\ddagger}/R_gT} \times e^{\Delta\Delta S^{\ddagger}/R_g} = e^{\Delta\Delta G^{\ddagger}/R_gT} \tag{14.16}$$

$$EIE = \left(\frac{K_{\text{H}}}{K_{\text{D}}}\right)_{\text{Gibbs}} = e^{\Delta\Delta H^{P}/R_gT} \times e^{\Delta\Delta S^{P}/R_g} = e^{\Delta\Delta G^{P}/R_gT} \tag{14.17}$$

$$\Delta\Delta G^{\ddagger/P} = \left(G_{\text{D}}^{\ddagger/P} - G_{\text{D}}^{R}\right) - \left(G_{\text{H}}^{\ddagger/P} - G_{\text{H}}^{R}\right) \tag{14.18}$$

Spreadsheet calculations of *H-S* IE contributions can be mechanically easier to perform than the aforementioned B-M calculations because enthalpy (*H*), entropy (*S*), and Gibbs free energy (*G*) values can be accessed directly from Gaussian output files, i.e., without manipulating lists of vibrational frequencies or inertial moments. The limitation with computing *H-S* contributions using this "cut and paste" approach is that certain Gaussian thermochemical values are printed with insufficient precision for studies requiring careful comparison of theory with experiment (such as when IEs are being compared at the thousandth place or beyond).

To address this shortcoming in precision, the extracted frequencies and rotational temperatures can be used to directly compute the isotope-dependent vibrational and rotational contributions to the enthalpy and entropy.[31,32] The enthalpic internal thermal energy, defined here as H_{thermal}, can be written as the sum of the zero-point energy and a temperature-dependent vibrational term:

$$H_{\text{thermal}} = H_{\text{ZPE}} + H_{\text{vib}} \tag{14.19}$$

$$H_{\text{ZPE}} = R_g \sum \frac{\theta_v}{2} \tag{14.20}$$

$$H_{\text{vib}} = R_g \sum \frac{\theta_v}{e^{\theta_v/T} - 1} \tag{14.21}$$

where $\theta_\nu = h\nu/k$ is the characteristic vibrational temperature.

Entropy has vibrational and rotational terms dependent upon isotopic substitution

$$S = S_{\text{vib}} + S_{\text{rot}} \tag{14.22}$$

$$S_{\text{vib}} = R_g \sum \left(\frac{\theta_v/T}{e^{\theta_v/T} - 1} - \ln\left(1 - e^{-\theta_v/T}\right) \right) \tag{14.23}$$

$$S_{\text{rot}} = R_g \left(\ln q_r + \frac{3}{2} \right) \tag{14.24}$$

$$q_r = \frac{\pi^{1/2}}{\sigma_r} \left(\frac{T^{3/2}}{(\theta_{r,x}\theta_{r,y}\theta_{r,z})^{1/2}} \right) \tag{14.25}$$

$$\theta_{r,x,y,z} = h^2/8\pi^2 I_{x,y,z} k \tag{14.26}$$

The vibrational entropy contribution (Eq. (14.23)) is a function of the aforementioned characteristic vibrational temperatures. The rotational entropy contribution (Eqs. (14.24)–(14.26)) is dependent on symmetry numbers and the characteristic rotational temperatures (θ_r), computed from the principal moments of inertia for a given isotopolog.

The relationship of the *H*-*S* terms to the B-M ZPE, EXC, MMI, and SYM components is shown in Fig. 14.2. Why bother to compute IE components? Interestingly, there are a number of examples in the literature where the overall IE is the sum of antagonistic contributions. For example, in the upcoming case study, we will see a secondary conformational KIE of 1.1257 that consists of $\Delta\Delta H$ and $\Delta\Delta S$ terms of 0.9601 and 1.1725, respectively. Clearly, this overall

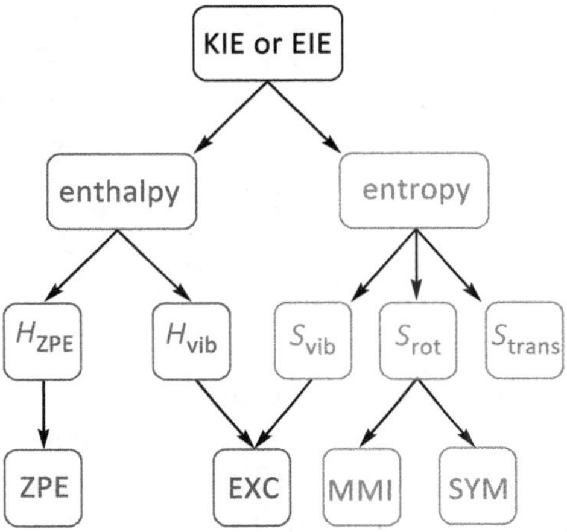

Figure 14.2. A flowchart showing how a KIE or EIE partitions into enthalpy and entropy contributions and their relationship to the B-M ZPE, EXC, MMI, and SYM contributions.

normal IE ($k_H/k_D > 1$) is dominated by entropy (making a normal contribution) and *not* enthalpy (making an inverse contribution). The computed ZPE and EXC terms are 1.0122 and 1.0915, respectively, and unlike the H-S decomposition, both are normal contributions. These values call into question two general assumptions about IEs: (1) that IEs are caused primarily by zero-point (enthalpic) vibrational energy differences and (2) that ZPE and EXC terms always correlate with enthalpy and entropy. But it is the case that these "exceptions to the rule" have been well documented in the literature and, somewhat ironically, date back to some of the earliest experimentally measured conformational KIEs.[16,24–26,33–35] The physical origins of IEs are in fact quite diverse, and the spreadsheet outlined in this chapter has been developed to explore and tally their contributions.

14.2.3 *Tunneling corrections*

The foregoing discussion of KIEs was developed on the premise of isotopomers moving *over* a barrier at different rates. Tunneling is a quantum correction for isotopomers moving *through* a barrier at different rates. The effect of tunneling can be quite dramatic, resulting in k_H/k_D ranging as large as 450–730,[36,37] in comparison with the semiclassical limit of approximately 10 for C–H bond rupture.

Equation (14.27) is known as the Bell tunneling correction[17]

$$Q_t(\text{Bell}) = \frac{\pi\alpha/\beta}{\sin(\pi\alpha/\beta)} = \frac{\frac{1}{2}u^{\ddagger}}{\sin\frac{1}{2}u^{\ddagger}} \tag{14.27}$$

where $\alpha = E/kT$, $\beta = 2\pi/h\nu^{\ddagger}$, and $u^{\ddagger} = hv/kT$. The u term should be familiar, as it was computed for the real frequencies in the B-M calculations of ZPE and EXC (see Eqs. (14.4) and (14.5)). The Bell correction is widely used and is generally applicable for evaluating small or moderate tunneling corrections, where those thresholds are governed by $\alpha < \beta$, or $u^{\ddagger} < 2\pi$. The beauty of the right-hand side of Eq. (14.27) is that the tunneling correction requires only the imaginary TS frequency for each isotopomer. The Bell KIE correction is expressed as:

$$\frac{Q_t(\text{H})}{Q_t(\text{D})} = \frac{\dfrac{\frac{1}{2}u(\text{H})^{\ddagger}}{\sin\frac{1}{2}u(\text{H})^{\ddagger}}}{\dfrac{\frac{1}{2}u(\text{D})^{\ddagger}}{\sin\frac{1}{2}u(\text{D})^{\ddagger}}} \tag{14.28}$$

In his treatise on tunneling,[17] Bell noted that the first two terms of Eq. (14.29), which is the series expansion of the right-hand side of Eq. (14.27), are equivalent to Wigner's expression[9] for small tunneling corrections (Eq. (14.30)).

$$Q_t = \frac{\frac{1}{2}u^{\ddagger}}{\sin\frac{1}{2}u^{\ddagger}} = 1 + \frac{(u^{\ddagger})^2}{24} + \frac{7\,(u^{\ddagger})^4}{5760} + \frac{31\,(u^{\ddagger})^6}{967,680} + \cdots \tag{14.29}$$

$$Q_t(\text{Wigner}) = 1 + \frac{(u^{\ddagger})^2}{24} \tag{14.30}$$

The behavior of Eqs. (14.27), (14.29), and (14.30) will be explored in this chapter's principal case study.

We have found that the Bell tunneling model works reasonably well for many applications, especially in the high temperature limit. At higher temperatures, most of the reactants cross over the classical barrier and the rate enhancement due to tunneling is an additional perturbation that can be accurately treated with the Bell model. If, however, one observes an unusually large KIE or a significant discrepancy between the experimental KIE and that computed with the Bell correction, it may be necessary to reach for more sophisticated tunneling models.

For example, one of the assumptions of the Bell model is that the classical barrier is parabolic with a width determined by the magnitude of the imaginary frequency. A more accurate potential energy barrier can be generated from a series of nonstationary points along the minimum energy path (MEP) that connects reactants to TS and then to products.[38] Computationally, the MEP is calculated by following the path of steepest descent from the TS to reactants and from the TS to products. With a well-quantified MEP, and the associated gradients and Hessians, the effects of tunneling are then incorporated into the transmission coefficient (κ) via an imaginary-action integral, which in essence describes the dampening of the wavefunction as the protium or deuterium enters the tunnel region. The tunneling that takes place along the MEP is the basis for the one-dimensional tunneling model, also known as the zero-curvature tunneling (ZCT) model.

In some cases, the imaginary vibrational mode corresponding to the reaction coordinate couples with some of the other $3n - 7$ vibrational modes orthogonal to the MEP, giving rise to multidimensional tunneling. More intuitively, tunneling mechanisms exist that lead to "corner cutting" along trajectories that do not lie on the MEP, but rather on the concave side of a curved MEP, which itself is but one of an infinite number of paths connecting reactants and products on a more complex multidimensional potential energy surface. The effects of a curved MEP on transmission coefficients can be modeled using a number of algorithms depending on the degree of curvature, e.g., small-curvature tunneling (SCT) or large-curvature tunneling (LCT).

Inclusion of these alternate tunneling models is beyond the scope of the methodology developed in this chapter, but programs are available if such corrections are needed.[39]

14.3 Secondary KIEs in the C–N rotation in formamide: A detailed case study

In 1992, Perrin *et al.*[40] reported NMR measurements of secondary conformational KIEs associated with C–N rotation in several amides, including formamide carrying one deuterium at the amide nitrogen, which exhibited a k_H/k_D of 1.16 ± 0.10 at 25°C. A computational study of conformational KIEs in simple amides, including formamide, was the focus of the 1995 Houk study mentioned earlier.[16] They reported a formamide potential energy surface consisting of a GS structure and two energetically different diastereomeric TS structures, TS-1 and TS-2 (Fig. 14.3). HF/6-31G(d) calculations, performed by us, indicate that TS-1 is lower in free energy than TS-2 by 2.7 kcal/mol. This means the KIE can be computed using harmonic frequencies and inertial moments calculated from the optimized GS structure and the optimized TS-1 structure, respectively. Despite formamide's structural simplicity, there are two unique motional trajectories for an N–D bond in this system, as it can begin its

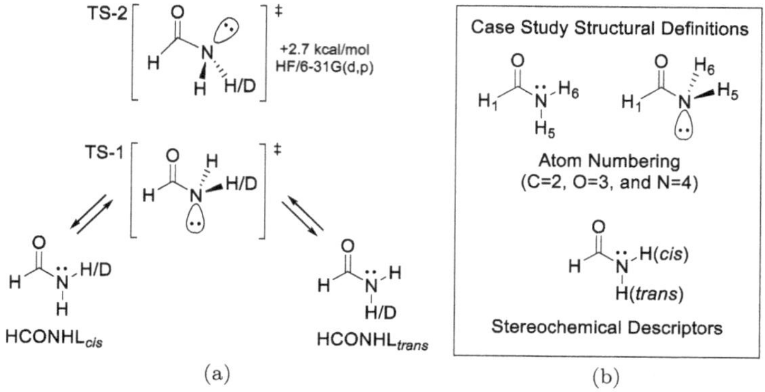

Figure 14.3. (a) Diastereomeric transition states TS-1 and TS-2 for C–N bond rotation in formamide containing an N–H/D bond. (b) Formamide atom labels and *cis*/*trans* descriptor definitions used in case study discussion.

Table 14.1. Calculated KIEs (k_H/k_D) for formamide C–N rotation via TS-1 at 298 K.

	RHF/ 6-31G(d)	RHF/ 6-311++G(d,p)	MP2(FC)/ 6-31G(d)	MP2(FC)/ 6-311++G(d,p)
$HCONHL_{cis}$	1.135	1.139	1.134	1.161
$HCONHL_{trans}$	1.138	1.148	1.121	1.163

GS rotation from positions 5 or 6. Houk *et al.* used a nomenclature system to define the starting point of these different trajectories from the diastereoisomeric and energetically inequivalent monodeuterated GS structures as $HCONHL_{cis}$ vs. $HCONHL_{trans}$, where L represents H or D (Fig. 14.3).

The thermochemical components of the formamide KIEs reported by Houk *et al.* will be discussed later in this case study. For now, we point out that the predicted KIEs, ranging from 1.121 to 1.163 depending upon level of theory and reaction trajectory (Table 14.1), were in good agreement with the experimental value (1.16 ± 0.10).

The qualitative agreement among theoretical approaches is typically observed in computational studies of isotope effects (IEs), underscoring the robustness of predictions based mainly or entirely upon computed harmonic vibrational frequencies. Frequency scaling factors, developed as corrections for computational approximations and anharmonicity, do not make a large difference in IE calculations. This is because IEs are calculated using ratios of frequency-dependent terms, and scaling effects tend to cancel. Houk *et al.* following this practice, used unscaled harmonic frequencies in their calculations. Later in the case study, we will explore the effect of frequency scaling on the formamide KIE.

14.3.1 *Geometry optimization and frequency calculations*

Formamide GS and TS-1 geometry optimizations and frequency calculations utilized HF/6-31G(d), the lowest level of theory used in the 1995 study. Representative input files are shown in Fig. 14.4.

As seen in Fig. 14.4, the HF geometry optimizations use the default convergence limits. In a later paper, Schaad *et al.*[30] showed that tightening the HF convergence force cutoffs and step sizes by including "*opt=tight*" or "*opt=vtight*" keywords made differences to only the fifth decimal place of the computed KIE. When

```
%chk=formamide_gs_hf631gd.chk
# opt freq rhf/6-31g(d)

formamide gs structure

0 1
 H            -0.15232500    1.47223500   -0.00005100
 C            -0.16464900    0.38155300   -0.00000200
 O            -1.17962600   -0.24514900   -0.00009200
 N             1.07327700   -0.15301300    0.00056300
 H             1.17595000   -1.14321900   -0.00115700
 H             1.88833700    0.41394500   -0.00198300

------------------------------------------------------
%chk=formamide_ts1_hf631gd.chk
# opt=(calcfc,ts) freq rhf/6-31g(d)

formamide ts-1 structure

0 1
 H             0.30991500    1.48807000    0.00006500
 C             0.20273500    0.40612500   -0.00003700
 O             1.14593300   -0.30802000   -0.00001900
 N            -1.15604600   -0.03095300    0.00000200
 H            -1.30036600   -0.62191300    0.80055600
 H            -1.30110100   -0.62207100   -0.80026900
```

Figure 14.4. Gaussian input files for formamide GS and TS-1 structures. Coordinates represent HF/6-31G(d) optimized structures.

density functional theory (DFT) calculations are used, however, it is our practice to employ "*opt=tight*" or "*opt=vtight*" in combination with "*scf=(conver=10)*" and "*integral=(grid=ultrafine)*" keywords.

14.3.2 *Generation of Gaussian freqchk outputs for isotopically substituted structures*

At this point in the process, one needs to generate an output of vibrational frequencies, moments of inertia, and Gaussian-generated thermochemistry values (H, S, and G) for both isotopically substituted and unsubstituted GS and TS-1 structures. Although there are several ways to generate such outputs in Gaussian, we use the freqchk utility program, run from the Linux command line environment, and direct its output to text files.

The freqchk utility requires a series of responses to queries such as temperature, pressure, scaling factor, and isotopic identity for each atom in the structure. To facilitate multiple analyses, text files containing freqchk responses in the query order are prepared for each isotopolog. Four such files are used for the calculation of the isomerization of $HCONHD_{trans}$ to $HCONHD_{cis}$ via TS-1 KIE, and their content is outlined in Table 14.2. This particular calculation is set up to provide thermochemical values computed at 298.15 K and 1 atm pressure, using unscaled vibrational frequencies. If a HF/6-31G(d) thermochemistry scaling factor, such as 0.9135, were to be used, "1" would be replaced by "0.9135" on line 4 of each text file. The isotopic identity of each atom position is defined by an integer for each atom, ordered by its position in the structure (1 = ^{1}H, 2 = ^{2}H, 12 = ^{12}C, 14 = ^{14}N, and 16 = ^{16}O).

The four text files are used as inputs for freqchk analyses of each isotopolog, and the output is generated with the following commands:

```
$ freqchk formamide_gs_hf631gd.chk <gs_298_0.frq >thermo_gs_298_0.txt
$ freqchk formamide_gs_hf631gd.chk <gs_298_5.frq >thermo_gs_298_5.txt
$ freqchk formamide_ts1_hf631gd.chk <ts_298_0.frq >thermo_ts_298_0.txt
$ freqchk formamide_ts1_hf631gd.chk <ts_298_5.frq >thermo_ts_298_5.txt
```

Table 14.2. Freqchk input file structures for $HCONHD_{trans}$ via TS-1 KIE calculation.

GS $HCONH_2$ `gs_298_0.frq`	GS $HCONHD_{trans}$ `gs_298_5.frq`	TS-1 $HCONH_2$ `ts_298_0.frq`	TS-1 HCONHD `ts_298_5.frq`	`Filename/notes`
N	N	N	N	line 1[a]
298.15	298.15	298.15	298.15	temp., K
1	1	1	1	pressure, atm
1	1	1	1	scale factor
N	N	N	N	Line 5[b]
1	1	1	1	atom 1
12	12	12	12	atom 2
16	16	16	16	atom 3
14	14	14	14	atom 4
1	1	1	1	atom 5
1	2	1	2	atom 6
N	N	N	N	line 12[c]

Notes: [a]Line 1 returns "No" to the query "Write Hyperchem file?"
[b]Line 5 returns "No" to the query "Do you want to use the principal isotope masses?"
[c]Line 12 returns "No" to the query "Project out gradient direction?"

These commands produce four new text files containing freqchk outputs, which include moments of inertia, vibrational frequencies, and thermochemical analyses. The thermochemical analysis is relevant for KIE calculation using the *H-S* approach, and we will return to these file components later in the case study, when the *H-S* KIE calculations are described.

For the B-M KIE spreadsheet calculation, the GS/TS-1 isotopolog moments of inertia are readily extracted from the freqchk output files by simply cutting and pasting the three eigenvalues into the spreadsheet. These values are found in the "–Thermochemistry–" section of the freqchk output (Fig. 14.5) and are reproduced for each formamide isotopomer in Table 14.3.

Extracting and tabulating frequencies from the freqchk output files is done from the Linux command line with an awk script (Fig. 14.6). The script is run four times to generate a file for each isotopolog tabulating its frequencies in descending order.

```
-------------------
- Thermochemistry -
-------------------
Temperature   298.150 Kelvin.  Pressure   1.00000 Atm.
Molecular mass:    45.02146 amu.
Principal axes and moments of inertia in atomic units:
                          1         2         3
    Eigenvalues --    23.86848 155.53446 179.40287
Rotational symmetry number  1.
Rotational temperatures (Kelvin)     3.62880     0.55688     0.48279
Rotational constants (GHZ):         75.61191    11.60348    10.05971
Zero-point vibrational energy     128588.6 (Joules/Mol)
                                   30.73342 (Kcal/Mol)

Zero-point correction=                           0.048977 (Hartree/Particle)
Thermal correction to Energy=                    0.052854
Thermal correction to Enthalpy=                  0.053798
Thermal correction to Gibbs Free Energy=         0.024032
                     E (Thermal)             CV                S
                      KCal/Mol        Cal/Mol-Kelvin    Cal/Mol-Kelvin
Total                   33.166             10.440            62.647
Electronic               0.000              0.000             0.000
Translational            0.889              2.981            37.339
Rotational               0.889              2.981            21.126
Vibrational             31.389              4.479             4.181
```

Figure 14.5. Partial Gaussian freqchk output for the unlabeled formamide GS structure $HCONH_2$ computed at 298.15 K using unscaled frequencies.

Table 14.3. Inertial moment eigenvalues (a.u.) for $HCONHD_{trans}$ via TS-1 B-M KIE calculation.

Isotopolog	I_x	I_y	I_z
GS $HCONH_2$	23.86848	155.53446	179.40287
GS $HCONHD_{trans}$	28.34383	160.92625	189.27000
TS-1 $HCONH_2$	27.49320	160.72960	178.99984
TS-1 HCONHD	31.25501	168.93926	187.03062

Representative awk commands are shown below, and the frequencies used for the case study are tabulated in Table 14.4.

```
$ awk -f vib.awk thermos_gs_298_0.txt >thermo_gs_298_0.txtf
$ awk -f vib.awk thermos_gs_298_5.txt >thermo_gs_298_5.txtf
$ awk -f vib.awk thermos_ts_298_0.txt >thermo_ts_298_0.txtf
$ awk -f vib.awk thermos_ts_298_5.txt >thermo_ts_298_5.txtf
```

It is worth noting that freqchk prints unscaled frequencies, irrespective of the scaling factor used for its input. All of the thermochemical quantities, such as ZPVE, G, H, and S are calculated using the scaled frequencies, but the awk-extracted frequencies must be scaled separately if such values are needed for spreadsheet-based IE calculations.

14.3.3 *Calculating MMI*

Substituting the inertial moments from Table 14.3 into Eq. (14.3) yields the MMI term (note the mass term in Eq. (14.3) is unity)

$$\text{MMI} = \left(\frac{I^{\ddagger}_{\text{H1}} I^{\ddagger}_{\text{H2}} I^{\ddagger}_{\text{H3}} I^{R}_{\text{D1}} I^{R}_{\text{D2}} I^{R}_{\text{D3}}}{I^{\ddagger}_{\text{D1}} I^{\ddagger}_{\text{D2}} I^{\ddagger}_{\text{D3}} I^{R}_{\text{H1}} I^{R}_{\text{H2}} I^{R}_{\text{H3}}} \right)^{\frac{1}{2}} = 1.01894$$

Using the frequency values in Table 14.4, the MMI term can be alternatively calculated by using the Redlich–Teller product rule in Eq. (14.6)

$$\left(\frac{v^{\ddagger}_{i(\text{H})}}{v^{\ddagger}_{i(\text{D})}} \right) \prod_{n}^{3n-6} \frac{v_{n(\text{D})^R}}{v_{n(\text{H})^R}} \Big/ \prod_{n}^{3n^{\ddagger}-7} \frac{v_{n(\text{D})^{\ddagger}}}{v_{n(\text{H})^{\ddagger}}} = \text{VP} = \text{MMI} = 1.01892$$

To familiarize readers with the overall spreadsheet design, we present a series of figures showing regions within the sheet and how

```
# This awk script extracts vibrational frequencies from Gaussian files
# frequency calculations or freqchk text outputs and arranges them in descending order.
# To run the script give the following awk command at the UNIX prompt
# for log files use awk -f vib.awk filename.log
# for freqchk text files use awk -f vib.awk filename.txt >filename.txtf
BEGIN { n = -3 }
{
# Search Gaussian output to find frequencies
if ($1 == "Frequencies")
  {
# Gaussian prints a maximum of 3 freq per line so every new frequency
# line requires we bump the counter by 3
  n=n+3
# The actual number of freq/line may be less than 3, e.g. NF-2
# where NF is the number of fields in the line.  The first two
# fields are formatting only, i.e. "Frequencies" and "--".
# Therefore the counter that controls the loop that reads the
# frequencies needs to run 1 to NF-2.  j is the variable assigned
# to the field number that contains the frequencies.  The values
# of these fields, $j, are stored in an array called v.
  for (i=1; i<=(NF-2); i++)
      {
      j=i+2
      v[n+i]=$j
      }
  }
}
END {
    for (k=1; k<=n+i-1; k++)
    printf ("%3i\t%10.4f\n", k, v[n+i-k])
#    to print in ascending order add # to line above
#    and remove # in line below
#    printf ("%3i\t%10.4f\n", k, v[k])
  }
```

Figure 14.6. Awk script for extracting and tabulating vibrational frequencies from either Gaussian frequency log files or from freqchk-generated text output.

Table 14.4. Unscaled HF/6-31G(d) frequencies (cm^{-1}) extracted from freqchk output files for spreadsheet calculation of $HCONHD_{trans}$ via TS-1 KIE using B-M approach.

Mode	GS $HCONH_2$	GS $HCONHD_{trans}$	TS-1 $HCONH_2$	TS-1 HCONHD
1	3972.9263	3925.3296	3773.4011	3735.6277
2	3838.2671	3214.3235	3694.5883	3258.8966
3	3214.2414	2843.1700	3258.8306	2723.7625
4	1998.7911	1993.0226	2032.5896	2029.9273
5	1789.1482	1654.6296	1786.8118	1604.3231
6	1563.4218	1560.8339	1548.6917	1543.9542
7	1378.8948	1277.5801	1381.1754	1297.8912
8	1182.7712	1179.2951	1214.1388	1187.9324
9	1160.1847	1090.6506	1026.4034	974.095
10	673.4572	583.3101	964.8177	856.2402
11	617.7182	553.0729	642.3170	592.6181
12	108.5189	99.6190	−505.1031	−442.9869

various KIE components are calculated. We find it useful to define commonly used physical constants at the top of each sheet. This region and the inertial MMI portion of the sheet are outlined in Fig. 14.7, where cell B14 is the location of the MMI term computed using the products of extracted inertial moments.

Calculation of the Redlich–Teller product-derived MMI value is shown in Fig. 14.8; the reader can also see how the spreadsheet manages the GS/TS frequency lists for each isotopolog. The frequencies are arranged numerically in descending order and grouped by light and heavy reactant and TS (R-H, TS-H, R-D, TS-D) in the "B" column. Any frequency- or temperature-dependent terms are computed alongside in proximal columns. For the MMI calculation using Eq. (14.6), the appropriate frequency terms are collected in the "Q" column (Fig. 14.8).

14.3.4 *Calculating ZPE*

The B-M ZPE term is readily computed by summing and computing the exponentials of the frequency- and temperature-dependent u

	A	B	C	D	E	F
1	h=	6.6262E-34	Js			
2	k=	1.3807E-23	J/K			
3	c=	2.9979E+10	cm/s			
4	R=	1.9872E-03	kcal/mol K			
5	1 kcal/mol	4184	Joule/mol			
6	R=	8.314	J/mol K			
7	R=	1.9872	cal/mol K			
8				Ix	Iy	Iz
9	IxIyIz R(H)	6.66E+05	<=D9*E9*F9	23.86848	155.53446	179.40287
10	IxIyIz TS(H)	7.91E+05	<=D10*E10*F10	27.49320	160.72960	178.99984
11						
12	IxIyIz R(D)	8.63E+05	<=D12*E12*F12	28.34383	160.92625	189.2700
13	IxIyIz TS(D)	9.88E+05	<=D13*E13*F13	31.25501	168.93926	187.03062
14	MMI=	1.01894	<=SQRT((B10*B12)/(B13*B9))			

Figure 14.7. Portion of spreadsheet showing physical constant definitions and MMI calculation for $HCONHD_{trans}$ via TS-1 KIE using Eq. (14.3). Cell B equations are defined by the <= symbol.

terms for each isotopomer, as described by Eq. (14.5). The u terms are collected in column "C" immediately adjacent to the frequencies (Fig. 14.9). Though not shown in the figure, it is useful to employ a set of neighboring columns (e.g., D-I) for u term calculations at different temperatures.

14.3.5 *Calculating EXC*

The calculation of the EXC contribution is done analogously to the ZPE calculation, with the u terms carried from the "C" to the "J" column to form the $1/1 - e^{-u}$ terms of Eq. (14.4) at the same temperature used for the "C" column (Fig. 14.10). The same temperature range/cell number used for the ZPE calculation (C–I) is used for the EXC calculation; the EXC terms calculated at different temperatures are therefore collected across columns J–P.

	A	B	Q	Q cell equations
18	mode	R-H ν	R-D ν/R-H ν	
19	1	3972.9263	0.9880	<=B59/B19
:	:	:	:	
30	12	108.5189	0.9180	<=B70/B30
31			0.4166	<=PRODUCT(Q19:Q30)
...	...	...	...	...
34		TS-H ν	TS-D ν/TS-H ν	
35	1	3773.4011	0.9900	<=B75/B35
:	:	:	:	
45	11	642.317	0.9226	<=B85/B45
46	12	−505.1031	0.8770	<=B86/B46
47			0.4088	<=PRODUCT(Q35:Q46)
...	...	...	...	...
58		R-D ν		
59	1	3925.3296		
:	:	:		
70	12	99.619		
...	...	...	...	...
74		TS-D ν		
75	1	3735.6277		
:	:	:		
85	11	592.6181		
86	12	−442.9869	-	
...	...	...	...	...
89		VP (MMI) =	1.018922	<=Q31/Q47

Figure 14.8. Spreadsheet region showing partial frequency tabulation and VP (MMI) calculation for $HCONHD_{trans}$ via TS-1 KIE using Eq. (14.6). Cells with dark highlight indicate a continuation of cells with complete frequency lists and calculated values, as well as formatting breaks. Cell Q equations are defined by the <= symbol.

14.3.6 *B-M KIE compilation and temperature calculations*

The overall B-M KIE values are compiled for each temperature at the base of each u-specific column, after reserving cells to compile values of MMI, ZPE, and EXC. Representative data for two temperatures (298.15 and 321.15 K) are shown in Fig. 14.11. Unlike the MMI term, the ZPE and EXC terms are temperature dependent,

	A	**B**	**C**	cell C equations
18	mode	R-H ν	*u*, 298.15 K	
19	1	3972.9263	19.1722	<=$B$1*$B$3*$B19/($B$2*C$18)
:	:	:	:	
30	12	108.5189	0.5237	<=$B$1*$B$3*$B30/($B$2*C$18)
31			103.7449	<=SUM(C19:C30)
...	...	...	...	...
34		TS-H ν		
35	1	3773.4011	18.2094	<=$B$1*$B$3*$B35/($B$2*C$18)
:	:	:	:	
45	11	642.317	3.0996	<=$B$1*$B$3*$B45/($B$2*C$18)
46	12	−505.1031	−2.4375	<=$B$1*$B$3*$B46/($B$2*C$18)
47			102.9024	<=SUM(C35:C45)
48		ZPE(‡H)/ZPE(RH) =	1.5238	<=EXP(−0.5*(C47))/EXP(-0.5*(C31))
...	...	...	...	...
58		R-D ν		
59	1	3925.3296	18.9425	<=$B$1*$B$3*$B59/($B$2*C$18)
:	:	:	:	
70	12	99.619	0.4807	<=$B$1*$B$3*$B70/($B$2*C$18)
71			96.3929	<=SUM(C59:C70)
...	...	...	...	...
74		TS-D ν		
75	1	3735.6277	18.0271	<=$B$1*$B$3*$B75/($B$2*C$18)
:	:	:	:	
85	11	592.6181	2.8598	<=$B$1*$B$3*$B85/($B$2*C$18)
86	12	−442.9869	−2.1377	<=$B$1*$B$3*$B86/($B$2*C$18)
87			95.5746	<=SUM(C75:C85)
88		ZPE(‡D)/ZPE(RD) =	1.5055	<=EXP(−0.5*(C87))/EXP(-0.5*(C71))
89		ZPE =	1.01215	<=C48/C88

Figure 14.9. Spreadsheet region showing partial frequency tabulation and ZPE calculation for $HCONHD_{trans}$ via TS-1 KIE at 298.15 K using Eq. (14.5). Dark highlighting indicates a continuation of cells with complete frequency lists and calculated values, as well as formatting breaks. Cell C equations are defined by the <= symbol.

and it can be seen that the ZPE term is decreasing with increasing temperature, with the opposite being true for the EXC term. Readers who are interested in these temperature dependencies are encouraged to consult the paper by Parkin.[41]

	A	B	C	J	cell J equations
18	mode	R-H ν	*u*, 298.15 K	298.15	
19	1	3972.9263	19.1722	1.0000	<=1/(1−EXP(−C19))
:	:	:	:	:	:
30	12	108.5189	0.5237	2.4530	<=1/(1−EXP(−C30))
31			103.7449	2.7129	<=PRODUCT(J19:J30)
...	...	...	...	...	...
34		TS-H ν			
35	1	3773.4011	18.2094	1.0000	<=1/(1−EXP(−C35))
:	:	:	:	:	:
45	11	642.317	3.0996	1.0472	<=1/(1−EXP(−C45))
46	12	−505.1031	−2.4375	−0.0957	<=1/(1−EXP(−C46))
47				1.0700	=<PRODUCT(J35:J45)
48			EXC(‡H)/EXC(RH) =	0.3944	=J47/J31
...	...	...		...	...
58		R-D ν			
59	1	3925.3296	18.9425	1.0000	<=1/(1−EXP(−C59))
:	:	:	:	:	:
70	12	99.619	0.4807	2.6201	<=1/(1−EXP(−C70))
71				3.0296	<=PRODUCT(J59:J70)
...	...	...		...	...
74		TS-D ν			
75	1	3735.6277	18.0271	1.0000	<=1/(1−EXP(−C75))
:	:	:	:	:	:
85	11	592.6181	2.8598	1.0608	<=1/(1−EXP(−C85))
86	12	−442.9869	−2.1377	−0.1337	<=1/(1−EXP(−C86))
87				1.0947	<=PRODUCT(J75:J85)
88			EXC(‡D)/EXC(RD) =	0.3613	<=PRODUCT(J75:J85)
89			EXC =	1.09152	<=C48/C88

Figure 14.10. Spreadsheet region showing partial frequency tabulation and EXC calculation for $HCONHD_{trans}$ via TS-1 KIE at 298.15 K using Eq. (14.4). Dark highlighting indicates a continuation of cells with complete frequency lists and calculated values, as well as formatting breaks. Cell J equations are defined by the <= symbol.

		C	D	
93	KIE / Temp	298.15	321.15	<=D18
94	VP(MMI)	1.01892	1.01892	<=Q89
95	MMI	1.01894	1.01894	<=D14
96	ZPE	1.01215	1.01128	<=D89
97	EXC	1.09152	1.09575	<=K89
98	VP(MMI)*ZPE*EXC	1.12569	1.12908	<=D94*D96*D97 = k_H/k_D
99	MMI*ZPE*EXC	1.12571	1.12910	<=D95*D96*D97 = k_H/k_D

Figure 14.11. Spreadsheet region for $HCONHD_{trans}$ via TS-1 KIE calculation at 298.15 and 321.15 K using unscaled frequencies and Eqs. (14.1) and (14.3). Cell D equations are defined by the <= symbol.

		C	D	
93	KIE / Temp	298.15	321.15	<=D18
94	VP(MMI)	1.01892	1.01892	<=Q89
95	MMI	1.01894	1.01894	<=D14
96	ZPE	1.01110	1.01030	<=D89
97	EXC	1.09665	1.10069	<=K89
98	VP(MMI)*ZPE*EXC	1.12980	1.13306	<=D94*D96*D97 = k_H/k_D
99	MMI*ZPE*EXC	1.12982	1.13308	<=D95*D96*D97 = k_H/k_D

Figure 14.12. Spreadsheet region for $HCONHD_{trans}$ via TS-1 showing KIE calculation at 298.15 and 321.15 K using scaled frequencies (0.9135) and Eqs. (14.1) and (14.3). Cell D equations are defined by the <= symbol.

The data shown in Fig. 14.12 was computed with scaling (0.9135) producing IE differences of approximately 4% compared to those without.

14.3.7 *Calculating enthalpy and entropy contributions*

As described earlier, we have used two approaches to calculating H-S IE contributions, and we will compare and contrast both in the context of this case study. The first approach simply uses the Gaussian-printed H, S, and ZPVE values for each isotopomer from its corresponding freqchk output file (Fig. 14.5). Those values, for the $HCONHD_{trans}$ via TS-1 KIE calculation, are listed in Table 14.5.

Table 14.5. Freqchk thermochemistry output for spreadsheet calculation of $HCONHD_{trans}$ via TS-1 KIE using the enthalpy–entropy (*H-S*) approach.

File/Header		*E* (Thermal) Kcal/mol	ZPVE Kcal/mol	*S* Cal/mol-K
GS-H	Total	33.166	30.73342	62.647
GS-D	Total	31.058	30.48385	63.424
TS-H	Total	32.414	28.55546	59.283
TS-D	Total	30.282	28.31305	59.745
GS-H	Vibrational	31.389		
GS-D	Vibrational	29.281		
TS-H	Vibrational	30.637		
TS-D	Vibrational	28.505		

The enthalpy terms relate to Eqs. (14.19)–(14.21) as follows: (1) H_{ZPE} as defined in Eq. (14.20) correlates to the Gaussian-printed ZPVE term, and (2) H_{vib} as defined in Eq. (14.21) correlates to the Gaussian-printed E(thermal) *Vibrational* term, minus the printed ZPVE term. H_{thermal} in Eq. (14.19) is therefore equal to the Gaussian-printed E(thermal) *Vibrational* term. This quantity is different from the E(thermal) *Total* value by two 0.889 kcal/mol increments, which are the constant $\left(\frac{3}{2}R_gT\right)$ translational and rotational enthalpy contributions for each isotopomer. As such, one can use either enthalpy value ("Total" or "Vibrational") from the Gaussian outputs if one is simply computing the total $\Delta\Delta H$ contribution to the KIE.

A spreadsheet module for computing the *H-S* KIE components for the formamide example, using the "cut and paste" method is shown in Fig. 14.13. The relevant H and S values are compiled at the top of the module, and the $\Delta\Delta H$ and $\Delta\Delta S$ terms are computed below these entries, using Eqs. (14.13) and (14.15), respectively. Their individual contributions to the KIE, computed with Eq. (14.16), are: 0.960 ($\Delta\Delta H$) and 1.172 ($\Delta\Delta S$). When combined using Eq. (14.17), these terms predict an overall KIE of 1.125($\Delta\Delta G$). When these values are compared with the normal ZPE and EXC*MMI terms computed in Fig. 14.9, it is apparent that this

	W	X	Y	Z	AA	BB
4	R=	0.0019872	kcal/mol K			
...	...	...		...	...	...
16		R_H	R_D	TS_H	TS_D	
17	enthalpy, H	33.166	31.058	32.414	30.282	kcal/mol
18	entropy, S	62.647	63.424	59.283	59.745	cal/mol-K
19			ΔΔH (thermal) calculation			
20	(TS_D-R_D) =	−0.776	<=AA17−Y17			
21	(TS_H-R_H) =	−0.752	<=Z17−X17			
22	ΔΔH =	−0.024	<=X20−X21			
23		ΔΔS calculation and KIE (H), KIE (S), and KIE (G) calculation				
24	(TS_H-R_H) =	−3.364	<=Z18−X18			
25	(TS_D-R_D) =	−3.679	<=(AA18−Y18)			
26	ΔΔS =	0.315	<=X24−X25			
27			KIE (H)=	0.9603	<=EXP((X22)/($W$29*$X$4))	
28	Temp		KIE (S)=	1.1718	<=EXP(X26/$X$4/1000)	
29	298.15		KIE (G)=	1.1253	<=Z27*Z28	
30	ΔΔH (thermal), ΔΔH (ZPVE), and ΔΔH (vib) calculation					
31		H(thermal)	H(ZPVE)	H(vib)		
32	R_H	31.389	30.73342			
33	TS_H	30.637	30.48385			
34	R_D	29.281	28.55546		X36=(X35−X34) − (X33−X32)	
35	TS_D	28.505	28.31305		Y36=(Y35−Y34) − (Y33−Y32)	
36	ΔΔH =	−0.024	0.00716	−0.03116	<=X36−Y36	

Figure 14.13. Spreadsheet for $HCONHD_{trans}$ via TS-1 enthalpy-entropy KIE calculation at 298.15 K using freqchk thermochemistry H/S output in Table 14.5 and Eqs. (14.12)–(14.16). Cells B and D equations are defined by the <= symbol.

particular formamide KIE is a case in which the B-M terms do not correlate with the H-S KIE decomposition. The lower portion of the spreadsheet in Fig. 14.13 illustrates the dissection of $\Delta\Delta H_{\text{thermal}}$ (Eq. (14.19)) into its constituent $\Delta\Delta H_{\text{ZPE}}$ (Eq. (14.20)) and $\Delta\Delta H_{\text{vib}}$ (Eq. (14.21)) terms. Although more will be said later about this breakdown, it is apparent from the data that the enthalpic IE contribution is controlled by a $\Delta\Delta H_{\text{vib}}$ term (−31 cal/mol) that is negative and larger than the $\Delta\Delta H_{\text{ZPE}}$ term (+7 cal/mol). Therefore, it is not surprising that the B-M ZPE and H-$S\Delta\Delta H$ terms lack correlation. Nor should we expect the B-M EXC and MMI terms to correlate with the $\Delta\Delta S$ term, as the latter will compensate for the contribution made by the $\Delta\Delta H_{\mathrm{vib}}$ term (Fig. 14.2).

As shown in Table 14.5 and Figs. 14.5 and 14.13, the Gaussian-derived H and S values are printed to three decimal places, limiting the precision of the overall KIE. Still, the agreement between the H-S derived KIE (1.125) and the value computed via the B-M approach (1.12571) is reasonably good, and accurate enough for many applications. Still, it is desirable to obtain the same precision irrespective of approach, and the spreadsheet can be easily programmed to compute the H and S values using the same primary input used for the B-M calculations, i.e., the vibrational frequencies.

Beginning with the enthalpy calculation, the H_{ZPE} terms (Eq. (14.20)) are indexed, computed, and summed alongside their contributing frequency values. The appropriate differences (Eq. (14.13)) are calculated and the $\Delta\Delta H_{\mathrm{ZPE}}$ term is reported at the bottom (column R in Fig. 14.14). The same strategy is used to tabulate and compute the $\Delta\Delta H_{\mathrm{vib}}$ term using Eq. (14.21) (Column T in Fig. 14.15). Computed this way, both terms (H_{ZPE}: 0.0072, H_{vib}: −0.0313; Figs. 14.14 and 14.15) are in good agreement with the values derived from the "cut and paste" method (H_{ZPE}: 0.0072, H_{vib}: −0.0312; Fig. 14.13). This is a consequence of the Gaussian-printed ZPVE values carrying five-digit precision and H_{vib} in Fig. 14.13 being calculated as a difference involving these ZPVE values.

Continuing with the entropy calculation, the S_{vib} terms (Eq. (14.23)) utilize the frequency- and temperature-dependent u terms in column C and are summed in a manner similar to H_{vib}. The appropriate differences (Eq. (14.15)) are calculated and the $\Delta\Delta S_{\mathrm{vib}}$ term is reported, in cal/mol K, at the bottom of the column (column V in Fig. 14.16).

Calculation of the $\Delta\Delta S_{\mathrm{rot}}$ term (Eqs. (14.24)–(14.26)) requires a temperature and the symmetry numbers and rotational temperatures for each isotopomer. As was done for the MMI calculation using the Gaussian-printed moments of inertia (Fig. 14.7), a spreadsheet

	B	R	cell R equations
18	R-H ν	H (ZPE)	kcal/mol
19	3972.9263	5.6793	=0.5*($B$1*$B$3*B19*$B$6)/($B$2*$B$5)
:	:	:	
30	108.5189	0.1551	=0.5*($B$1*$B$3*B30*$B$6)/($B$2*$B$5)
31	R-H ZPE	30.7319	=SUM(R19:R30)
...	...	...	...
34	TS-H ν		
35	3773.4011	5.3941	=0.5*($B$1*$B$3*B35*$B$6)/($B$2*$B$5)
:	:	:	
45	642.317	0.9182	=0.5*($B$1*$B$3*B45*$B$6)/($B$2*$B$5)
46	−505.1031		
47	TS-H ZPE	102.9024	<=SUM(R35:R45)
...	...	...	...
58	R-D ν		
59	3925.3296	5.6113	=0.5*($B$1*$B$3*B59*$B$6)/($B$2*$B$5)
:	:	:	
70	99.619	0.1424	=0.5*($B$1*$B$3*B70*$B$6)/($B$2*$B$5)
71	R-D ZPE	28.5541	=SUM(R59:R70)
...	...	...	...
74	TS-D ν		
75	3735.6277	5.3401	=0.5*($B$1*$B$3*B75*$B$6)/($B$2*$B$5)
:	:	:	
85	592.6181	0.8471	=0.5*($B$1*$B$3*B85*$B$6)/($B$2*$B$5)
86	−442.9869		
87	TS-D ZPE	28.3117	<=SUM(R75:R85)
...	...	...	...
89	ΔΔH (ZPE) =	0.0072	<=(R87−R71) − (R47−R31)

Figure 14.14. Spreadsheet region showing partial frequency tabulation and $\Delta\Delta H_{\mathrm{ZPE}}$ calculation (kcal/mol) for $\mathrm{HCONHD_{trans}}$ via TS-1 KIE at 298.15 K using Eq. (14.20). Cells highlighted in grey indicate continuation of cells with complete frequency lists and calculated values.

region compiling this information is situated below column V, so that the $\Delta\Delta S_{\mathrm{rot}}$ and $\Delta\Delta S_{\mathrm{vib}}$ terms can be readily summed (cell V99 in Fig. 14.17). Readers will recognize that computing $\Delta\Delta S_{\mathrm{rot}}$ this way is akin to the "cut and paste" approach described earlier for the MMI term using Eq. (14.8), except that the rotational temperatures are being pasted from the freqchk outputs. The freqchk-derived

	T	cell V equations
18	H (vib)	kcal/mol
19	0.00000	<=$B$4*(($B$1*$B$3*B19)/($B$2))/(EXP((($B$1*$B$3*B19)/($B$2))/$C$18)−1)
:	:	
30	0.45082	<=$B$4*(($B$1*$B$3*B30)/($B$2))/(EXP((($B$1*$B$3*B30)/($B$2))/$C$18)−1)
31	0.65524	<=(SUM(T19:T30))
...	...	...
35	0.00000	<=$B$4*(($B$1*$B$3*B35)/($B$2))/(EXP((($B$1*$B$3*B35)/($B$2))/$C$18)−1)
:	:	
45	0.08667	<=$B$4*(($B$1*$B$3*B45)/($B$2))/(EXP((($B$1*$B$3*B45)/($B$2))/$C$18)−1)
...	...	...
47	0.15274	<=SUM(T35:T45)
...	...	...
59	0.00000	<=$B$4*(($B$1*$B$3*B59)/($B$2))/(EXP((($B$1*$B$3*B59)/($B$2))/$C$18)−1)
:	:	
70	0.46143	<=$B$4*(($B$1*$B$3*B70)/($B$2))/(EXP((($B$1*$B$3*B70)/($B$2))/$C$18)−1)
71	0.72524	<=SUM(T59:T70)
...	...	...
75	0.00000	<=$B$4*(($B$1*$B$3*B75)/($B$2))/(EXP((($B$1*$B$3*B75)/($B$2))/$C$18)−1)
:	:	
85	0.10295	<=0.5*($B$1*$B$3*B85*$B$6)/($B$2*$B$5)
...	...	...
87	0.19145	<=SUM(T75:T85)
...	...	...
89	−0.03129	<=(T87−T71) − (T47−T31) = ΔΔH(vib)

Figure 14.15. Spreadsheet region showing partial frequency tabulation and $\Delta\Delta H_{\text{vib}}$ calculation (kcal/mol) for $\text{HCONHD}_{\text{trans}}$ via TS-1 KIE at 298.15 K using Eq. (14.21). Cells highlighted in grey indicate continuation of cells with complete frequency lists and calculated values.

rotational temperatures and inertial moments are printed with five digits of precision, which is one digit better than the frequencies printed with default precision. So when these values are used, there is no loss in precision. On the other hand, the $\Delta\Delta S_{\text{rot}}$ and MMI terms are typically small for most medium- and large-sized molecules. As described earlier for the MMI term, their contribution to the formamide KIE is a consequence of its small molecular size.

Symmetry numbers (σ_r) are used to calculate a portion of S_{rot}, in same manner described earlier for the B-M SYM term (Eq. (14.2)). As shown in the spreadsheet module in Fig. 14.17, each isotopomer's σ_r value can be input and used in the calculation of S_{rot}. In the

	C	V	cell V equations
18	*u*, 298.15 K	S (vib)	cal/mol K
19	19.1722	0.00000	<=$B$7*(((C19/(EXP(C19) −1)))−(LN(1−EXP(−C19))))
:	:		
30	0.5237	3.29517	<=$B$7*(((C30/(EXP(C30) −1))) − (LN(1-EXP(−C30))))
31	R-H S(vib)	4.18095	<=SUM(V19:V30)
...	...		...
35	18.2094	0.00000	<=$B$7*(((C35/(EXP(C35) −1))) − (LN(1−EXP(−C35))))
:	:		
45	3.0996	0.38231	<=$B$7*(((C45/(EXP(C45) −1))) − (LN(1−EXP(−C45))))
46	−2.43748		<=[skip imaginary frequency]
47	TS-H S(vib)	0.64679	<=SUM(V35:V45)
...	...		...
59	18.9425	0.00000	<=$B$7*(((C59/(EXP(C59) −1))) − (LN(1−EXP(−C59))))
:	:		
70	0.4807	3.46168	<=$B$7*(((C70/(EXP(C70) −1))) − (LN(1−EXP(−C70))))
71	R-D S(vib)	4.63514	<=SUM(V59:V70)
...	...		...
75	18.0271	0.00000	=$B$7*(((C75/(EXP(C75) −1))) − (LN(1−EXP(−C75))))
:	:		
85	2.8598	0.46251	=$B$7*(((C85/(EXP(C85) −1))) − (LN(1−EXP(−C85))))
86	−2.1377		<=[skip imaginary frequency]
87	TS-D S(vib)	0.82200	<=SUM(V75:V85)
...	...		...
89	ΔΔS(vib) =	0.27898	=(V47−V31) − (V87−V71)

Figure 14.16. Spreadsheet region showing partial frequency tabulation and $\Delta\Delta S_{\text{vib}}$ calculation for $\text{HCONHD}_{\text{trans}}$ via TS-1 KIE at 298.15 K using Eq. (14.5). Cells highlighted in grey indicate continuation of cells with complete frequency lists and calculated values.

monodeuterated formamide system, three of the isotopomers have $C_s(\sigma_r = 1)$ symmetry, whereas the monodeuterated TS-1 structure has C_1 symmetry ($\sigma_r = 1$). Accordingly, rotational symmetry numbers make no contribution to $\Delta\Delta S_{\text{rot}}$ or to SYM.

The $\Delta\Delta S$ value calculated using the approach outlined in Figs. 14.16 and 14.17 (0.31626 cal/mol K) can be compared with the value obtained by using the lower-precision freqchk-printed entropy values (0.315 cal/mol K, Fig. 14.13). The discrepancy in the third decimal place may seem insignificant, but KIEs are computed using the exponential of $\Delta\Delta S$, and small differences matter. If optimal

	Q	R	S	T	U	V	cell V equations
91			rotational temperatures, K				
92		σ_r	$\theta_{r,x}$	$\theta_{r,y}$	$\theta_{r,z}$	S(rot)	cal/mol K
93	R-H	1	3.62880	0.55688	0.48279	21.12580	<=see V93 below
94	TS-H	1	3.15037	0.53888	0.48388	21.29668	<=see V94 below
95	R-D	1	3.05583	0.53822	0.45762	21.38361	<=see V95 below
96	TS-D	1	2.77120	0.51269	0.46310	21.51721	<=see V96 below
97					ΔΔS(rot) =	0.03728	<=(V94−V93)−(V96−V95)
98							
99					ΔΔS(vib) + ΔΔS(rot) =	0.31626	<=V89+V97

```
V93=$B$7*(LN((SQRT(PI())/(R93))*(($C$18^1.5)/(SQRT(S93*T93*U93))))+1.5)
V94=$B$7*(LN((SQRT(PI())/(R94))*(($C$18^1.5)/(SQRT(S94*T94*U94))))+1.5)
V95=$B$7*(LN((SQRT(PI())/(R95))*(($C$18^1.5)/(SQRT(S95*T95*U95))))+1.5)
V96=$B$7*(LN((SQRT(PI())/(R96))*(($C$18^1.5)/(SQRT(S96*T96*U96))))+1.5)
```

Figure 14.17. Spreadsheet region showing $\Delta\Delta S_{\text{rot}}$ and $\Delta\Delta S_{\text{vib}} + \Delta\Delta S_{\text{rot}}$ calculation for $\text{HCONHD}_{\text{trans}}$ via TS-1 KIE at 298.15 K using user-defined symmetry numbers and Gaussian-printed rotational temperatures and Eqs. (14.15) and (14.24)–(14.26).

agreement between B-M and *H-S* IE calculations is desired, it is necessary to recompute the $\Delta\Delta H$ and $\Delta\Delta S$ terms as shown in Figs. 14.14–14.18.

The spreadsheet module compiling all of the B-M and *H-S* terms at a given temperature is shown in Fig. 14.18. This region is an extension of the B-M region described in Fig. 14.11, and its placement allows for easy comparison of all contributing B-M and *H-S* terms. The overall B-M KIE (1.12571) agrees with the overall *H-S* KIE (1.12571) to five decimal places. Both calculations used the vibrational frequencies printed with four-digit precision and inertial moments or rotational temperatures with five-digit precision. The B-M KIE (1.2369) computed using the VP (MMI) term can be reported with similar precision. By comparison, the "cut and paste" *H-S* approach yielded an overall KIE of 1.125; its precision is limited by the Gaussian-printed entropy values.

	B	C	cell C equations
	Bigeleisen-Mayer KIE calculation with VP(MMI), MMI, ZPE, and EXC term breakdown		
93	term	KIE 298.15 K	<=C18
94	VP(MMI)	1.01892	<= Q89
95	MMI	1.01894	<=C14
96	ZPE	1.01215	<=C89
97	EXC	1.09152	<=J89
98	VP(MMI)*ZPE*EXC	1.12569	<=C94*C96*C97 = k_H/k_D
99	MMI*ZPE*EXC	1.12571	<=C95*C96*C97 = k_H/k_D
100	Enthalpy-Entropy KIE calculation with ΔΔH, ΔΔS, and ΔΔG term breakdown		
101	Term	KIE 298.15 K	<=C18
102	ΔΔH(ZPVE)	1.01215	<=EXP(R89/($B$4*$C$18))
103	ΔΔH(vib)	0.94855	<=EXP(T89/($B$4*$C$18))
104	ΔΔH(thermal)	0.96008	<=C102*C103
105	ΔΔS(vib)	1.15073	<=EXP(V89/($B$7))
106	ΔΔS(rot)	1.01894	<=EXP(V97/($B$7))
107	ΔΔS	1.17252	<=C105*C106
108	ΔΔG	1.12571	<=C104*C107 = k_H/k_D

Figure 14.18. Spreadsheet region for $HCONHD_{trans}$ via TS-1 KIE calculation at 298.15 K using unscaled frequencies, comparing B-M with enthalpy–entropy calculation and term breakdown for each, using Eqs. (14.1)–(14.18). Cell C equations are defined by the <= symbol.

Examination of the contributing B-M and H-S terms in Fig. 14.18 reveals several things, some obvious and some not so obvious. First, as expected, the B-M ZPE and $\Delta\Delta H_{ZPVE}$ terms and the MMI and $\Delta\Delta S_{\text{rot}}$ terms are pairwise equivalent. But, as alluded to earlier, it is also true that there is little correlation between the B-M ZPE term (1.01215, a normal IE contribution) and the aggregated $\Delta\Delta H_{\text{thermal}}$ term (0.96008, an inverse IE contribution). Nor does the magnitude of the B-M EXC term (1.09152) come close to the aggregated $\Delta\Delta S$ term (1.17252). The complicating factor in the formamide KIE, of course, is the unusually large and inverse $\Delta\Delta H_{\text{vib}}$ term (0.94855). This term, and $\Delta\Delta S_{\text{vib}}$, appear to be important in cases where IEs are governed or influenced by thermally excited low-frequency vibrational modes.[24–26] Indeed, the Perrin and Houk groups had identified the importance of thermally excited vibrational states as a determining factor in the formamide KIE,[16,40] so the calculations reported in this chapter serve to verify their conclusions and recast the IE contributions in terms of enthalpy and entropy.

14.3.8 *Tunneling corrections*

The spreadsheet module for incorporating tunneling corrections for KIE predictions is meant to explore the interplay of the Bell[17] and Wigner[9] corrections described in Eqs. (14.27)–(14.30). To this end, a series of constants used for the series expansion (Eq. (14.29)) of the Bell correction (Eq. (14.27)) are defined in a grid situated just above the region where the tunneling corrections are calculated (Fig. 14.19).

Below this section are a series of columns, arranged by temperature as was done for the B-M calculations, tabulating the constituent Q_t(H) and Q_t(D) elements of Eq. (14.29) for light and heavy transition structures, respectively (Fig. 14.20). These terms are formed from the TS $u^{\ddagger}$(H) and $u^{\ddagger}$(D) terms (computed, but not used in the B-M ZPE and EXC calculation, Figs. 14.9 and 14.10) and combined with the series expansion constants defined in the prior step.

	A	B	C	D	cell D equations
	Bell tunneling correction, Q_t, in series-expanded form:				
	Q_t =1+a(u‡^2)+b(u‡^4)+c(u‡^6)+d(u‡^8)+e(u‡^10)+f(u‡^12)+g(u‡^14)+h(u‡^16)				
117	a=	1	24	0.04166667	<=B117/C117
118	b=	7	5760	0.00121528	<=B118/C118
119	c=	31	967680	3.2035E−05	<=B119/C119
120	d=	127	154828800	8.2026E−07	<=B120/C120
121	d=	73	3503554560	2.0836E−08	<=B121/C121
122	e=	1414477	2.67812E+15	5.2816E−10	<=B122/C122
123	f=	8191	6.12141E+14	1.3381E−11	<=B123/C123
124	g=	16931177	4.99507E+19	3.3896E−13	<=B124/C124

Figure 14.19. Spreadsheet region showing definition of constants for Bell and Wigner tunneling corrections. Cell D equations are defined by the <= symbol.

	A	B	C	D	E
			Q_t(H)	Q_t(H)	
126	TS-H	T=	298.15	321.15	<=$D$18
127	u‡		2.437	2.263	<=ABS(D46)
128	Bell0		1.298	1.250	<=(D127/2)/SIN(D127/2)
129	Bell2		1.248	1.213	<=1+$D$117*POWER(D127,2)
130	Bell3		1.290	1.245	<= see D130 below
131	Bell4		1.297	1.250	<= see D131 below
132	Bell5		1.298	1.250	<= see D132 below
133	Bell6		1.298	1.250	<= see D133 below
134	Wigner		1.248	1.213	<=1+$D$117*POWER(D127,2)

D130 =1+$D$117*POWER(D127,2)+$D$118*POWER(D127,4)

D131 =1+$D$117*POWER(D127,2)+$D$118*POWER(D127,4)+$D$119*POWER(D127,6)

D132=1+$D$117*POWER(D127,2)+$D$118*POWER(D127,4)+$D$119*POWER(D127,6)+$D$120*POWER(D127,8)

D133 =1+$D$117*POWER(D127,2)+$D$118*POWER(D127,4)+$D$119*POWER(D127,6)+$D$120*POWER(D127,8)+$D$121*POWER(D127,10)

			Q_t(D)	Q_t(D)	
136	TS-D	T=	298.15	321.15	<=$D$18
137	u‡		2.138	1.985	<=ABS(D86)
138	Bell0		1.219	1.185	<=(D137/2)/SIN(D137/2)
139	Bell2		1.190	1.164	<=1+$D$117*POWER(D137,2)
140	Bell3		1.216	1.183	<= see D140 below
141	Bell4		1.219	1.185	<= see D141 below
142	Bell5		1.219	1.185	<= see D142 below
143	Bell6		1.219	1.185	<= see D143 below
144	Wigner		1.190	1.164	<=1+$D$117*POWER(D137,2)

D140 =1+$D$117*POWER(D137,2)+$D$118*POWER(D137,4)

D141 =1+$D$117*POWER(D137,2)+$D$118*POWER(D137,4)+$D$119*POWER(D137,6)

D142 =1+$D$117*POWER(D137,2)+$D$118*POWER(D137,4)+$D$119*POWER(D137,6)+$D$120*POWER(D137,8)

D143 =1+$D$117*POWER(D137,2)+$D$118*POWER(D137,4)+$D$119*POWER(D137,6)+$D$120*POWER(D137,8)+$D$121*POWER(D137,10)

Figure 14.20. Spreadsheet region showing Q_t(H) and Q_t(D) calculations for Bell and Wigner tunneling corrections. Cell D equations are defined by the <= symbol.

	A	B	C	D	Cell D equations
			Qt(H)/ Qt(D)*KIE	Qt(H)	
146	TS-H/D	T=	298.15	321.15	<=$D$18
147	KIE (uncorr.)		1.126	1.129	<=D111
...	...	...	...	...	...
149	Bell0		1.199	1.191	<=(D128/D138)*D111
150	Bell2		1.180	1.177	<=(D129/D139)*D111
151	Bell3		1.195	1.189	<=(D130/D140)*D111
152	Bell4		1.198	1.191	<=(D131/D141)*D111
153	Bell5		1.199	1.191	<=(D132/D142)*D111
154	Bell6		1.199	1.191	<=(D133/D143)*D111
155	Wigner		1.180	1.177	<=(D134/D144)*D111

Figure 14.21. Spreadsheet region showing uncorrected KIE values and Bell and Wigner tunneling-corrected KIE values at 298.15 K and 321.15 K for $HCONHD_{trans}$ via TS-1 KIE study. Cell D equations are defined by the <= symbol.

The tunneling corrections are tabulated immediately below this section (Fig. 14.21). The uncorrected KIE is recalled for sake of comparison at the top of each column. It is then multiplied by the appropriate $Q_t(\mathrm{H})/Q_t(\mathrm{D})$ term. The commonly used Bell tunneling correction (Eq. (14.28)) is listed first. This elevates the KIE from 1.126 to 1.199 at 298.15 K, a 6.5% increase, and indicative of moderate tunneling.[16]

The series expansion terms are tabulated below these entries (Fig. 14.21). Here it is evident that that the series expansion of the Bell correction converges to the same result when five terms are used to calculate $Q_t(\mathrm{H})/Q_t(\mathrm{D})$. The Wigner tunneling correction, described earlier as being equivalent to the first two terms of the Bell series expansion, is also listed. It provides a marginally smaller tunneling correction at 298.15 K (1.180, a 4.8% increase).

14.4 Applications of isotope effects in the literature

In this section of the chapter, we point the reader to five studies that have featured KIE or EIE estimations to elucidate reaction mechanisms and conformational preferences. First on the list is a study from O'Leary and coworkers looking at the origins of conformational KIEs in Mislow's dihydrophenanthrenes and doubly-bridged biphenyl derivatives. Here, the reader will appreciate the importance of calculating KIEs using both B-M and *H-S* approaches, and of close analysis of the components of both approaches. Second is a seminal study by Houk and Singleton using computed and experimental KIEs to decisively delineate the mechanism of a classic Diels–Alder reaction, effectively putting to rest a long-standing question in organic chemistry. We then highlight the work of Saunders, Wolfsberg, Anet, and Kronja in using deuterium EIEs to understand the conformational preferences of CH_3/CD_3 groups in tetra-substituted cyclohexanes. We conclude with a more recent example where Grubbs, Houk, and O'Leary investigated the mechanism of C(sp^3)-H activation at ruthenium(II) alkylidenes. It is our hope that with these studies, the reader will appreciate the wide utility of IE calculations.

14.4.1 *Origins of conformational KIEs*

O'Leary *et al.*[25] and Fong *et al.*[26] utilized the IEs decomposition method described earlier in this chapter to investigate the origins of conformational KIEs in Mislow's 9,10-dihydro-4,5-dimethylphenanthrene (Table 14.6). In delineating the ZPE, EXC, and MMI terms from the B-M approach, and the enthalpy and entropy contributions from the *H-S* approach, they verified that antagonistic enthalpy–entropy effects contribute to the overall KIE observed in deuterated variants of this molecule.

For example, in the species where only the methyl groups are deuterated (i.e., X = D, Y = H), a k_H/k_D, of 0.888 is computed, in excellent agreement with the experimentally measured value of 0.880. The faster racemization rate of this deuterated isotopolog (i.e., origins of the CKIE) is usually attributed to the fact that C–D bonds are shorter than C–H bonds, and hence $-CD_3$ groups smaller than $-CH_3$. This is consistent with the observed reduction in the inverse KIE when only the backbone methylene groups are deuterated.

A steric IE such as this is expected to be purely enthalpic in nature, and hence governed exclusively by the difference in ZPE

Table 14.6. $^{1}H/^{2}H$ CKIEs observed in dihydrophenanthrene isotopologs.

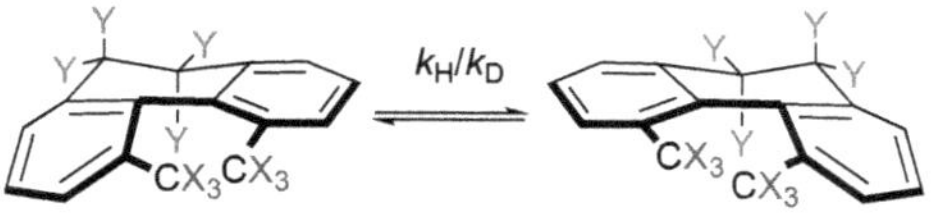

k_H/k_D	X = D, Y = H	X = H, Y = D	X = D, Y = D
Experimental	**0.880**	**0.952**	**0.847**
B-M	**0.888**	**0.953**	**0.846**
ZPE	0.755	0.943	0.712
EXC	1.182	1.009	1.193
MMI	0.995	1.001	0.997
H-S	**0.888**	**0.953**	**0.846**
$\Delta\Delta H^{\ddagger}$	0.743	0.927	0.689
$\Delta\Delta S^{\ddagger}$	1.193	1.028	1.227

Note: Comparisons between B-M and *H-S* approaches are shown. Quantum mechanical structures and vibrational frequencies were computed using B3LYP/6 − 31G** level of theory.

between the isotopologs. However, a computational breakdown of the enthalpic and entropic contributions in the deuterated methyl isotopologs reveals two features: (1) a larger inverse KIE would be observed if contributions were solely enthalpic in nature and (2) a normal KIE entropic contribution serves to diminish the enthalpic inverse KIE. A comparison with the B-M terms shows that the excitation (EXC) term is largely correlated with vibrational contributions to the entropy (*vide supra*). The results shown here support the recommendation that enthalpy and entropy contributions should be investigated in these systems to fully understand the origins of CKIEs. This particular example was studied by Mislow *et al.*[35] in the early 1960s, and they provided experimental evidence for the antagonistic enthalpy–entropy contributions revealed by the calculations described here.

An entirely different sort of KIE is observed in the racemization of doubly-bridged biphenyl derivatives (Table 14.7). This class of molecules was also studied by Mislow in the early 1960s, but in this case, the racemization rate of the deuterated diketone was

Table 14.7. $^1H/^2H$ CKIEs observed and computed at 368 K in doubly bridged biphenyl isotopologs.

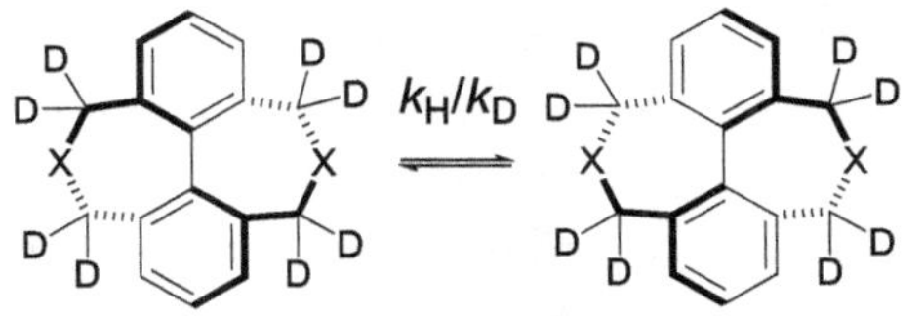

k_H/k_D	X = C = O	X = CH_2/CD_2
Experimental	1.06	—
B-M	**1.075**	**1.172**
ZPE	1.026	1.022
EXC	1.050	1.148
MMI	0.998	0.998
H-S	**1.075**	**1.171**
$\Delta\Delta H$	0.973	0.935
$\Delta\Delta S$	1.105	1.252
$\Delta\Delta H_{ZPVE}$ (kcal/mol)	0.018	0.016
$\Delta\Delta H_{vib}$ (kcal/mol)	−0.038	−0.065
$\Delta\Delta H_{thermal}$ (kcal/mol)	−0.020	−0.049

Note: Comparisons between B-M and *H-S* approaches are shown. Quantum mechanical structures and vibrational frequencies were computed using B3LYP/6 − 31G** level of theory.

found to be *slower* than the unlabeled compound.[33,34] O'Leary *et al.* found this KIE (1.075) to consist of normal ZPE (1.026) and EXC (1.050) terms, with the latter roughly twice as large as the former. Conversely, the *H-S* decomposition revealed inverse $\Delta\Delta H$ (0.973) and normal $\Delta\Delta S$ (1.105) contributions, with the latter dominating the overall IE. Importantly, this example highlights one major point made earlier in Sec. 14.2.2: the enthalpic and entropic contributions of the *H–S* approach do not always correlate with the ZPE and EXC components, respectively, of the B-M approach. As discussed earlier in the formamide case study (Fig. 14.13), the entropically driven IE in Mislow's diketone is a consequence of having small overall zero-point vibrational energy differences and an unusually large $\Delta\Delta H_{vib}$ term of the opposite sign. A computational study of the hydrocarbon-bridged analog revealed even larger such effects (Table 14.7).[26]

14.4.2 *Unambiguous delineation of a Diels–Alder mechanism*

A seminal report by Beno *et al.*[42] showcases the use of IEs to settle the long-standing quest to reveal the mechanism of the Diels–Alder reaction between isoprene and maleic anhydride (Fig. 14.22). Prior to this report, both theory and experiments concluded that the concerted mechanism is favored over a stepwise diradical mechanism by 2–7 kcal/mol.[43] However, the synchronicity or asynchronicity of the concerted reaction had not yet been elucidated.

Storer *et al.*[43] reported, for the first time, the use of precise IE measurements to unambiguously show that the Diels–Alder reaction occurs *via* a concerted, asynchronous mechanism. Both *endo* and *exo* Diels–Alder TSs were quantum mechanically computed — the *endo* was favored over the *exo* by 1.6 kcal/mol. The first clue to the asynchronicity is readily seen in the asymmetrical C–C bond-forming distances (2.225 Å vs. 2.333 Å). ^{13}C isotopic substitution at the two isoprene carbons directly involved in the C–C bond formation reveal a predicted $^{12}C/^{13}C$ KIE of 1.022 and 1.018, respectively, strongly indicating asynchronicity of the bond-forming process in TS. The KIEs were computed with both B3LYP and RHF vibrational frequencies (only B3LYP results shown in Fig. 14.22) using the B-M method as implemented by QUIVER, including Bell and Wigner tunneling corrections. Overall, the computed results were in excellent agreement with experiments.

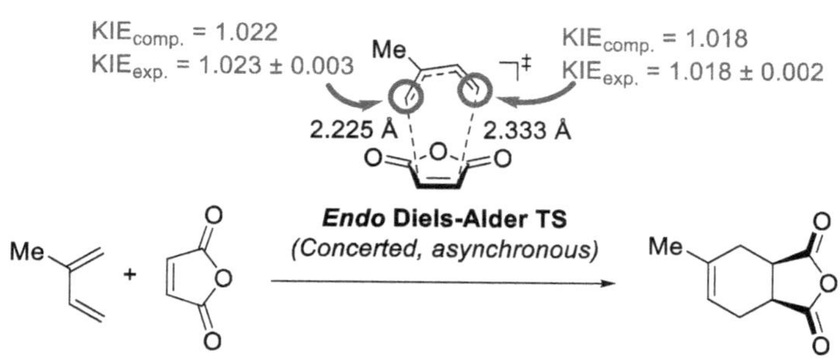

Figure 14.22. Precise computational and experimental measurements of $^{12}C/^{13}C$ KIEs used to elucidate the mechanism of Diels–Alder reactions.

14.4.3 *Origins of the EIE in 1,1,3,3-tetramethylcyclohexane*

Saunders *et al.*[44] computationally studied the origins of the steric EIEs in 1,1,3,3-tetramethylcyclohexane (Fig. 14.23). In the unlabeled compound, rapid interconversion (i.e., ring flip) between the two equivalent isomers yields an averaged frequency on an NMR time scale. In the deuterated isotopolog (i.e., $X_1 = X_2 = X_3 = D$), rapid interconversion still occurs. However, the equilibrium constant is altered from unity and shifts toward the right where the $-CD_3$ group is positioned axial to the ring, yielding an isotopic shift difference on the NMR timescale.

The authors calculated IEs using the THERMISTP program, implemented from QUIVER but including the Redlich–Teller product rule to derive the MMI value. Computations predict an EIE of 1.0417, which is in excellent agreement with experiment. To investigate the origins of the EIE, a series of isotopologs were computed with deuterium placed at the three different positions on the axial methyl group. Deuterium at the most sterically hindered position resulted in an equilibrium constant of 1.0370, much larger than at any of the other positions, indicating that steric compression at that location is a strong contributor to the observed EIE.

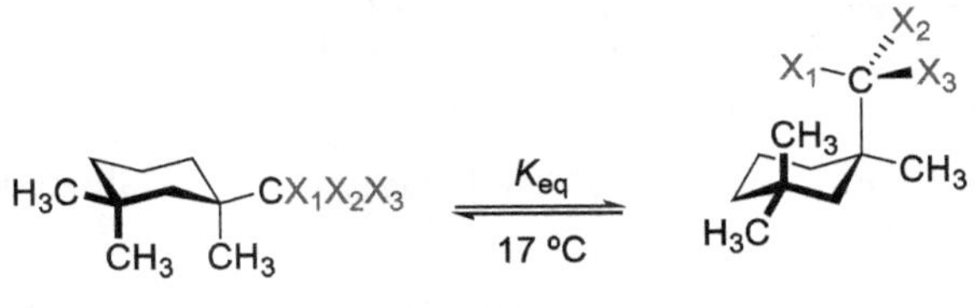

$K_{eq,\,exp.} = 1.042 \pm 0.001$ for $X_1 = X_2 = X_3 = D$
$K_{eq,\,comp.} = 1.0417$ for $X_1 = X_2 = X_3 = D$
$K_{eq,\,comp.} = 1.0370$ for $X_1 = D$, $X_2 = X_3 = H$
$K_{eq,\,comp.} = 1.0017$ for $X_2 = D$, $X_1 = X_3 = H$ OR $X_3 = D$, $X_1 = X_2 = H$

Figure 14.23. Steric $^1H/^2H$ EIEs observed in 1,1,3,3-tetramethylcyclohexane. Quantum mechanical structures and vibrational frequencies were computed using B3LYP/6-31G* level of theory.

14.4.4 *Elucidating the mechanism of a carboxylate-assisted C–H activation by ruthenium (II) complexes*

Cannon *et al.*[45] recently investigated the mechanism of C(sp^3)-H activation by ruthenium (II) alkylidenes. Through a combination of experimental kinetic studies and computational exploration of mechanistic pathways, important features of the C–H activation step were elucidated. The resting state of the reaction is the dicarboxylate intermediate. However, C–H activation could occur *via* a monopivalate or dipivalate TS as shown in Fig. 14.24.

Computations show that the latter is favored over the former by ~9 kcal/mol. Moreover, the computed free-energy barrier for the dipivalate pathway is in excellent agreement with experiments ($(\Delta G^{\ddagger}_{\text{comp}} = 23.5$ kcal/mol, $(\Delta G^{\ddagger}_{\text{exp.}} = 22.2 \pm 0.1$ kcal/mol). A unique combination of features is observed in the lowest energy transition structure: (1) dissociation of the Ru–O bond between the ruthenium center and *o*-isopropoxyphenyl group, and rotation of the *o*-isopropoxyphenyl group, and (2) deprotonation of the adamantyl C–H by a bottom-bound pivalate *via* a six-membered ring concerted metalation–deprotonation mechanism. The other computed dipivalate TS mechanisms were shown to be disfavored by ≥ 6.9 kcal/mol. In further support of the predicted lowest energy mechanism, a ^{1}H/^{2}H KIE of 8.1 ± 1.7 was obtained by measuring the rates of C–H and C–D activation in separate NMR tubes at 25°C. Computations comparing the C–H and C–D activation between the resting state and the lowest energy transition structure predict a ^{1}H/^{2}H KIE of 6.19. This KIE value was calculated using a spreadsheet-based B-M approach on B3LYP/LANL2DZ-6-31G* structures and scaled (0.97) vibrational frequencies. It is worth noting here that the computed KIE alone would not be sufficient to distinguish between the six- or four-membered ring dipivalate mechanisms shown in Fig. 14.24. It was important to couple the KIE results with the computed barriers to delineate the favored pathway in this reaction.

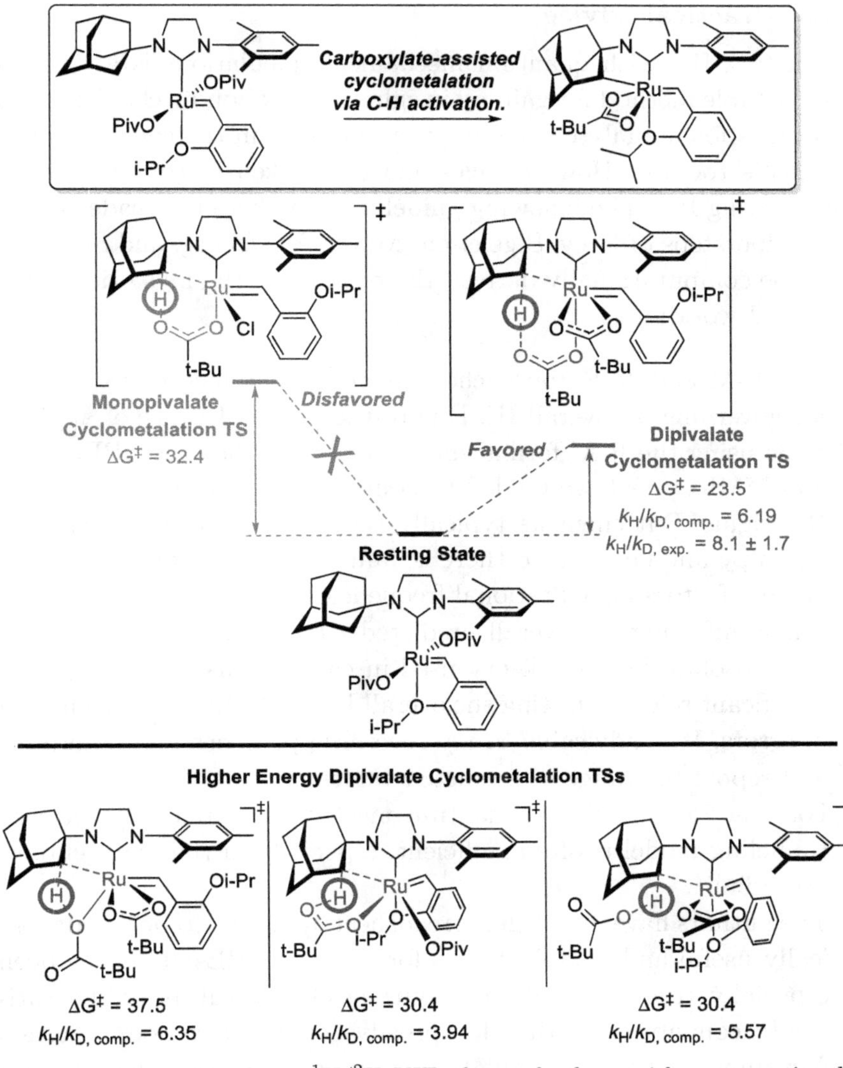

Figure 14.24. The large $^{1}H/^{2}H$ KIE observed, along with computationally derived reaction coordinate pathway, delineates the major mechanism for the carboxylate-assisted C–H activation in olefin metathesis-relevant ruthenium (II) complexes. Quantum mechanical structures and vibrational frequencies were computed using B3LYP/LANL2DZ-6-31G* level of theory.

14.5 Practical advice

Predicting IEs with quantum mechanical calculations is a remarkably reliable method for gaining insights on reaction mechanisms and conformational analyses, making it a mainstay in the computational chemist's toolbox. However, care must be taken in computing and interpreting IEs. The following guidelines may help the reader follow in the footsteps of Urey, Bigeleisen, Mayer, Wolfsberg, and Saunders and use computationally derived IEs to uncover the mysteries of the chemical world:

- The B-M and *H-S* approaches can reveal the factors responsible for governing an overall IE. Due to the preponderance of applications using the B-M formalism, it is natural that the ZPE, EXC, and MMI terms have tended to dominate literature discussions of IE origins. But chemists typically think and measure in terms of enthalpy and entropy, so there is value in the *H-S* approach.
- Scaling factors for vibrational frequencies do not play a significant role in affecting the overall predicted KIE/EIE.
- The absolute level of theory used in calculations does not play a significant role in affecting the overall KIE/EIE due to cancellation of errors. It is advisable to try several approaches and determine and report the range of predicted values.
- For reactions involving quantum mechanical tunneling, the Bell tunneling model is often sufficient, especially in the high temperature limit.
- The authors have found the spreadsheet approach to be a pedagogically useful and flexible means for examining IEs. This has been especially true in situations requiring close scrutiny of the various IE contributions. But the spreadsheet approach does require a fair amount of file manipulation effort, and in response to this, a new program *Onyx* has been developed that utilizes both B-M and *H-S* approaches, decomposes the IE results into the respective contributions, and recreates a spreadsheet similar to that presented in this chapter.[46] This program enables the user to couple the luxury of automation with the rigor of computing IEs using the dual approach described in this chapter.

References

1. Lindemann, F. A. and Aston, F. W. (1919). The possibility of separating isotopes. *Philos. Mag.*, 37, pp. 523–534.
2. Lindemann, F. A. (1919). Note on the vapour pressure and affinity of isotopes, *Philos. Mag.*, 38, pp. 173–181.
3. Keesom, W. H. and van Dijk, H. (1931). On the possibility of separating neon into its isotopic components by rectification, *Physica*, 34, pp. 42–50.
4. Urey, H. C., Brickwedde, F. G. and Murphy, G. M. (1932). A hydrogen isotope of mass 2, *Phys. Rev.*, 39, pp. 164–165.
5. Urey, H. C. and Rittenberg, D. (1933). Some thermodynamic properties of the H^1H^2, H^2H^2 molecules and compounds containing the H^2 atom, *J. Chem. Phys.*, 1, pp. 137–143.
6. Urey, H. C. and Greiff, L. J. (1935). Isotopic exchange equilibria, *J. Am. Chem. Soc.*, 57, pp. 321–327.
7. Urey, H. C. (1947). The thermodynamic properties of isotopic substances, *J. Chem. Soc.*, pp. 562–581.
8. Bigeleisen, J. and Mayer, M. G. (1947). Calculation of equilibrium constants for isotopic exchange reactions, *J. Chem. Phys.*, 15, pp. 261–267.
9. Bigeleisen, J. (1949). The relative reaction velocities of isotopic molecules, *J. Chem. Phys.*, 17, pp. 675–678.
10. Wigner, E. (1932). Concerning the excess of potential barriers in chemical reactions, *Zeitschrift Fur Physikalische Chemie-Abteilung B-Chemie Der Elementarprozesse Aufbau Der Materie*, 19, pp. 203–216.
11. Hout, R. F., Wolfsberg, M. and Hehre, W. J. (1980). Direct calculation of equilibrium constants for isotopic exchange reactions by ab initio molecular-orbital theory, *J. Am. Chem. Soc.*, 102, pp. 3296–3298.
12. Williams, I. H. (1986). Calculated conformational equilibrium isotope effect for [$^2H_1$]cyclohexane, *J. Chem. Soc., Chem. Commun.*, pp. 627–628.
13. Anet, F. A. L. and Kopelevich, M. (1986). Deuterium isotope effects on the ring inversion equilibrium in cyclohexane: The A value of deuterium and its origin, *J. Am. Chem. Soc.*, 108, pp. 1355–1356.
14. Kopelevich, M. (1989). Theoretical and experimental studies of deuterium isotope effects on degenerate conformational equilibria. Ph.D. Dissertation, University of California, Los Angeles.
15. Saunders, M., Laidig, K. E. and Wolfsberg, M. (1989). Theoretical calculation of equilibrium isotope effects using ab initio force constants: Application to NMR isotopic perturbation studies, *J. Am. Chem. Soc.*, 111, pp. 8989–8994.
16. Olson, L. P., Li, Y., Houk, K. N., Kresge, A. J. and Schaad, L. J. (1995). Theoretical analysis of secondary kinetic isotope effects in C-N rotation of amides, *J. Am. Chem. Soc.*, 117, pp. 2992–2997.
17. Bell, R. P. (1980). *The Tunnel Effect in Chemistry* (Chapman and Hall, London; New York).
18. Kwan, E. E., Park, Y., Besser, H. A., Anderson, T. L. and Jacobsen, E. N. (2017). Sensitive and accurate ^{13}C kinetic isotope effect measurements enabled by polarization transfer, *J. Am. Chem. Soc.*, 139, pp. 43–46.

19. Sims, L. B., Burton, G. W. and Lewis, D. E. (1997). BEBOVIB-IV, QCPE No. 337 (Quantum Chemistry Program Exchange, Department of Chemistry, University of Indiana: Bloomington, IN, USA).
20. Anisimov, V. and Paneth, P. (1999). ISOEFF98. A program for studies of isotope effects using Hessian modifications, *J. Math. Chem.*, 26, pp. 75–86.
21. Peles, D. N. and Thoburn, J. D. (2008). Multidimensional tunneling in the [1,5] shift in (*Z*)-1,3-pentadiene: How useful are Swain-Schaad exponents at detecting tunneling? *J. Org. Chem.*, 73, pp. 3135–3144.
22. Allen, B. D. and O'Leary, D. J. (2003). Fomenting proton anisochronicity in the CH_2D group, *J. Am. Chem. Soc.*, 125, pp. 9018–9019.
23. Allen, B. D., Cintrat, J. C., Faucher, N., Berthault, P., Rousseau, B. and O'Leary, D. J. (2005). An isosparteine derivative for stereochemical assignment of stereogenic (chiral) methyl groups using tritium NMR: Theory and experiment, *J. Am. Chem. Soc.*, 127, pp. 412–420.
24. O'Leary, D. J., Hickstein, D. D., Hansen, B. K. and Hansen, P. E. (2010). Theoretical and NMR studies of deuterium isotopic perturbation of hydrogen bonding in symmetrical dihydroxy compounds, *J. Org. Chem.*, 75, pp. 1331–1342.
25. O'Leary, D. J., Rablen, P. R. and Meyer, M. P. (2011). On the origin of conformational kinetic isotope effects, *Angew. Chem. Int. Ed.*, 50, pp. 2564–2567.
26. Fong, A., Meyer, M. P. and O'Leary, D. J. (2013). Enthalpy/entropy contributions to conformational KIEs: Theoretical predictions and comparison with experiment, *Molecules*, 18, pp. 2281–2296.
27. Frisch, M. J. *et al.* (2017). Gaussian 09, Revision C.01, Gaussian, Inc., Wallingford, CT, USA.
28. Eliel, E. L., Wilen, S. H. and Doyle, M. P. (2001). *Basic Organic Stereochemistry* (Wiley-Interscience, New York).
29. Parkin, G. (2007). Applications of deuterium isotope effects for probing aspects of reactions involving oxidative addition and reductive elimination of H-H and C-H bonds, *J. Label. Comp. Radiopharm.*, 50, pp. 1088–1114.
30. Schaad, L. J., Bytautas, L. and Houk, K. N. (1999). Ab initio test of the usefulness of the Redlich-Teller product rule in computing kinetic isotope effects, *Can. J. Chem.- Revue Canadienne de Chimie*, 77, pp. 875–878.
31. Ochterski, J. W. (2000). Thermochemistry in gaussian. See: http://gaussian.com/thermo/.
32. Laurendeau, N. M. (2005). *Statistical Thermodynamics: Fundamentals and Applications* (Cambridge University Press, New York).
33. Mislow, K., Simon, E. and Hopps, H. B. (1962). A secondary kinetic isotope effect in conformational racemization, *Tetrahedron Lett.*, 3, pp. 1011–1014.
34. Mislow, K., Simon, E., Glass, M. A. W., Wahl, G. H. and Hopps, H. B. (1964). Stereochemistry of doubly bridged biphenyls: Synthesis, spectral properties, and optical stability, *J. Am. Chem. Soc.*, 86, pp. 1710–1733.

35. Mislow, K., Wahl, G. H., Gordon, A. J. and Graeve, R. (1964). Conformational kinetic isotope effects in racemization of 9,10-dihydro-4,5-dimethylphenanthrene, *J. Am. Chem. Soc.*, 86, pp. 1733–1741.
36. Huynh, M. H. V. and Meyer, T. J. (2004). Colossal kinetic isotope effects in proton-coupled electron transfer, *Proc. Natl. Acad. Sci. USA*, 101, pp. 13138–13141.
37. Hu, S. S., Sharma, S. C., Scouras, A. D., Soudackov, A. V., Carr, C. A. M., Hammes-Schiffer, S., Alber, T. and Klinman, J. P. (2014). Extremely elevated room-temperature kinetic isotope effects quantify the critical role of barrier width in enzymatic C-H activation, *J. Am. Chem. Soc.*, 136, pp. 8157–8160.
38. For reviews of variational transition state theory and multidimensional tunneling see: Truhlar, D. G., Isaacson, A. D. and Garrett, B. C. (1985). Generalized transition state theory. In *The Theory of Chemical Reaction Dynamics*, ed. Baer, M., Vol. 4 (CRC Press, Boca Raton, USA), pp. 65–137; Tucker, S. C. and Truhlar, D. G., (1989). Dynamical formulation of transition state theory: Variational transition states and semiclassical tunneling. In *New Theoretical Concepts For Understanding Organic Reactions*, eds. Bertran, J. and Csizmadia, I. G. (Kluwer, Dordrecht), pp. 291–346. Truhlar, D. G. (2006). Variational transition state theory and multidimensional tunneling for simple and complex reactions in the gas phase solids, liquids, and enzymes. In *Isotope Effects in Chemistry and Biology*, eds. Kohen, A. and Limbach, H.-H. (CRC Press: Boca Raton), pp. 579–619.
39. Zheng, J., Bao, J. L., Meana-Pañeda, R., Zhang, S., Lynch, B. J., Corchado, J. C., Chuang, Y.-Y., Fast, P. L., Hu, W.-P., Liu, Y.-P., Lynch, G. C., Nguyen, K. A., Jackels, C. F., Ramos, A. F., Ellingson, B. A., Melissas, V. S., Villà, J., Rossi, I., Coitiño, E. L., Pu, J., Albu, T. V., Ratkiewicz, A., Steckler, R., Garrett, B. C., Isaacson, A. D. and Truhlar, D. G. (2016). https://comp.chem.umn.edu/polyrate/.
40. Perrin, C. L., Thoburn, J. D. and Kresge, A. J. (1992). Secondary kinetic isotope effects in C-N rotation of amides, *J. Am. Chem. Soc.*, 114, pp. 8800–8807.
41. Parkin, G. (2009). Temperature-dependent transitions between normal and inverse isotope effects pertaining to the interaction of H-H and C-H bonds with transition metal centers, *Acc. Chem. Res.*, 42, pp. 315–325.
42. Beno, B. R., Houk, K. N. and Singleton, D. A. (1996). Synchronous or asynchronous? An "experimental" transition state from a direct comparison of experimental and theoretical kinetic isotope effects for a Diels-Alder reaction, *J. Am. Chem. Soc.*, 118, pp. 9984–9985.
43. Storer, J. W., Raimondi, L. and Houk, K. N. (1994). Theoretical secondary kinetic isotope effects and the interpretation of transition state geometries. 2. The Diels-Alder reaction transition state geometry, *J. Am. Chem. Soc.*, 116, pp. 9675–9683.
44. Saunders, M., Wolfsberg, M., Anet, F. A. L. and Kronja, O. (2007). A steric deuterium isotope effect in 1,1,3,3-tetramethylcyclohexane, *J. Am. Chem. Soc.*, 129, pp. 10276–10281.

45. Cannon, J. S., Zou, L. F., Liu, P., Lan, Y., O'Leary, D. J., Houk, K. N. and Grubbs, R. H. (2014). Carboxylate-assisted C(sp^3)-H activation in olefin metathesis-relevant ruthenium complexes, *J. Am. Chem. Soc.*, 136, pp. 6733–6743.
46. Brueckner, A. C., Cevallos, S. L., Ogba, O. M., Walden, D. M., Meyer, M. P., O'Leary, D. J. and Cheong, P. H.-Y. (2016). Onyx, version 1.0; Oregon State University, Corvallis, OR, USA & Pomona College, Claremont, CA, USA.

Chapter 15

Stereoelectronic Effects: Analysis by Computational and Theoretical Methods

Gabriel dos Passos Gomes and Igor Alabugin

Department of Chemistry and Biochemistry
Florida State University, USA

15.1 Introduction

One of the important roles of computational techniques in organic chemistry is in translating "fuzzy" qualitative concepts developed for rationalizing the vast diversity of empirical observations into the rigorous and quantitative language of quantum chemistry. The goal of this chapter is to illustrate the utility of computational methods for quantitative analysis of stereoelectronic effects.

"Stereoelectronic" is occasionally confused with "steric + electronic"! This is inaccurate. Stereoelectronic effects are stabilizing interactions originating from increased delocalization in a favorable conformation.[1] Repulsive steric interactions also depend on the spatial arrangement of orbitals but have a different electronic origin and, generally, are not considered as a part of stereoelectronic effects family.

More rigorously, stereoelectronic effects are defined as stabilizing electronic interactions maximized by a particular geometric arrangement which can be traced to a favorable orbital overlap. These effects are ubiquitous in chemistry, and such orbital interactions can stabilize required conformations and reactive intermediates, deliver electron density to electron-deficient centers, synchronize bond forming and bond breaking, differentiate between alternative transition states (TSs), etc. Furthermore, such interactions can involve orbitals from different molecules or from different parts of the same molecule (i.e., be either inter- or intramolecular). Often, different layers of stereoelectronic interactions co-exist and co-evolve by coming into play at different stages of a multistep transformation.

The difficulty in applying the concept of stereoelectronic effects is associated with the flip side of its generality. Broad variations exist in the properties of donor and acceptor orbitals that can be stereoelectronically coupled. It is hard to have confidence with the relative donor ability of C–H and C–C bonds, especially if hybridization of carbon atoms in these bonds are different — such discussions have been heated and controversial.[2] Again, hybridization in particular can have confusing consequences that counteract electronegativity. For example, intuition can be misleading when comparing a p-orbital at oxygen with a sp^5 orbital at nitrogen where electronegativity and hybridization effects are directed in opposite ways.

15.2 Types of overlap in two-orbital interactions

In the language of molecular orbital (MO) theory, the most common situation corresponds to a stabilizing two-electron interaction involving two orbitals: a filled bonding orbital and an empty antibonding orbital (Fig. 15.1).

For the intermolecular formation of a chemical or supramolecular bond between two interacting fragments, the co-linear overlap of interacting orbitals is preferred, leading to the textbook description of supramolecular stereoelectronic effects such as H-bond formation. This orbital interaction pattern resembles direct-overlap in a σ-bond (Fig. 15.2(a)).

For molecules where the interacting atoms are already connected via a σ-bond, the dominant orbital interaction pattern corresponds to the π-type overlap. For example, the π-overlap is important

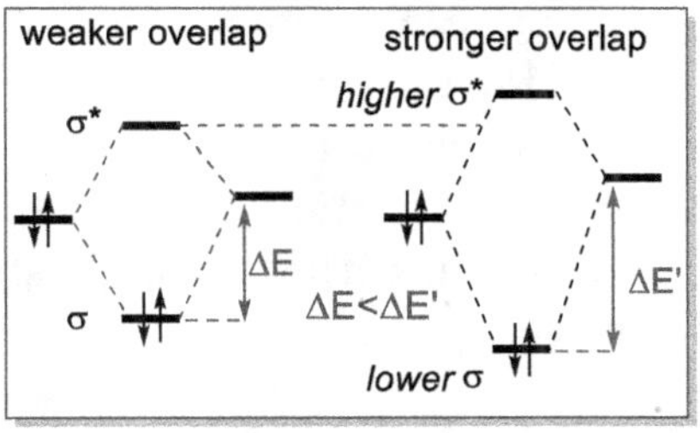

Figure 15.1. Formation of two-center two-electron chemical bonds and role of overlap in bond strength.

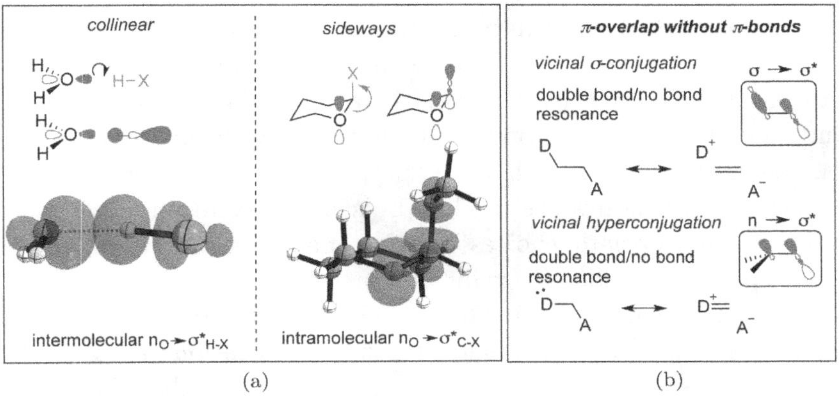

Figure 15.2. (a) Comparison of intermolecular and intramolecular overlap patterns for interaction between lone pairs and antibonding orbitals. (b) Examples of interactions using π-overlap in systems lacking formal double bonds in the main Lewis structure.

in vicinal hyperconjugative interactions (Fig. 15.2(b)), providing stereoelectronic basis to such phenomena as the anomeric effect, gauche effect, and *cis*-effect (*vide infra*).

15.3 Computational and theoretical approaches for studies of stereoelectronic effects

Even though stereoelectronic effects manifest themselves in a variety of conformational preferences and reactivity effects,[1a] their precise experimental measurement can be challenging because structure, properties, and reactivity can originate from a combination of multiple effects. Gauging the relative importance of coexisting effects can be greatly facilitated with the help of modern theoretical and computational advances. Section 15.3.1 will provide a general description of theoretical approaches to understanding, detecting, and quantifying stereoelectronic effects.

15.3.1 *Conformational changes*

The simplest and historically most important approach to quantifying stereoelectronic effects is based on conformational analysis (see Chapter 5). Conformational changes provide a convenient way to change orbital overlap responsible for the interaction in question. More importantly, conformational analysis unites experiment and theory because the quality of computed conformational profiles can be cross-checked with the experimental measurements. When experimental energies are used to benchmark computational methods, the appropriate choices can be made for the level of theory that can provide the right compromise between the accuracy and speed of calculations. High level conformational profiles also provide the necessary geometries for quantification of stereoelectronic effects with computational dissections described in the following sections on natural bond orbital (NBO) analysis and related methods.

The dependence of energies from rotation around single bonds is very general, as illustrated by the selection of examples provided in Fig. 15.3. The conformational preferences illustrated there involve a variety of functional groups with σ-bonds, π-bonds, and lone pairs that adopt well-defined geometries, very often with the antiperiplanar arrangement of the best donor and the best acceptor.[3]

However, the differences in conformer energies can potentially originate from multiple sources, leaving room for controversies and discussions regarding the relative importance of multiple components. Even when specific interactions can be "switched off" by a conformational change, more often than not, turning off one interaction activates another one.

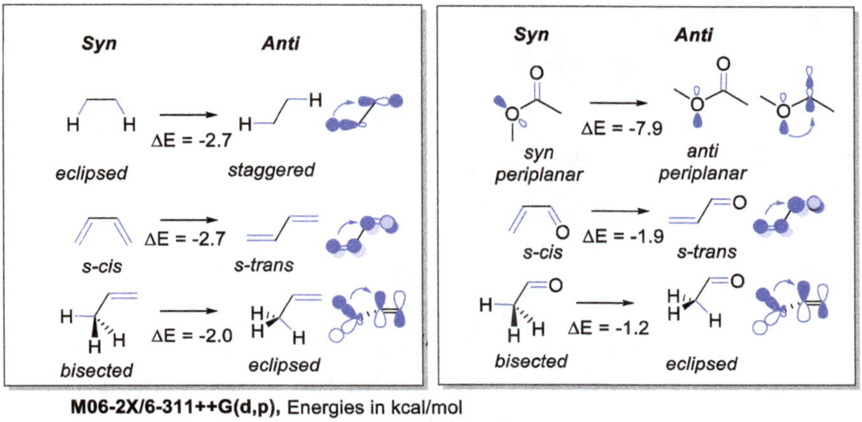

Figure 15.3. Energy differences for conformers originating from rotation around a single bond, from an anti to a syn arrangement between key interacting orbitals (shown in blue).

15.4 Isogyric, isodesmic, hypohomodesmotic, homodesmotic, and hyperhomodesmotic equations

In order to "isolate" a desired electronic effect, chemists often design hypothetical reactions of different degrees of sophistication (see Chapter 7). The advantage of these equations is that, in many cases, the thermochemical data can either be obtained experimentally or calculated with a high degree of accuracy. The challenge lies within zooming in on the key electronic effect without introducing additional structural and electronic perturbations. An "ideal" reaction for the analysis of a delocalizing electronic effect would involve no changes in hybridization and bond types. In addition, it should also have negligible changes in steric and electrostatic factors. Meeting all of these requirements is often a challenge.

The hierarchy of these equations was comprehensively summarized in a recent work by Wheeler *et al.*[4] We will discuss this order only briefly but will provide several examples to illustrate the possible caveats.

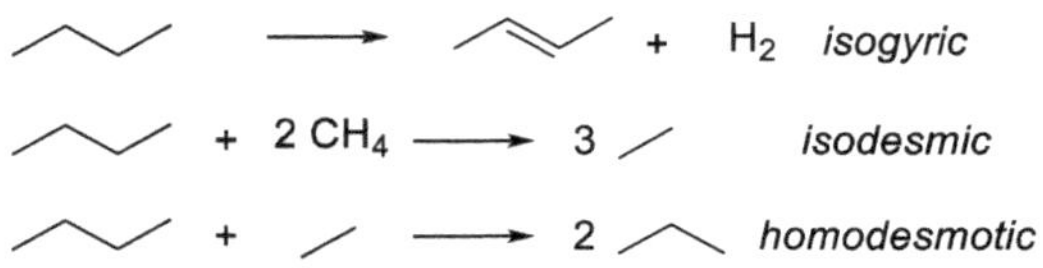

Figure 15.4. The simplest isogyric, isodesmic, and homodesmotic equations for butane.

Such equations can be divided into the categories: isogyric, isodesmic, and homodesmotic (further divided into hypohomodesmotic, homodesmotic, or hyperhomodesmotic). The level of accuracy and sophistication increases in the following order, where each next category is a subclass of the previous one: isogyric (RC1) $\supseteq$ isodesmic (RC2) $\supseteq$ hypohomodesmotic (RC3) $\supseteq$ homodesmotic (RC4) $\supseteq$ hyperhomodesmotic (RC5).

An illustration of the differences between simple isogyric, isodesmic, and homodesmotic equations is given in Fig. 15.4. The detailed definitions are given below.

15.4.1 *Isogyric equations*

This is the least restrictive comparison. In isogyric reactions (RC1), the total number of electron pairs is conserved but the number of specific bonds of a given type does not need to be balanced. Many reactions used in undergraduate chemistry textbooks (e.g., hydrogenation reactions) are isogyric.

15.4.2 *Isodesmic equations*

For hydrocarbons, the subset of isogyric reactions in which the number of C–C bonds of a given formal type (single, double, and triple) is conserved was defined as isodesmic (from Greek, *desmos* = bond) by Pople *et al.* in 1970.[5] Isodesmic reactions (RC2) are defined by preserving the number of bonds of each formal type (i.e., the single, double, or triple C–C bonds) in both reactants and products, and they include "bond separation reactions," a term for equations in which each bond between nonhydrogen atoms is separated into the simplest two-heavy-atom fragments with the same formal bond types.

- Difference from isogyric equations: Formal type of bonds (single, double, triple, etc.) is conserved.

15.4.3 *Homodesmotic equations*

Homodesmotic ("equal bonds") equations are separated further into hypohomodesmotic (RC3), homodesmotic (RC4), and hyperhomodesmotic equations (RC4).

15.4.4 *Hypohomodesmotic reactions (RC3)*

(1) Equal numbers of carbon atoms in their various states of hybridization in reactants and products;
(2) Equal number of carbon atoms (regardless of hybridization state) with zero, one, two, and three hydrogens attached in reactants and products (i.e., CH, CH_2, CH_3 groups).

- Difference from isodesmic equations: The types of bonds are preserved, including heavy-atom hydrogen bonds. Carbon hybridizations are conserved.

15.4.5 *Homodesmotic reactions (RC4)*

(1) Equal numbers of carbon–carbon [Csp^3–Csp^3, Csp^3–Csp^2, Csp^3–Csp, Csp^2–Csp^2, Csp^2–Csp, Csp–Csp, Csp^2=Csp^2, Csp^2=Csp, Csp=Csp, Csp≡Csp] bond types in reactants and products;
(2) Equal number of each type of carbon atom (sp^3, sp^2, sp) with zero, one, two, and three hydrogens attached in reactants and products.

- Difference from hypohomodesmotic equations: The hybridization of partners in the individual C–C and C–H bonds is preserved.

15.4.6 *Hyperhomodesmotic reactions (RC5)*

(1) Equal number of carbon–carbon [H_3C–CH_2, H_3C–CH, H_2C–CH_2, H_3C–C, H_2C–CH, H_2C–C, HC–CH, HC–C, C–C, H_2C=CH, HC=CH, H_2C=C, HC=C, C=C, HC≡C, and C≡C] bond types in reactants and products;
(2) Equal numbers of each type of carbon atom (sp^3, sp^2, sp) with zero, one, two, and three hydrogens attached in reactants and products.

- Difference from homodesmotic equations: Same number of C–H bonds at identically hybridized fragments.

15.5 Examples

An example of an instructive and accurate equation is provided by that for biradical stabilization energies for the estimation of electronic interaction between the two radical centers in *p*-benzynes.[6] Although the two radicals do not overlap in space, they couple the σ^* orbitals of the C–C bridges ("through-bond coupling"). This interaction not only renders the Bergman cyclization to be a symmetry-allowed process[7] but also has important consequences for the reactivity of these species. All radical reactions of *p*-benzynes (i.e., H-abstraction, addition, etc.) have to pay a penalty for uncoupling the electrons, i.e., losing the through-bond interaction (estimated as $\sim 2-5$ kcal/mol).[8] This interaction can be conveniently evaluated by the "radical separation" equation in Fig. 15.5. Note that all bond types and hybridizations are balanced perfectly in this hyperhomodesmotic equation.

Although accurate, homodesmotic equations are often avoided due to their seeming complexity. However, useful information can also be obtained when isodesmic or isogyric equations are combined. As an example, let us analyze equations designed to evaluate the stabilization provided to a carbonyl group by a cyclopropyl substituent.

Because bent C–C bonds of cyclopropanes are good hyperconjugative donors, they are expected to lead to significant

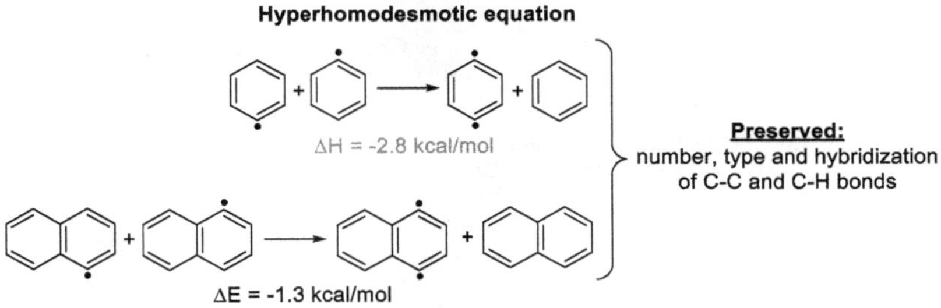

Figure 15.5. Appropriate hyperhomodesmotic equations display stabilization due to the through-bond coupling between the two radical centers of p-benzyne and its naphthyl derivative.

$\sigma_{C-C} \rightarrow \pi^*_{C=O}$ donation. This stabilizing effect accounts for the decreased reactivity of cyclopropyl esters in reactions where such donation is lost, like the formation of the tetrahedral intermediate of the acid- and base-promoted hydrolysis. Based on this hyperconjugative effect, cyclopropyl esters have been suggested as potential prodrugs with enhanced hydrolytic stability.[9] "Replacement energies" from the equations in Fig. 15.6 suggest that a cyclopropyl group provides ~2 kcal/mol more stabilization to the carbonyl than a methyl.

Note that one can convert the above equations into a more accurate form by converting two isogyric equations into one in a way that can partially cancel the errors. For example, one can subtract the two equations as shown in Fig. 15.7.

The resulting equation is hypohomodesmotic (where the types of bonds are conserved, but hybridization partners in bonds can be different), and thus is better balanced than the parent equations. Furthermore, the new equation provides a direct evaluation of the phenomenon in question (i.e., the relative conjugative donor ability of cyclopropane vs. methyl).

15.6 The limitations of reaction equations: Caveat emptor

The importance of careful design of reaction equations aimed to evaluate a theoretical concept cannot be overstated. Otherwise, such

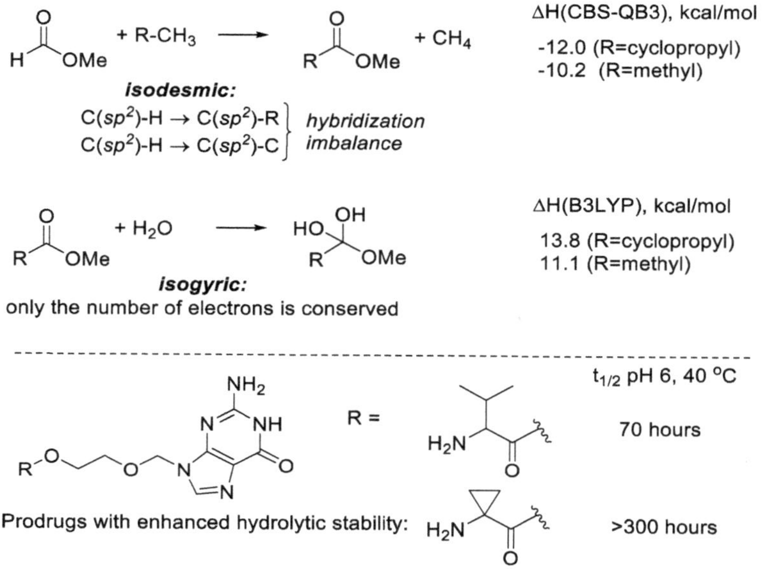

Figure 15.6. Control of the stability and reactivity of esters via $\sigma_{C-C} \rightarrow \pi^*_{C=O}$ interactions with strained cyclopropyl C–C bonds.

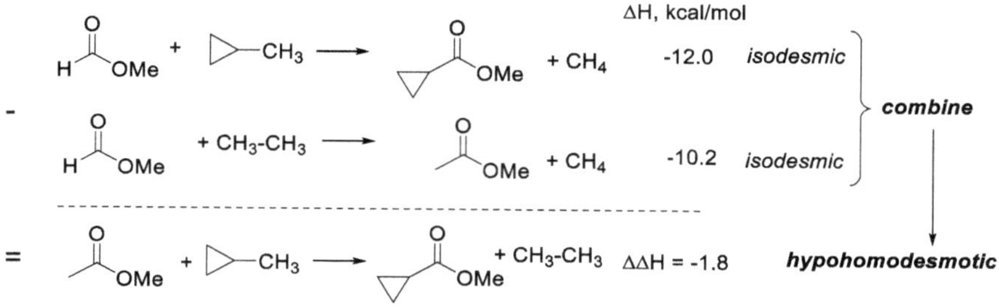

Figure 15.7. Converting a system of two isodesmic equations into a hypohomodesmotic equation by subtraction.

equations can lead to puzzling results such as the recent claim that "the conjugation stabilization of 1,3-butadiyne is zero".[10] Let us discuss this interesting example in detail.

The straightforward comparison of hydrogenation energies in two parent conjugated molecules, 1,3-butadiyne and 1,3-butadiene, revealed interesting differences (Fig. 15.8). Whereas the first

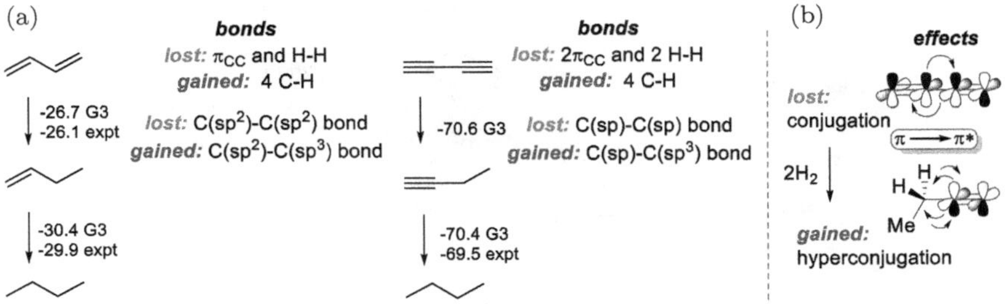

Figure 15.8. (a) Isogyric hydrogenation equations for butadiyne and butadiene. (b) Stereoelectronic origins of "disappearing" conjugative stabilization in butadiynes.

hydrogenation of a double bond in the diene is ~3.8 kcal/mol less exergonic than the second hydrogenation, the complete hydrogenation of each of the triple bonds of butadiyne is equally exergonic. Because the difference between the two hydrogenation energies for the diene corresponds to the textbook example of conjugative stabilization, the absence of such difference in alkynes was suggested to indicate the lack of such conjugation in diynes. Clearly, this result is puzzling because one would expect resonance in conjugated diynes to be approximately twice of that in conjugated dienes. Why is it not so? Can the answer lie in the nature of equations used to describe this system?

It is immediately obvious that the hydrogenation equations are isogyric (not only are the bond types not conserved, but more importantly, the types of atoms are not conserved) and hence are not properly balanced. However, the lack of balance is not the main reason why these equations do not provide a physically reasonable estimate for the resonance stabilization of 1,3-diynes. This notion can be readily illustrated by converting the above equations into their more balanced version by subtracting equations for the first and second hydrogenation steps. Note that the resulting hypohomodesmotic equation conserves bond types but still provides the same surprising answer: "conjugation in butadiyne is close to zero!"

So, why does a better equation still not solve the problem? The real issue with the above equations is much larger. The key

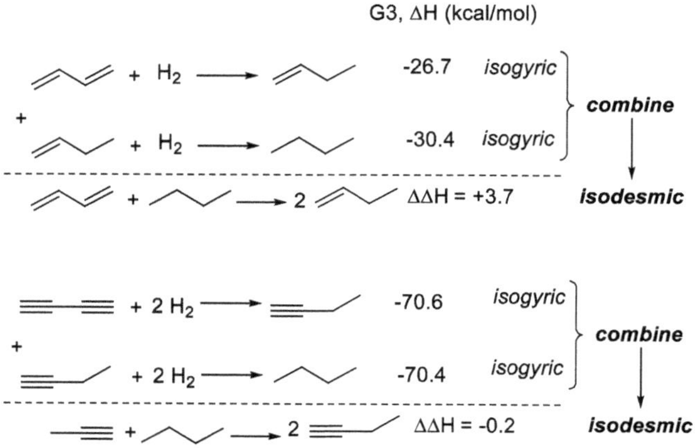

Figure 15.9. More balanced equations for butadiyne and butadiene still seem to suggest the "disappearance" of conjugative stabilization in butadiynes.

omission of Fig. 15.9 is its neglect of an important physical effect, i.e., hyperconjugation. The importance of hyperconjugation in alkynes is illustrated by the more appropriate isodesmic equations suggested by Jarowski *et al.*[11] For example, the difference between the heats of hydrogenation of ethylene and 1-butene provides an estimate for stabilization of ethylene (in kcal/mol) by an ethyl substituent (2.4 G3; 2.2 G3-(MP2); 2.7 experimental). Likewise, the hyperconjugative stabilization of acetylene by an ethyl group (4.9 G3; 4.8 G3-(MP2); 4.7 experimental) is evaluated by the difference between the heat of hydrogenation of acetylene and 1-butyne. Equivalently, the hyperconjugative stabilization can also be described by reactions in Fig. 15.9 that produce data consistent with the above evaluation.

Equations described in Fig. 15.10 illustrate that hyperconjugative stabilization of 1-butene and 1-butyne compensates for the loss of conjugative stabilization for butadiene only partially, but this compensation is complete in butadiyne. In other words, hyperconjugation in alkynes is twice as large as hyperconjugation in alkenes. This difference is sufficient for fully obscuring the conjugative stabilization in 1,3-butadiyne!

The take-home message here is the importance of designing meaningful, good-quality equations. Such equations must reflect the

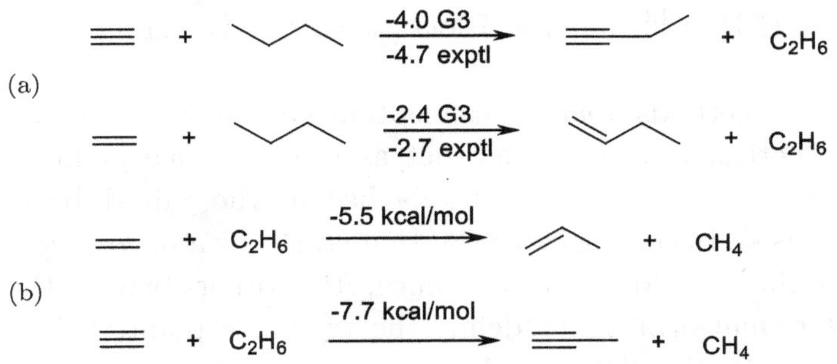

Figure 15.10. (a) Conventional equations for the evaluation of hyperconjugation. (b) Bond separation energy (BSE) values for alkene and alkyne hyperconjugation, corrected for protobranching.

Source: Reprinted with permission from Ref. [12].

answers raised by a particular problem. Not only should the types of bonds and atoms should be taken into account but specific physical phenomena (i.e., the interplay between different delocalization patterns) need to be properly balanced.

15.7 Dissecting electronic interactions

15.7.1 *Localized orbitals from delocalized wavefunctions*

The previous sections provided several indirect schemes for analyzing orbital interactions. One can see that isolating the desired effect is often difficult and sometimes impossible. Quantifying such electronic effects via a direct computational approach is seemingly straightforward. Conceptually, all one needs to do is calculate the energy penalty for removing this interaction from the delocalized wavefunction. The difference in energy between the noninteracting, localized state (sometimes called a diabatic state) and the full state (sometimes called adiabatic) can be taken then as the interaction energy. The main challenge lies in defining the appropriate localized state to serve as a reference point. Three approaches have emerged as popular tools for isolating delocalizing interactions (see Chapters 8 and 10): NBO analysis, Energy Decomposition

Analysis (EDA),[13] and the Block-Localized Wavefunction method (BLW).[14]

These methods share a conceptual similarity by starting with a hypothetical localized construct as the reference point. The key difference between these methods lies in the initial basis set of orbitals used. In contrast to NBO that starts with *orthogonal* orbitals to describe the localized reference, the other two methods use *nonorthogonal* orbitals to define the reference point.[15] This difference leads to significant variations in the estimated magnitude of delocalizing interactions. It also increases the apparent contribution from steric effects for the methods based on nonorthogonal orbitals.

The nonorthogonal initial orbitals cannot be the eigenfunctions of any physical (Hermitian) Hamiltonian corresponding to the reference "unperturbed system." Although the overlap contamination effects do not change energies evaluated on the basis of the *overall* molecular wavefunctions (whether orbitals of a determinate wavefunction are orthogonal or not has no effect on the overall expectation value), orbitals (and charge density) attributed to one group have overlap with (and thus could equally well be attributed to) orbitals of the other group. If the "bond" of one group overlaps with the "antibond" of the other group, such overlap will automatically be labeled "exchange repulsion" in a scheme based on nonorthogonal orbitals.[16]

The ambiguity about which nonorthogonal subunits receive credit for unaccounted density in the overlap region is the source of many reported differences between alternative computational dissections. The associated overlap density can be assigned to the filled orbital (and counted toward steric effects) or to the unfilled orbital (and counted toward hyperconjugative charge-transfer). All methods that harbor such overlap ambiguities are expected to differ sharply from NBO-based assessments of intramolecular or intermolecular interactions.

Furthermore, Fig. 15.11(a) illustrates the origin of "four-electron destabilization" between two nonorthogonal filled orbitals that is often taken as the physical origin of the steric destabilization.

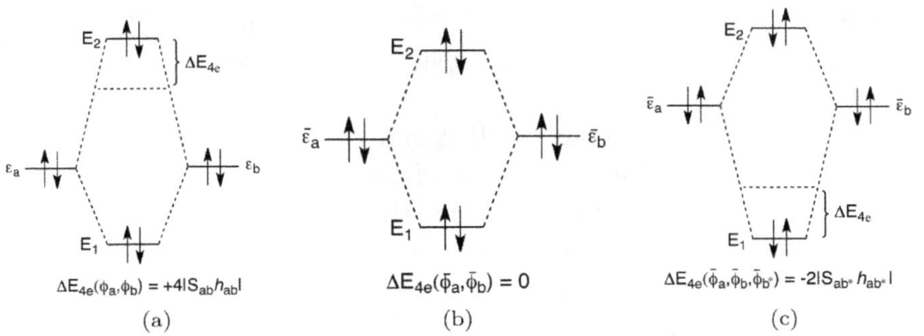

Figure 15.11. (a) Four-electron destabilizing interaction expressed in terms of nonorthogonal "unperturbed" orbitals (for which there is no imaginable Hermitian perturbation theory). (b) Four-electron nonstabilizing interaction expressed in terms of orthogonalized unperturbed orbitals (for which there exists a valid Hermitian). (c) Four-electron stabilizing interaction for a proper three-term description of orbital energies in terms of Löwdin-orthogonalized basis orbitals.

Source: Reprinted with permission from Ref. [17(b)].

According to Weinhold *et al.*, such destabilization is a mathematical artifact of nonorthogonality and does not, in fact, correspond to a physical interpretation of any imaginable physical process.[17] Once orbitals are orthogonalized, the "four-electron destabilization" disappears (Fig. 15.11(b)). When at least one unoccupied orbital is added to the system, the overall interaction becomes stabilizing (Fig. 15.11(c)).[17]

After this general preface, let us describe the three popular approaches in more detail.

15.7.2 *Natural bond orbital analysis*

The NBO analysis transforms the canonical delocalized Hartree–Fock (HF) MOs and nonorthogonal atomic orbitals (AOs) into the sets of localized "natural" atomic orbitals (NAOs), hybrid orbitals (NHOs), and bond orbital (NBOs). Each of these localized basis sets is complete, orthonormal, and describes the wavefunction with the minimal amount of filled orbitals in the most rapidly convergent fashion. Filled NBOs describe the hypothetical, strictly localized Lewis structure. Natural Population Analysis (NPA) charge assignments based on NBO analysis correlate well with empirical charge measures.[18]

The interactions between filled and antibonding orbitals represent the deviation from the localized Lewis structure and can be used as a measure of delocalization. Since the occupancies of filled NBOs are highly condensed, the delocalizing interactions can be treated by a standard second-order perturbation approach (Eq. (15.1)) or by deletion of the corresponding off-diagonal elements of the Fock matrix in the NBO basis and recalculating the energy (referred to as E_{del} energies)[19,20] where $\langle\sigma|F|\sigma^*\rangle$, or F_{ij} is the Fock matrix element between the orbitals (NBOs) i and j, ε_σ and ε_{σ^*} are the energies of the σ and σ^* NBOs, and n_σ is the population of the donor σ-orbital.[21] Usually, there is a good linear correlation between the deletion (E_{del}) and perturbation ($E(2)$) energies.[22] Deviation from such correlation often reflects cooperativity between the individual delocalizing interactions.

$$E(2) = -n_\sigma \frac{\langle\sigma|F|\sigma^*\rangle^2}{\varepsilon_{\sigma^*} - \varepsilon_\sigma} = -n_\sigma \frac{F_{i,j^2}}{\Delta E} \tag{15.1}$$

Natural Steric Analysis[23] in the NBO procedure is based on the model of Weiskopf where orbital orthogonalization leads to the "kinetic energy pressure" that opposes interpenetration of matter.[24] As the orbitals begin to overlap, the physically required orthogonalization leads to additional oscillatory and nodal features in the orbital waveform, which correspond to increased wavefunction curvature and kinetic energy, the essential "destabilization" that opposes interpenetration. The overlap-type analysis of Pauli interactions can be introduced to the NBO framework through interactions of nonorthogonalized pre-NBOs.

The NBO procedure is not the only localization technique for transforming delocalized MOs into the intuitive Lewis structure description. Foster and Boys,[25] Edmiston and Ruedenberg,[26] and Pipek and Mezey[27] reported alternative localization procedures that provide additional bridges between MO and valence bond (VB) theories.

From the organic chemist's point of view, these approaches are conceptually similar to NBO and, for the sake of brevity, will not be discussed herein.

The set of localized bonding Lewis systems can be extended to *3c-2e* bonds, and even to electron pairs localized over bigger fragments, via adaptive natural density partitioning (AdNDP).[28] When lone pairs are truly delocalized over all of the available atoms in a cyclic system, it can be taken as a sign of aromaticity (see Chapter 9).

15.7.3 *Energy decomposition analysis*[29]

This analysis starts with a "zeroth-order" wavefunction from the overlapping orbitals of the isolated molecular fragments (see Chapter 7). In EDA, the interactions between these fragments are divided into three steps. In the first step, the fragments, which are calculated with the frozen geometry of the entire molecule, are superimposed without electronic relaxation; this yields the quasi-classical electrostatic attraction ΔE_{elstat}. In the second step, the product wavefunction becomes antisymmetrized and renormalized, which gives the repulsive term ΔE_{Pauli}, termed Pauli repulsion. In the third step, the MOs relax to their final form to yield the stabilizing orbital interaction ΔE_{orb}. The latter term can be divided into contributions of orbitals having different symmetry which is useful for separation of σ- and π-effects. The sum of the three terms $\Delta E_{\text{elstat}} + \Delta E_{\text{Pauli}} + \Delta E_{\text{orb}}$ gives the total interaction energy ΔE_{int}.

The utility of EDA is illustrated by comparison of conjugation and hyperconjugation by Fernandez and Frenking,[30] who suggested that hyperconjugation is roughly half as strong as π-conjugation between two multiple bonds and that the hyperconjugative stabilization of C–H and C–C bonds with double bonds is half as strong as such stabilization by the triple bonds. As a result, hyperconjugative stabilization of the two π-systems in alkyl substituted alkynes (20.1 kcal/mol) such as 1-propyne and 4,4-dimethyl-1-butyne is almost identical as the conjugative stabilization in 1,3-butadiene (19.5 kcal/mol).

15.7.4 *Block localized wavefunction method*[31]

The electron delocalization to the cationic carbon and neutral boron center can be estimated by removing the vacant *p*-orbitals from the expansion space of MOs. Although this simple orbital deletion procedure (ODP) technique is limited to the analysis of positive hyperconjugation in carbocations and boranes, it has been generalized and extended to the BLW method.[32–34]

The BLW method combines the MO and VB theories. In this method, the wavefunction for a localized (diabatic) state is defined by limiting the expansion of each MO (called block-localized MO) to a predefined subspace. Block-localized MOs belonging to different subspaces are generally nonorthogonal. A conceptual advantage of this method is that the BLWs for diabatic states are optimized self-consistently, and the adiabatic state is a combination of a few (usually two or three) diabatic state wavefunctions.

For example, for propene, the delocalized and localized (BLW) wavefunctions can be expressed as $\Psi(\text{del}) = \hat{\text{A}}(\sigma 1a"^2 2a"^2)$ and $\Psi(\text{loc}) = \hat{\text{A}}(\sigma\pi^2_{\text{c=c}}\pi^2_{\text{CH3}})$, where $\pi_{\text{C=C}}$ and π_{CH3} are group orbitals expanded in $CH_2{=}CH$ and CH_3 groups, and are nonorthogonal. In contrast, canonical MOs 1a" and 2a" are delocalized for the whole system and orthogonal. In this example, the energy difference between these two wavefunctions, which are independently optimized self-consistently, is taken as the vicinal hyperconjugative interaction between the π-double bond and the adjacent methyl group.

15.7.5 *Other approaches to wavefunction analysis*

The electronic structure and chemical bonding of molecular systems can also be described via Electron Sharing Indexes (ESI)[34] and other approaches that evaluate electronic sharing between two atoms using the concept of bond order (bond index).[35]

Alternatively, a variety of approaches to the analysis of chemical bonding avoids the notion of "chemical bonds" altogether and replaces them with the analysis of charge density, such as the topological properties of the Laplacian of the electron density (quantum theory of atoms in molecules QTAIM),[36] or electron localization function (ELF) that defines "localization attractors" of bonding, nonbonding, and core type.[37]

In this part, we will provide several examples of practical evaluation of different types of delocalization. We will use NBO analysis for this purpose due to its close connection to classic concepts of chemical bonding.

15.8 Illustrative examples of NBO analysis of stereoelectronic interactions

15.8.1 *Types of delocalization: Neutral, negative, positive*

Hyperconjugation is the most common of the stereoelectronic effects. It can manifest itself in a number of physical observables such as bond lengths, bond angles, and quantifiable parameters such as orbital population, charges and orbital interaction energies. Let us start in a simple way and analyze three parent H_3C–X systems of different electronic nature (X = NH_2, CH_3, BH_2) to calibrate our perception of negative, neutral, and positive hyperconjugation (see Fig. 15.12).

MeNH$_2$: Methyl amine presents a negative hyperconjugative $n_N \rightarrow \sigma^*_{CH}$ interaction, estimated by NBO analysis as worth 8.6 kcal/mol. This interaction has structural consequences as well. In particular, it elongates the adjacent C–H bond relative to the same C–H bond in ethane, and, at the same time, it shortens the C–N bond. These geometric changes are consistent with the "no-bond/double-bond" resonance structure that describes this delocalization. The geometric consequences of such interactions can be more accurately evaluated with optimization with such interactions "switched off" using NBO deletion procedures.

NBO analysis also allows one to analyze effects on orbital populations. For example, donation of electronic density from the N's lone pair to a vacant σ^*_{CH} orbital decreases the population of the former

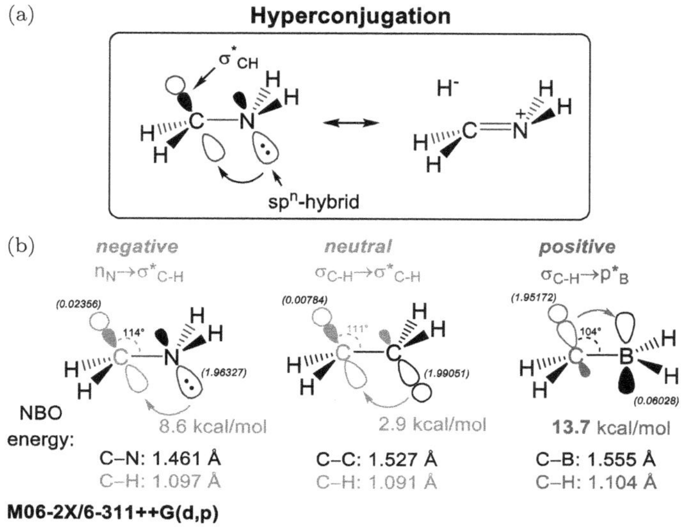

Figure 15.12. (a) General (negative) hyperconjugation with the "double bond/no bond" resonance of methylamine. (b) Examples of negative, neutral, and positive hyperconjugation. Note that neutral hyperconjugation needs to be multiplied by two since it is bidirectional.

(from 2 to 1.963 e, or, in other words, 0.037 e is missing from the lone pair) and increases population of the latter (from 0 to 0.023 e). The stereoelectronic connection between these effects is illustrated by the similarity of the two values and by the significantly greater populations of the "misaligned" vicinal C–H bonds (1.991 e).

$MeBH_2$: Delocalization in methylborane is dominated by positive hyperconjugation with a large $\sigma_{CH} \rightarrow p_B^*$ interaction (13.7 kcal/mol). The resonance description of such interaction as "double bond/no bond" is consistent with the elongation of the interacting C–H bond. The decrease of the H–C–B angle (when compared to H–C–H and even H–C–N angles) is associated with rehybridization of the donor C–H bond, which acquires much more *p*-character (79.91%) in comparison with the other two C–H bonds (76.51%). By aligning the C–H bond better with the boron's empty p-orbital, the molecule maximizes the donor–acceptor hyperconjugative interaction, thus gaining a greater stabilization.

Ethane: The neutral hyperconjugation observed in ethane is perfectly bidirectional, meaning that all antiperiplanar C–H bonds interact with each other both as a donor and an acceptor. By symmetry, these interactions are identical in magnitude. This situation is conceptually different from negative hyperconjugation, where a lone pair can only act as a donor, and from positive hyperconjugation, where an empty orbital can only act as an acceptor. However, neutral hyperconjugation can be rather imbalanced when different sets of donors and acceptors interact.

15.8.2 *Evolution of neutral hyperconjugation in polar molecules*

As an illustration of polar effects in hyperconjugation, let us compare fluoroethane, ethyllithium, and ethane (Fig. 15.13). The $\sigma_{\mathrm{CH}} \rightarrow \sigma^*_{\mathrm{CH}}$ interaction in ethane is estimated by NBO as worth 2.9 kcal/mol and, as it mentioned earlier, it is perfectly balanced by an identical interaction in the opposite direction. There is no net charge transfer between the two parts of the molecule. In ethyllithium, the situation is different. The $\sigma_{\mathrm{CLi}} \rightarrow \sigma^*_{\mathrm{CH}}$ donation is much stronger than the $\sigma_{\mathrm{CH}} \rightarrow \sigma^*_{\mathrm{CLi}}$ interaction (8.5 vs. 1.8 kcal/mol, respectively), so the C–Li bond is the net hyperconjugative donor in the C–H/C–Li pair. As expected, the balance in the C–F/C–H tug-of-war for electron density is shifted toward the C–F bond ($\sigma_{\mathrm{CH}} \rightarrow \sigma^*_{\mathrm{CF}} > \sigma_{\mathrm{CF}} \rightarrow \sigma^*_{\mathrm{CH}}$, 5.1 vs. 1.0 kcal/mol, respectively). Electronic density in the C–H bond antiperiplanar to the σ^*_{CF} (1.982 e) is depleted relative to the C–H bonds of ethane (1.990 e) and the *gauche* C–H bonds of fluoroethane (1.988 e).

Although the above trends are, of course, not surprising, the value of NBO dissection comes from the additional information that it provides. First, it gives an estimate of how large the differences in energies are. Second, NBO gives useful insight into the two factors that are responsible for the magnitude of orbital interactions for a pair of the donor and the acceptor orbitals, i.e., the contributions from the energy gap and the Fock matrix element (proportional to the orbital overlap). In this particular comparison of

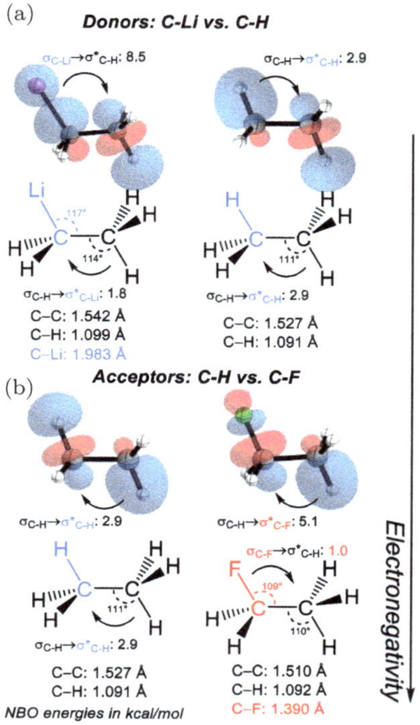

(c)

	orbital	Energy (a.u.)	Occupancy (electrons)
Li	$\sigma_{C\text{-}H}$	-0.51026	1.99250
	$\sigma_{C\text{-}Li}$	-0.26622	1.96399
	$\sigma^*_{C\text{-}H}$	+0.52019	0.02557
	$\sigma^*_{C\text{-}Li}$	+0.18888	0.00408
H	$\sigma_{C\text{-}H}$	-0.55974	1.99051
	$\sigma^*_{C\text{-}H}$	+0.48440	0.00784
F	$\sigma_{C\text{-}H}$	-0.58286	1.98269
	$\sigma_{C\text{-}F}$	-1.03228	1.99707
	n_F	-0.48800	1.97484
	$\sigma^*_{C\text{-}H}$	+0.46421	0.00451
	$\sigma^*_{C\text{-}F}$	+0.30841	0.02158

Figure 15.13. (a) Comparison of donor ability between C–Li and C–H bonds. (b) Comparison of acceptor ability between C–H and C–F bonds. (c) Energy and population of the orbitals in ethyllitium, ethane, and fluoroethane.

the three substituted "ethanes," the NBO dissection clearly show that the difference in the energies of delocalizing interactions comes from both sources, but the relative contributions are different. From the second-order perturbation energy equation, it is clear that the Fock matrix element $F_{i,j}$ contribution is dominant, especially in ethyllithium (Fig. 15.14).

In addition to evaluating individual orbital contributions, NBO can also evaluate the "global" importance of resonance for the whole molecule. This can be done by comparing the total amount of "non-Lewis density" in the NBO output. Interestingly, this parameter shows that, overall, ethyllithium is slightly less delocalized than ethyl fluoride ("valence non-Lewis density" for EtLi is 0.367% vs. 0.381%

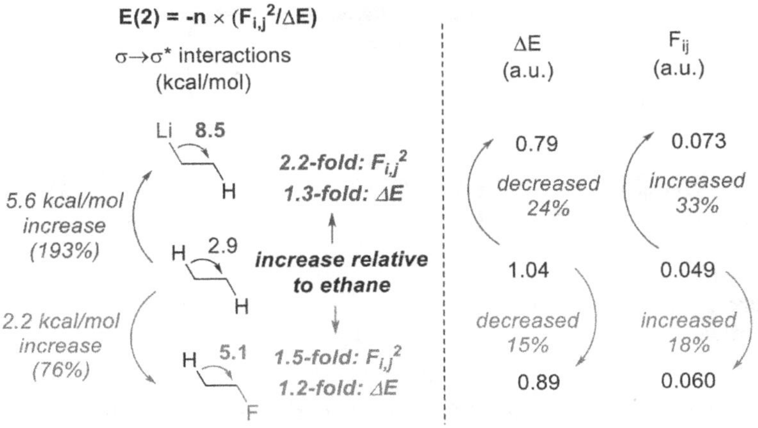

Figure 15.14. Analysis of $F_{i,j}$ and ΔE terms in the hyperconjugative interactions. The second-order perturbation energy equation illustrates that the $F_{i,j}$ term is the dominant component in the σ-conjugation strengthening observed for more polar bonds.

for EtF). This finding may look surprising considering the great magnitude of the σC–Li $\rightarrow$ σ^*C–H interaction discussed above. However, one should not forget the presence of multiple lone pairs at the fluorine atom. These lone pairs are strongly delocalized through negative hyperconjugation interactions (nF $\rightarrow$ σ^*C–H and nF $\rightarrow$ σ^*C–C) that are not present in EtLi.

15.8.3 *Overlap effects on hyperconjugative interactions in the absence of polarization*

When analyzing the concentration of interactions between bonds of different polarity, one should not forget that simple geometric factors have a large effect on the magnitude of delocalizing interactions. For example, even the neutral C–H/C–H hyperconjugation can be increased by ~80% in magnitude when the interacting orbitals are separated by the shorter C=C bond of ethene. It is noteworthy that this large change is a consequence of just a 0.2 Å decrease in the distance between two carbon atoms.

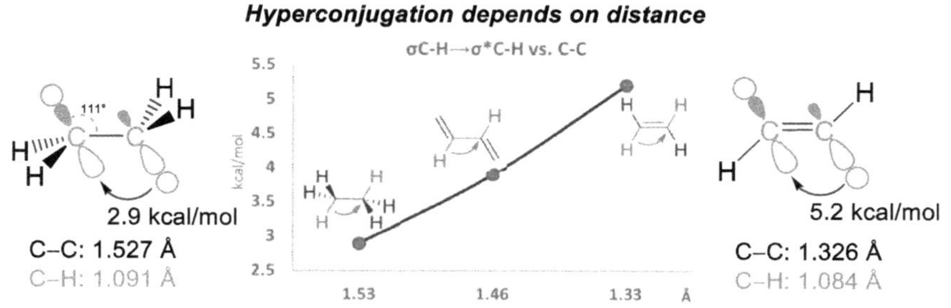

Figure 15.15. The dependence of hyperconjugation from the distance between two interacting orbitals in representative hydrocarbons.

The dominating role of C–C distance, rather than hybridization, is further illustrated by the comparison of such interaction in 1,3-butadiene where the antiperiplanar C–H bonds at the central atom are separated by a C(sp^2)–C(sp^2) bond. Although the C–C bond-forming hybrids have the same hybridization as in the C–C bond of ethane, there is only partial double bond order and the C–C distance is longer. As a result, the $\sigma_{\text{C-H}} \rightarrow \sigma^*_{\text{C-H}}$ interaction energy lies between that in ethane and ethene (Fig. 15.15).

15.8.4 *Directionality of stereoelectronic effects*

The above examples provided quantitative insights but for the systems that are relatively simple and intuitively understandable. In the present section, we will provide analysis of less obvious, yet important, effects associated with directionality of stereoelectronic interactions. In order to illustrate this phenomenon, let us compare two similar interactions in ethyl methyl ether: $\sigma_{\text{CH}} \rightarrow \sigma^*_{\text{C-O}}$ and $\sigma_{\text{CH}} \rightarrow \sigma^*_{\text{O-C}}$. The interactions involve the same σ^* acceptor that overlaps, at its opposite ends, with two analogous $\sigma_{\text{C-H}}$ donors. Despite these similarities, the magnitude of these effects is noticeably different: acceptor ability of the bridge bond increased by 38% (from 3.2 to 4.4 kcal/mol) at the carbon end relative to the oxygen end (Fig. 15.16).

NBO analysis readily answers why C–O bonds are ~40% better acceptors than O–C bonds in $\sigma_{\text{C-H}} \rightarrow \sigma^*_{\text{X-Y}}$ interactions

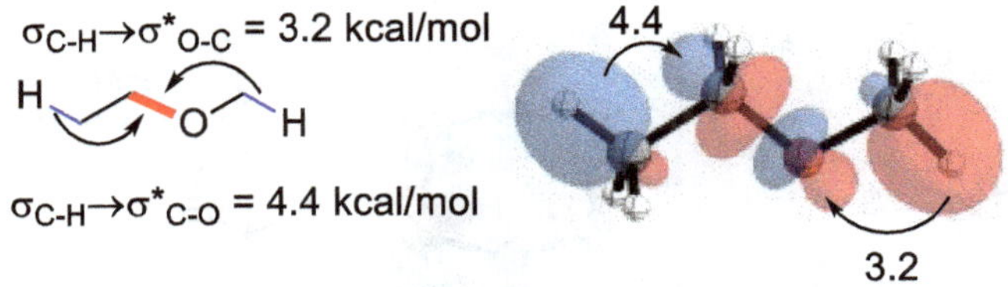

Figure 15.16. Polarization of C–O bond dramatically changes the magnitude of interactions depending on the C or O end of its antibonding orbital.

(X, Y=O, C) in ethers. The origin of this directionality lies in polarization of σ^*_{CO} orbital.[24] Due to the greater electronegativity of carbon, the carbon end of the σ^*_{CO} orbital has the larger orbital coefficient, leading to a larger $F_{i,j}$ term for the $\sigma_{\mathrm{C-H}} \rightarrow \sigma^*_{\mathrm{C-O}}$ interaction.

However, the expectations based on electronegativity fail to provide an explanation for an even stronger directionality of hyperconjugative interactions in the analogous sulfur compounds. This observation is not consistent with bond polarization. Not only is the difference in electronegativity between carbon and sulfur considerably smaller than between carbon and oxygen, but a C–S bond is polarized toward C and therefore its antibonding orbital has a greater orbital coefficient at S.

So what is the origin of this surprising behavior? Figure 15.17 illustrates that the 1,3-dithiane ring is distorted because S–C bonds are longer than C–C bonds. This geometric feature imposes very different orbital overlaps,[38] leading to a larger difference in magnitude between $\sigma_{\mathrm{C-H}} \rightarrow \sigma^*_{\mathrm{C-S}}$ (5.6 kcal/mol) vs. $\sigma_{\mathrm{C-H}} \rightarrow \sigma^*_{\mathrm{S-C}}$ (1.3 kcal/mol).

Directionality and anisotropy of stereoelectronic effects manifests itself often in more complex systems, especially when stronger donors such as lone pairs are present. Recently, such effects were shown to contribute to the structural and electronic differences between peroxides and ethers (and between acetals and bis-peroxides).[39]

In these systems, the directionality of $n_{\mathrm{O}} \rightarrow \sigma^*_{\mathrm{X-Y}}$ interactions is pronounced even further. At the same level of theory, the NBO energy for $n_{\mathrm{O}} \rightarrow \sigma^*_{\mathrm{C-O}}$ interaction in the acetal is worth 14.9 kcal/mol,

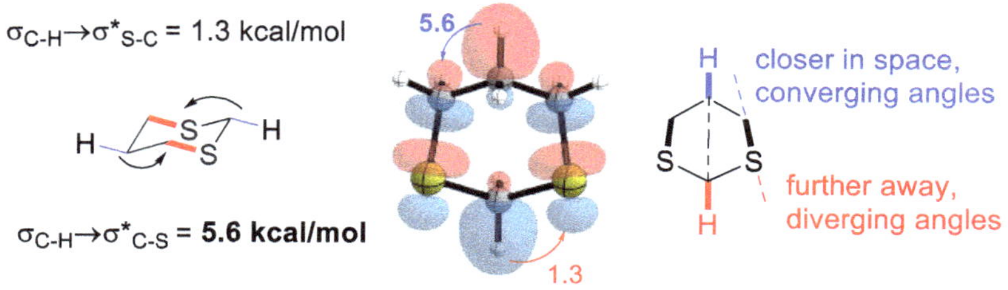

Figure 15.17. The difference in energy of the hyperconjugative interactions in 1,3-dithiane stem from different orbital overlaps rather than (lack of thereof) polarization of the C–S bond.

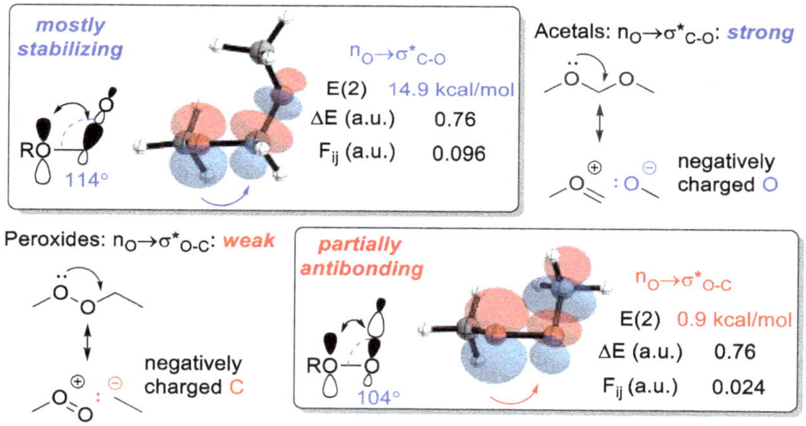

Figure 15.18. The difference between $n_O \rightarrow \sigma^*_{C-O}$ interactions in acetals and $n_O \rightarrow \sigma^*_{O-C}$ interactions in peroxides is amplified by more than 16-fold (>1600% difference!).

whereas the energy $n_O \rightarrow \sigma^*_{O-C}$ interaction in the peroxide is worth ~0.9 kcal/mol (Fig. 15.18). In other words, the stereoelectronic difference increased from 40% to more than 1600% in the σ^* interactions with a stronger donor (σ_{C-H} vs. n_O, see Fig. 15.17 vs. 15.18).

The reason for the greatly increased hyperconjugative anisotropy lies in the intricate combination of effects that control the resulting orbital overlap. When a σ-orbital serves as a donor, most of the stabilizing orbital overlap in the $\sigma \rightarrow \sigma^*$ interaction originates from overlap of this σ-orbital with the *back* lobe of an antiperiplanar

σ^*-acceptor. In contrast, when a p-orbital serves as a donor in an anomeric interaction, the notions of syn- and antiperiplanarity vanish. In such systems, the n_p/σ^* overlap is significant with both the back lobe of the σ^* orbital (e.g., from the O–C bond in the peroxide in Fig. 15.18) and the antibonding region between the two atoms (e.g., O and C). In peroxides, the unusually small O–O–C angle brings the σ^*_{OC} node closer to the p-orbital. The destabilizing interaction with the out-of-phase hybrid at carbon largely offsets the in-phase stabilizing interaction of the p-donor with the oxygen part of the $\sigma^*_{\mathrm{O-C}}$ orbital (Fig. 15.18).

The greater than 16-fold decrease in the magnitude of $n_{\mathrm{O}} \rightarrow \sigma^*_{\mathbf{O-C}}$ interactions in peroxides in comparison to $n_{\mathrm{O}} \rightarrow \sigma^*_{\mathbf{C-O}}$ interactions in acetals is striking. Taken together with the abovementioned structural effects, the nonsymmetric nature of σ-acceptors explains why the anomeric effect is dramatically diminished in peroxides in comparison with acetals. This stereoelectronic analysis reveals the weakening of anomeric hyperconjugative interactions as an additional source of thermodynamic instability of dialkyl peroxides.

A logical expansion of this stereoelectronic analysis explains why bis-peroxides can possess surprising stability. When the second peroxide moiety in the same molecule is separated by from the first peroxide by a one-atom bridge, such bis-peroxides can be considered as acetals. In cyclic molecules of this family, the two peroxide moieties stabilize each other via strong anomeric $n_{\mathrm{O}} \rightarrow \sigma^*_{\mathbf{C-O}}$ interactions (Fig. 15.19). Note that the NBO dissection and isogyric and isodesmic equations agree in identifying the increased stability of bis-peroxides.

15.8.5 *Multilayered orbital interactions in carbonyl compounds*

In this section, we show how NBO analysis can be used for the dissection of hyperconjugative interactions in organic functional groups using ketones as a molecular guinea pig. A pedagogically interesting feature of this important class of organic molecules is that

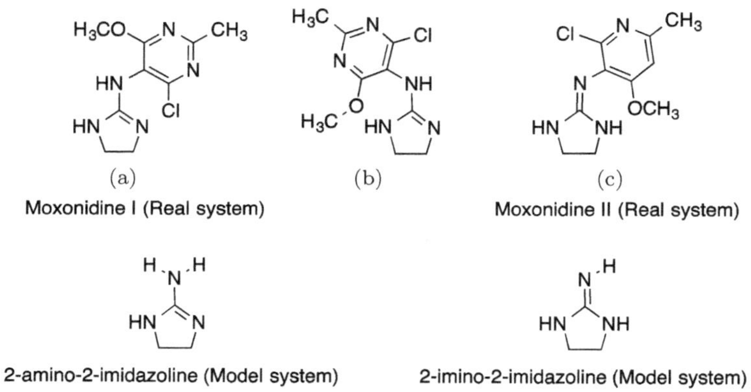

Figure 15.19. Relative weakness of hyperconjugative donation from oxygen lone pairs to vicinal σ^*-acceptors in peroxides vs. acetals (top) and stereoelectronic transformation of bis-peroxides to bis-acetals (bottom).

they contain all major types of orbitals: σ-bonds, π-bonds, and lone pairs. Let us discuss the key orbital interactions revealed by NBO analysis.

The "first level" of interactions involve delocalization of C–H bonds with $\pi_{C=O}$. Note that this interaction increases when C–H bonds are anti to a carbonyl π-bond (i.e., in the eclipsed conformation). The respective combined NBO energies for the π-component of donation to and from the carbonyl increase from 13.2 vs. 14.9 kcal/mol. One could stop here, because this answer does capture the most important change in the magnitude of delocalization. Furthermore, the greater stabilizing effect is observed in the more stable (eclipsed) conformation (*vide infra*). However, it is instructive to take a moment and examine the vicinal effects in greater detail, getting a closer view of the full complexity of this system, which originates form the interplay between multiple delocalizing interactions of different nature (Fig. 15.20).[40]

The "second level" of interactions that are also different in the bisected and eclipsed conformations of ethanal includes the vicinal interactions between anti- and syn-periplanar σ-bonds. Note that changes in the individual interactions are substantial. For example, the 0.9 kcal/mol interaction of the in-plane C–H bond with

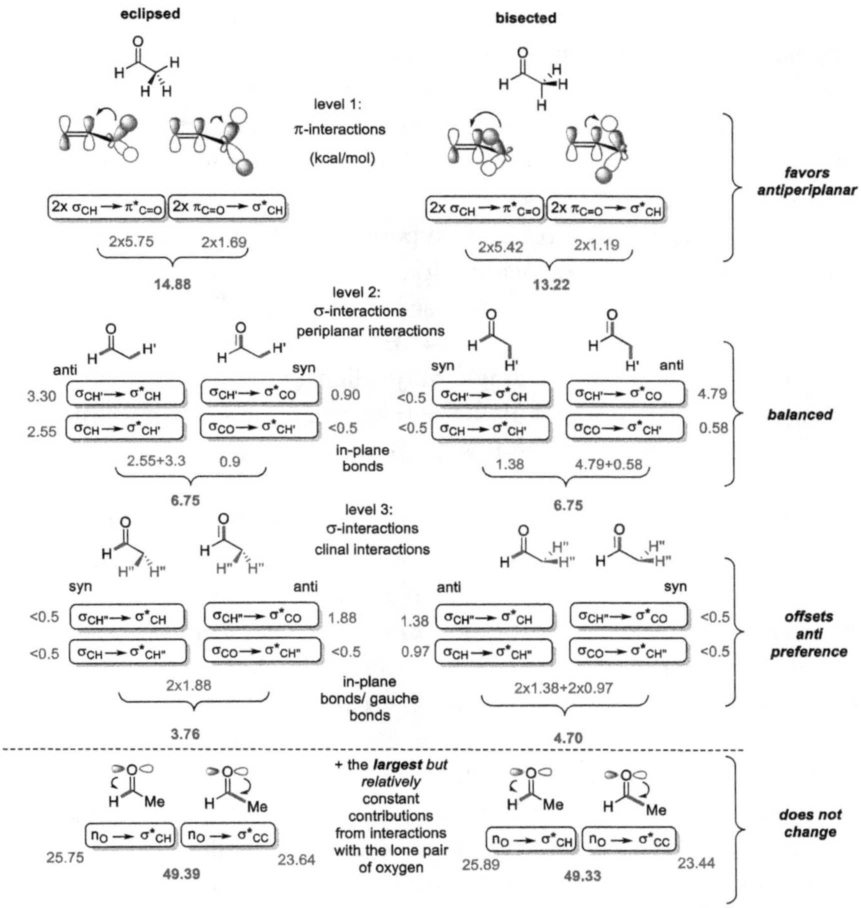

Figure 15.20. NBO analysis of stereoelectronic interactions involved in the conformational profile of ethanal. The combined energies should be treated as approximate because interaction with energies below the default NBO threshold of 0.5 kcal/mol was not used in determining the overall balance.

the $\sigma^*_{\mathrm{C-O}}$ orbital increases to 4.8 kcal/mol due to the change from syn- to anti-periplanar orbital arrangement. Note that the change in energy of $\sigma_{\mathrm{C-H}} \rightarrow \sigma^*_{\mathrm{C-H}}$ interactions between the two vicinal in-plane C–H bonds partially compensates for the difference in C–H/$\pi_{\mathrm{C-O}}$ interactions.

The "third level" of resonance delocalization involves the imperfectly aligned syn- and anticlinal vicinal σ-bonds. Here, the effect of $\sigma_{C-H} \rightarrow \sigma^*_{C-H}$ interactions favors the bisected conformation where these orbitals are anticlinal. However, this effect is smaller and cannot overcome the π-effects that impose a greater bias toward the eclipsed conformation.

Interestingly, the strongest hyperconjugative effect in this system, the $n_O \rightarrow \sigma^*$ delocalization, remains a bystander that it is fully impartial to the "eclipsed vs. bisected" conformational tug-of-war. Although the $n_O \rightarrow \sigma^*$ interactions are very large (23–26 kcal/mol), they change only slightly upon conformational change. They are important for determining the overall stability of carbonyl compounds but impose hardly any effect on the conformational equilibrium.

There are several take-home messages from this analysis. First, the several layers of conjugative interactions illustrate the electronic richness that can hide even in a relatively simply system. A subset of these interactions can play a determining role in a specific conformational or reactivity feature whereas the other factors may play a secondary role that can be either cooperative or anticooperative. Figure 15.20 illustrates how the qualitative analysis based on the quantum-mechanical dissection of the high-quality wavefunctions (such as NBO analysis) can help to untangle this complexity. Furthermore, the balance between different "levels" of interactions can change upon physical or chemical changes in the environment.

Let us now summarize conformational profiles of carbonyl compounds and show how the interplay between the first three layers of hyperconjugation in Fig. 15.20 can explain the relative stabilities of aldehyde conformations. The eclipsed conformation is favored over the bisected in straight chain aldehydes and ketones.[41] Contrary to the expectations based on steric reasoning, linear alkyl groups prefer to eclipse the carbonyl. For example, the doubly methyl-eclipsed conformation is the most stable rotamer of 3-pentanone. For the larger alkyl groups (i.e., t-butyl), the preference is for the hydrogen group to

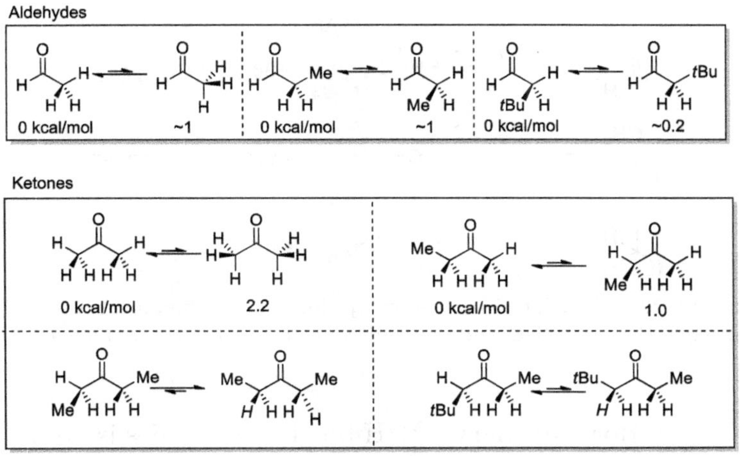

Figure 15.21. Stereoelectronic preference for the eclipsed conformation in carbonyl compounds, analogous to the preference in alkenes.

eclipse the carbonyl (Fig. 15.21). These conformational preferences are general. For example, similar trends, not conforming to a model based on sterics, are observed for alkenes.[42] This conformational effect can be traced back to the slightly higher donor ability of σ_{CH} bonds over that of σ_{CC} bonds in $\sigma \rightarrow \pi^*$ interactions. Delocalization in carbonyls is less balanced and more unidirectional than it is in alkenes: the $\sigma_{CH} \rightarrow \pi^*_{C=O}$ interactions become more important in carbonyls, as the $\pi^*_{C=O}$ acts as a better acceptor due to polarization and the higher electronegativity of oxygen.

It is important to mention that effects that are invisible and "unimportant" for conformational equilibria (e.g., the "fourth level" of delocalization in Fig. 15.20) can play key roles in reactivity. For example, the large $n_O \rightarrow \sigma^*_{C-H}$ interaction in aldehydes evolves, upon the C–H bond scission, into a $2c{-}3e$ bond in acyl radical. The latter effect manifests itself as the source of dramatic weakening of the aldehyde C–H bond dissociation energy (BDE) (~88 kcal/mol) — much smaller than the BDE for a C–H bond in ethane (~111 kcal/mol; Fig. 15.22).[43] The difference is especially striking, since both carbon atoms are sp^2 hybridized and expected to have relatively strong

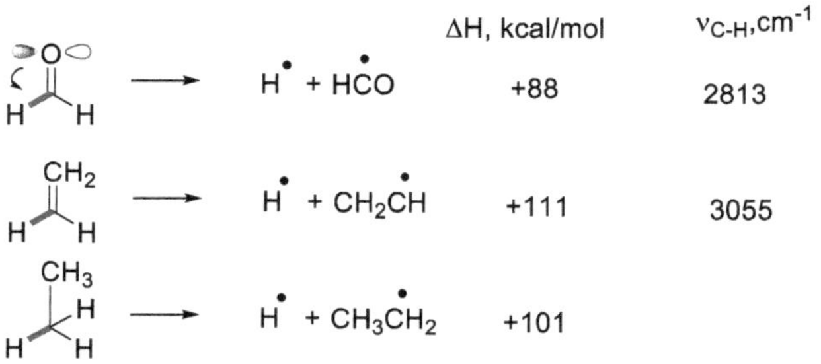

Figure 15.22. The large $n_O \rightarrow \sigma^*_{C-H}$ interactions manifest themselves as decreased C–H bond BDE and red-shifted C–H stretching frequencies in aldehydes.

C–H bonds. However, the C(O)–H bond in aldehydes is even weaker than a typical C–H bond in alkanes. This structural feature and the resulting ease of acyl radical formation has important consequences for the stability and reactivity of aldehydes under radical-forming conditions.

15.8.6 *Competition between hyperconjugation and (re)hybridization/polarization*

Stereoelectronic effects compete with other factors, such as direct changes in bond polarization and hybridization. Although the interplay between conjugation and hybridization/polarization effects can be complex, methods such as NBO analysis allow one to evaluate the relative variations in such effects, identifying the regions where either hybridization or hyperconjugation can dominate. Such analysis can be very helpful in understanding structural and spectroscopic phenomena. For example, the tug-of-war between hybridization and hyperconjugation was used to rationalize the peculiar properties of blue-shifting H-bonds (also referred to as "improper" H-bonds). These supramolecular interactions display the reversal of the usual red-shift behavior of H-bonds in the seemingly paradoxical C–H bond shortening and the blue-shift in the respective IR stretching frequency.[44] This behavior reflects the importance of rehybridization as a structural force controlling X–H bond length in the process of X–H$\cdots$Y bond formation (Fig. 15.23).[45]

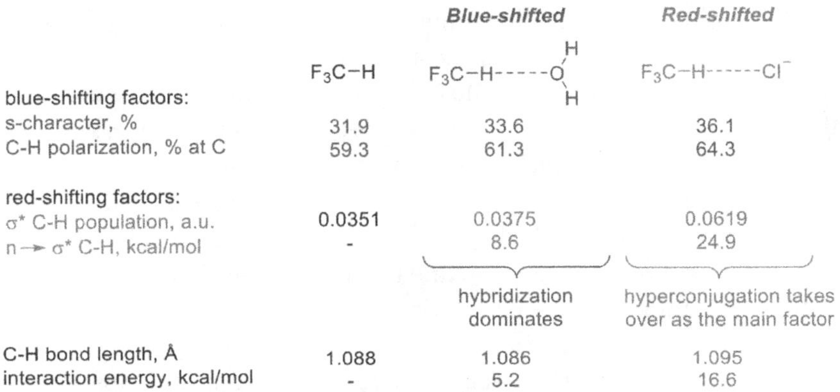
Blue-shifted
Red-shifted
F3C–H
F3C–H-----OH2
F3C–H------Cl−
blue-shifting factors:
s-character, %
31.9
33.6
36.1
C-H polarization, % at C
59.3
61.3
64.3
red-shifting factors:
σ* C-H population, a.u.
0.0351
0.0375
0.0619
n → σ* C-H, kcal/mol
-
8.6
24.9
hybridization dominates
hyperconjugation takes over as the main factor
C-H bond length, Å
1.088
1.086
1.095
interaction energy, kcal/mol
-
5.2
16.6

Figure 15.23. Structures, energies, and NBO analysis of blue- and red-shifted H-bonds at the MP2/6–31+G* level of theory.
Source: From Ref. [35].

The delicate balance of charge transfer and rehybridization becomes apparent when examining the H-bonds between F_3C–H and water (blue-shifted) and F_3C–H and Cl^- (red-shifted, Fig. 15.23).[35] Relative to the "free" C–H bond, the C–H-bonds in the two complexes show an increase in s-character, C–H bond polarization, and σ^*_{C-H} population. Interestingly, the F_3C–H$\cdots$$Cl^-$ H-bond actually shows a greater percentage of s-character (36.1% vs. 33.6%) and a more polarized bond (64.3% vs. 61.3%), relative to the blue-shifted F_3C–H$\cdots$$OH_2$ H-bond. Despite these larger changes in hybridization and polarization, the hyperconjugative charge transfer acts as a dominating force leading to the red-shifted F_3C–H$\cdots$$Cl^-$ H-bond. Interestingly, at greater F_3C–H$\cdots$$Cl^-$ separations, the C–H bond initially undergoes a contraction (that should lead to a blue-shift in the respective IR stretching frequency) and only at the distances where the direct orbital overlap with the chloride lone pair is sufficiently strong to make the $n_{Cl} \rightarrow \sigma^*_{C-H}$ interaction to be sufficiently strong, the C–H bond contraction is replaced by C–H bond elongation. This analysis illustrates that the same factors are involved in the two H-bond types and it is the balance of them that controls the observed shift in the C–H stretching frequency.

To further illustrate the generality of hybridization/hyperconjugation competition, let us discuss the relative strength of C–H bonds in halogen-substituted methanes (Fig. 15.24). These simple molecules display surprisingly rich structural reorganization patterns, depending on the type and number of halogen atoms.

If only vicinal stereoelectronic interactions are considered, one would expect that the C–H bond in fluoromethane would be longer than the C–H bond of methane due to the contribution of $n_F \rightarrow \sigma^*_{C-H}$ interaction. Indeed, such elongation is observed (1.089 vs. 1.091 Å) along with the red-shift of the respective IR-stretching C–H frequency (2992 vs. 2976 cm^{-1}). However, this simple explanation stops working when the second fluorine is introduced or when fluorine is changed to a heavier halogen. In both cases, the C–H bond gets shorter and a blue-shift is observed for the C–H stretching frequencies.

What can explain the red-shift of these frequencies in fluoroalkanes when compared to methane? Furthermore, how can one explain

molecule	ν(CH)[a]	C-H (Å)[b]	C-H hybridization[b]
CH_4	**2992**	1.0892	$sp^{3.00}$
FCH_3	2976	1.0906	$sp^{2.76}$
F_2CH_2	2984	1.0901	$sp^{2.53}$
F_3CH	2990	1.0885	$sp^{2.27}$
$ClCH_3$	3012	1.0862	$sp^{2.76}$
Cl_2CH_2	3025	1.0841	$sp^{2.61}$
Cl_3CH	3034	1.0829	$sp^{2.51}$
$BrCH_3$	3027	1.0852	$sp^{2.70}$
Br_2CH_2	3040	1.0832	$sp^{2.52}$
Br_3CH	3050	1.0815	$sp^{2.42}$

b: calculated at M06-2X/6-311++G(d,p)

(a)

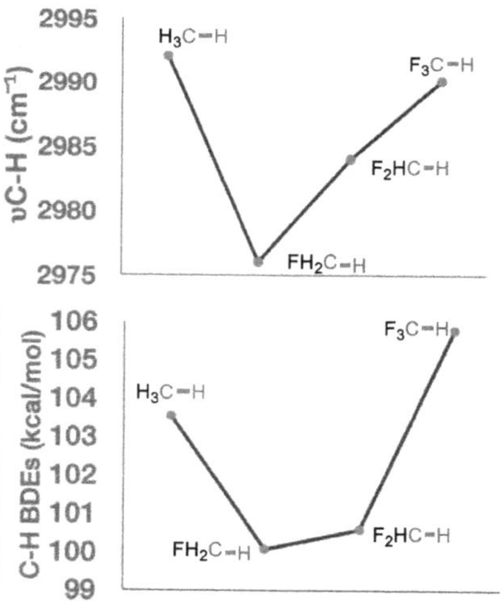

(b)

Figure 15.24. (a) C–H frequencies, lengths, and hybridizations for methane and a series of mono-, di- and tri-substituted halomethanes. (b) Trends for C–H bond BDEs and stretching frequencies for methane and its fluoro-substituted analogues.

the fact that trend is just the *opposite* for chloro- and bromoalkanes? This is a case where a detailed analysis of the interplay of the stereoelectronic effects and hybridization is crucial and only the power of computational analysis can provide an answer.

Below are presented the experimental IR C–H frequencies alongside C–H bond lengths and hybridizations for methane and its halo-substituted analogues. In parallel, we show the trends in C–H BDEs. Note that the energies follow the same general trend as the respective v(C–H) values: decreasing at first and progressively increasing the further fluorine atoms are added. The unusual behavior is unique for fluorine. For the other halogens, progressive blue-shift in the C–H stretching frequency and bond contraction are observed.

A closer analysis of NBO data illustrates how substituting hydrogen for fluorine causes a series of perturbations in the other three C–H bonds (Fig. 15.25). First, due to Bent's rule,[46] carbon has to put more *p*-character into the C–F bond. Because the overall *p*-character is conserved (i.e., there are only 3 p-AOs), the C–H bonds at the same atom have to rehybridize (from sp^3 to $sp^{2.76}$, according to NBO analysis). The increase in *s*-character is expected to make this C–H bond shorter. Thus, the observed elongation relative to the C–H bonds of methane is even more impressive, suggesting the dominating structural effect of hyperconjugative $n_F \rightarrow \sigma^*_{C-H}$ interactions. Indeed,

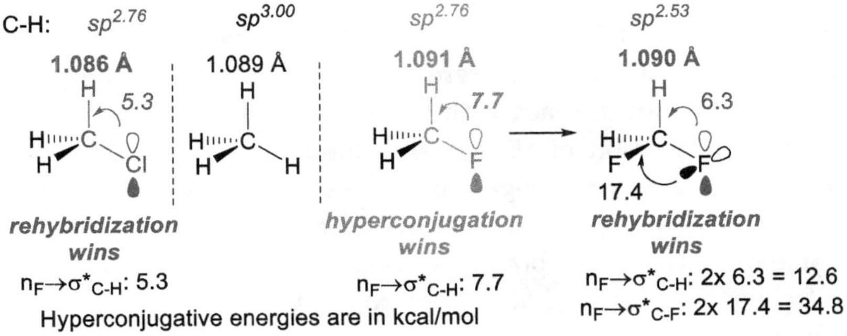

Figure 15.25. Tug-of-war between (re)hybridization and hyperconjugation. Hyperconjugative interactions are given in kcal/mol.

NBO analysis suggests that these are quite strong — worth up to 7.7 kcal/mol. In this case, hyperconjugation wins over hybridization, leading to C–H bond elongation and weakening. The structural effects are fully consistent with the energetic and spectroscopic effects: i.e., decrease in the C–H BDE and red-shift in the IR stretching frequency.

On the other hand, chloromethane follows what is expected based on Bent's rule: the C–H bond is shortened and it shows by being blue-shifted in the IR spectra when compared to methane. This is a case where rehybridization overcomes hyperconjugation. NBO analysis clearly shows that hyperconjugation is 2.4 kcal/mol weaker in comparison to fluoromethane: the energy of $n_{\mathrm{Cl}} \rightarrow \sigma^*_{\mathrm{C-H}}$ interaction is 5.3 kcal/mol. This decrease is due to the greater length of the C–Cl bond which diminishes the π-type $n_{\mathrm{Cl}} \rightarrow \sigma^*_{\mathrm{C-H}}$ vicinal overlap. Because the $\sim$sp$^{2.76}$ hybridization of C–H bonds in fluoro- and chloromethane is almost identical (but different from methane!), the weaker hyperconjugation in chloromethane allows the rehybridization effect to manifest itself fully (chlorine's lone pairs are not as good donors as fluorine's), in making the C–H bond shorter and stronger.

Further replacement of H atoms by fluorines increases the frequency of the C–H stretch, a sign that the bond is getting stronger and shorter. Indeed, the expected bond shortening is reproduced in the calculated structures. This trend is consistent with rehybridization imposed by Bent's rule. For example, carbon uses sp$^{2.27}$ hybrid in the C–F bonds of CHF_3. Furthermore, the additional F atoms introduce another stereoelectronic effect into the picture: the lone pairs can interact with the much stronger $\sigma^*_{\mathrm{C-F}}$ acceptors (i.e., a manifestation of the general anomeric effect). The competition between hyperconjugation ($n_{\mathrm{F}} \rightarrow \sigma^*_{\mathrm{C-H}}$) and the anomeric effect ($n_{\mathrm{F}} \rightarrow \sigma^*_{\mathrm{C-F}}$) combined with further increase on the *s*-character in the C–H bond (rehybridization) bring the stretching frequency for this bond close to that for the C–H bonds of methane.

15.9 Marcus analysis

In this section, we will introduce Marcus theory as a tool for detecting stereoelectronic stabilization of TS structures and show how it was used to identify a new effect that facilitates radical fragmentations (Fig. 15.26). After a sequence of reactions that convert aromatic enynes into α-Sn-substituted naphthalenes into a six-membered cyclic radical, the penultimate species of the overall cascade, the final step involves a C–C bond scission. The efficiency of fragmentation can be enhanced by stabilizing the rational design of radical leaving groups (Fig. 15.26).[47] The stabilization, conveniently evaluated by the equation given in Fig. 15.27, illustrates that the presence of a lone pair adjacent to the radical center stabilizes the radical.

However, fragmentation leading to the formation of the *more* stable α-oxy radical ($\bullet CH_2OMe$) was calculated to be *less* exergonic that the analogous fragmentation that forms the propyl radical. The paradoxical lower thermodynamic driving force for the formation of a *more* stable O-containing radical can be explained by the presence of a selective reactant stabilization. A through-bond (TB) interaction

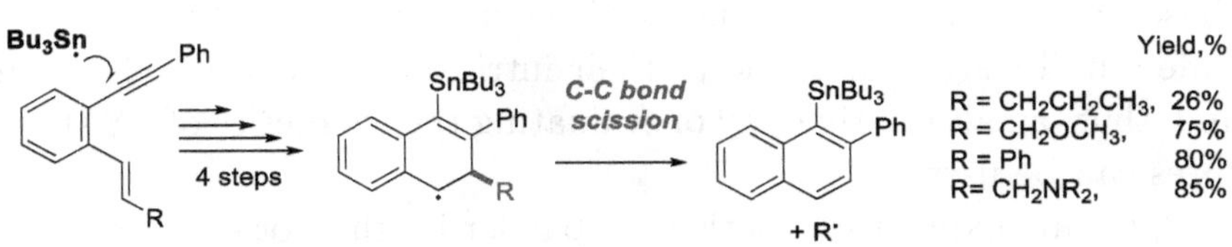

Figure 15.26. Efficiency of fragmentation can be increased by proper substitution at the alkene terminus.

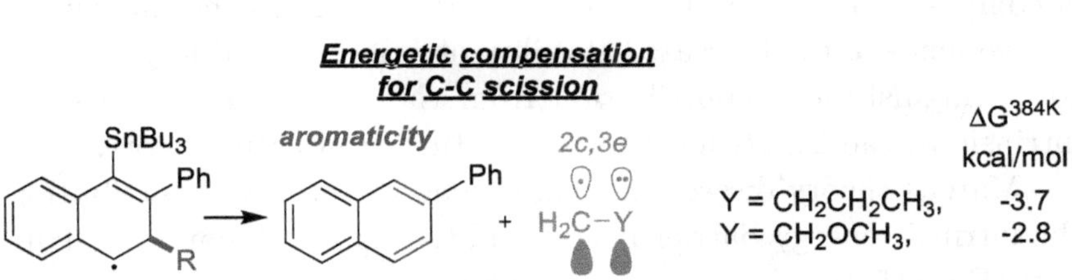

Figure 15.27. Two ways to compensate a C–C scission: aromaticity and the radical stabilizing 2*c*,3*e* interaction.

between the benzylic radical and the lone pair at the δ-position was suggested as the source of this reactant stabilization.[47]

Such TB coupling between two nonbonding orbitals populated with three electrons (the lone pair of X and the radical center) can rationalize the calculated fragmentation exergonicities by providing additional stabilization to the benzylic radical with a β-X group. From a practical point of view, such stabilization should deactivate the reactant and make this fragmentation less efficient. However, in another seeming paradox, the experimental data clearly suggest the opposite — the fragmentations were clearly assisted by the presence of a lone pair next to the radical center of the departing species.

These observations can only be resolved if the presence of oxygen leads to selective TS stabilization that exceeds such stabilization of the reactant. It was suggested that the odd-electron TB communication between the radical and the lone pair through the σ-bridge is increased at the TS and, thus, can serve as the source of additional TS stabilization. NBO analysis of orbital interactions supported this hypothesis.[47] However, the quantitative accuracy of NBO dissections (based on localized dominant Lewis structures as a starting point) suffers in TS structures due to their intrinsically delocalized nature and, thus, an alternative way of evaluating the energetics of TS structures can be useful.

Of course, direct comparison of barrier heights for different reactions is the most straightforward way for making such evaluations. However, for reactions with different exothermicity, such comparison is complicated by thermodynamic contributions to the barrier (the consequence of the increased stability of reaction products). In order to distinguish such contribution from the direct stabilizing effects intrinsic to the TS structure, one can turn to Marcus theory.[48,49]

Marcus theory dissects activation energy into two contributions: the intrinsic energy barrier height and the thermodynamic contribution (Eq. (15.2)).

$$\Delta E^{\ddagger} = \Delta E_0^{\ddagger} + \frac{1}{2}\Delta E_{rxn} + \Delta E_{rxn}^2/16\Delta E_0^{\ddagger} \qquad (15.2)$$

Stereoelectronic differences in the TS can be identified by examining the intrinsic barrier $\Delta E_0^{\ddagger}$, i.e., the barrier of a thermoneutral process lacking the thermodynamic contributions. The intrinsic barrier can be estimated from the rearranged Eq. (15.3) if both the activation and reaction energies are known

$$\Delta E_0^{\ddagger} = \frac{\Delta E^{\ddagger} - \frac{1}{2}\Delta E_{rxn} + \sqrt{(\Delta E^{\ddagger})^2 - \Delta E^{\ddagger}\Delta E_{rxn}}}{2} \tag{15.3}$$

To eliminate the complication associated with the difference in the entropic penalties, one can focus on reaction energies ΔE (rather than free energies, ΔG). When thermodynamic contributions to the barrier are removed, significant differences in the TS energies still remain (1–2 kcal/mol for X = CH_2OR and *ca.* 6 kcal/mol for X = CH_2NMe_2). These large effects on the fragmentation barrier originate from electronic communication between the nonbonding orbitals, which weakens the bridging σ-bond in the TS.

The increase in TB interaction through stretched bonds is documented by NBO orbital interaction energies. In the fragmentation process, the energy of the σ^*-antibonding bridge orbital is lowered, decreasing the ΔE_{ij} term for the stabilizing interaction that couples the nonbonding orbitals (i.e., the radical and lone pair). In addition, as the fragmentation progresses, the $\sim$sp^3 σ-bond is transformed into two *p*-orbitals (one π-bonded in naphthalene and the other in a 2*c*–3*e* "half-bond"), increasing overlap between interacting orbitals. Together these interactions are responsible for selective TS stabilization for the fragmentation process.

An analysis of the Marcus theory shows that intrinsic reaction barriers for the fragmentation are decreased by the presence of heteroatoms with lone pairs. This finding illustrates the utility of Marcus dissection for finding and evaluating selective TS stabilization effects. It does not identify what the interaction may be. Other theoretical methods such as MO and NBO analyses suggested that such stereoelectronic effect is the TB interaction described above. This interaction serves as a stereoelectronic conduit for the observed acceleration of the bond scission.

In summary, one can apply the Marcus theory to analyze stereoelectronic effects in TS structures with the following sequence of steps:

(1) Calculate the activation barrier $\Delta E^{\ddagger}$ and the reaction energy of the process, ΔE_{rxn}.
(2) Use the $\Delta E^{\ddagger}$ and ΔE_{rxn} values to calculate Marcus intrinsic barriers $\Delta E_0^{\ddagger}$ using Eq. (15.3).
(3) Because $\Delta E_0^{\ddagger}$ values are free from thermodynamic and entropic contributions, they reflect the electronic component of reactivity. Comparison of intrinsic barriers for different substituents allows the detection of the stereoelectronic effects in the TSs.
(4) This approach can be complemented with NBO analysis of reactants, TSs, and products to gauge the relative importance of individual interactions and evaluate their evolution along the reaction path.

The caveat in using such Marcus analysis is that the bonds broken and formed in reactions should be similar. For two bonds of drastically different nature with different shapes of potential energy surfaces, the accuracy of simple Marcus description of such surfaces as two parabolas may become inadequate. For example, the use of Marcus theory for comparison cycloaddition to alkenes vs. alkynes is expected to suffer from a systematic error associated with the different intrinsic strengths of π-bonds in these two functionalities.[50,51]

15.10 Distortion analysis — Interaction energies

Another useful method for detecting stereoelectronic effects is distortion analysis (see Chapter 13).[52] Distortion analysis dissects the activation barrier for cycloadditions (or other bimolecular reactions) into distortion and interaction energies. Distortion describes the energy penalty for adopting the TS geometry by the reactants, whereas interaction energy reflects energy lowering due to covalent and noncovalent interaction between the reactants in the TS geometry. Here we will only summarize the key points needed for the application of this approach toward stereoelectronic analysis.

The Distortion–Interaction model is based on the dissection of the activation barrier, $\Delta E^{\ddagger}$, in two components:

(1) The energy to distort reactants to the geometry they adopt in the TS, ΔE_{dist}. This can be interpreted as the ***penalty*** the reactants have to pay to adopt the TS geometry $\Delta E_{\text{dist}} > 0$.
(2) The interaction energy, ΔE_{int}, that ***lowers*** the overall energy due to covalent and noncovalent interactions during the TS $\Delta E_{\text{int}} < 0$.

For bimolecular reactions, stereoelectronic effects that lead to TS stabilization are often manifested via an increase of stabilizing interaction energy between the fragments. Such effects can be detected as considerable deviations in the correlation of distortion energy with full activation barriers. The activation barrier $\Delta^{E\ddagger}$ can be calculated as:

$$\Delta E^{\ddagger} = \Delta E_{\text{dist}} + \Delta E_{\text{int}} \tag{15.4}$$

In summary, one can apply the Distortion–Interaction model to analyze stereoelectronic effects in TS structures with the following methodology for bimolecular processes (i.e., cycloadditions):

(1) One needs to calculate the geometries of reactants and TS using an appropriate theoretical method.
(2) One calculates the energy of each reactant separately in their TS specific geometries with two single-point calculations (one for each distorted reactant).
(3) ΔE_{dist} is calculated by subtracting the energy of the two fragments of the TS by the energy of the separated reactants in the ground state:

$$\Delta E_{\text{dist}} = \Delta E_{\text{reactants}}^{\text{TSgeometry}} - \Delta E_{\text{reactants}}^{\text{GSgeometry}} \tag{15.5}$$

For example, one can lower the activation barrier of a system by destabilizing one of the reactants, therefore paying less distortion penalties.

(4) Now one can simply calculate ΔE_{int} by the following difference: $\Delta E_{\text{int}} = \Delta E^{\ddagger} - \Delta E_{\text{dist}}$.
(5) Since ΔE_{int} directly reflects the electronic interactions in the TS, the impact of different stereoelectronic effects can be evaluated by this term. A straightforward approach is to vary different substituents in both reactants and see how that changes the activation barrier. Then one can compare how much of those variations come from stereoelectronics and design solutions for a particular problem.

As an example, let us illustrate how distortion analysis can be used to identify specific orbital stabilizing effects in the TS of the noncatalyzed alkyne–azide cycloaddition (Fig. 15.28).[53]

Figure 15.28(a) illustrates that activation energy increases linearly with the amount of total distortion energy require to "bend" the alkyne to its geometry in the TS. Increased strain activates the alkyne by forcing it to a geometry that is closer to the one it would have during the TS. In other words, the alkyne can "pre-pay" the distortion penalty to reach the TS geometry. However, the simple

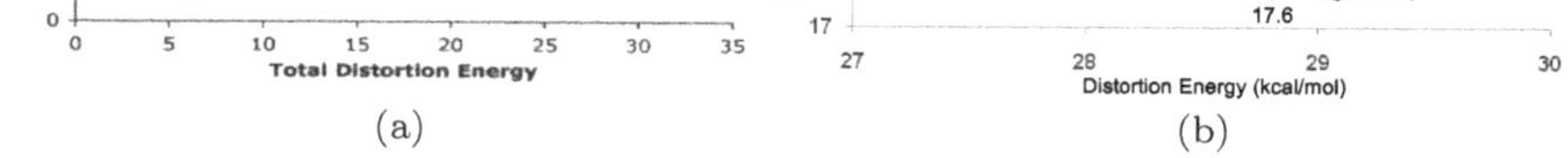

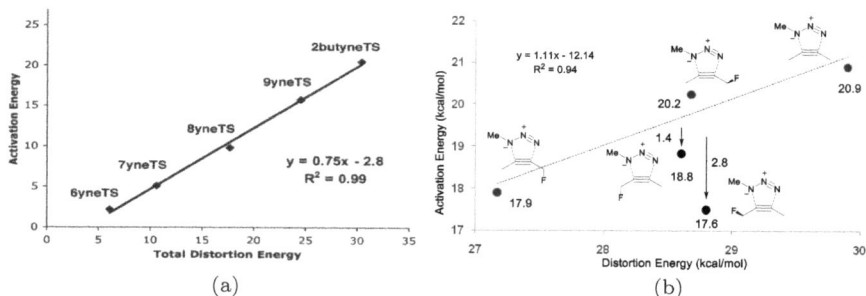

Figure 15.28. (a) Correlation between activation and distortion energies of (cyclo)alkynes in cycloaddition reactions. Energies in kcal/mol. (b) Correlation between activation and distortion energies of (α-fluorosubstitued)-2-butynes in cycloaddition reactions. Deviations from the straight line indicate additional stabilizing interactions in the TS. (Calculations at the B3LYP/6-31G(d) level.)

Source: Reprinted with permission from Refs. [53a, 54], respectively.

correlations of distortion and activation energies become more complicated when stereoelectronic effects are present.

As an example, let us analyze Fig. 15.28(b), where the same cycloaddition reaction shows different behavior. It is clear that data for several substrates with a σ-acceptor at the propargylic carbon do not follow the linear correlation. Two points deviate from the linear trend most significantly, with activation energies much smaller than expected from the respective distortion energies. This behavior is evidence of stereoelectronic effects that stabilize the TS. We will discuss the specific nature of these effects later in this section.

When viewed through the prism of the distortion analysis, these strategies for TS stabilization can be quantified (Table 15.1) as increases in "interaction energy." Increase in this interaction relative to the isolated alkyne is mostly due to an increase in the population of donor π^* orbital which is empty in the reactant but gains some electronic density in the TS.[53] Because the above interaction is manifested to a greater degree in the TS, this effect fits the classic definition of TS stabilization — the ideal scenario for selective reaction acceleration.

However, the utility of distortion analysis does not stop here. Table 15.1 also illustrates the decrease in the "distortion energy" of the alkyne component for the fluoro-substituted substrates. So, which stereoelectronic effects are manifested in the increased interaction energy and in the decreased distortion penalty?[53a]

15.10.1 *Stereoelectronic origin of decrease in distortion energies*

The formation of both 1,4- and 1,5-isomers of the triazole product is facilitated via hyperconjugative assistance to alkyne bending and C$\cdots$N bond formation provided by antiperiplanar σ-acceptors at the propargylic carbons, diminishing the distortion energy. As the alkyne is bent, the in-plane π-bond becomes a better donor, therefore making hyperconjugative interactions with the appropriately positioned substituents stronger. This reinforcement is illustrated by the symmetric bending of 2-butyne relative to fluoro-2-butyne (Fig. 15.29), with a decreasing energy cost of bending in the following

Table 15.1. Activation, reaction, distortion, and interaction energies for 2-butyne and 1-fluoro-2-butyne TSs at the B3LYP/6-31G(d) level of theory.

	Distortion energy-alkyne (kcal/mol)	Distortion energy-azide (kcal/mol)	Distortion energy total (kcal/mol)	Interaction energy, kcal/mol)	Activation energy (kcal/mol)	Reaction energy (kcal/mol)
2-butyne Gas phase	10.12	19.78	29.90	−9.01	20.89	−67.8
CPCM $(H_2O)^{a,b}$	10.11	20.14	30.26	−7.66	22.62	−70.4
1,5-*app* Gas phase	9.25	19.41	28.66	−9.83	18.83	−69.8
CPCM $(H_2O)^{a,b}$	9.50	19.75	29.25	−8.41	20.84	−71.2
1,5-*gauche* Gas phase	10.23	18.52	28.75	−11.20	17.55	−69.8
CPCM $(H_2O)^{a,b}$	10.10	18.57	28.67	−8.52	20.15	−71.2
1,4-*app* Gas phase	8.74	18.43	27.17	−9.27	17.90	−70.8
CPCM $(H_2O)^{a,b}$	8.93	18.75	27.68	−7.77	19.91	−72.3
1,4-*gauche* Gas phase	10.16	18.52	28.68	−8.44	20.24	−70.8
CPCM $(H_2O)^{a,b}$	10.06	18.85	28.91	−7.71	21.20	−72.3

Notes: [a]B3LYP/6-31G(d) geometry.
[b]Radii = UA0.

order: 2-butyne > 1-fluoro-2-butyne (gauche) > 1-fluoro-2-butyne (synperiplanar) > 1-fluoro-2-butyne (antiperiplanar).

15.10.2 *Stereoelectronic origin of increase in interaction energies*

The above stabilizing effect increases even further in the full TS, where the azide is close to the bent alkyne. In the TS, the propargylic

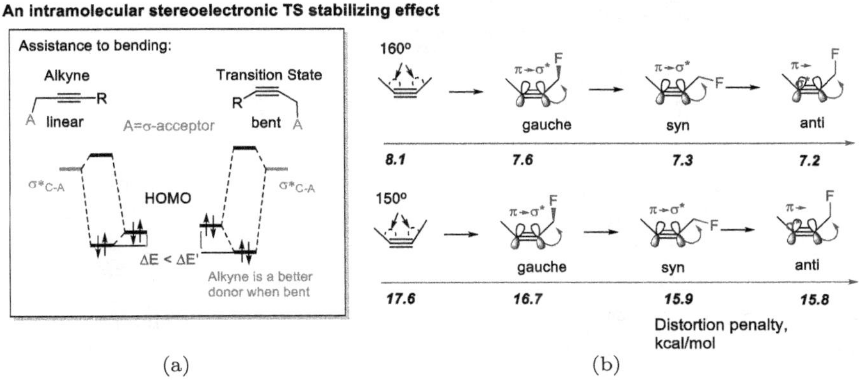

Figure 15.29. (a) Stereoelectronic basis for assistance to alkyne bending utilized in TS stabilization in azide-alkyne cycloadditions. (b) Symmetric bending scan of butyne and 2-fluorobutyne in the gauche, synperiplanar, and antiperiplanar conformations.

acceptor facilitates bond formation (Fig. 15.30) due to the interaction of a strategically placed antiperiplanar C–F bond with the reacting π-system, σ^*_{C-F}. The donor ability of the distorted "incipient" bond (i.e., the "hyperconjugative assistance to bond formation"[54]) decreases the activation energy by 1.4 kcal/mol. NBO analysis suggests that π^* of the alkyne (i.e., its LUMO) becomes populated as the azide approaches. The high-energy π^*-orbital is usually empty, but in the cycloaddition TS, it has sufficient electron density (transferred from the azide in the bond forming interaction) to serve as a very strong donor. Figure 15.30 illustrates that orbital mixing in the TS leads to larger stabilization when delocalization into a propargylic acceptor ($\pi + \pi^* \rightarrow \sigma^*_{CF}$) is conceivable. This qualitative analysis is fully supported by NBO dissection of the individual stereoelectronic contributions.

In the other case, where an even greater barrier lowering is observed in Fig. 15.29 ($\sim$2.8 kcal/mol), fluorine plays a different stereoelectronic role: a 1,5-gauche fluoro substituent can interact with the C–H bond of the methyl azide moiety via a stabilizing Me$\cdots$F interaction that can be considered as a weak C–H$\cdots$X hydrogen bond with a considerable electrostatic component.

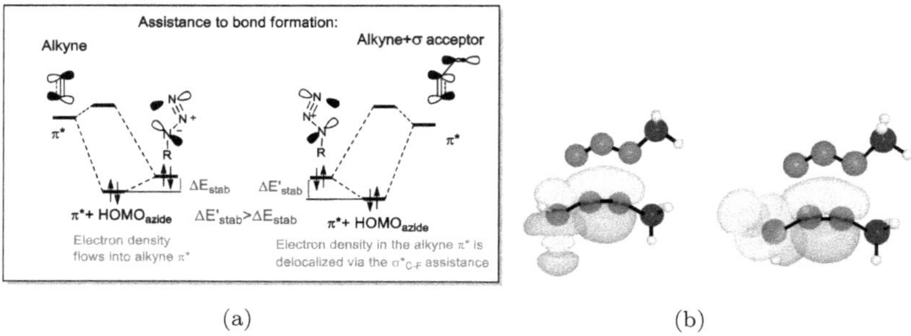

Figure 15.30. (a) Electronic basis for TS stabilization. The increase in the alkyne π^* population due to the C$\cdots$N bond-forming interaction complements the effect of propargylic acceptor on alkyne bending. (b) NBO plots for orbital interactions between the propargylic σ-acceptors and the reacting in-plane alkyne π-bond in the cycloadditions TS for the antiperiplanar (left) and gauche (right) orbital arrangements.

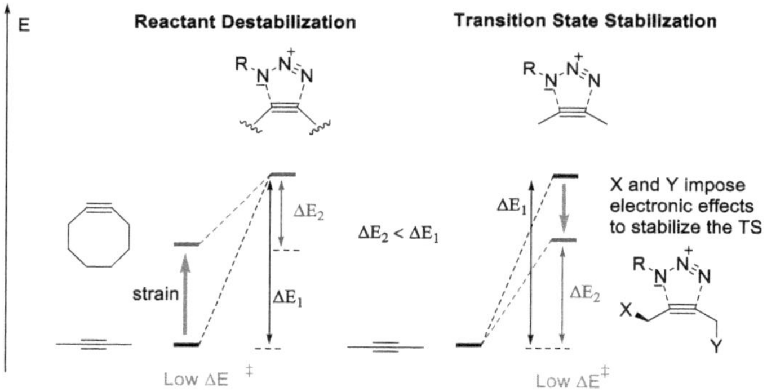

Figure 15.31. Comparison of alternative strategies for acceleration of copper-free click reactions.

Comparison of reactant destabilization by strain and TS stabilization by stereoelectronic effects suggest two complementary strategies toward faster noncatalyzed click cycloadditions (Fig. 15.31).

Combined with strain activation, the stereoelectronic strategy can incorporate several components. As a result, one can design a system that exploits both distortion and stereoelectronic effects for interaction in the TS. The combination of several strategies was estimated to provide greater than 10,000-fold acceleration, upon the transition from the regular cyclooctyne to its electronically activated derivatives.[53b]

15.11 Conclusions and practical considerations

We have described several computational approaches for the detection and quantification of stereoelectronic orbital interactions using computational methods: from conformational analysis to thermochemistry, from wavefunction dissections to global energy components revealed by distortion analysis and Marcus theory. We hope that with the help of these quantitative techniques, the large variety of stereoelectronic effects suggested in the literature can be critically evaluated, bringing a new level of computational rigor in this field.

References

1. (a) Deslongchamps, P. (1984). *Stereoelectronic Effects in Organic Chemistry* (Pergamon Press, Oxford); (b) Kirby, A. J. (1983). *The Anomeric Effect and Related Stereoelectronic Effects at Oxygen* (Springer-Verlag, Berlin, New York).
2. Alabugin, I. V. and Manoharan, M. (2004). Effect of double hyperconjugation on the apparent donor ability of σ-bonds: Insights from the relative stability of δ-substituted cyclohexyl cations, *J. Org. Chem.*, 69, pp. 9011–9024.
3. Kirby, A. J. (1996). *Stereoelectronic Effects* (Oxford University Press, Oxford).
4. For definitions see: Wheeler, S. E., Houk, K. N., Schleyer, P. v. R. and Allen, W. D. (2009). A hierarchy of homodesmotic reactions for thermochemistry, *J. Am. Chem. Soc.*, 131, pp. 2547–2560.
5. (a) Hehre, W. J., Ditchfield, R., Radom, L. and Pople, J. A. (1970). Molecular orbital theory of the electronic structure of organic compounds. V. Molecular theory of bond separation, *J. Am. Chem. Soc.*, 92, pp. 4796–4801; (b) Radom, L., Hehre, W. J. and Pople, J. A. (1971). Molecular orbital theory of the electronic structure of organic compounds. VII. Systematic study of energies, conformations, and bond interactions, *J. Am. Chem. Soc.*, 93, pp. 289–300; (c) Hehre, W. J., Radom, L., Schleyer, P. v. R. and Pople, J. A. (1986). *Ab Initio Molecular Orbital Theory* (Wiley-Interscience, New York).
6. Hoffmann, R. (1971). Interaction of orbitals through space and through bonds, *Acc. Chem. Res.*, 4, pp. 1–9.
7. Mohamed, R. K., Peterson, P. W. and Alabugin, I. V. (2013). Concerted reactions that produce diradicals and zwitterions: Electronic, steric, conformational, and kinetic control of cycloaromatization processes, *Chem. Rev.*, 113, pp. 7089–7129.

8. (a) Schottelius, M. J. and Chen, P. (1996). 9,10-Dehydroanthracene: *p*-Benzyne-type biradicals abstract hydrogen unusually slowly, *J. Am. Chem. Soc.*, 118, pp. 4896–4903; (b) Pickard, F. C. IV, Shepherd, R. L., Gillis, A. E., Dunn, M. E., Feldgus, S., Kirschner, K. N., Shields, G. C., Manoharan, M. and Alabugin, I. V. (2006). Ortho effect in the Bergman cyclization: Electronic and steric effects in hydrogen abstraction by 1-substituted naphthalene 5,8-diradicals, *J. Phys. Chem. A*, 110, pp. 2517–2526.
9. Bender, D. M., Peterson, J. A., McCarthy, J. R., Gunaydin, H., Takano, Y. and Houk, K. N. (2008). Cyclopropanecarboxylic acid esters as potential prodrugs with enhanced hydrolytic stability, *Org. Lett.*, 10, pp. 509–511.
10. Rogers, D. W., Matsunaga, N., Zavitsas, A. A., McLafferty, F. J. and Liebman, J. F. (2003). The conjugation stabilization of 1,3-butadiyne is zero, *Org. Lett.*, 5, pp. 2373–2375.
11. Jarowski, P. D., Wodrich, M. D., Wannere, C. S., Schleyer, P. V. R. and Houk, K. N. (2004). How large is the conjugative stabilization of diynes? *J. Am. Chem. Soc.*, 126(46), pp. 15036–15037.
12. Alabugin, I. V., Gilmore, K. and Peterson, P. (2011). Hyperconjugation, *WIREs Comput. Mol. Sci.*, 1, pp. 109–141.
13. (a) Bickelhaupt, F. M. and Baerends, E. J. (2003). The case for steric repulsion causing the staggered conformation of ethane, *Angew. Chem. Int. Ed.*, 42, pp. 4183–4188; See also: (b) Cappel, D., Tullmann, S., Krapp, A. and Frenking, G. (2005). Direct estimate of the conjugative and hyperconjugative stabilization in diynes, dienes, and related compounds, *Angew. Chem. Int. Ed.*, 44, pp. 3617–3620; (c) Fernandez, I. and Frenking, G. (2007). Direct estimate of conjugation and aromaticity in cyclic compounds with the EDA method, *Faraday Discuss.*, 135, pp. 403–421; (d) Fernandez, I. and Frenking, G. (2006). Direct estimate of the strength of conjugation and hyperconjugation by the energy decomposition analysis method, *Chem. Eur. J.*, 12, pp. 3617–3629.
14. (a) Mo, Y., Wu, W., Song, L., Lin, M., Zhang, Q. and Gao, J. (2004). The magnitude of hyperconjugation in ethane: A perspective from *ab initio* valence bond theory, *Angew. Chem. Int. Ed.*, 43, pp. 1986–1990; (b) Mo, Y. and Gao, J. (2007). Theoretical analysis of the rotational barrier of ethane, *Acc. Chem. Res.*, 40, pp. 113–119.
15. Weinhold, F. and Carpenter, J. E. (1988). Some remarks on non-orthogonal orbitals in quantum chemistry, *J. Mol. Struct. (THEOCHEM)*, 165, pp. 189–202.
16. Corcoran, C. T. and Weinhold, F. (1980). Antisymmetrization effects in bond-orbital models of internal rotation barriers, *J. Chem. Phys.*, 72, pp. 2866–2868.
17. (a) Weinhold, F. (2001). Chemistry: A new twist on molecular shape, *Nature*, 411, pp. 539–540; (b) Weinhold, F. (2003). Rebuttal to the Bickelhaupt-Baerends case for steric repulsion causing the staggered conformation of ethane, *Angew. Chem. Int. Ed.*, 42, pp. 4188–4194.

18. Gross, K. C. and Seybold, P. G. (2001). Substituent effects on the physical properties and pKa of phenol, *Int. J. Quantum Chem.*, 85, pp. 569–579.
19. Weinhold, F. and Schleyer, P. v. R. (1998). *Encyclopedia of Computational Chemistry* (Wiley, New York), p. 3, 1792.
20. Reed, A. E. and Weinhold, F. (1985). Natural localized molecular-orbitals, *J. Chem. Phys.*, 83, pp. 1736–1740.
21. Reed, A. E., Curtiss, L. A. and Weinhold, F. (1988). Intermolecular interactions from a natural bond orbital, donor–acceptor viewpoint, *Chem. Rev.*, 88, pp. 899–926.
22. Alabugin, I. V. and Zeidan, T. A. (2002). Stereoelectronic effects and general trends in hyperconjugative acceptor ability of s bonds, *J. Am. Chem. Soc.*, 124, pp. 3175–3185.
23. (a) Badenhoop, J. K. and Weinhold, F. (1997). Natural bond orbital analysis of steric interactions, *J. Chem. Phys.*, 107, pp. 5406–5422; (b) Badenhoop, J. K. and Weinhold, F. (1999). Natural steric analysis of internal rotation barriers, *Int. J. Quantum Chem.*, 72, pp. 269–280.
24. Weisskopf, V. F. (1975). Of atoms, mountains, and stars: A study in qualitative physics, *Science*, 187, pp. 605–612.
25. Foster, J. M. and Boys, S. F. (1960). Canonical configurational interaction procedure, *Rev. Mod. Phys.*, 32, pp. 300–302.
26. Edmiston, C. and Ruedenberg, K. (1963). Localized atomic and molecular orbitals, *Rev. Mod. Phys.*, 35, pp. 457–465.
27. Pipek, J. and Mezey, P. G. (1989). A fast intrinsic localization procedure applicable for *ab initio* and semiempirical linear combination of atomic orbital wave functions, *J. Chem. Phys.*, 90, pp. 4916–4926.
28. Zubarev, D. Y. and Boldyrev, A. I. (2008). Developing paradigms of chemical bonding: Adaptive natural density partitioning, *Phys. Chem. Chem. Phys.*, 10, pp. 5207–5217.
29. (a) Bickelhaupt, F. M., Baerends, E. J. and Evert, J. (2000). Kohn-Sham density functional theory: Predicting and understanding chemistry, *Rev. Comput. Chem.*, 15, pp. 1–86; (b) te Velde, G., Bickelhaupt, F. M., Baerends, E. J., Van Gisbergen, S. J. A., Snijders, J. G. and Ziegler, T. (2001). Chemistry with ADF, *J. Comp. Chem.*, 22, pp. 931–967; See also (c) Ziegler, T. and Rauk, A. (1977). On the calculation of bonding energies by the Hartree Fock Slater method. I. The transition state method, *Theo. Chim. Acta*, 46, pp. 1–10; (d) Morokuma, K. (1971). Molecular orbital studies of hydrogen bonds III. C=O...H-O hydrogen bond in $H_2CO...H_2O$ and $H_2CO...2H_2O$, *J. Chem. Phys.*, 55, pp. 1236–1244.
30. Fernandez, I. and Frenking, G. (2007). Direct estimate of the conjugative and hyperconjugative stabilization in diynes, dienes, and related compounds, *Faraday Discuss.*, 135, pp. 403–421.
31. Mo, Y. (2006). Intramolecular electron transfer: Computational study based on the orbital deletion procedure (ODP), *Curr. Org. Chem.*, 10, pp. 779–790.
32. (a) Mo, Y. and Peyerimhoff, S. D. (1998). Theoretical analysis of electronic delocalization, *J. Chem. Phys.*, 109, pp. 1687–1697; (b) Mo, Y., Zhang, Y. and

Gao, J. (1999). A simple electrostatic model for trisilylamine: Theoretical examinations of the n $\rightarrow$ σ^* negative hyperconjugation, pπ $\rightarrow$ dπ bonding, and stereoelectronic interaction, *J. Am. Chem. Soc.*, 121, pp. 5737–5742; (c) Mo, Y., Gao, J. and Peyerimhoff, S. D. (2000). Energy decomposition analysis of intermolecular interactions using a block-localized wave function approach, *J. Chem. Phys.*, 112, pp. 5530–5538; (d) Mo, Y., Subramanian, G., Ferguson, D. M. and Gao, J. (2002). Cation-π interactions: An energy decomposition analysis and its implication in δ-opioid receptor-ligand binding, *J. Am. Chem. Soc.*, 124, pp. 4832–4837; (e) Mo, Y., Song, L., Wu, W. and Zhang, Q. (2004). Charge transfer in the electron donor–acceptor complex BH_3NH_3, *J. Am. Chem. Soc.*, 126, pp. 3974–3982; (f) Mo, Y. and Gao, J. (2000). An *ab initio* molecular orbital-valence bond (MOVB) method for simulating chemical reactions in solution, *J. Phys. Chem.*, 104, pp. 3012–3020; (g) Mo, Y. and Gao, J. (2000). *Ab initio* QM/MM simulations with a molecular orbital-valence bond (MOVB) method: Application to an SN2 reaction in water, *J. Comp. Chem.*, 21, pp. 1458–1469; (h) Mo, Y. (2004). Resonance effect in the allyl cation and anion: A revisit, *J. Org. Chem.*, 69, pp. 5563–5567.

33. Mo, Y. (2010). Computational evidence that hyperconjugative interactions are not responsible for the anomeric effect, *Nat. Chem.*, 2, pp. 666–671.
34. Fulton, R. L. (1993). Sharing of electrons in molecules, *J. Phys. Chem.*, 97, pp. 7516–7529.
35. (a) Coulson, C. A. (1939). The electronic structure of some polyenes and aromatic molecules. VII. bonds of fractional order by the molecular orbital method, *Proc. R. Soc. A*, 158, pp. 413–428; (b) Wiberg, K. B. (1968). Application of the Pople-Santry-Segal CNDO method to the cyclopropylcarbinyl and cyclobutyl cation and to bicyclobutane, *Tetrahedron*, 24, pp. 1083–1096; (c) Mayer, I. (1983). Charge, bond order and valence in the *ab initio* SCF theory, *Chem. Phys. Lett.*, pp. 270–274.
36. Bader, R. F. W. (1990). *Atoms in Molecules: A Quantum Theory* (Oxford University Press, Oxford).
37. (a) Becke, A. D. and Edgecombe, K. E. (1990). A simple measure of electron localization in atomic and molecular systems, *J. Chem. Phys.*, 92, pp. 5397–5403; (b) Silvi, B. and Savin, A. (1994). Classification of chemical bonds based on topological analysis of electron localization functions, *Nature*, 371, pp. 683–686.
38. For a more detailed analysis of saturated heterocycles, see Alabugin, I. V. (2000). Stereoelectronic interactions in cyclohexane, 1,3-dioxane, 1,3-oxathiane, and 1,3-dithiane: W-effect, $\sigma_{C-X} \leftrightarrow \sigma^*_{C-H}$ interactions, anomeric effect — what is really important? *J. Org. Chem.*, 65, pp. 3910–3919.
39. Gomes, G. P., Vil', V., Terent'ev, A. and Alabugin, I. V. (2015). Stereoelectronic source of the anomalous stability of bis-peroxides, *Chem. Sci.*, 6, pp. 6783–6791.
40. For a similar analysis, see Rzepa, H. The conformation of acetaldehyde: A simple molecule, a complex explanation? http://doi.org/10.15200/winn.142795.56009.

41. (a) Guirgis, G. A., Drew, B. R., Gounev, T. K. and Durig, J. R. (1998). Conformational stability and vibrational assignment of propanal, *Spectrochim. Acta Mol. Biomol.*, 54, pp. 123–143; (b) Karabastos, G. J. and Hsi, N. (1965). Structural studies by nuclear magnetic resonance. X. conformations of aliphatic aldehydes, *J. Am. Chem. Soc.*, 87, pp. 2864–2870.
42. (a) Jalbouta, A. F., Basso, E. A., Pontes, R. M. and Das, D. (2004). Hyperconjugative interactions in vinylic systems: The problem of the barrier to methyl rotation in acetone, *THEOCHEM*, 677, pp. 167–171; (b) Allinger, N. L., Hirsch, J. A., Miller, M. A. and Tyminski, I. J. (1969). Conformational analysis. LXV. Calculation by the Westheimer method of the structures and energies of a variety of organic molecules containing nitrogen, oxygen, and halogen, *J. Am. Chem. Soc.*, 91, pp. 337–343.
43. Blanksby, S. J. and Ellison, G. B. (2003). Bond dissociation energies of organic molecules, *Acc. Chem. Res.*, 36, pp. 255–263.
44. (a) Hobza, P. and Havlas, Z. (2000). Blue-shifting hydrogen bonds, *Chem. Rev.*, 100, pp. 4253–4264; (b) Budesinsky, M., Fiedler, P. and Arnold, Z. (1989). Triformylmethane: An efficient preparation, some derivatives, and spectra, *Synthesis*, 1989, pp. 858–86; (c) Boldeskul, I. E., Tsymbal, I. F., Ryltsev, E. V., Latajka, Z. and Barnes, A. J. (1997). Reversal of the usual v(C-H/D) spectral shift of haloforms in some hydrogen-bonded complexes, *J. Mol. Struc.*, 436, pp. 167–171; (d) Hobza, P., Spirko, V., Selzle, H. L. and Schlag, E. W. (1998). Anti-hydrogen bond in the benzene dimer and other carbon proton donor complexes, *J. Phys. Chem. A*, 102, pp. 2501–2504.
45. Alabugin, I. V., Manoharan, M., Peabody, S. and Weinhold, F. (2003). Electronic basis of improper hydrogen bonding: A subtle balance of hyperconjugation and rehybridization, *J. Am. Chem. Soc.*, 125, pp. 5973–5987.
46. For a more detailed review on orbital hybridization, see: Alabugin, I. V., Bresch, S. and dos Passos Gomes, G. (2014). Orbital hybridization: A key electronic factor in control of structure and reactivity, *J. Phys. Org. Chem.*, 28(2), pp. 147–162. See also: Alabugin, I. V., Bresch, S. and Manoharan, M. (2014). Hybridization trends for main group elements and expanding the Bent's rule beyond carbon: More than electronegativity, *J. Phys. Chem.*, 118, pp. 3663–3677.
47. (a) Mondal, S., Gold, B., Mohamed, R. K. and Alabugin, I. V. (2014). Design of leaving groups in radical C–C fragmentations: Through-bond 2c–3e interactions in self-terminating radical cascades, *Chem. Eur. J.*, 20(28), pp. 8664–8669; (b) Mohamed, R. K., Mondal, S., Gold, B., Evoniuk, C. J., Banerjee, T., Hanson, K. and Alabugin, I. V. (2015). Alkenes as alkyne equivalents in radical cascades terminated by fragmentations: Overcoming stereoelectronic restrictions on ring expansions for the preparation of expanded polyaromatics, *J. Am. Chem. Soc.*, 137(19), pp. 6335–6349.
48. (a) Marcus, R. A. (1956). On the theory of oxidation-reduction reactions involving electron transfer. I. *The J. Chem. Phys.*, 24(5), pp. 966–978; (b) Marcus, R. A. (1964). Chemical and electrochemical electron-transfer

theory, *Annu. Rev. Phys. Chem.*, 15(1), pp. 155–196; (c) Marcus, R. A. (1968). Theoretical relations among rate constants, barriers, and Broensted slopes of chemical reactions, *J. Phys. Chem.*, 72(3), pp. 891–899. For alternative models, see also: (d) Evans, M. G., Polanyi, M. (1938). Inertia and driving force of chemical reactions. *Trans. Faraday Soc.*, 34, pp. 11–24; (e) Koeppl, G. W., Kresge, A. J. (1973). Marcus rate theory and the relationship between Brønsted exponents and energy of reaction, *J. Chem. Soc., Chem. Comm.*, 11, pp. 371–373.

49. For the application of Marcus theory to radical reactions, see: (a) Alabugin, I. V. and Manoharan, M. (2005). 5-endo-dig radical cyclizations: "The poor cousins" of the radical cyclizations family, *J. Am. Chem. Soc.*, 127(26), pp. 9534–9545. For potential caveats, see: (b) Osuna, S. and Houk, K. N. (2009). Cycloaddition reactions of butadiene and 1,3-dipoles to curved arenes, fullerenes, and nanotubes: Theoretical evaluation of the role of distortion energies on activation barriers, *Chem. Eur. J.*, 15(47), pp. 13219–13231.
50. Alabugin, I. V. and Gold, B. (2013). "Two functional groups in one package": Using both alkyne π-bonds in cascade transformations, *J. Org. Chem.*, 78, pp. 7777–7784.
51. Osuna, S. and Houk, K. N. (2009). Cycloaddition reactions of butadiene and 1,3-dipoles to curved arenes, fullerenes, and nanotubes: Theoretical evaluation of the role of distortion energies on activation barriers, *Chem. Eur. J.*, 15, pp. 13219–13231.
52. (a) Ess, D. H. and Houk, K. N. (2007). Distortion/interaction energy control of 1,3-dipolar cycloaddition reactivity, *J. Am. Chem. Soc.*, 129, pp. 10646–10647. For examples in cycloadditions: (b) Ess, D. H., Jones, G. O. and Houk, K. N. (2008). Transition states of strain-promoted metal-free click chemistry: 1,3-dipolar cycloadditions of phenyl azide and cyclooctynes, *Org. Lett.*, 10, pp. 1633–1636; (c) Schoenebeck, F., Ess, D. H., Jones, G. O. and Houk, K. N. (2009). Reactivity and regioselectivity in 1,3-dipolar cycloadditions of azides to strained alkynes and alkenes: A computational study, *J. Am. Chem. Soc.*, 131, pp. 8121–8133.
53. (a) Gold, B., Shevchenko, N., Bonus, N., Dudley, G. B. and Alabugin, I. V. (2012). Selective transition state stabilization via hyperconjugative and conjugative assistance: Stereoelectronic concept for copper-free click chemistry, *J. Org. Chem.*, 77, pp. 75–89; (b) Gold, B., Dudley, G. B. and Alabugin, I. V. (2013). Moderating strain without sacrificing reactivity: Design of fast and tunable noncatalyzed alkyne–azide cycloadditions via stereoelectronically controlled transition state stabilization, *J. Am. Chem. Soc.*, 135, pp. 1558–1569.
54. Schoenebeck, F., Ess, D. H., Jones, G. O. and Houk, K. N. (2009). Reactivity and regioselectivity in 1,3-dipolar cycloadditions of azides to strained alkynes and alkenes: A computational study, *J. Am. Chem. Soc.*, 131(23), pp. 8121–8133.

Chapter 16

pK_a Prediction

Yijie Niu and Jeehiun K. Lee

Department of Chemistry and Chemical Biology
Rutgers, The State University of New Jersey, USA

16.1 Introduction

Acidity is a key molecular feature that figures extensively in a wide range of organic reactivity and our understanding of it. In solution, acidity is represented most commonly by its pK_a value. For an acid HA in water, the equilibrium for dissociation can be expressed as

$$HA(aq) + H_2O(aq) \rightarrow A^-(aq) + H_3O^+(aq) \tag{16.1}$$

The equilibrium constant, K_{eq}, is therefore

$$K_{eq} = [A^-][H_3O^+]/[HA][H_2O] \tag{16.2}$$

Generally, since the concentration of water is ~55.5 M, the K_a used in pK_a calculations is defined as

$$K_a = 55.5K_{eq} = [A^-][H_3O^+]/[HA] \tag{16.3}$$

The measurement of pK_a is often straightforward, but for many organic molecules of interest, with pK_a values higher than that of water, the experimental determination can be more complex. Therefore, much work has gone into pK_a prediction.[1–9]

Here, we will focus on practical methods by which experimental organic chemists might calculate pK_a with some accuracy. First, we will review methodology and then we will highlight some organic chemical examples from the recent literature. Finally, we will proffer some practical advice on using some pK_a methods.

16.2 Methodology

16.2.1 *First-principles methods*

"First-principles" computations refer to those which utilize the relationship between the Gibbs free energy change of the acid dissociation reaction and K_a

$$\Delta G^0 = -\mathrm{RT}\ln K_a = -2.303\mathrm{RT}\log_{10} K_a \tag{16.4}$$

We will review two main strategies within the first-principles method, with and without thermodynamic cycles.

16.2.1.1 *First-principles computations using thermodynamic cycles*

Most often, this type of calculation utilizes a thermodynamic cycle, a common example of which is shown in Fig. 16.1.[10]

Thermodynamic cycles are often invoked because accessible methodology (Hartree–Fock (HF) or Density Functional Theory (DFT); continuum solvation) does not yield accurate ΔG_{aq} values.[11] The gas-phase values of $G(\mathrm{HA}_{(g)})$ and $G(\mathrm{A}^-_{(g)})$ can be calculated using quantum mechanical methods. Most commonly, $G(\mathrm{HA}_{(aq)})$ and $G(\mathrm{A}^-_{(aq)})$ are calculated using an "implicit solvation" method, or more specifically, continuum solvation models[1,2,6,7] Explicit solvation can also be used, but we will focus on the use of implicit solvation, which

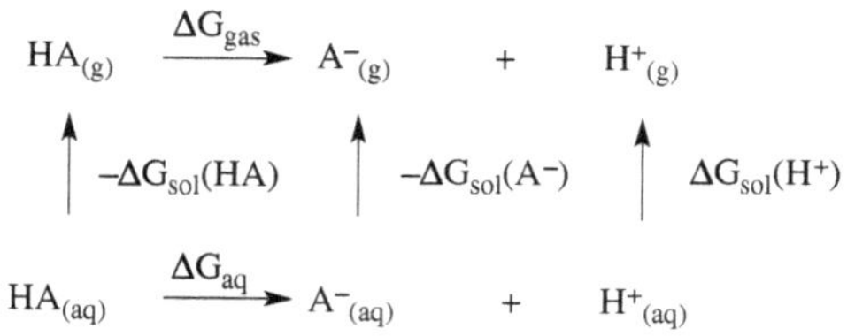

Figure 16.1. Typical thermodynamic cycle used for pK_a calculations.

is practically simpler than explicit methods (see Chapter 4 for a detailed description of solvation methods).

The only value that cannot be calculated with quantum mechanical methods is $\Delta G_{sol}(H^+)$. This problem has two main solutions: a relative pK_a method and an absolute pK_a method.

The relative pK_a method uses a known experimentally determined pK_a value of a "reference molecule." In this method, because the pK_a of the reference acid (HA') is known, then two thermodynamic cycles can be set up and subtracted to eliminate $\Delta G_{sol}(H^+)$. The pK_a of the unknown HA then is simply[1,9,12]

$$\begin{aligned} pK_a(HA) = pK_a(HA') + [&G_g(A^-) - G_g(A'^-) - G_g(HA) + G_g(HA') \\ &+ \Delta G_{sol}(A^-) - \Delta G_{sol}(A'^-) - \Delta G_{sol}(HA) \\ &+ \Delta G_{sol}(HA')]/RT\ln(10) \end{aligned} \tag{16.5}$$

"Relative pK_a" calculations are known for their accuracy (within a few tenths of a kcal/mol) due to the cancellation of errors in the calculated gas- vs. solution-phase values.[1,2,6,7,9,12] The method, however, is limited by the need for a reference molecule that is ideally close in acidity and structure to the unknown.[13,14] Therefore, sometimes a second "thermodynamic cycle" method is used, called the "absolute pK_a calculation." In this, there is no reference acid, so one needs the $G_g(H^+)$ and $\Delta G_{sol}(H^+)$ values. The former appears to be quite well-known, at -6.28 kcal/mol.[9] However, much discussion surrounds the correct value for $\Delta G_{sol}(H^+)$. Recent papers point to a "best value" of -265.6 ± 1 kcal/mol (at 1 M standard state).[15–24]

For the thermodynamic cycle shown in Fig. 16.1, the gas-phase values are, at this point, easily calculated with reasonable accuracy using quantum mechanical methods. Many gas-phase acidities are known (and available on the NIST website) such that assessment of a computational method is relatively straightforward.[25] While high levels of theory such as CBS-QB3 have been used, even MP2 and DFT methods can be quite accurate.[3,9,13,14,19,26–35]

The errors in absolute pK_a calculations thus do not arise from gas-phase calculations, but from the solution-phase calculations.

The simplest way to calculate the solution-phase part of the thermodynamic cycle is to use implicit solvation calculations. These methods do not take explicit solvent molecules into account, but rather model a dielectric continuum to mimic a bulk solvent effect. Dielectric continuum calculations are reasonably accurate for the neutral HA, but not for the anion A^-.[1–3,9,14,15,36–39] This appears to be both due to the fact that the anion solvation is more poorly modeled with bulk solvent than is neutral solvation, and because experimental Gibbs energies of solvation for ions also have error bars of roughly 2–5 kcal/mol, so any models have the same uncertainty.[2,13]

Because of these shortcomings, explicit solvation is sometimes used. One approach is to use an implicit model, but to add discrete solvent molecules ("implicit-explicit" model). These methods have been explored since the late 1970s and have been heretofore reviewed.[2,15,40–43] One of the more common methods is the "cluster-continuum," where the first solvent shell is treated quantum mechanically while the outer shell is treated using dielectric continuum models.[2,16,17,22,40,44,45] Explicit solvent methods that are completely explicit are relatively rare due to the significantly higher computational cost. These methods include DFT and quantum mechanics/molecular mechanics (QM/MM) molecular dynamics simulations.[2,46] Prominent among these is DFT-based molecular dynamics, pioneered by Car and Parrinello.[47]

In terms of continuum solvation models,[2,3,6,7] the most versatile and accurate currently appear to be the "SM" series (SMD, SM8, SM12) of Cramer and Truhlar.[48–50] These methods can be used alone (for implicit solvation), but also give the user the ability to add explicit water molecules for "implicit-explicit" modeling. These methods are also not limited to water; any solvent can be used such that pK_as in organic solvents could be calculated.

16.2.1.2 *First-principles computations without using thermodynamic cycles*

A more recent development is the calculation of pK_as without using a thermodynamic cycle. Such calculations have not historically been accurate because of the difficulty of calculating ΔG_{aq}. Ho's[2,11] group is at the forefront of this trend, having recently shown that MP2 or G3 with SMD, for 117 acids, yielded mean unsigned errors within 0.5 kcal/mol.

One other "first-principles, no cycle" method that has been used to estimate pK_a values is to perform first-principles computations on a data set of known molecules, then to correlate those values to experimental data to yield a relationship that can then be used to calculate the pK_a of an unknown. For example, Bartberger, Fukuto and Houk calculated the pK_a of HNO by first correlating computed gas-phase deprotonation energies plus aqueous solvation energy differences (using continuum solvation) between an acid and conjugate base with experimental pK_a values.[51] This yielded a linear relationship that had a correlation coefficient of 0.95 and standard deviation of 1.0 pK_a units, which they then used to estimate the pK_a of HNO.

16.2.2 *QSAR methods*

Despite considerable progress in solvent modeling, first-principles methods are still often inaccurate, since an error of just 1.4 kcal/mol in the calculation of $\Delta G_{(aq)}$ will result in an error of 1 pK_a unit. To gain greater accuracy, pK_a can alternatively be calculated in a "parametrized" fashion, using a quantitative structure-activity reactivity (QSAR) approach.[1,4,8,52,53]

The QSAR method involves first collecting known, experimental pK_a values for a series of compounds that are related to one's unknown. One then establishes — often using computations — the "features" or "molecular descriptors" that appear to be related to the variations in the pK_a values among the series.

In the simplest form, a single molecular descriptor is used, allowing one to conduct a linear free energy analysis. For organic chemists, an easily understood example would be a Hammett plot, where the molecular descriptor would be the σ value of a substituent on the aromatic ring of a compound whose pK_a is of interest.[54–56] For example, one could examine the pK_a values of a series of *para*-substituted

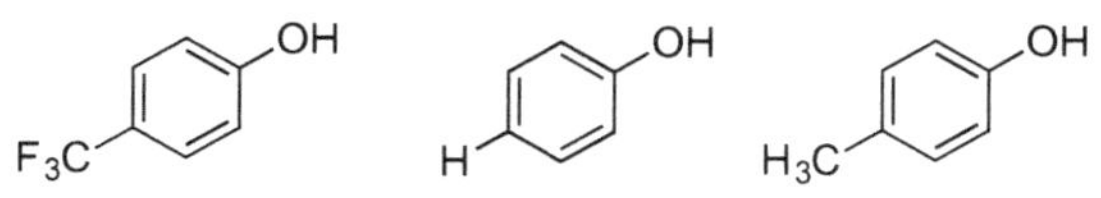

Figure 16.2. *Para*-substituted phenols.

phenols, relative to the parent phenol (Fig. 16.2). Each substituent ($-NO_2$, $-OCH_3$) has a different, unique σ_p value, which was long ago established by Hammett for the effect of each substituent on the ionization of benzoic acids. These values reflect the different effects of the substituents; $-CF_3$ is electron-withdrawing and will lower the pK_a relative to the parent phenol (by stabilizing the phenoxide ion), while methyl is overall electron donating, resulting in a higher pK_a value than that of the parent phenol. A plot of pK_a vs. σ_p should yield a straight line.

Use of multiple molecular descriptors requires a multiple linear regression equation:

$$pK_a = c_0 + c_1X_1 + c_2X_2 + c_3X_3 + \cdots \tag{16.6}$$

where the c_i is constant.[1] While Hammett constants (and the related ones of Taft and Hansch) are used most commonly and widely,[1,54–56] computational methods have made it possible to find many other useful descriptors, including partial atomic charges, atomic properties, and electrostatic potential maxima and minima.[57–64]

The QSAR method also involves a fair amount of statistics, in order to establish how reliable the pK_a linear regression equation is[1,4] Ultimately, the QSAR approach can be quite accurate, even to the point of being more accurate than the experimental data, since the linear regression can even out random errors in that data![65]

16.3 Selected recent examples (survey of literature from last five years)

Next, we cover a few recent examples from the literature highlighting the various methods, applied to problems of interest to organic chemists.

16.3.1 pK_a *in ionic liquids*

Xu *et al.*[66] recently used an implicit-explicit method (cluster continuum) to calculate absolute pK_a values in a commonly used ionic liquid, [Bmim][NTf_2]. The novelty is the ability to calculate acidity in an ionic liquid environment. The authors were able to calculate the pK_a values for a series of benzoic acids and benzenethiols with a mean unsigned error of ≤ 0.5 pK_a units.

16.3.2 pK_a *of amino acids using MD and metadynamics sampling*

Tummanapelli and Vasudevan[67] used *ab initio* Car–Parrinello molecular dynamics simulations to estimate the pK_a values for the 20 amino acids. Because a weak acid does not undergo much dissociation, long simulation times are required before a dissociation event is seen. In situations such as this, metadynamics sampling can be used, which handles infrequent events in molecular dynamics simulations by allowing the system to explore the free-energy surface without "revisiting" regions to which it has already been. The authors utilized metadynamics sampling and were able to calculate the amino acid pK_a values with a mean relative error of 0.2 pK_a units.

16.3.3 pK_a *of fluorinated alkylsulfonylpentafluoroanilides*

Kogel *et al.*[68] used a cluster continuum method (COSMO) to ascertain the pK_a values of fluoro- and perfluoroalkylsulfonylpentafluoroanilides in DMSO and acetonitrile.[68] Interestingly, their computational results reveal that the known experimental pK_a of $HN(SO_2CF_3)_2$ should be interpreted as the pK_a of a $Me_2SO\text{-}H^+\cdot N(SO_2CF_3)_2^-$ contact ion pair.

16.3.4 *Chiral bronsted acids*

Using a first-principles approach with a thermodynamic cycle, Yang *et al.*[69] calculated the acidities of a series of chiral Bronsted acids, including BINOL-derived phosphoric and disulfonic acids and disulfonimides in DMSO. SMD and M06-2X were utilized. Their interest was in providing insight into the catalytic and stereoselective ability of these acids. Since more acidic catalysts are often more effective, the authors also designed/proposed possible new catalysts based on calculated acidities of their parent acids.

16.3.5 *Acidity in nonpolar solvents*

The calculation of acidity in nonpolar solvents (hexane and toluene) was carried out using M06-2X//B3LYP and SMD by Xue *et al.*[70] Specifically, the pK_a shifts of proline when assembled with various hydrogen-bond–donating catalysts were assessed. Assembly was found to increase pK_a by up to over 9 pK_a units. The authors also found a correlation between the shifts and the catalyst acidity. Through this work, the authors hope to lend insight into co-catalyst understanding and development.

16.3.6 *Organometallic complex acidity*

Estrada-Montano *et al.*[71] carried out a combined experimental-theoretical study that included an interesting pK_a calculation.[71] These authors studied the reactions of $[Fe_3(CO)_{12}]$ with the alkyne $HCCSiR_3$ (where R = Me, Ph). In the course of their studies, they wished to characterize the pK_a values of $[(\mu\text{-H})Fe_3(CO)_9CCSi(CH_3)_3]$ and $[(\mu\text{-H})Fe_3(CO)_9CCC(CH_3)_3]$ in acetonitrile. To this end, they conducted a relative pK_a study with a thermodynamic cycle. They first calculated the pK_a values of two known organometallic complexes and found the computations to be reasonably accurate (one was 0.27 pK_a units off, while the other was 1.75 pK_a units different from experiment). Their calculations allowed them to establish that both of their trinuclear iron complexes have similar pK_a values.

16.3.7 pK_a *studies of "green" catalysts*

M06-2X and SMD were used with a thermodynamic cycle to calculate the relative pK_a values of 1,2-benzenedisulfonimide and some of its derivatives.[72] 1,2-Benzenedisulfonimide was found previously by these authors to be a safe, nonvolatile, noncorrosive, and excellent Bronsted acid catalyst for a variety of organic reactions. In an effort to find other derivatives that might be even more effective and acidic, the authors calculated pK_a values and confirmed those calculations with potentiometric titrations of some of the compounds. The calculations indicated some good candidates for improved catalysts. Based on the calculated pK_a values, the authors synthesized a nitro derivative and found it to be a more effective catalyst than the parent 1,2-benzenedisulfonimide.

16.3.8 *QSAR based on one bond length*

Alkorta *et al.*[73] conducted a study comparing the experimental pK_a values of a series of bicyclo[2.2.2]octane and cubane carboxylic acids and computed bond lengths. Their goal was to ascertain whether there would be a linear relationship between pK_a and bond length, so that bond length could be used as a single molecular descriptor for a potential QSAR-type study. They examined the C=O, C–O, and O–H bond lengths at HF/6-31G(d), in the gas phase, and in water polarizable continuum model (PCM). Linear correlations were found between bond length and pK_a; different bond lengths appeared to be better depending on the particular set of conditions (gas phase vs. water; neutral vs. anions). This work builds upon earlier work by the same authors in trying to stablish a relationship between gas-phase bond length and pK_a values in aqueous solution.

16.3.9 *Properties of Fe^{+3} therapeutic chelating agents*

Chen *et al.*[74] conducted a detailed study centered on predicting the hydroxyl pK_a of 3-hydroxy-4-pyridin-4ones (HPOs), which are of interest as therapeutic Fe^{+3} chelating agents. The authors first assessed the various pK_a prediction methods with a training set of 15 HPOs with known pK_a values. The most reliable method was then tested for a further 48 HPOs with known pK_a values (including 11 measured by the authors, and 2 novel HPOs they synthesized).

Both first-principles approaches using a thermodynamic cycle were tested: a relative pK_a, or "proton exchange" method using a reference acid, and an absolute pK_a, or "direct" method, using a $\Delta G_{sol}(H^+)$ value of -265.9 kcal/mol. Several solvent models (CPCM, IEF-PCM, I-PCM, SCI-PCM, SMD) with various radii were also tested. Two commercial QSAR programs were also used; ACD/pK_a DB 12.0 is a linear free energy method while Marvin 5.4.0.1 is a quantitative structure–property relationship method that correlates calculated structural descriptors with experimental pK_a values. Neither QSAR method provided accurate pK_a prediction across all the HPOs. The authors attribute this failure to intramolecular hydrogen bonding and electronic effects. However, using B3LYP/6-31+G(d)/CPCM (Pauling radii, water solvent) with a thermodynamic cycle and reference acid (relative pK_a method) was quite reliable in predicting the pK_as of HPOs.

16.3.10 *Effect of Cu^{+2}-imidazole coordination on* pK_a

These authors tackled the challenging problem of the influence of Cu^{+2} coordination on the pK_a of imidazole.[75] This study was quite exhaustive, covering multiple thermodynamic cycles, with and without reference acids, and with and without explicit solvent molecules, to calculate the pK_a of isolated imidazole and imidazole in Cu^{+2}. All the methods predicted the same trend (coordination would decrease the pK_a) but the shift ranged from -2 to -7 units. The pK_a shifts are much smaller when explicit water is not considered, highlighting the importance of chemical environment on the acidity. Biologically it is also of interest that coordination of Cu^{+2} could potentially lead to a more acidic imidazole that might be deprotonated at physiological pH.

16.4 Practical considerations

For the calculation of aqueous pK_a values, a reasonable approach that should balance accuracy and time would be to use a DFT/SMD method, with a thermodynamic cycle like that shown in Fig. 16.1.

Better accuracy would be achieved with a "relative pK_a" calculation that uses a reference acid. Otherwise, if an absolute pK_a method is used, then one has to be aware of the uncertainty of the $\Delta G_{sol}(H^+)$ value. Suggested approaches would be to either use the "best" value quoted herein for $\Delta G_{sol}(H^+)$ [-265.6 ± 1 kcal/mol] or to use a value that gives a "best match" between calculated and experimental values for a "training set" of similar molecules.[1,19]

References

1. Seybold, P. G. (2015). Computational estimation of pK_a values, *WIREs Comp. Molec. Sci.*, 5, pp. 290–297.
2. Ho, J. (2014). Predicting pK_a in implicit solvents: Current status and future directions, *Aust. J. Chem.*, 67, pp. 1441–1460 and references therein.
3. Alongi, K. S. and Shields, G. C. (2010). Theoretical calculations of acid dissociation constants: A review article, *Annu. Rep. Comput. Chem.*, 6, pp. 113–138.
4. Seybold, P. G. (2012). Quantum chemical-qspr estimation of the acidities and basicities of organic compounds, *Adv. Quant. Chem.*, 64, pp. 84–104.
5. Lee, A. C. and Crippen, G. M. (2009). Predicting pK_a, *J. Chem. Inf. Model.*, 49, pp. 2013–2033.
6. Ho, J. and Coote, M. L. (2010). A universal approach for continuum solvent pK_a calculations: Are we there yet? *Theor. Chem. Acc.*, 125, pp. 3–21.
7. Ho, J. and Coote, M. L. (2011). First-principles prediction of acidities in the gas and solution phase, *WIREs Comp. Molec. Sci.*, 1, pp. 649–660.
8. Zevatskii, Y. E. and Samoilov, D. V. (2011). Modern methods for estimation of ionization constants of organic compounds in solution, *Russ. J. Org. Chem.*, 47, pp. 1445–1467.
9. Shields, G. C. and Seybold, P. (2014). *Computational Approaches for the Prediction of pK_a Values* (CRC Press, Boca Raton, FL, USA).
10. Casanovas, R., Ortega-Castro, J., Frau, J., Donoso, J. and Munoz, F. (2014). Theoretical pK_a calculations with continuum model solvents, alternative protocols to thermodynamic cycles, *Int. J. Quantum Chem.*, 114, pp. 1350–1363.
11. Ho, J. (2015). Are thermodynamic cycles necessary for continuum solvent calculation of pK_as and reduction potentials? *Phys. Chem. Chem. Phys.*, 17, pp. 2859–2868.
12. Toth, A. M., Liptak, M. D., Phillips, D. L. and Shields, G. C. (2001). Accurate relative pK_a calculations for carboxylic acids using complete basis set and Gaussian-n models combined with continuum solvation methods, *J. Chem. Phys.*, 114, pp. 4595–4606.
13. Klicic, J. J., Friesner, R. A., Liu, S.-Y. and Guida, W. C. (2002). Accurate prediction of acidity constants in aqueous solution via density functional

theory and self-consistent reaction field methods, *J. Phys. Chem. A*, 106, pp. 1327–1335.
14. Chipman, D. M. (2002). Computation of pK_a from dielectric continuum theory, *J. Phys. Chem. A*, 106, pp. 7413–7422.
15. Kelly, C. P., Cramer, C. J. and Truhlar, D. G. (2006). Adding explicit solvent molecules to continuum solvent calculations for the calculation of aqueous acid dissociation constants, *J. Phys. Chem. A*, 110, pp. 2493–2499.
16. Bryantsev, V. S., Diallo, M. S. and Goddard III, W. A. (2008). Calculation of solvation free energies of charged solutes using mixed cluster/continuum models, *J. Phys. Chem. B*, 112, pp. 9709–9719.
17. Zhan, C.-G. and Dixon, D. A. (2001). Absolute hydration free energy of the proton from first-principles electronic structure calculations, *Theor. Model. Simul.*, 105, pp. 11534–11540.
18. Noyes, R. M. (1962). Thermodynamics of ion hydration as a measure of effective dielectric properties of water, *J. Am. Chem. Soc.*, 84, pp. 513–522.
19. Jang, Y. H., Goddard III, W. A., Noyes, K. T., Sowers, L. C., Hwang, S. and Chung, D. S. (2003). pK_a values of guanine in water: Density functional theory calculations combined with Poisson-Boltzmann continuum-solvation model, *J. Phys. Chem. B*, 107, pp. 344–357.
20. Liptak, M. D. and Shields, G. C. (2001). Accurate pK_a calculations for carboxylic acids using complete basis set and Gaussian-n models combined with cpcm continuum solvation methods, *J. Am. Chem. Soc.*, 123, pp. 7314–7319.
21. Marcus, Y. (1985). *Ion Solvation* (John Wiley & Sons, Ltd., New York).
22. Tawa, G. J., Topol, I. A., Burt, S. K., Caldwell, R. A. and Rashin, A. A. (1998). Calculation of the aqueous solvation free energy of the proton, *J. Chem. Phys.*, 109, pp. 4852–4863.
23. Reiss, H. and Heller, A. (1985). The absolute potential of the standard hydrogen electrode: A new estimate, *J. Phys. Chem.*, 89, pp. 4207–4213.
24. Tissandier, M. D., Cowen, K. A., Feng, W. Y., Gundlach, E., Cohen, M. H., Earhart, A. D., Coe, J. V. and Tuttle Jr, T. R. (1998). The proton's absolute aqueous enthalpy and Gibbs free energy of solvation from cluster-ion solvation data, *J. Phys. Chem. A*, 102, pp. 7787–7794.
25. Linstrom, P. J. and Mallard, W. G. (2017). NIST Chemistry WebBook, NIST Standard Reference Database Number 69, National Institute of Standards and Technology, Gaithersburg MD, 20899. See: http://webbook.nist.gov. Last accessed October 3, 2017.
26. Richardson, W. H., Peng, C. Y., Bashford, D., Noodelman, L. and Case, D. A. (1997). Incorporating solvation effects into density functional theory: Calculation of absolute acidities, *Int. J. Quantum Chem.*, 61, pp. 207–217.
27. Fu, Y., Liu, L.-P., Li, R.-Q., Liu, R. and Guo, Q.-X. (2004). First-principle predictions of absolute pK_a's of organic acids in dimethyl sulfoxide solution, *J. Am. Chem. Soc.*, 126, pp. 814–822.

28. Merrill, G. N. and Kass, S. R. (1996). Calculated gas-phase acidities using density functional theory: Is it reliable? *J. Phys. Chem.*, 100, pp. 17465–17471.
29. Smith, B. J. and Radom, L. (1995). Gas-phase acidities: A comparison of density functional, MP2, MP4, F4, G2(MP2, SVP), G2(MP2) and G2 procedures, *Chem. Phys. Lett.*, 245, pp. 123–128.
30. Pokon, E. K., Liptak, M. D., Feldgus, S. and Shields, G. C. (2001). Comparison of CBS-QB3, CBS-APNO, and G3 predictions of gas phase deprotonation data, *J. Phys. Chem. A.*, 105, pp. 10483–10487.
31. Pickard, F. C., Dunn, M. E. and Shields, G. C. (2005). Comparison of model chemistry and density functional theory thermochemical predictions with experiment for formation of ionic clusters of the ammonium cation complexed with water and ammonia; atmospheric implications, *J. Phys. Chem. A*, 109, pp. 4905–4910.
32. Pickard, F. C., Griffith, D. R., Ferrara, S. J., Liptak, M. D., Kirschner, K. N. and Shields, G. C. (2006). Ccsd(t), W1 and other model chemistry predictions for gas-phase deprotonation reactions, *Int. J. Quantum Chem.*, 106, pp. 3122–3128.
33. Liptak, M. D. and Shields, G. C. (2005). Comparison of density functional theory predictions of gas-phase deprotonation data, *Int. J. Quantum Chem.*, 105, pp. 580–587.
34. Zhao, Y. and Truhlar, D. G. (2008). The M06 suite of density functionals for main group thermochemistry, thermochemical kinetics, noncovalent interactions, excited states, and transition elements: Two new functionals and systematic testing of four M06 functionals and twelve other functionals, *Theor. Chem. Acc.*, 120, pp. 215–241.
35. Zhao, Y. and Truhlar, D. G. (2008). Density functionals with broad applicability in chemistry, *Acc. Chem. Res.*, 41, pp. 157–167.
36. Cramer, C. J. (2004). *Essentials of Computational Chemistry: Theories and Models* (John Wiley & Sons, Ltd., Chichester).
37. Kelly, C. P., Cramer, C. J. and Truhlar, D. G. (2006). Aqueous solvation free energies of ions and ion-water clusters based on an accurate value for the absolute aqueous solvation free energy of the proton, *J. Phys. Chem. B*, 110, pp. 16066–16081.
38. Saracino, G. A. A., Improta, R. and Barone, V. (2003). Absolute pK_a determination for carboxylic acids using density functional theory and the polarizable continuum model, *Chem. Phys. Lett.*, 373, pp. 411–415.
39. Klamt, A., Eckert, F., Diedenhofen, M. and Beck, M. E. (2003). First principles calculations of aqueous pK_a values for organic and inorganic acids using COSMO-RS reveal an inconsistency in the slope of the pK_a scale, *J. Phys. Chem. A*, 107, pp. 9380–9386.
40. Pliego, J. R. J. and Riveros, J. M. (2001). The cluster-continuum model for the calculation of the solvation free energy of ionic species, *J. Phys. Chem. A*, 105, pp. 7241–7247.

41. Jia, Z.-k., Du, D.-m., Zhou, Z.-y., Zhang, A.-g. and Hou, R.-y. (2007). Accurate pK_a determinations for some organic acids using an extended cluster method, *Chem. Phys. Lett.*, 439, pp. 374–380.
42. Pliego, J. R. Jr. and Riveros, J. M. (2002). Theoretical calculation of pK_a using the cluster-continuum model, *J. Phys. Chem. A*, 106, pp. 7434–7439.
43. Pham, H. H., Taylor, C. D. and Henson, N. J. (2014). Acidity constants and its dependence on solvent selection from first-principles calculations using cluster-continuum models, *Chem. Phys. Lett.*, 610–611, pp. 141–147.
44. Asthagiri, D., Pratt, L. R. and Ashbaugh, H. S. (2003). Absolute hydration free energies of ions, ion–water clusters, and quasichemical theory, *J. Chem. Phys.*, 119, pp. 2702–2708.
45. Mejias, J. A. and Lago, S. (2000). Calculation of the absolute hydration enthalpy and free energy of H^+ and OH^-, *J. Chem. Phys.*, 113, pp. 7306–7316.
46. Sulpizi, M. and Sprik, M. (2010). Acidity constants from DFT-based molecular dynamics simulations, *J. Phys. Condens. Matter.*, 22, pp. 284116–284124.
47. Car, R. and Parrinello, M. (1985). Unified approach for molecular dynamics and density-functional theory, *Phys. Rev. Lett.*, 55, pp. 2471–2474.
48. Marenich, A. V., Cramer, C. J. and Truhlar, D. G. (2009). Universal solvation model based on solute electron density and on a continuum model of the solvent defined by the bulk dielectric constant and atomic surface tensions, *J. Phys. Chem. B*, 113, pp. 6378–6396.
49. Miertuš, S., Scrocco, E. and Tomasi, J. (1981). Electrostatic interaction of a solute with a continuum. A direct utilization of AB initio molecular potentials for the prevision of solvent effects, *Chem. Phys.*, 55, pp. 117–129.
50. Pascual-Ahuir, J. L., Silla, E. and Tuñón, I. (1994). GEPOL: An improved description of molecular surfaces. III. A new algorithm for the computation of a solvent-excluding surface, *J. Comput. Chem.*, 15, pp. 1127–1138.
51. Bartberger, M. D., Fukuto, J. M. and Houk, K. N. (2001). On the acidity and reactivity of HNO in aqueous solution and biological systems, *Proc. Natl. Acad. Sci. USA*, 98, pp. 2194–2198.
52. Tehan, B. G., Lloyd, E. J., Wong, M. G., Pitt, W. R., Montana, J. G., Manallack, D. T. and Gancia, E. (2002). Estimation of pK_a using semiempirical molecular orbital methods. Part 1: Application to phenols and carboxylic acids, *Quant. Struct-Act. Relat.* 21, pp. 457–472 and references therein.
53. Tehan, B. G., Lloyd, E. J., Wong, M. G., Pitt, W. R., Manallack, D. T. and Gancia, E. (2002). Estimation of pK_a using semiempirical molecular orbital methods. Part 2: Application to amines, anilines and various nitrogen containing heterocyclic compounds, *Quant. Struct-Act. Relat.*, 21, pp. 473–485.
54. Hammett, L. P. (1937). The effect of structure on the reactivity of organic compounds. Benzene derivatives, *J. Am. Chem. Soc.*, 59, pp. 96–103.
55. Taft, R. W., Taagepera, M., Abboud, J. L. M., Wolf, J. F., DeFrees, D. J., Hehre, W. J., Bartmess, J. E. and McIver, R. T., Jr. (1978). Regarding the

separation of polarizability and inductive effects in gas- and solution-phase proton-transfer equilibria, *J. Am. Chem. Soc.*, 100, pp. 7765–7767.
56. Hansch, C., Leo, A. and Taft, R. W. (1991). A survey of hammett substituent constants and resonance and field parameters, *Chem. Rev.*, 91, pp. 165–195.
57. Adam, K. R. (2002). New density functional and atoms in molecules method of computing relative pK_a values in solution, *J. Phys. Chem. A*, 106, pp. 11963–11972.
58. Murray, J. S. and Politzer, P. (2011). The electrostatic potential: An overview, *WIREs Comp. Molec. Sci.*, 1, pp. 153–163.
59. Brink, T., Murray, J. S. and Politzer, P. (1991). A relationship between experimentally determined pK_as and molecular surface ionization energies for some azines and azoles, *J. Org. Chem.*, 56, pp. 2934–2936.
60. Dixon, S. L. and Jurs, P. C. (1993). Estimation of pK_a for organic oxyacids using calculated atomic charges, *J. Comput. Chem.*, 14, pp. 1460–1467.
61. Gross, K. C., Hadad, C. M. and Seybold, P. G. (2002). Comparison of different atomic charge schemes for predicting the pK_a variations in substituted anilines and phenols, *Int. J. Quantum Chem.*, 90, pp. 445–458.
62. Ma, Y., Gross, K. C., Hollingsworth, C. A. and Seybold, P. G. (2004). Relationships between aqueous acidities and computed surface-electrostatic potentials and local ionization energies of substituted phenols and benzoic acids. *J. Molec. Model.*, 10, pp. 235–239.
63. Zhang, J., Kleinoder, T. and Gasteiger, J. (2006). Prediction of pK_a values for aliphatic carboxylic acid and alcohols with empirical atomic charge descriptors, *J. Chem. Inf. Model.*, 46, pp. 2256–2266.
64. Varekova, R. S., Geidl, S., Ionescu, C.-M., Skrehota, O., Kudera, M., Sehnal, D., Bouchal, T., Abagyan, R., J., H. H. and Koca, J. (2011). Predicting pK_a values of substituted phenols from atomic charges: Comparison of different quantum mechanical methods and charge distribution schemes, *J. Chem. Inf. Model.*, 51, pp. 1795–1806.
65. Gauch, H. G. J. (1993). Prediction, parsimony, and noise, *Am. Sci.*, 81, pp. 468–478.
66. Xue, X.-S., Wang, Y., Yang, C., Ji, P. and Cheng, J.-P. (2015). Toward prediction of the chemistry in ionic liquids: An accurate computation of absolute pK_a values of benzoic acids and benzenethiols, *J. Org. Chem.*, 80, pp. 8997–9006.
67. Tummanapelli, A. K. and Vasudevan, S. (2015). Ab initio molecular dynamics simulations of amino acids in aqueous solutions: Estimating pK_a values from metadynamics sampling, *J. Phys. Chem. B*, 119, pp. 12249–12255.
68. Kogel, J. F., Linder, T., Schroder, F. G., Sundermeyer, J., Goll, S. K., Himmel, D., Krossing, I., Kutt, K., Saame, J. and Leito, I. (2015). Fluoro- and perfluoralkylsulfonylpentafluoroanilides: Synthesis and characterization of NH acids for weakly coordinating anions and their gas-phase and solution acidities, *Chem. Eur. J.*, 21, pp. 5769–5782.

69. Yang, C., Xue, X.-S., Li, X. and Cheng, J.-P. (2014). Computational study on the acidic constants of chiral Brønsted acids in dimethyl sulfoxide, *J. Org. Chem.*, 79, pp. 4340–4351.
70. Xue, X.-S., Yang, C., Li, X. and Cheng, J.-P. (2014). Computational study on the pK_a shifts in proline induced by hydrogen-bond-donating cocatalysts, *J. Org. Chem.*, 79, pp. 1166–1173.
71. Estrada-Montano, A. S., Leyva, M. A., Grande-Aztatzi, R. G., Vela, A. and Rosales-Hoz, M. J. (2014). The reaction of $[Fe_3(CO)_{12}]$ with $HCCSiR_3$ (R =Me, Ph) and reactivity of $[HFe_3(C))_9(CCSiMe_3)]$ with amines. Theoretical studies on NMR 1H and 13C chemical shifts and some advances in the theoretical determinations of pK_a in cluster compounds, *J. Organomet. Chem.*, 751, pp. 420–429.
72. Barbero, M., Berto, S., Cadamuro, S., Daniele, P. G., Dughera, S. and Ghigo, G. (2013). Catalytic properties and acidity of 1,2-benzenedisulfonimide and some of its derivatives. An experimental and computational study, *Tetrahedron*, pp. 3212–3217.
73. Alkorta, I., Griffiths, M. Z. and Popelier, P. L. A. (2013). Relationship between experimental pK_a values in aqueous solution and a gas phase bond length in bicyclo[2.2.2]octane and cubane carboxylic acids, *J. Phys. Org. Chem.*, 26, pp. 791–796.
74. Chen, Y.-L., Barlow, D. J., Kong, X.-L., Ma, Y.-M. and Hider, R. C. (2012). Prediction of 3-hydroxypyridin-4-one (HPO) hydroxyl pK_a values, *Dalton Trans.*, 41, pp. 6549–6557.
75. Ali-Torres, J., Rodriguez-Santiago, L. and Sodupe, M. (2011). Computational calculations of pK_a values of imidazole in Cu(ii) complexes of biological relevance, *Phys. Chem. Chem. Phys.*, 2011, pp. 7852–7861.

Chapter 17

Issues Particular to Organometallic Reactions

Gang Lu, Huiling Shao, Humair Omer and Peng Liu

Department of Chemistry, University of Pittsburgh, USA

17.1 Introduction

Concurrently with the rapid experimental advancement in the field of transition metal catalysis, there has been growing interest in performing computational studies on the mechanisms, reactivity, and selectivity of these systems.[1–3] The recent developments of effective density functional theory (DFT) methods has made it possible to calculate thermochemistry data and barrier heights of organometallic reactions with reasonable accuracy.[4] Now, computations have been almost routinely used to study all types of transition-metal–catalyzed reactions[5–10] to predict the most favorable reaction pathways, determine the origins of the effects of ligands and substituents, and provide useful insights for the design of new reactions and improved catalysts. The computational studies are especially valuable in cases where the experimental mechanistic investigations are challenging or impractical due to the difficulty in characterizing the reactive organometallic intermediates.[11–13]

The unique properties of transition metal catalysts pose a few distinct challenges for the computational studies on these systems.[14–16] First, there are often multiple possible reaction pathways in transition-metal–catalyzed reactions. These competing mechanisms often involve metal complexes at different oxidation states or different coordination geometries. Each reaction pathway usually has several elementary steps involving a number of possible intermediates and transition states. In many organometallic reactions, especially in the recently developed catalytic systems, little is known in terms of the exact reaction mechanism, the structures and reactivities of the intermediates, and the rate- and selectivity-determining steps. Thus, a computational study often relies on a careful investigation of possible intermediates and transition-state structures in all potential pathways. The relatively large size of the organometallic compounds and the number of isomers and conformers also dramatically complicate the computations.

While many of the computational studies aim to provide chemically meaningful explanations and predictions of experimental results, the interpretation of the computational data is often another significant challenge. The reactivities and regio- and stereoselectivities in many organometallic reactions are often determined by rather complex interactions between the catalyst and the substrate.[17,18] Although computations have been shown to provide reliable predictions of reactivities and selectivities in many cases,[5–10] the identification of the underlying interactions that account for the observed reactivity and selectivity is not always straightforward. The catalyst–substrate interaction is often governed by a combination of a few factors, such as the electron donicity of the ligand, the distortion of the auxiliary ligand and the substrate, and the through-space interactions between the ligand and the substrate, which include steric repulsions, electrostatic interactions, and dispersion interactions. As the focus of many recent computational studies moves from simply predicting the most favorable mechanism to providing insights about the effects of ligand, reagents, and additives,[19–23] better strategies to derive general and predictive models to help rationalize the underlying effects that govern reactivity and selectivity are highly desirable.

In this chapter, we will discuss the general approaches to address these challenges in modeling organometallic reactions. Rather than providing an extensive overview of the entire field, we focus on a few recent examples from our research group on the carbon–carbon and carbon–hydrogen bond functionalization. The capability and limitations of the state-of-the-art computational methods to solve practical problems in transition-metal–catalyzed reactions will be highlighted in these examples.

17.2 Computational studies of multistep organometallic reaction mechanisms

A distinct feature of the mechanisms of transition-metal–catalyzed reactions is the complexity of the multistep reaction pathways. In the computational mechanistic studies of these processes, it is often essential to investigate the entire catalytic cycle to identify the key step(s) that controls the outcome of the reactions. For example, the transition-metal–catalyzed functionalization of C–H and C–C bonds usually initiates via the "activation" step that forms a metal–carbon bond via cleavage of the C–H or C–C bond in the substrate (Fig. 17.1). A subsequent "functionalization" step then converts the organometallic intermediate to the functionalized products. While many early computational studies focused on the elementary steps of C–H and C–C bond cleavage,[24–28] novel pathways in reactions of the M–C intermediate may also play a key role in controlling reactivity and selectivity.

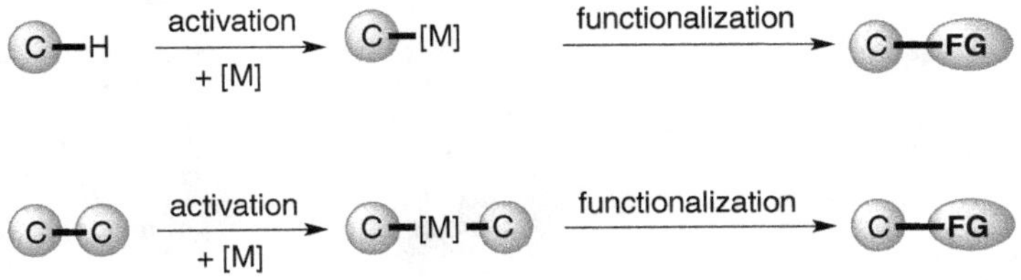

Figure 17.1. General strategy of transition metal-catalyzed C–H and C–C bond functionalization reactions.

The importance of understanding the mechanistic features in the functionalization step is highlighted in our recent computational study of Pd-catalyzed C–H amination reactions.[29] The Pd-catalyzed intramolecular C–H amination of benzylamines with a picolinamide-directing group offered the first practical synthesis of the highly strained benzazetidine derivatives (Fig. 17.2). The choice of oxidant is critical for the highly selective formation of the benzazetidine product. Under the typical reaction condition with a Pd(II) catalyst and using $PhI(OAc)_2$ as the oxidant, only the undesired C–O coupling product is formed. Using PhI(DMM) in place of $PhI(OAc)_2$ as the oxidant, the product selectivity is completely reversed to favor the formation of benzazetidine. DFT calculations indicated that the reactions under both conditions occur via the same C–H cleavage mechanism to form a common palladacycle intermediate (**2**, Fig. 17.3). In the subsequent C–N and C–O bond-formation (functionalization) steps, the mechanism and product selectivity are completely different using different oxidants. In the reaction with $PhI(OAc)_2$, the Pd(III)/Pd(III) dimer **3** is formed, which undergoes facile C–O reductive elimination with the axial OAc ligand via a five-membered cyclic transition state **6-TS**. In contrast, using the dicarboxylate DMM in place of acetate, there is no anionic ligand at the axial

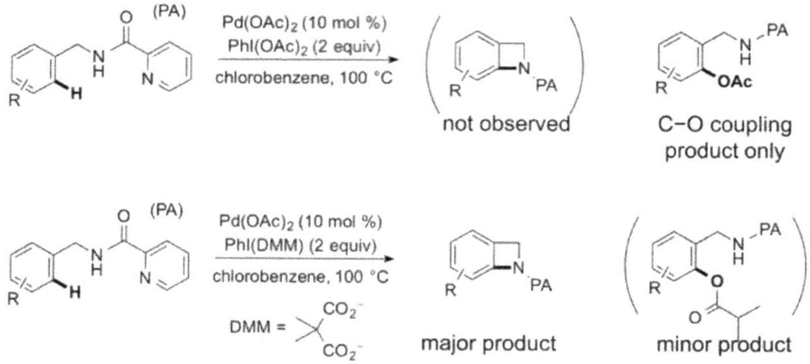

Figure 17.2. Experimental studies of Pd-catalyzed C–H amination to synthesize benzazetidine.

Source: Adapted from Ref. [29].

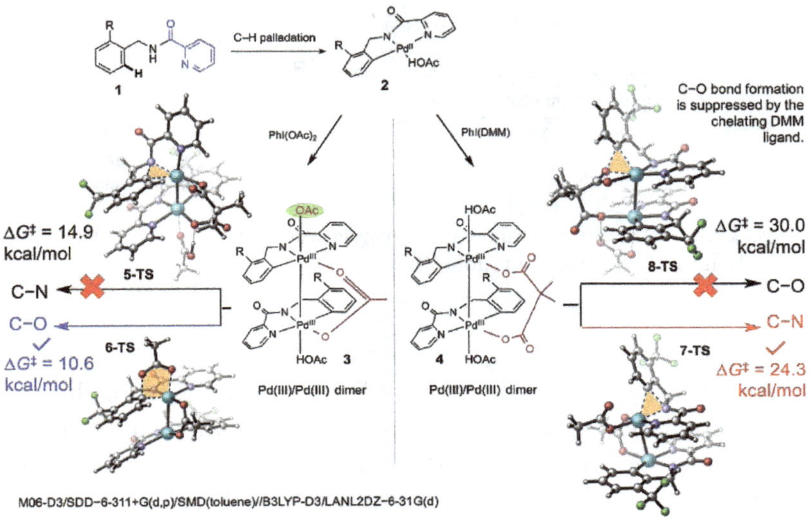

Figure 17.3. DFT-predicted product selectivity in Pd-catalyzed C–H amination of **1** (R = CF_3). The Gibbs free energies are with respect to the catalyst resting state, **3** and **4** in reactions with $PhI(OAc)_2$ and PhI(DMM), respectively.

position of the Pd(III)/Pd(III) dimer **4**. Thus the C–O reductive elimination (**8-TS**) is suppressed due to the high activation energy of the three-membered transition state. As a result, the C–N reductive elimination (**7-TS**) becomes favorable and selectively forms the benzazetidine product.

Careful investigation of the *entire* catalytic cycle is also important in mechanistic studies of C–C bond activation and functionalization reactions. Although C–C bonds are often thought to be thermodynamically stable and kinetically inert, the cleavage of C–C bonds, especially those in a strained molecule, could be a facile process in the presence of an appropriate transition metal catalyst.[30–32] For example, a recent computational study by Ding *et al.*[33] indicated the C–C bond cleavage in the Rh-catalyzed ring-opening of benzocyclobutenol only requires an activation free energy of 10 kcal/mol. This suggests the functionalization of the M–C bonded intermediate formed after the C–C bond cleavage may become rate- and selectivity-determining in many C–C bond functionalization reactions.

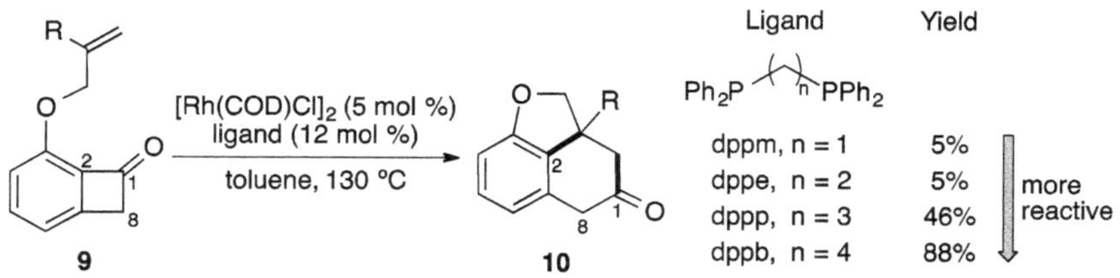

Figure 17.4. Rhodium-catalyzed carboacylation of olefins via functionalization of C–C bond in benzocyclobutenone **9**.

Source: Adapted from Ref. [35].

We recently investigated the Rh-catalyzed carboxylation of olefins, a reaction that occurs through activation of C–C bond in benzocyclobutenones (Fig. 17.4).[34] This reaction allows for the insertion of an alkene or alkyne into the four-membered carbocycles to form more complex fused carbocycles.[35–37] An interesting feature of this reaction is that the final product (**10**) is formed via insertion of an olefin to the relatively strong C1–C2 bond in benzocyclobutenone **9**, while the much weaker C1–C8 bond remains intact. The DFT-computed potential energy profile of the complete catalytic cycle is shown in Figs. 17.5–17.7. The DFT calculations indicated that the initial C–C bond cleavage is neither rate-determining nor selectivity-determining. Although only the C1–C2 bond cleavage product is formed, the initial C–C bond cleavage occurs at the C1–C8 bond of **9** via **13-TS** to form rhodacycle intermediate **14** (Fig. 17.5). A rather unusual mechanism then follows in the functionalization of this metallacycle intermediate, which first undergoes decarbonylation (**15-TS**) and CO insertion (**17-TS**) to isomerize to metallacycle **12** (Fig. 17.6), followed by alkene insertion (**18-TS**) and reductive elimination (**20-TS**) to form the final product (Fig. 17.7).

Probably the most important mechanistic insight from the computational investigation of a catalytic cycle is the identification of the rate- and selectivity-determining steps. This knowledge is particularly useful to understand and predict the effects of catalysts on reactivity and selectivity, and thus to help the rational design of catalytic reactions. From the reaction energy profile shown in Figs. 17.5–17.7,

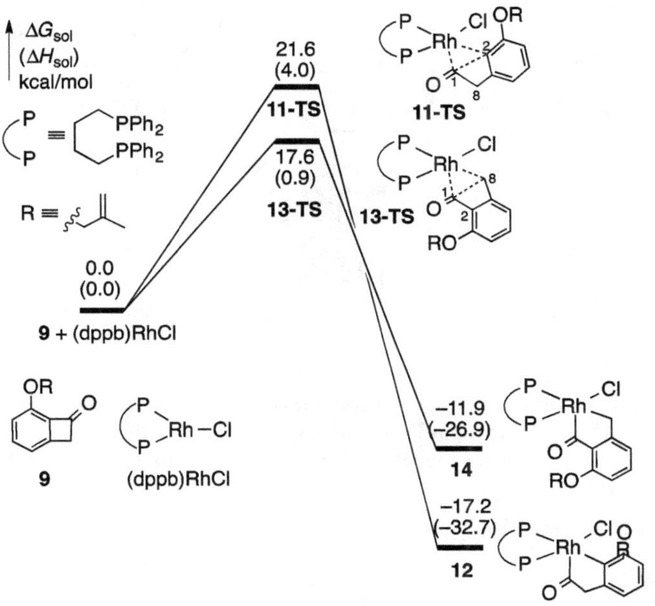

Figure 17.5. Computed energy profiles of the oxidative addition pathways. All energies are in kcal/mol and are with respect to the separate substrate **9** and the rhodium catalyst (dppb)RhCl.

Source: Adapted from Ref. [34].

one could calculate the activation barrier for each step. According to the energy span model,[38] the overall rate of a catalytic reaction (i.e., turnover frequency) is determined by the energy difference between the rate-determining transition state (RDTS) and the catalyst resting state that is the lowest energy intermediate before the RDTS. Here, the RDTS is the CO insertion (**17-TS**) and the catalyst resting state is the metallacycle intermediate **14**. Thus, the overall barrier of the catalytic cycle is 28.1 kcal/mol.

The computed reaction energy profile can also be used to predict whether each individual step is reversible. The reversibility is determined by the relative barriers of the forward vs. reverse reactions starting from an intermediate. For example, after the formation of **14**, the forward decarbonylation (**14**→**15-TS**, $\Delta G^{\ddagger} = 24.2$ kcal/mol, Fig. 17.6) occurs faster than the reverse C–C activation (**14**→**13-TS**,

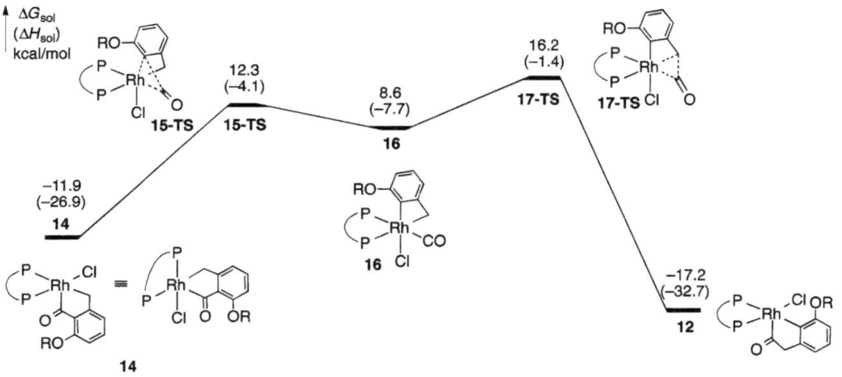

Figure 17.6. Computed energy profile of the metallacycle isomerization pathway via decarbonylation (**15-TS**) and CO insertion (**17-TS**). All energies are in kcal/mol and are with respect to the separate substrate **9** and the rhodium catalyst (dppb)RhCl.

Source: Adapted from Ref. [34].

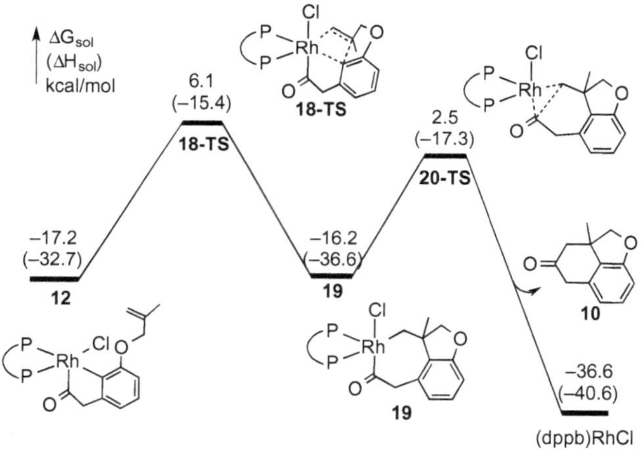

Figure 17.7. Computed energy profile of the olefin migratory insertion (**18-TS**) and reductive elimination (**20-TS**) steps. All energies are in kcal/mol and are with respect to the separate substrate **9** and the rhodium catalyst (dppb)RhCl.

Source: Adapted from Ref. [34].

$\Delta G^{\ddagger} = 29.5\,\text{kcal/mol}$, Fig. 17.5), and thus the C–C activation step is irreversible. In addition, it should be noted that the selectivity-determining step may differ from the rate-determining step. Since the alkene insertion step (**18-TS**) is irreversible, it determines the enantioselectivity in reactions with chiral bisphosphine ligands. Here the enantioselectivity-determining step is different from the rate-determining CO insertion step (**17-TS**).

17.3 Computational studies of the effects of auxiliary ligand on reactivity and selectivity

One of the most significant challenges in the development of transition-metal–catalyzed reactions is the proper choice of ancillary ligands, such as phosphines[39,40] and *N*-heterocyclic carbenes.[41–43] Modification of the ligand structure often dramatically alters the steric and electronic properties of the catalyst and affects the outcome of the catalytic reaction.[44] Thus, understanding and predictions of how the structure of the ligand controls the reactivity and selectivity of the catalyst have gained extensive interest in the computational studies of transition metal catalysis.

The bisphosphine ligand plays an important role on the reactivity in the Rh-catalyzed carboxylation of benzocyclobutenone.[35] As shown in Fig. 17.4, this reaction is clearly promoted by ligands with more methylene groups on the backbone of the bisphosphine (e.g., dppp and dppb). The computational investigation of ligand effects on reactivity is often achieved through calculating the overall activation barriers of the catalytic cycle using different ligands. Based on the mechanism revealed by the DFT calculations (see above), the overall barriers in reactions with four bisphosphine ligands, dppm, dppe, dppp, and dppb, were calculated and are summarized in Table 17.1.

Based on transition state theory and the Eyring equation, an activation free energy difference of 1.4 kcal/mol corresponds to a 10-fold difference in reaction rate at room temperature. The computationally predicted activation barriers indicated the

Table 17.1. Ligand steric parameters and computed activation free energies of the rate-determining CO insertion step with different ligands.

Ligand	Bite angle[a] (°)	Buried volume[b] ($\%V_{bur}$)	Activation energy[c] ($\Delta G^{\ddagger}$, kcal/mol)
dppm	72	38.3	32.8
dppe	85	42.6	32.3
dppp	91	46.3	31.7
dppb	98	47.3	28.1

Notes: [a]Average bite angle in X-ray crystal structures.[49]
[b]Buried volume of the bisphosphine ligand in rate-determining CO insertion transition state **17-TS** computed using the SambVca program[48] based on the B3LYP/LANL2DZ–6-31G(d) optimized geometry.
[c]Activation free energy of **17-TS** with respect to **14**.

dppb-based catalyst will react significantly faster than catalysts with other bisphosphine ligands. This agreement with the experimentally observed reactivity trend not only indicated the satisfactory accuracy of the computational method but also provided support to the computationally derived rate-determining step. However, a more important question has yet to be answered: what is the fundamental cause of the much higher reactivity with the dppb ligand?

The most straightforward rationalization of ligand effects is typically based on the steric and electronic ligand parameters. Over the past a few decades, many types of ligand parameters have been developed and widely used to understand ligand effects on rates and selectivity.[45] For example, Tolman cone angle[46] and bite angle[47] are probably the most commonly used steric parameters for mono- and bisphosphine ligands, respectively. The percent buried volume ($\%V_{bur}$) developed by Poato *et al.*[48] is a universal steric parameter for both phosphine and *N*-heterocyclic carbene ligands. The literature bite angle values[49] and the computed buried volume of the four bisphosphine ligands listed in Table 17.1 indicated a qualitative correlation of the ligand steric parameters and reactivity. The reaction is clearly promoted when a bulkier ligand with greater bite angle and buried volume is employed.

Quantum mechanical calculations also enabled us to obtain more insights into the origin of the steric effects on reactivity. Careful analysis of the steric interactions in the resting state and the rate-determining transition state can reveal the exact ligand–substrate repulsion that accounts for the steric effects on reactivity. For reactions between two simple organic molecules, steric repulsions are typically attributed to a short distance between two nonbonded atoms. A rule of thumb is that the steric interaction becomes repulsive when the interatomic distance is shorter than the sum of the van der Waals radii of the two atoms. Unfortunately, analysis of the steric interactions between a structurally complex ancillary ligand and a substrate can be much more complicated. It is often not clear which particular interatomic distance is responsible for the suspected steric effects. For example, one's chemical intuition may suggest that bulkier ligand such as dppb promotes the carboxylation reaction due to destabilizing steric repulsions in the catalyst resting state, metallacycle **14**. However, it is not straightforward to identify which parts of the ligand and which substituents on the substrate have the greatest contribution to such steric repulsions.

A solution to this challenge is by mapping and visualizing the noncovalent interactions between the ligand and the substrate. There have been many applications that use the NCIPLOT software developed by Contreras-García *et al.*[50] to plot and analyze the noncovalent interaction (NCI) index for the ligand–substrate interactions in organometallic complexes.[51–53] The NCI index, derived from electron density, illustrates the nature and strength of the inter- and intramolecular through-space interactions, and can be efficiently applied to study very large molecules, such as organometallic complexes and even protein–ligand interactions.[54–56]

A different approach to visualize the ligand–substrate steric interactions is the ligand steric contour plot.[57,58] The ligand steric contour is derived from the van der Waal surface of the ligand (Fig. 17.8(a)) and is typically created using optimized geometry of ligand in key transition state or intermediate structures to illustrate the exact shape of the ancillary ligand and the relative orientation

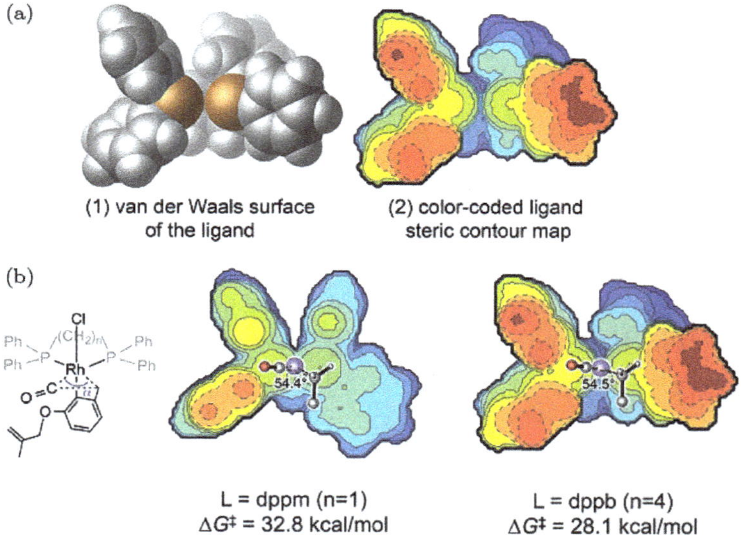

Figure 17.8. (a) Ligand steric contour map derived from the van der Waals surface of the bisphosphine ligand. (b) Ligand steric contour maps of the CO insertion transition state (**17-TS**) with dppm and dppb ligand. The CO ligand and the a CH_2 group on the substrate are overlaid in the contour maps to illustrate the relative orientation of the ligand with respect to the substrate.

Source: Adapted from Ref. [34].

with respect to the substrate and other ligands on the metal catalyst. The red color on the contour map indicates regions of the ligand that is placed closer to the substrate. The contour plots of the dppm and dppb ligands shown in Fig. 17.8(b) clearly indicate the dppb ligand is much bulkier than the dppm ligand. The increased bulkiness of dppb results from the shorter distance between the Ph substituents on the ligand and the substrate. The majority of the steric clashes with the dppb ligand is expected near the equatorial plane, where the CO ligand and the α CH_2 group are located. In the rate-determining CO insertion transition state, both the CO and CH_2 groups are moving away from the ligand, as the substrate bite angle (α, Fig. 17.8(b)) between CO and CH_2 decreases. Thus, the steric clash with the dppb ligand is diminished in the transition state. This explains why the dppb ligand is particularly effective to promote this reaction.

17.4 Practical considerations

The level of complexity of a computational investigation of organometallic reaction depends on many factors. Before each computational project, one should first consider the following questions:

- What chemical structures will be calculated?
- What level of theory will be used?
- How will the computational results be analyzed?

At first glance, these are all fairly straightforward questions that do not seem to differ from any other computational chemistry project. However, as discussed in the beginning of this chapter, there are unique challenges associated with each of these three questions when performing calculations on organometallic systems.

First, computational investigation of many experimentally observed phenomena requires a thorough understanding of the reaction mechanism that elucidates the rate- and selectivity-determining step in the most favorable pathway. Again, taking the Rh-catalyzed carboxylation of benzocyclobutenone as an example, although there are only two general strategies for the activation of C–C bonds using transition metal catalysts[59,60]: (a) oxidative addition of the C–C bond and (b) coordination of the metal center to a heteroatom followed by β-carbon elimination (Fig. 17.9), in practice, a much greater number of possible pathways (Fig. 17.10) need to be considered by computations.

The typical computational approach to investigate the reaction mechanism is to calculate the energies of all the possible intermediates and transition states in each pathway and construct the

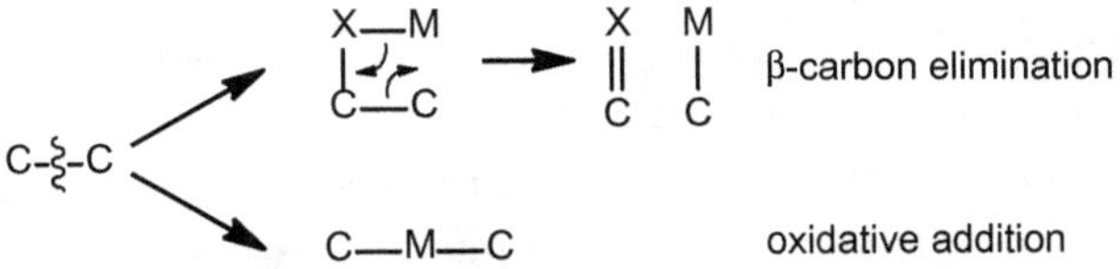

Figure 17.9. General strategies to activate C–C bonds using transition metal catalysts.

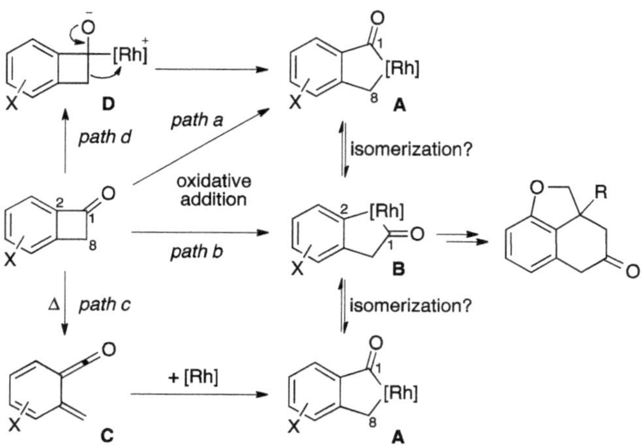

Figure 17.10. Possible C–C bond cleavage pathways that need to be considered in the computational mechanistic study.

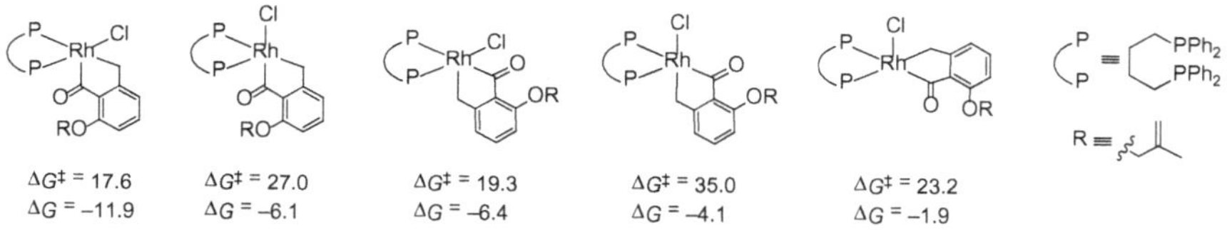

Figure 17.11. Relative Gibbs free energies (ΔG) of the possible coordination isomers of metallacycle intermediate **A**. The barriers ($\Delta G^{\ddagger}$) for the corresponding oxidative addition transition state isomer are also provided. All energies are in kcal/mol and are with respect to the separate substrate **9** and the rhodium catalyst (dppb)RhCl.

energy profile for the reaction. This requires the calculations of all possible isomers for each transition state and intermediate shown in Fig. 17.10. For example, the five-coordinated metallacycle intermediate **A** has a square-based pyramidal geometry, and at least five isomers are possible (Fig. 17.11). For each isomer of the metallacycle, there is a corresponding oxidative addition transition state structure. Since conformational search of the intermediates and transition states are often time-consuming and tedious (see Chapter 5), there has been great interest in developing automatic conformational search algorithms for organometallic complexes. Because of the lack of reliable force field parameters for transition metals, force-field–based conformational search software programs usually cannot be readily applied to organometallic systems. Nonetheless, promising research has started to emerge in this area. Foscato *et al.*[61,62] and Zimmerman[63,64] have developed new algorithms that allows for the automatic construction and evaluation of isomers of organometallic complexes. Careful conformational analysis of intermediates and transition states is extremely useful and important not only to provide accurate and reliable energetics but also to predict the relative arrangements of the catalyst and the substrate. This structural information can then be applied to investigate ligand effects on reactivity and selectivity.

In terms of the computational method, the vast majority of modern computations on metal-catalyzed reactions were performed using DFT. A recent review by Sperger *et al.*[2] surveyed the most popular DFT methods and pseudopotential basis sets in modern computational studies on organometallic systems. Although a large number of DFT methods are available, only a handful of popular functionals are actually used in most applications. Among the most popular choices are M06,[65] M06L,[66] ωB97xD,[67] and other modern functionals that account for dispersion interactions,[68] which are especially important for ligand-binding energy and ligand–substrate interactions in transition metal complexes. The performances of these DFT methods have been rigorously benchmarked for geometries and energies of various organometallic systems. Since a large number of review articles and benchmark studies of the performances and limitations of different DFT methods are available in the literature,[69–72] these discussions will not be repeated here. It should be noted although using DFT has become common practice in studies of transition-metal–catalyzed reactions, it is essential to understand the limitations of the different functionals to choose an appropriate method for the computational project and to avoid overinterpretation of the computational results.

Last but not the least, it often requires a great amount of effort to interpret computational data on organometallic systems.

Calculations provide energies and structures of specific model compounds. How can these data be used to rationalize and predict experiment? Classical concepts, models, and empirical parameters in physical organic chemistry and organometallic chemistry, such as steric and electronic ligand parameters, *trans* effects, and field theory, are all valuable tools to provide chemically meaningful explanation of the computational data. In addition, the raw data from quantum mechanical calculations (i.e., optimized structures and energies) are often augmented with more in-depth analysis, including calculating molecular orbitals and partial atomic charges to understand electronic effects, and using NCI plots and steric contour plots to study the nature of ligand–substrate steric interactions. Comparing to merely reproducing experimental reactivity or selectivity trends, the development of straightforward and chemically meaningful models to understand and predict a broad range of reactions is clearly a more desirable goal for the computational investigation of organometallic systems.

References

1. Cramer, C. J. and Truhlar, D. G. (2009). Density functional theory for transition metals and transition metal chemistry, *Phys. Chem. Chem. Phys.*, 11, pp. 10757–10816.
2. Sperger, T., Sanhueza, I. A., Kalvet, I. and Schoenebeck, F. (2015). Computational studies of synthetically relevant homogeneous organometallic catalysis involving Ni, Pd, Ir, and Rh: An overview of commonly employed DFT methods and mechanistic insights, *Chem. Rev.*, 115, pp. 9532–9586.
3. Cheng, G.-J., Zhang, X., Chung, L. W., Xu, L. and Wu, Y.-D. (2015). Computational organic chemistry: bridging theory and experiment in establishing the mechanisms of chemical reactions, *J. Am. Chem. Soc.*, 137, pp. 1706–1725.
4. Pribram-Jones, A., Gross, D. A. and Burke, K. (2015). DFT: A theory full of holes? *Annu. Rev. Phys. Chem.*, 66, pp. 283–304.
5. Sunoj, R. B. (2016). Transition state models for understanding the origin of chiral induction in asymmetric catalysis, *Acc. Chem. Res.*, 49, pp. 1019–1028.
6. Santoro, S., Kalek, M., Huang, G. and Himo, F. (2016). Elucidation of mechanisms and selectivities of metal-catalyzed reactions using quantum chemical methodology, *Acc. Chem. Res.*, 49, pp. 1006–1018.

7. Lam, Y.-h., Grayson, M. N., Holland, M. C., Simon, A. and Houk, K. N. (2016). Theory and modeling of asymmetric catalytic reactions, *Acc. Chem. Res.*, 49, pp. 750–762.
8. Balcells, D., Clot, E., Eisenstein, O., Nova, A. and Perrin, L. (2016). Deciphering selectivity in organic reactions: A multifaceted problem, *Acc. Chem. Res.*, 49, pp. 1070–1078.
9. Zhang, X., Chung, L. W. and Wu, Y.-D. (2016). New mechanistic insights on the selectivity of transition-metal-catalyzed organic reactions: The role of computational chemistry, *Acc. Chem. Res.*, 49, pp. 1302–1310.
10. Park, Y., Ahn, S., Kang, D. and Baik, M.-H. (2016). Mechanism of Rh-catalyzed oxidative cyclizations: Closed versus open shell pathways, *Acc. Chem. Res.*, 49, pp. 1263–1270.
11. Lin, Z. (2010). Interplay between theory and experiment: computational organometallic and transition metal chemistry, *Acc. Chem. Res.*, 43, pp. 602–611.
12. Ananikov, V. P. (2014). *Understanding Organometallic Reaction Mechanisms and Catalysis: Computational and Experimental Tools* (Wiley-VCH, Germany).
13. Sperger, T., Sanhueza, I. A. and Schoenebeck, F. (2016). Computation and experiment: A powerful combination to understand and predict reactivities, *Acc. Chem. Res.*, 49, pp. 1263–1270.
14. Cundari, T. R. (2001). *Computational Organometallic Chemistry* (Marcel Dekker, Switzerland).
15. Wiest, O. and Wu, Y. (2012). *Computational Organometallic Chemistry* (Springer-Verlag, Berlin).
16. Macgregor, S. A. and Eisenstein, O. (2016). *Computational Studies in Organometallic Chemistry* (Springer International Publishing, Switzerland).
17. Carboni, S., Gennari, C., Pignataro, L. and Piarulli, U. (2011). Supramolecular ligand-ligand and ligand-substrate interactions for highly selective transition metal catalysis, *Dalton Trans.*, 40, pp. 4355–4373.
18. Dydio, P. and Reek, J. N. H. (2014). Supramolecular control of selectivity in transition-metal catalysis through substrate preorganization, *Chem. Sci.*, 5, pp. 2135–2145.
19. Jindal, G., Kisan, H. K. and Sunoj, R. B. (2015). Mechanistic insights on cooperative catalysis through computational quantum chemical methods, *ACS Catal.*, 5, pp. 480–503.
20. Anand, M., Sunoj, R. B. and Schaefer, H. F. (2014). Non-innocent additives in a palladium(II)-catalyzed C–H bond activation reaction: Insights into multimetallic active catalysts, *J. Am. Chem. Soc.*, 136, pp. 5535–5538.
21. Figg, T. M., Wasa, M., Yu, J.-Q. and Musaev, D. G. (2013). Understanding the reactivity of Pd^0/PR_3-catalyzed intermolecular $C(sp^3)$–H bond arylation, *J. Am. Chem. Soc.*, 135, pp. 14206–14214.

22. Haines, B. E., Xu, H., Verma, P., Wang, X.-C., Yu, J.-Q. and Musaev, D. G. (2015). Mechanistic details of Pd(II)-catalyzed C–H iodination with molecular I_2: Oxidative addition vs electrophilic cleavage, *J. Am. Chem. Soc.*, 137, pp. 9022–9031.
23. Haines, B. E., Berry, J. F., Yu, J.-Q. and Musaev, D. G. (2016). Factors controlling stability and reactivity of dimeric Pd(II) complexes in C–H functionalization catalysis, *ACS Catal.*, 6, pp. 829–839.
24. Siegbahn, P. E. M. (1996). Comparison of the C−H activation of methane by $M(C_5H_5)(CO)$ for M = cobalt, rhodium, and iridium, *J. Am. Chem. Soc.*, 118, pp. 1487–1496.
25. Sundermann, A., Uzan, O., Milstein, D. and Martin, J. M. L. (2000). Selective C−C vs C−H bond activation by rhodium(I) PCP pincer Complexes. a computational study, *J. Am. Chem. Soc.*, 122, pp. 7095–7104.
26. Rybtchinski, B., Oevers, S., Montag, M., Vigalok, A., Rozenberg, H., Martin, J. M. L. and Milstein, D. (2001). Comparison of steric and electronic requirements for C−C and C−H bond activation. Chelating vs nonchelating case, *J. Am. Chem. Soc.*, 123, pp. 9064–9077.
27. Sundermann, A., Uzan, O. and Martin, J. M. L. (2001). Exclusive C−C activation in the rhodium(I) PCN pincer complex. A computational study, *Organometallics*, 20, pp. 1783–1791.
28. Cundari, T. R., Jimenez-Halla, J. O. C., Morello, G. R. and Vaddadi, S. (2008). Catalytic tuning of a phosphinoethane ligand for enhanced C−H activation, *J. Am. Chem. Soc.*, 130, pp. 13051–13058.
29. He, G., Lu, G., Guo, Z., Liu, P. and Chen, G. (2016). Benzazetidine synthesis via palladium-catalyzed intramolecular C−H amination, *Nat. Chem.*, 8, pp. 1131–1136.
30. Dong, G. (2014). *C−C Bond Activation* (Springer-Verlag, Berlin).
31. Souillart, L. and Cramer, N. (2015). Catalytic C–C bond activations via oxidative addition to transition metals, *Chem. Rev.*, 115, pp. 9410–9464.
32. Murakami, M. and Chatani, N. (2016). *Cleavage of Carbon–Carbon Single Bonds by Transition Metals* (Wiley-VCH, Germany).
33. Ding, L., Ishida, N., Murakami, M. and Morokuma, K. (2014). sp^3–sp^2 vs sp^3–sp^3 C–C site selectivity in Rh-catalyzed ring opening of benzocyclobutenol: A DFT study, *J. Am. Chem. Soc.*, 136, pp. 169–178.
34. Lu, G., Fang, C., Xu, T., Dong, G. and Liu, P. (2015). Computational study of Rh-catalyzed carboacylation of olefins: Ligand-promoted rhodacycle isomerization enables regioselective C–C bond functionalization of benzocyclobutenones, *J. Am. Chem. Soc.*, 137, pp. 8274–8283.
35. Xu, T. and Dong, G. (2012). Rhodium-catalyzed regioselective carboacylation of olefins: A C-C bond activation approach for accessing fused-ring systems, *Angew. Chem., Int. Ed.*, 51, pp. 7567–7571.
36. Xu, T., Ko, H. M., Savage, N. A. and Dong, G. (2012). Highly enantioselective Rh-catalyzed carboacylation of olefins: Efficient syntheses of chiral poly-fused rings, *J. Am. Chem. Soc.*, 134, pp. 20005–20008.

37. Xu, T. and Dong, G. (2014). Coupling of sterically hindered trisubstituted olefins and benzocyclobutenones by C-C activation: Total synthesis and structural revision of cycloinumakiol, *Angew. Chem. Int. Ed.*, 53, pp. 10733–10736.
38. Kozuch, S. and Shaik, S. (2011). How to conceptualize catalytic cycles? The energetic span model, *Acc. Chem. Res.*, 44, pp. 101–110.
39. Hayashi, T. (2000). Chiral monodentate phosphine ligand MOP for transition-metal-catalyzed asymmetric reactions, *Acc. Chem. Res.*, 33, pp. 354–362.
40. Tang, W. and Zhang, X. (2003). New chiral phosphorus ligands for enantioselective hydrogenation, *Chem. Rev.*, 103, pp. 3029–3070.
41. Cesar, V., Bellemin-Laponnaz, S. and Gade, L. H. (2004). Chiral N-heterocyclic carbenes as stereodirecting ligands in asymmetric catalysis, *Chem. Soc. Rev.*, 33, pp. 619–636.
42. Diez-Gonzalez, S. and Nolan, S. P. (2007). Stereoelectronic parameters associated with N-heterocyclic carbene (NHC) ligands: A quest for understanding, *Coord. Chem. Rev.*, 251, pp. 874–883.
43. Hopkinson, M. N., Richter, C., Schedler, M. and Glorius, F. (2014). An overview of N-heterocyclic carbenes, *Nature*, 510, pp. 485–496.
44. Gorin, D. J., Sherry, B. D. and Toste, F. D. (2008). Ligand effects in homogeneous Au catalysis, *Chem. Rev.*, 108, pp. 3351–3378.
45. Fey, N. (2010). The contribution of computational studies to organometallic catalysis: Descriptors, mechanisms and models, *Dalton Trans.*, 39, pp. 296–310.
46. Tolman, C. A. (1977). Steric effects of phosphorus ligands in organometallic chemistry and homogeneous catalysis, *Chem. Rev.*, 77, pp. 313–348.
47. Birkholz, M.-N., Freixa, Z. and van Leeuwen, P. W. N. M. (2009). Bite angle effects of diphosphines in C-C and C-X bond forming cross coupling reactions, *Chem. Soc. Rev.*, 38, pp. 1099–1118.
48. Poater, A., Cosenza, B., Correa, A., Giudice, S., Ragone, F., Scarano, V. and Cavallo, L. (2009). SambVca: A web application for the calculation of the buried volume of N-heterocyclic carbene ligands, *Eur. J. Inorg. Chem.*, 13, pp. 1759–1766.
49. Dierkes, P. and W. N. M. van Leeuwen, P. (1999). The bite angle makes the difference: A practical ligand parameter for diphosphine ligands, *J. Chem. Soc., Dalton Trans.*, pp. 1519–1530.
50. Contreras-García, J., Johnson, E. R., Keinan, S., Chaudret, R., Piquemal, J.-P., Beratan, D. N. and Yang, W. (2011). NCIPLOT: A program for plotting noncovalent interaction regions, *J. Chem. Theory Comput.*, 7, pp. 625–632.
51. Bhaskararao, B. and Sunoj, R. B. (2015). Origin of stereodivergence in cooperative asymmetric catalysis with simultaneous involvement of two chiral catalysts, *J. Am. Chem. Soc.*, 137, pp. 15712–15722.
52. Bhattacharjee, R., Nijamudheen, A., Karmakar, S. and Datta, A. (2016). Strain control: Reversible H_2 activation and H_2/D_2 exchange in Pt complexes, *Inorg. Chem.*, 55, pp. 3023–3029.

53. Lepetit, C., Poater, J., Alikhani, M. E., Silvi, B., Canac, Y., Contreras-García, J., Solà, M. and Chauvin, R. (2015). The missing entry in the agostic–anagostic series: Rh(I)–η1-C interactions in P(CH)P pincer complexes, *Inorg. Chem.*, 54, pp. 2960–2969.
54. Johnson, E. R., Keinan, S., Mori-Sánchez, P., Contreras-García, J., Cohen, A. J. and Yang, W. (2010). Revealing noncovalent interactions, *J. Am. Chem. Soc.*, 132, pp. 6498–6506.
55. Wu, P., Chaudret, R., Hu, X. and Yang, W. (2013). Noncovalent interaction analysis in fluctuating environments, *J. Chem. Theory Comput.*, 9, pp. 2226–2234.
56. Campomanes, P., Neri, M., Horta, B. A. C., Röhrig, U. F., Vanni, S., Tavernelli, I. and Rothlisberger, U. (2014). Origin of the spectral shifts among the early intermediates of the rhodopsin photocycle, *J. Am. Chem. Soc.*, 136, pp. 3842–3851.
57. Wucher, P., Caporaso, L., Roesle, P., Ragone, F., Cavallo, L., Mecking, S. and Goettker-Schnetmann, I. (2011). Breaking the regioselectivity rule for acrylate insertion in the Mizoroki-Heck reaction, *Proc. Natl. Acad. Sci. USA*, 108, pp. 8955–8959.
58. Liu, P., Montgomery, J. and Houk, K. N. (2011). Ligand steric contours to understand the effects of N-heterocyclic carbene ligands on the reversal of regioselectivity in Ni-catalyzed reductive couplings of alkynes and aldehydes, *J. Am. Chem. Soc.*, 133, pp. 6956–6959.
59. Aissa, C. (2011). Transition-metal-catalyzed rearrangements of small cycloalkanes: Regioselectivity trends in beta-carbon elimination reactions, *Synthesis*, pp. 3389–3407.
60. Ruhland, K. (2012). Transition-metal-mediated cleavage and activation of C-C single bonds, *Eur. J. Org. Chem.*, pp. 2683–2706.
61. Foscato, M., Venkatraman, V., Occhipinti, G., Alsberg, B. K. and Jensen, V. R. (2014). Automated building of organometallic complexes from 3D fragments, *J. Chem. Inf. Model.*, 54, pp. 1919–1931.
62. Foscato, M., Occhipinti, G., Venkatraman, V., Alsberg, B. K. and Jensen, V. R. (2014). Automated design of realistic organometallic molecules from fragments, *J. Chem. Inf. Model.*, 54, pp. 767–780.
63. Zimmerman, P. M. (2013). Automated discovery of chemically reasonable elementary reaction steps, *J. Comput. Chem.*, 34, pp. 1385–1392.
64. Zimmerman, P. M. (2015). Navigating molecular space for reaction mechanisms: An efficient, automated procedure, *Mol. Simul.*, 41, pp. 43–54.
65. Zhao, Y. and Truhlar, D. G. (2008). The M06 suite of density functionals for main group thermochemistry, thermochemical kinetics, noncovalent interactions, excited states, and transition elements: Two new functionals and systematic testing of four M06-class functionals and 12 other functionals, *Theor. Chem. Acc.*, 120, pp. 215–241.
66. Zhao, Y. and Truhlar, D. G. (2006). A new local density functional for main-group thermochemistry, transition metal bonding, thermochemical kinetics, and noncovalent interactions, *J. Chem. Phys.*, 125, pp. 194101–194118.

67. Chai, J.-D. and Head-Gordon, M. (2009). Long-range corrected double-hybrid density functionals, *J. Chem. Phys.*, 131, pp. 174105–174117.
68. Grimme, S., Antony, J., Ehrlich, S. and Krieg, H. (2010). A consistent and accurate ab initio parametrization of density functional dispersion correction (DFT-D) for the 94 elements H-Pu, *J. Chem. Phys.*, 132, pp. 154104–154123.
69. Quintal, M. M., Karton, A., Iron, M. A., Boese, A. D. and Martin, J. M. L. (2006). Benchmark study of DFT functionals for late-transition-metal reactions, *J. Phys. Chem. A*, 110, pp. 709–716.
70. Gusev, D. G. (2013). Assessing the accuracy of M06-L organometallic thermochemistry, *Organometallics*, 32, pp. 4239–4243.
71. Hansen, A., Bannwarth, C., Grimme, S., Petrovic, P., Werle, C. and Djukic, J.-P. (2014). The thermochemistry of London dispersion-driven transition metal reactions: Getting the 'right answer for the right reason,' *ChemistryOpen*, 3, pp. 177–189.
72. Goerigk, L. (2015). Treating London-dispersion effects with the latest minnesota density functionals: Problems and possible solutions. *J. Phys. Chem. Lett.*, 6, pp. 3891–3896.

Chapter 18

Computationally Modeling Nonadiabatic Dynamics and Surface Crossings in Organic Photoreactions

Arthur Winter

Department of Chemistry
Iowa State University, USA

18.1 Introduction

Photochemical reactions delight in the suffering of molecules: generation of strained rings and high-energy reactive intermediates, difficult to access by thermal chemistry, are readily provided by photochemical means. Consequently, synthetic photochemistry continues to attract interest due to the frequent differences between the chemistries induced by light and heat. Furthermore, understanding the subtleties of important life processes such as photosynthesis, vision, and bioluminescence has always been an important objective, and renewed interest in photochemistry has resulted from emerging areas such as photocatalysis, biological probes, solar energy storage, and photomedicine. Photons are a cheap renewable reagent and the push for green energy sources as alternatives to fossil fuels will likely involve reactions induced by the sun. Because photochemistry touches on many of the big problems society faces, such as new medicines, understanding biology, sustainability, and renewable resources and energy, it has never been more important to understand

and control the consequences of the interaction of molecules with light.

Given the excitement about exploiting light for synthesis and probing and manipulating chemical and biological processes, it is highly desirable to be able to computationally model these photochemical events. The objective of this chapter is to provide an overview of the most recent developments in computational chemistry for modeling organic photochemical reactions and to provide practical advice on the kinds of models available for a given photochemical problem. We focus in particular on the most recent developments in the modeling of photochemical organic reactions and the problem of surface touchings and nonadiabatic dynamics.

Ideally, one would like to be able to predict photochemical or photophysical outcomes for new molecules (e.g., prediction of product ratios and quantum yields). Regrettably, for reasons described below, modeling photochemical reactions is far more difficult than modeling thermal reactions, and the current state of the art in computational photochemistry can be characterized as being more in the understanding and development phase rather than the predicting phase. Outside of a few simple cases, the current ability to predict photochemistry comes more from the extension of past empirical observations (e.g., photolysis of aryl azides provides aryl nitrenes) rather than *a priori* prediction. This inability to make predictions outside of past experience is obviously highly limiting for the computational discovery or prediction of new photoreactions.

While difficulties remain, tremendous progress in the development of computational models for understanding photochemical reactions has been made in the last few years, and new ideas and paradigms for understanding photochemistry are surely yet to be uncovered. This chapter highlights some of the most noteworthy developments in the last few years. Given the incredible volume of interest in this area, it is far from exhaustive, and a number of important and meritorious studies are not highlighted here.

In this chapter, we begin by providing an overview of photochemical mechanisms and the importance of surface crossings and dynamics on understanding the fate of the electronically excited molecule. We then discuss some of the methods that can be used for modeling excited states. For the remainder of the chapter, we focus on the development of methods for performing dynamic simulations of photochemical reactions, including methods that include the possibility of intersystem crossing, as well as automated tools that can search for conical intersections without requiring any kind of special chemical insight or *a priori* knowledge. In the last section, we give some practical advice on performing such calculations. In all sections, we made the decision to focus on case studies so that the discussion of the computational methods always remains grounded in interesting problems.

18.2 Background

Prior to discussing computational models, we first describe the different mechanisms available for photochemistry reactions. Until relatively recently, it was believed that most photochemical mechanisms proceeded primarily through an avoided crossing mechanism, wherein molecules jump from the excited-state potential energy surface (PES) to the ground-state PES at a geometry where there was an energy gap between the two surfaces. The historical basis for belief in the avoided crossing mechanism is that in diatomic molecules, surface touchings between PESs of the same spin are forbidden, leading to the so-called *noncrossing rule* for diatomics. This noncrossing rule can be derived in a relatively simple way by considering the interaction of two states (Φ_1 and Φ_2) through the secular equation:

$$\begin{matrix} \mathrm{H}_{11} - \mathrm{E}\mathrm{H}_{12} \\ \\ \mathrm{H}_{21}\mathrm{H}_{22} - \mathrm{E} \end{matrix} = 0 \quad \mathrm{H}_{ij} = \textstyle\int \Phi_1 \mathrm{H} \Phi_2 d\tau$$

Which gives two energies by solving the quadratic:

$$\mathrm{E}_{12} = 1/2(\mathrm{H}_{11} + \mathrm{H}_{22} \pm ([\mathrm{H}_{11} - \mathrm{H}_{22}]^2 + 4\mathrm{H}_{12}^2)^{1/2})$$

For the energies of the two states to be equal, $\mathrm{H}_{11} = \mathrm{H}_{22}$ and $\mathrm{H}_{12} = 0$. Satisfying both constraints generally requires at least two coordinates unless the two configurations are of different electronic symmetries. Although Teller recognized fairly early on that in polyatomic chemical systems such surface touchings were possible, for many years, they were considered to be a curiosity and not particularly relevant for most organic photoreactions except in rare cases where surface crossings were enforced by symmetry arguments.

However, early suggestions by Zimmerman, Michl, and Turro, as well as more recent computational investigations by the groups of Robb, Bernardi, Olivucci, Klessinger, Domcke, Bearpark, Martinez, Serrano-Andres, Yarkony, and other researchers, has led to a change in this view.[1–18] The emergence of ultrafast spectroscopy experiments demonstrating that photoisomerization reactions (such as those involved in vision) occur with blazing sub-picosecond rates became impossible to reconcile with the rates predicted by the avoided crossing hypothesis. Within the typical avoided crossing mechanism, now associated with Van der Lugt and Oosteroff,[19–21] decay from the excited state to the ground state is governed by the Fermi Golden Rule (overlap of vibrational wavefunctions and density of states) and occurs on approximately the same time scale as fluorescence. It is now known that many photoreactions decay from the excited state to the ground state at (or near) a geometry where the ground-state and excited-state surfaces have degenerate energies. Such geometries, referred to as funnels or surface touchings or conical intersections, in order of increasing specificity, play a role in many photochemical reactions similar to that of the transition state for thermal reactions. A conical intersection connects at least two minima, and so the molecular dynamics of the molecule around a conical intersection can influence quantum yields and product ratios.

Before delving into different computational methods used for modeling photoreactions, it is worth stepping back and appreciating some of the difficulties in modeling chemistry occurring in excited states. For most ground-state molecules, the separation between the ground electronic state and the first excited state is large and the single-reference approximation is typically a reasonable one. This is not the case for excited states, where excited-state energies are frequently bunched together and are not well defined by a single determinant and nondynamical correlation effects are large. Even the intuitive idea of electronic orbitals, which are generated by invoking the separation of electron motion and nuclear motion (the Born–Oppenheimer approximation), become invalid: rapid motion of nuclei means that instantaneous relaxation of electrons to adapt to the new nuclear positions cannot be assumed, particularly at surface crossings.

Furthermore, photochemistry and photophysics is a zero-sum game involving a competition of rates, with the fastest decay channels back to the ground-state winning. Consequently, exact *a priori* prediction of the efficiency of photochemical reactions requires the ability to predict relative rates for *all* competitive decay channels. Predicting rates is a challenge even for ground–state reactions, and the problem is exponentially harder in the excited state where there are numerous competitive channels (intersystem crossing (ISC), internal conversion (IC), photochemistry, fluorescence, etc.), each of which is difficult to model accurately. Simplifying assumptions for estimating ground state rates (e.g., transition state theory), are frequently not applicable for excited state photochemistry, where excited states barriers can be small or nonexistent and nonstatistical dynamics can govern reaction rates and relative product distributions.

As mentioned earlier, geometries where the excited state and ground state (or another excited state) have the same energy have a special importance for photochemistry. At these geometries, the probability of a molecule surface hopping from one state to the other increases. With surfaces separated by more than a few kcal/mol, the probability of a surface crossing approaches zero. In particular, for a polyatomic system, the surface touching frequently resembles a double cone where the two surfaces meet at a single point, called a conical intersection. Remarkably, moving away from this single touching point in all coordinates except for two (the remaining $3N - 8$ coordinates) maintains the degeneracy between the two surfaces, giving rise to a seam or intersection space of coordinates with an infinite number of degenerate surface energies. At the conical intersection, the two coordinates that break the degeneracy between the two surfaces define the so-called branching space. Typical plots depicting the double-cone topology of conical intersections plot these two coordinates. These two coordinates that lift the degeneracy are related to the so-called g and h vectors. The g vector is the maximum gradient between the two surfaces (sometimes called the derivative coupling vector) and the h vector is the coordinate that involves maximal mixing of the two adiabatic wavefunctions (sometimes called the nonadiabatic coupling vector).

For a one-dimensional system, the probability of a surface hopping can be described approximately by the Landau–Zener equation:

$$P = \exp[(-\pi^2/\mathrm{h})(\Delta E^2/v\Delta s)]$$

where P is the probability of a surface hopping, h is Planck's constant, ΔE is the energy difference between the surfaces, v is the velocity of the nuclear wavepacket, and Δs is the difference in the slopes between the two surfaces. It follows from this equation that the probability of a surface jump is highest when the ΔE term approaches zero or the velocity of the nuclear wavepacket and the difference in slopes between the surfaces becomes large. The spirit of this equation also applies for nonlinear systems, with the added complication that the entire wavepacket cannot fit within the tip of the cone, so the probability of surface hopping is diminished from unity because some of the wavepacket experiences a region where there is an energy separation between the states.

Conical intersections provide a path between two minima, so computational models must include methods for accurately determining the branching plane to allow for the complete prediction of all the potential products available to the excited state. Theoretical methods that have been developed to understand organic photochemistry fall into two principle categories: methods that involve the use of molecular dynamics simulations to model surface hopping rates, and those that do not use molecular dynamics. It is clear that as the development of dynamic simulations improves in both accuracy and speed, excited state molecular dynamics will improve our ability to not only understand but also predict the outcomes of photoreactions.

The following examples are intended to be informative about current methods used for organic photochemistry and the size of the systems that are possible for study using these methods. Ideally, using computational models to understand organic photochemistry would be as simple as using any of the widely used computational chemistry packages; however, development of computational software is an area of active research and the ability to modify existing software packages may be needed. Ultimately, as the computational methods become more robust and the demand to study organic photochemistry increases, it is very likely these methods will become more routinely useable to the average chemist in widely distributed software packages.

The examples are presented by the model theory used to study the problem, with the first presented as a good tutorial for understanding Complete Active Space Self-Consistent Field (CASSCF) or Restricted Active Space Self-Consistent Field (RASSCF) methods, which does not include dynamic simulations calculations (see also Chapter 12). Recent work has made clear that correctly modeling organic photoreaction outcomes requires dynamic simulations of the surface hoping. Consequently, the remainder of the examples all involve dynamic simulations.

18.3 CASSCF/RASSCF

One of the principal reasons photochemistry is challenging to model is because reactions can involve changes from closed-shell to open-shell electronic states, or from covalent to ionic character (within the valence bond depictions).[22] Typically, open-shell singlet states and ionic states are more difficult to model, leading to imbalances in accuracy for the different surfaces and the possibility of large errors. The CASSCF method includes nondynamical correlation by allowing the mixing of all excited state configurations from a user-defined active space into the desired wavefunction. If the user selects the complete active space available to the molecule, within the infinite basis set limit, the method is an exact one. However, the number of configurations increases very rapidly, making such complete active spaces intractable for most chemically interesting systems.

Because it is challenging to achieve complete active spaces above about 14 electrons in 14 orbitals, RASSCF methods can be used that modify and expand upon the CASSCF method. This method incorporates additional electron correlation by expanding the number of active orbitals and electrons. Energies can then be corrected using perturbation theory (e.g., RASPT2) to capture dynamic correlation outside the active space. RASSCF is an extension of the CASSCF method with the active space divided into subspaces. In the example below, three subspaces are used: RAS1, RAS2, and RAS3. All possible electronic configurations of the RAS2 orbitals are considered, but only a limited number of vacancies (typically one) for the RAS1 orbitals and a limited number of occupancies (typically single and double occupancies) for the RAS3 orbitals are possible. Similar to CASSCF methods, these active spaces can be varied depending on the size of the molecular system being studied and the desired accuracy of the calculations.

To perform these calculations in a computationally tractable way, RASSCF typically starts with a reasonable CASSCF active space, for which all excitation types are computed, and adds an additional RASSCF layer, which restricts the number of holes (in occupied orbitals) or electrons (in unoccupied orbitals) into a user-selected restricted active space.[23–25] It is known that CASSCF with small basis sets and incomplete active spaces often inadequately model ionic-like structures in preference for covalent bonding.[22] However, RASSCF makes fully optimized orbitals with analytic gradient calculations possible and should be better able to model both ionic and covalent bonding, given the expanded consideration of electron correlation.

While it is expected that RASSCF will not be capable of reproducing accurate numerical values of spectroscopic data, due to the absence of consideration of dynamical correlation outside the active space, it may provide a reasonable hypersurface that can yield insights from on-the-fly dynamic simulations of photochemical trajectories.[22] For example, recently, Santolini *et al.*[22] used the RASSCF method on the highly studied *cis*-butadiene (BD) and cyclopentadiene (CPD) as examples of a covalent to ionic photoexcitation. The reaction path is fairly straightforward from a planar ground covalent state excited to a nonplanar ionic state that transits through an ionic–covalent conical intersection (CIX) followed by a final nonplanar covalent–covalent CIX yielding the ground-state surface. In this example, Bearpark *et al.* focused only on the reaction paths of the S_1 surface, therefore simplifying the problem to something that is tractable and informative. As to be expected, CASSCF using a minimal (4,4) active space fails in the ordering of the excited states for BD because of the importance of the electron correlation energy to correctly predict/calculate a lowering of the excited B_2 state compared to the $2A_1$ state. An expanded RASSCF calculation obtains the correct ordering of the vertical excitation energies.

Similar to all MCSCF models, a significant error can be introduced by the choice of the orbitals used in the active space. In this example, to improve on the CASSCF(4,4) model, the authors utilize correlating orbital pairs combined with the semi-internal correlation developed by Sinanoglu[26] and Oksuz and Sinanoglu.[27] Choosing orbitals in this way has the advantage of including electron correlation and are developed using the natural bond orbitals (NBO) model of Foster and Weinhold.[28] The orbitals are shown in Fig. 18.1. Expanding the basis set to include penetrating 3p-type orbitals facilitates building higher energy orbitals, having more nodes, which occupy the same space as the π-system. This method better models electron correlation and provides the correct ordering of excited states. This procedure allows for two main points: (1) σ-π semi-internal electron correlation with $[\sigma,\sigma^*]$ orbital pairs and (2) dynamic correlation of the π-electrons using the 3p-type oscillator orbitals in the restricted active space (RAS). On a technical note, the NBO formalism localizes orbitals and therefore they may not have the same molecular symmetry. Before carrying out the RASSCF calculation, it is typical to canonicalize the NBO orbitals, via block diagonalizing a

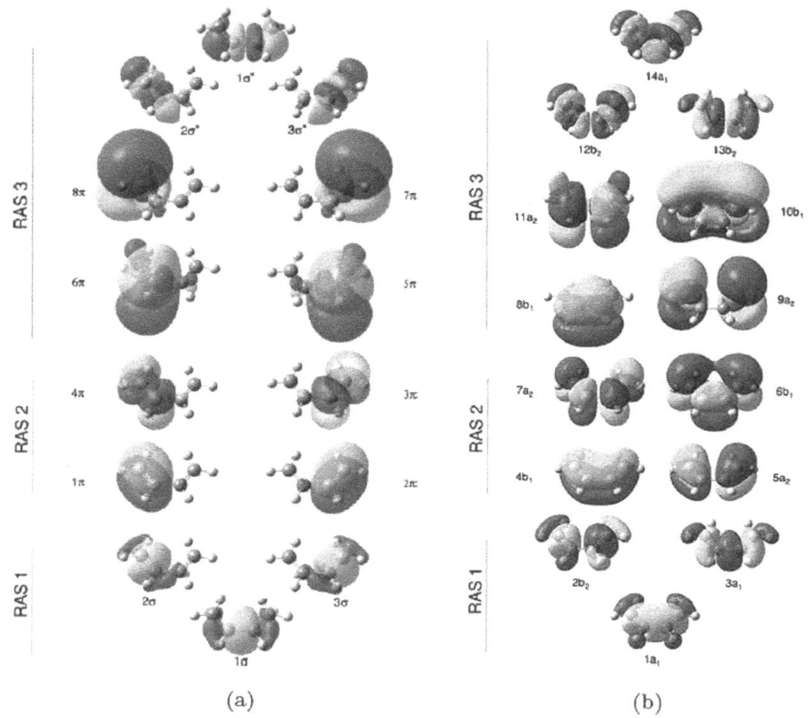

Figure 18.1. NBO (a) used to determine the largest RASSCF active space (b) for BD.

Source: From Ref. [22].

one-electron operator or other suitable method. Ground-state equilibrium geometries were determined by using C_{2v} symmetry, even though CASSCF calculations with larger active spaces calculate this geometry to be slightly higher in energy than a slight twist about the s-cis bond in BD. It is clear from Fig. 18.2 that increasing the active space to include more σ-π semi-internal electron correlation and using the +3p basis set to model the dynamic correlation of the π-electrons yields results that predict the correct orientation of the states — $^1B_2$ lower than $2A_1$ — and is getting qualitatively close, 0.8 eV (18 kcal/mol), to the experimental energy associated with the vertical excitation.

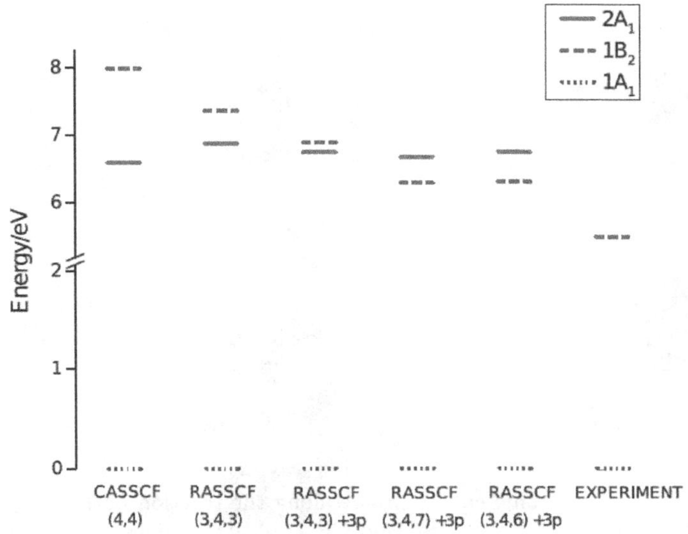

Figure 18.2. Vertical energies for the singlet electronic states of CB at the ground-state equilibrium geometry. There were two basis sets used in this study, that are of a relatively small size, 6-31G* and then a larger basis set that incorporates the 3p functions from the 6-31G basis set for silicon is used for the unsaturated carbons indicated by the "+3p" notation.

Source: From Ref. [22].

With increased confidence that the RASSCF method can yield the correct relative energies for the different states, structure optimizations were carried out using analytical gradient techniques to understand the photochemistry and PESs of these molecules, shown in Fig. 18.3.

Following the steepest decent path, the S_1 energy surface crosses the S_2 energy surface at the S_2/S_1 CIX, leading to reduced-symmetry structures that have decreased energies. The S_2/S_1C_{2v} CIX is shown in detail in Fig. 18.4.

Ultimately, using this method, the authors follow the energy surfaces to map the photochemistry/relaxation of these molecules through CIXs. Bearpark *et al.* demonstrate the importance of electron correlation for molecular structures of excited states, and more importantly photochemistry in general. Using a modest basis set and

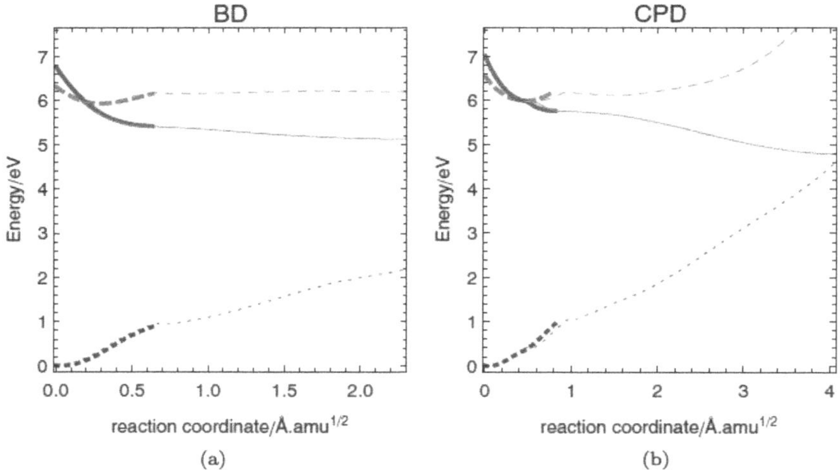

Figure 18.3. Electronic energy profile along the reaction path on S_1, starting from the FC geometry, for BD (a) and CPD (b). Thick lines follow the gradient and conserve the C_{2v} symmetry, with the $2A_1$ shown as a solid line, $1B_2$ shown as a dashed line, and $1A_1$ shown as a dotted line. The thin lines correspond to the reaction paths that deviate from the C_{2v} symmetry, where the electronic states start to mix.

Source: From Ref. [22].

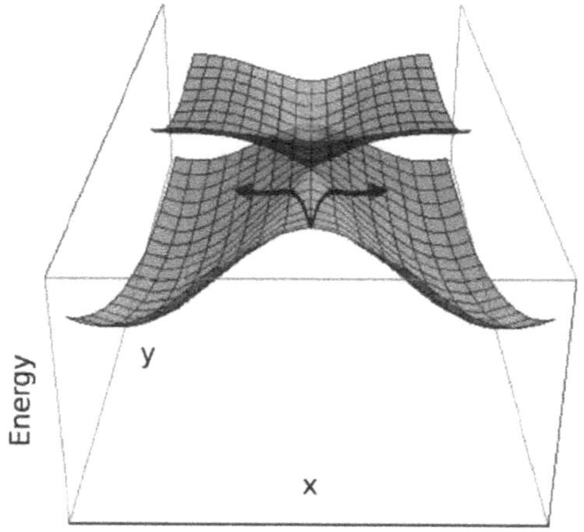

Figure 18.4. Schematic representation of the S_2 and S_1 PESs in the region of the S_2/S_1 C_{2v} CIX. The black arrows represent the branching of the system's trajectories/nuclear wavefunction in the approach to the CIX.

Source: From Ref. [22].

adding larger and more penetrating 3p orbitals with the RASSCF method provides a framework for optimizing and searching excited molecular energy surfaces required to map out the photochemical reaction. These reactions on the excited state surfaces and relaxation often occur through a minimum energy CIX (MECIX).

18.4 Nonadiabatic dynamics

Photochemistry has become a test bed for the development of *ab initio* molecular dynamics (AIMD) since the reactions are sometimes completed with sub-picosecond time scales, making such simulations computationally tractable.[29,30] Ben-Nun *et al.*[29] noted, in a useful review, that AIMD was poised to tackle problems in photochemistry because the combined advances of computational power and the development of experimental techniques to probe these very short timescale reactions with femtosecond spectroscopy. Now, with the establishment of attosecond spectroscopy, there is again a new frontier that can be a powerful tool to test the development of molecular dynamics that involve nonadiabatic processes. There are many methods for computing nonadiabatic dynamics; however, we will be limiting further discussions to a few recent examples or methods that have received broad use.

18.5 SHARC combined with CASSCF

An interesting aspect of photochemistry is the propensity of some systems, typically those containing atoms that are heavier than neon, to undergo intersystem crossing (ISC) during the deactivation process. While many excited chemical systems return to the ground state very rapidly, ISC provides a mechanism to achieve triplet excited states which can have lifetimes that are nanoseconds or longer. Nucleic bases, deoxyribonucleic acid and ribonucleic acid (DNA and RNA), are an interesting system to study this kind of photochemistry as there are a variety of excited state relaxation pathways available to DNA and RNA that span from femtosecond (fs) to nanosecond (ns) time scales.[31] Furthermore, from a biological perspective, it is an important property of DNA and RNA to relax rapidly from excited-state surfaces, dissipating the absorbed energy, thereby avoiding potentially damaging excited-state reactions.

Interest in this field arose from the observation of dimerized pyrimidines, commonly found in photoinduced damaged DNA and RNA, which are thought to occur on the triplet state surface. However, it has been a challenge for the theoretical field to develop methods that can carry out dynamical simulations which include both internal conversion and ISC. Recently, Richter *et al.*[31] used the surface-hopping algorithm, SHARC, a semi-classical *ab initio* molecular calculation, combined with on-the-fly CASSCF calculations for uracil. The CASSCF calculations used a 6-31G* basis set with two different active spaces, one containing 12 electrons in 9 orbitals, CASSCF (12,9), and other active space adding the nonbonding electrons and orbitals of the oxygen atoms yielding 14 electrons in 10 orbitals, CASSCF (14,10) (see Fig. 18.5). CASPT2 single-point energies calculated from the optimized CASSCF geometries were used to benchmark the CASSCF calculations and determine vertical excitation energies. The CASSCF optimized ground state (S_0) geometries

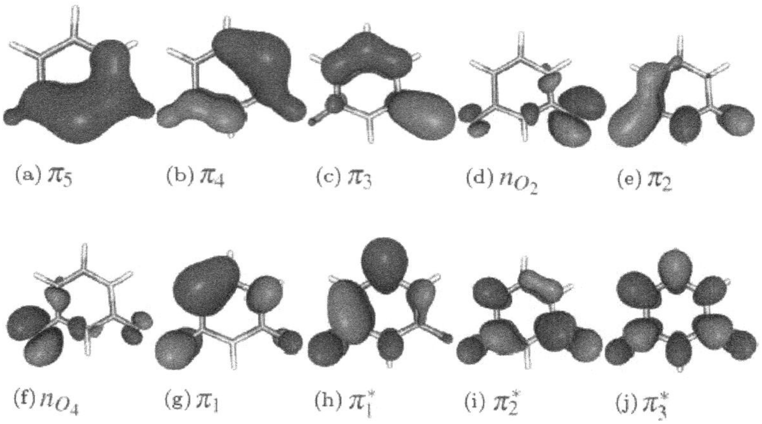

Figure 18.5. Active space orbitals used in the (14,10) CASSCF calculations. For the (12,9) CASSCF calculations the n_{O2} orbital (d) is omitted.
Source: From Ref. [31].

were used to determine initial conditions for 1 ps dynamic simulations. Only trajectories that completed the entire 1 ps or had relaxed to the S_0 or T_1 states for at least 15 fs were considered.

Richter *et al.*[31] found that the level of theory, CASPT2 vs. CASSCF, has dramatic effects (more that 1 eV) on predicted vertical excitation energies, with the largest variation between CASPT2 and CASSCF occurring for the singlet states compared to the triplet states. These differences in predicted vertical energies would be expected to have significant effects on the dynamic simulations, changing the probabilities of surface hopping during the computed trajectories. While it would be ideal to use the more accurate CASPT2 method for the on-the-fly *ab initio* calculations, unfortunately uracil is too large a system for the picosecond time scale to be practical. The faster but presumably less accurate CASSCF method was used. The predicted absorption spectra, CASSCF(14,10) and CASSCF(12,9), are shown in Fig. 18.6. The effect of the size of the active space is also large, as can be seen by the dominance of the S_2 state using the (12,9) active space; compare panels (a) and (b) of Fig. 18.7. The CASSCF(14,10) active space calculates a difference of 0.03 eV for the separation of the S_2 and S_3 states, and both states are similarly populated in the initial excitations, as shown in Fig. 18.7(a).

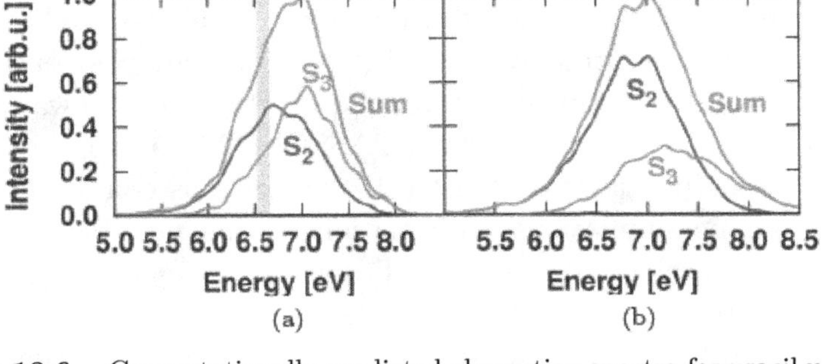

Figure 18.6. Computationally predicted absorption spectra for uracil using (a) CASSCF(14,10) and (b) CASSCF(12,9). The vertical grey bar, in (a), represents the excitation energy range used to model the experimental excitation energy. *Source*: From Ref. [31].

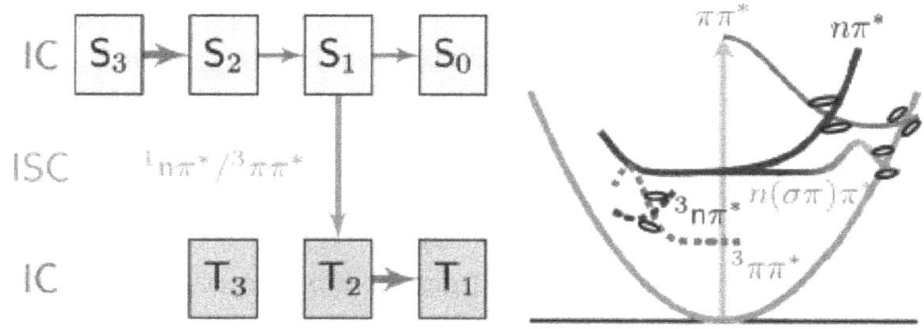

(a) (b)

Figure 18.7. Predicted model for the relaxation of uracil excited with UV light. (a) Scheme showing the predicted electronic states involved, with the thickness of the arrows representing the extent of population transfer. (b) Model showing a spectroscopic representation using molecular orbital transitions and showing the predicted conical intersections for surface hopping. The model has been simplified to a single dimension for ease of drawing, however the relaxation of uracil would occur in multidimensional space.

Source: From Ref. [31].

Using the predicted absorption spectra, Gonzalez *et al.*[13] tested four distinct sets of possible conditions in the dynamics calculations. To determine the importance of including ISC, populations using the same initial excitations to the S_2 and S_3 states were propagated for 600 fs. One set of 26 trajectories, excited to S_2, and 23 trajectories, excited to S_3, were not allowed to undergo ISC. These simulations were compared with another set of 64 trajectories excited to S_2 and 56 trajectories excited to S_3 which could both relax via IC and ISC. A third set of conditions were used to approximate previously published gas phase experimental data. The fourth condition was used to test the active space size by using CASSCF(12,9) as opposed to the CASSCF(14,10) calculations used in the previous conditions. From these dynamic simulations, information on the possible relaxation pathways for uracil, initially excited to either the S_2 or the S_3 states, were generated. Interestingly, Gonzalez *et al.* found that the model including both IC and ISC during the dynamic simulations gives computed decay rates that are in agreement with the experimental observed decay rates. This model, presented in Fig. 18.7, suggests a biexponential decay, with the initial fast, roughly 30 fs, relaxation process arising from IC and a slower, of about 2.4 ps, process arising from ISC. The importance of this work is that, for many photochemical reactions, the excited states may relax through both IC and ISC, and completely correct dynamic simulations should make allowances for both processes.

18.6 Calls for a computational benchmarking protocol using *cis-trans* photoisomerization

There have been many studies investigating the mechanism of the photoinduced isomerization of the retinal protonated Schiff base (rPSB) chromophore.[32–34] The particular features of this photoreaction make it useful for benchmarking the accuracy of new methods. This photoreaction presents a particularly challenging problem for theoreticians studying CIXs because this is a very large system if one includes the protein surrounding the rPSB. Furthermore, the rPSB chromophore itself is of moderate size, and many studies focus on using a reduced model of the rPSB so that higher computational methods are possible. In an interesting publication by Gozem *et al.*,[32] which establishes a benchmarking protocol to test theoretical methods, penta-2,4-dieniminium cation (PSB3) is used as a model compound for rPSB, as seen in Fig. 18.8. In a fairly comprehensive test, they included a variety of methods investigating if the method could correctly describe the shape and dimensionality of the CIX.[32]

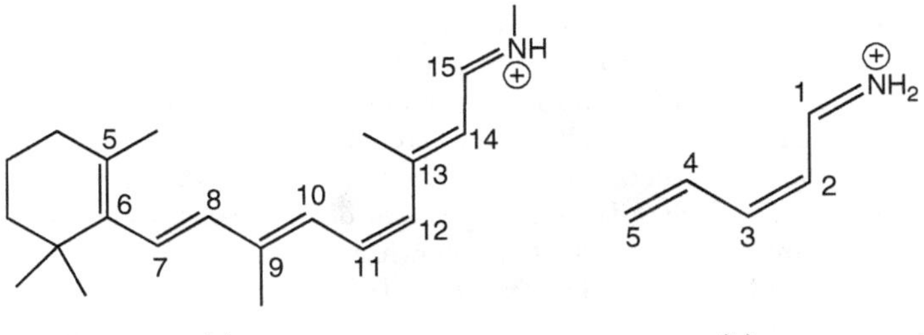

(a) (b)

Figure 18.8. Structure of rPSB in pigments used for (a) vision and (b) the reduced model version, penta-2,4-dieniminium cation.

Source: From Ref. [32].

This chromophore undergoes a *cis-trans* photoisomerization, arising from rotation of the π-bond between C2 and C3. The photoisomerization reaction is thought to proceed through a CIX from the S_1 to S_0 states. Gozem *et al.*[32] have found, using PSB3, that the nature of the PES consists of either significant charge transfer (CT) or diradical (DIR) character depending up the region of the PES surrounding the CIX. This feature makes this system useful as a model because of the difficulties in correctly modeling these two types of states. Furthermore they have found that the two axes of the branching plane (BP) leads to changes in bond lengths, vs. an electronic character change from more DIR to more CT character, shown in top part of Fig. 18.9(b).

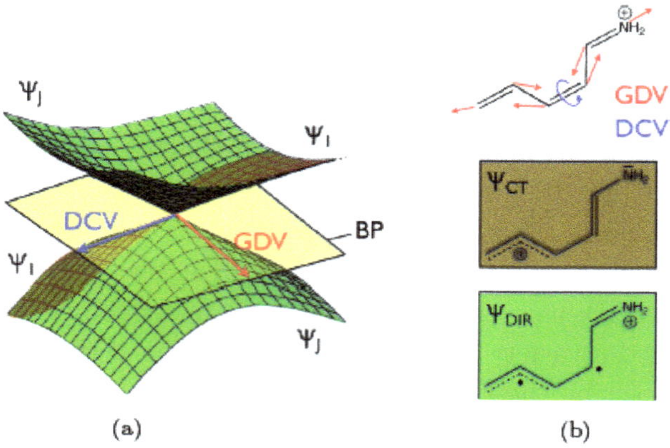

Figure 18.9. (a) Peaked representation of the PESs of the S1 and S0 electronic states at the CIX. The electronic character of the PESs are given by adiabatic wave functions Ψ_I shown in brown and having charge transfer character and Ψ_J shown in green and having diradical character. Bond length changes corresponding to the gradient difference vector vs. rotation of the π-bond between C2–C3 corresponding to the derivative coupling vector of the branching plane are shown as the yellow plane. (b) (Top) Diagram showing the dominant molecular changes of the branching plane. Bond length alternation of the gradient difference vector: red arrows indicating the lengthening of the π-bonds and shortening of the sigma bonds, blue arrow indicates the rotation of the π-bond corresponding to the derivative coupling vector. The Lewis structure representations of the charge transfer and diradical are shown in brown and green boxes, respectively.

Source: From Ref. [32].

Beyond providing a balanced depiction of covalent and diradical states, as described by Gozem *et al.* and references therein,[32] it is crucial for the theoretical method employed to correctly depict the BP of a CIX. Specifically, the BP should exist only in two degrees of freedom: the gradient difference vector (GDV) and the derivative coupling vector, DCV. To discover which methods — multiconfigurational, equations of motion couple cluster, and density functional theory (DFT) — give the CIX the expected shape and dimensionality, these authors computed the S_1 and S_0 energies for a ring centered on the crossing point for each method (see Fig. 18.10). By calculating the difference of the S_0 and S_1 energies around the ring shown in Fig. 18.10, it is possible to determine if a method incorrectly predicts the intersection to be a line, meaning there are multiple points of energy degeneracy. This line crossing is representative of a method that is predicting a $3N - 7$ degrees of freedom for the CIX, whereas it should be $3N - 8$ degrees of freedom. Unfortunately, from this procedure, it is not possible to determine if the method predicts a CIX with greater than $3N - 8$ degrees of freedom. Olivucci *et al.* found that a variety of methods

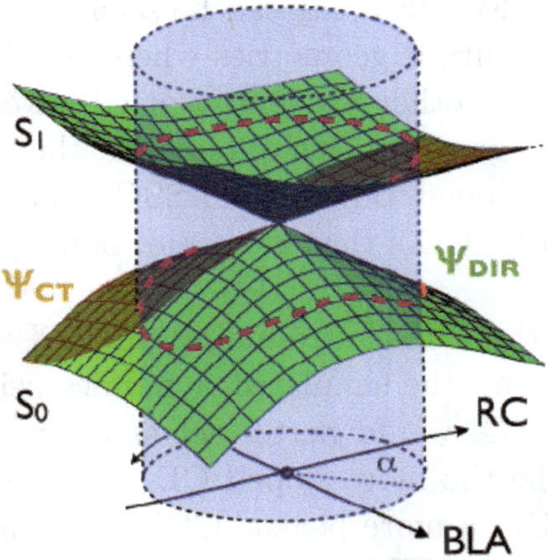

Figure 18.10. The CIX is illustrated as a peaked, double cone, crossing and the ring of structures (blue cylinder and red dotted lines on the PESs).
Source: From Ref. [32].

predict the correct dimensionality of the CIX and the BP. These methods include CASSCF, MRCISD, MS-CASPT2 with a symmetric Hamiltonian matrix or a large active space, MS-CASPT2 with IPEA = 0.25, XMCQDPT2 and QD-NEVPT2 with symmetric Hamiltonian matrices, SI-SA-REKS, and SF-TDDFT without including spin projection. Methods that predicted a linear crossing in the BP, having $3N - 7$ dimensions of the CIX, predict degenerate energies for two points along ring of structures. These methods include MRCISD+Q, SS-CASPT2, SS-CASPT2 with IPEA = 0.25, EOM-SF-CCSD(dT), and TD-mPW2PLYP. Methods that predicted the correct, nondegenerate energies for the computed ring of structures, however, are expected to yield a $3N - 9$ dimensional space of the CIX. These included MS-CASPT2, QD-NEVPT2, and the EOM-CC method.

For any new method in need of benchmarking, determining the S_1 and S_0 energies for a ring centered on the crossing point for the photoreaction is expected to yield a plot of the CIX. The plots generated by scanning the BP should produce CIXs that are linear, having more than one degenerate energy geometry, or peaked/sloped CIXs which will have only one degenerate energy geometry. Some of the plots are shown in Fig. 18.11. It can be seen that SS-CASPT2 with IPEA = 0.25 has multiple geometries where the S_1 and S_0 PESs are degenerate, while the other methods plotted have the correct single degenerate energy geometry. Of interest is the changing shape of the CIX: most are sloped, but with CASSCF and MS-CASPT2 with IPEA = 0 predicting CIXs that are somewhere between peaked and sloped PESs.

The shape of the PESs surrounding the CIX would be expected to have an impact on the simulated dynamics using these surfaces: the more peaked CIX would better act as a funnel facilitating surface crossing, while the more sloped CIX might slow surface crossing. Using CASSCF, a more peaked CIX was computed, while with the SS-CASPT2 with IPEA = 0.25, a more sloped CIX was computed. Then, Olivucci *et al.* ran semiclassical dynamics of 100 trajectories using the *trans*-3-MePSB3 model structure. Interestingly,

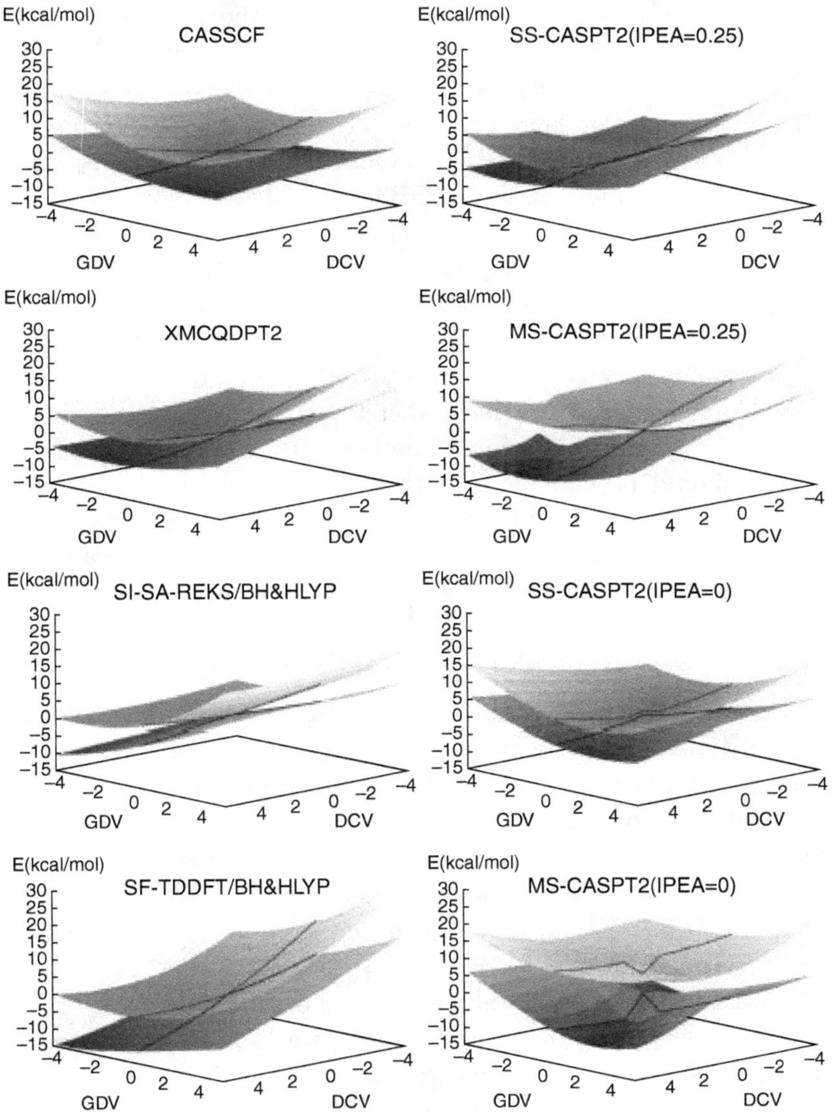

Figure 18.11. S_0 and S_1 PES scans along the branching plane. Plots were generated using an 11×11 grid of points centered on the CIX.

Source: From Ref. [32].

these initial results suggest that even though the shape of the PESs is fairly different, the two methods still predicted very similar crossing dynamics. The authors emphasize that further work needs to be done on this system. They find that, at the CASSCF level, nine trajectories remained unreacted on the excited S_1 PES, 58 trajectories cross to the product, *cis*-3Me-PSB3, and 33 trajectories return to the *trans*-3Me-PSB3 starting material, at the end of 200 fs simulation. This result is very similar to the SS-CASPT2 level, where 17 trajectories stay on the excited S_1 surface, 60 trajectories cross to from products, and 23 trajectories returned to the starting material. Overall this work establishes a protocol to be used to benchmark a method's ability to correctly predict the shape of the PESs around a CIX. For this benchmarking photoisomerization reaction, any new method should be capable of correctly modeling the PESs at and around the CIX, yielding PESs that have only one degenerate geometry at the CIX and no other PESs crossing points are calculated around the circle about the CIX.

18.7 *Ab initio* multiple spawning dynamics

While a static depiction of CIXs are useful, ultimately determining the dynamics of photochemical reactions is required in order to understand the rates of photochemical reactions and understand how photoreactions partition to different minima accessible at funnels of the PESs. This problem requires simultaneously determining the PESs of both the ground and excited states, while also calculating how the reaction will evolve according to the landscape of the excited and ground PESs. While the theory to carry out these dynamic calculations is available, it is unfortunately very computationally expensive to do so in a rigorous way, except for systems of a few atoms. Some approximations must be made to allow for the study of desired chemical systems.

Ab initio multiple spawning (AIMS) is a method that uses simplifications for determining the trajectories of the dynamics via the full multiple spawning (FMS) method while still solving the electronic and nuclear Schrödinger equations simultaneously.[29,35,36] Liu *et al.*[37] has recently used AIMS with MS-CASPT2 to study the *cis-trans* isomerization of PSB3, the model compound mentioned previously. Similar to Olivucci *et al.* discussed earlier, Liu *et al.* stress the importance of ensuring that the CIX possesses the correct dimensional space, namely exactly two molecular degrees of freedom that break the energy degeneracy of the excited and lower energy states. If the method determines a CIX with an artificial topography, then it is possible that "sloshing" between the excited and lower energy states may arise.[38] This is expected if the PESs surrounding the CIX allow for molecules to hop back and forth between the energy states multiple times.[37,38] Dynamic simulations have established that surface hopping can occur at points away from the CIX, so to address this issue, Mori and Martinez[39] utilized the seam space nudged elastic band (SS-NEB) method. This method searches for these nonadiabatic transitions along the intersection seam by finding the minimum energy path between two CIX while maintaining the electronic state degeneracy. Liu *et al.* use SA3-CASSCF(6,6)/6-31G, SA3-MS-CASPT2(6,6)/6-31G, and AS3-MS-CASPT2(6,6)/6-31G* methods to optimize the S_0 and S_1 energy states. From these methods, they find that there are significant differences between the CASSCF and MSPT2 PESs. Both the CASSCF and MSPT2 methods identify the optically bright S_1 excited minima; however, the CASSCF method overestimates this vertical energy by roughly 16.5 kcal/mol, relative to the same vertical energy calculated via the MSPT2 method. Furthermore, the CASSCF method also calculates three distinct minima with geometries that are twisted about each of the π-bonds, while the MSPT2 method only identifies a single minimum on the excited surface. These results demonstrate the importance of including dynamic electronic correlation effects when calculating excited electronic states.[14,33,40] Both methods find three CIXs, corresponding to twisting about the individual three π-bonds, and both methods find that twisting around the central C–C π-bond to be the lowest energy CIX. However, CASSCF also predicts two other CIXs corresponding twisting about multiple π-bonds simultaneously, the so-called hula-twist and bicycle-pedal types of motion.[41,42] Furthermore, the lowest energy intersection seam of the PESs connecting the two lowest energy CIXs are different for

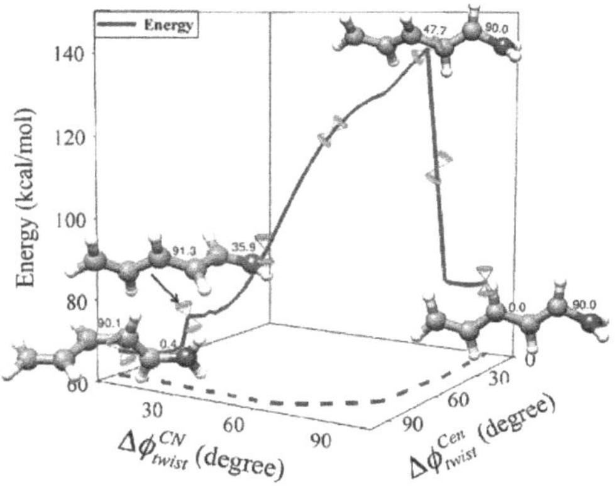

(a)

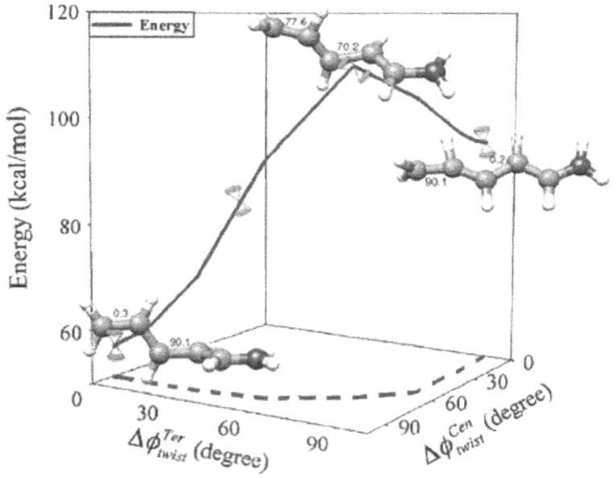

(b)

Figure 18.12. (a) CASSCF-calculated minimum energy intersection seam connecting the two lowest energy CIXs. (b) MSPT2-calculated minimum energy intersection seam connecting the two lowest energy CIXs. All energies are relative to the optimized *trans* S_0 Frank–Condon point.

Source: From Ref. [37].

CASSCF, which connects the CIXs twisting about the C–N π-bond and the central C–C π-bond, compared to MSPT2, which connects the CIXs twisting about the terminal C–C π-bond and the central C–C π-bond, shown in Fig. 18.12. Interestingly, energetically allowed intersections on the CIX seam have a more peak topography, while intersections in the higher energy forbidden regions have a more sloped topography. Also the topography of the MSPT2-calculated CIXs is more peaked than the CASSCF-determined CIXs.

These predicted PESs were then utilized in dynamics simulations using the AIMS method. It is found that, by analyzing the population densities that spawn to the S_0 surface, there is a broad distribution, indicating that many trajectories are crossing outside of the CIXs. This result suggests that simply finding the CIX may not provide an accurate description of the nonadiabatic process and that dynamic calculations are required. More importantly, the reaction path and outcome predicted by the two methods is significantly different, as shown in Fig. 18.13. The CASSCF method predicts the major product of the photodynamics to return to the starting *trans* geometry while the MSPT2 method predicts the major product to be the isomerized *cis* geometry, giving a much higher predicted quantum yield for the process.

18.8 TD-DFT on-the-fly surface-hopping method

While the CASPT2/CASSCF methods are anticipated to yield reliable results, it is becoming increasingly important to identify computational methods that will utilize reasonable levels of theory for chemically interesting systems combined with on-the-fly dynamics to understand the landscape and nature of the surface crossing. In this section, we will be presenting some recent reports of methods that carry out these types of calculations.

JADE, developed by Du and Lan[43] is a dynamics simulation that utilizes Tully's trajectory surface-hopping strategy with TD-DFT methods for determining molecular geometries and energies. This methodology was successfully used to study the lifetimes of the excited states of a range of molecular sizes from very small, methaniminium cation ($CH_2NH_2^+$) to a moderate-sized molecule, fullerene C_{20}. Of interest was the ability of the method to provide an explanation for the dual luminescence observed for *p*-dimethylaminobenzonitrile, which is not observed in absence of the methyl groups, p-aminobenzonitrile. Importantly, these structures, which have conjugated electron-donating and -accepting groups, imply that charge transfer effects will be critical. The researchers

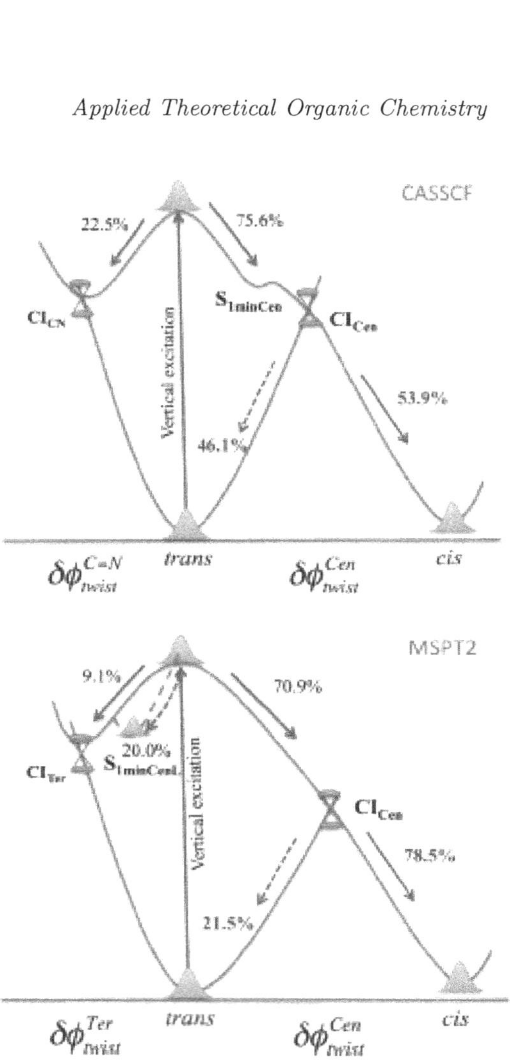

Figure 18.13. Reaction schemes predicted for the photoisomerization of PSB3 from the dynamics simulations, AIMS/CASSF upper panel and AIMS/MSPT2 lower panel.

Source: From Ref. [37].

recommend that the use of range-separated functionals, i.e., CAM-B3LYP, are needed to better model this phenomena.

18.9 Global reaction route mapping

On a more human level, computations can only give answers to questions posed by the chemist. Frequently, possible mechanisms of reactions are overlooked due to a lack of imagination. This problem manifests itself in the calculation of thermal reaction mechanisms by computational chemists failing to calculate all possible mechanisms available to a reaction. This problem is exacerbated in photochemical mechanisms because it is very challenging to rely on a chemical intuition to guide the calculations of excited states. For most people, it is easier to predict the location of thermal transition states than the locations of seam crossings or excited-state transition states. In the long term, solutions to the human lack of imagination are needed.

The global reaction route mapping (GRRM) is a method developed by Maeda *et al.*[44] (see Chapter 11) that tries to resolve this human problem by identifying minimum energy conical intersections (MECIXs) and minimum energy seam crossings (MESXs) in an automated fashion that does not require *a priori* reasonable guesses of the minimum energy geometries. At the core of GRRM is minimizing a penalty function, called the seam model function (SMF) of the computed PESs. The GRRM is agnostic to the model theory but is typically performed using a modest quantum mechanical method such as CASSCF with a small active space, which will lead to minima in regions where the PESs approach and or cross (see Fig. 18.14(a)).[45] Furthermore, the GRRM code can be used to search for a minimum energy reaction path from an excited state using an Avoiding Model Function (AMF). This function can be minimized and, by using a coupling term, the discontinuity at the MECIX or MESX is smoothed, allowing the use of well-established PES search methods to find the minima and saddle points (see Fig. 18.14(b)). All minima, including MECIX and MESX, and saddle points found using GRRM must then be reoptimized to find more accurate geometries and energies, typically with higher levels theory. Critically the reoptimization step is stressed, as minima and saddle points found using GRRM at lower levels of theory have been reported to disappear upon using higher levels of theory.[44] To date, the GRRM method has been used on a variety of molecules ranging from very small molecules, such as the photodissociation of formaldehyde,[45] to investigating the quenching pathways of moderately large aceneimides, with the largest consisting of nine fused rings.[46]

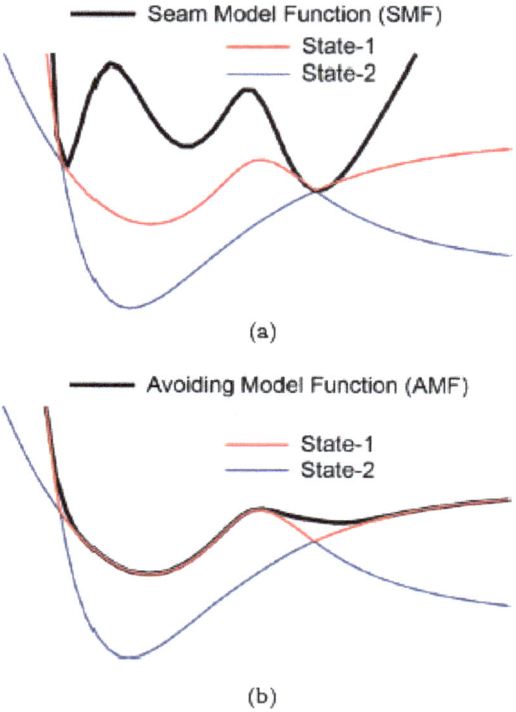

Figure 18.14. Schematic showing how the (a) SMFs and the (b) AMFs as black lines for the two PESs.
Source: From Ref. [44].

An example of the usefulness of the GRRM method is the photolysis of formaldehyde which can involve three PESs: the singlet ground electronic state (S_0), the singlet first excited electronic state (S_1), and the lowest triplet electronic state (T_1). Furthermore, in 2004, it was shown experimentally and theoretically that there exists an alternate reaction path called roaming that does not follow typical transition state theories, where one of the H atoms partially dissociates and then moves along the HCO fragment to eventually abstract the other H atom, thus generating the products CO plus H_2.[47] Using the GRRM method followed by CASPT2, Maeda *et al.*[44] were able to produce the data in Fig. 18.15. This demonstrates the versatility of

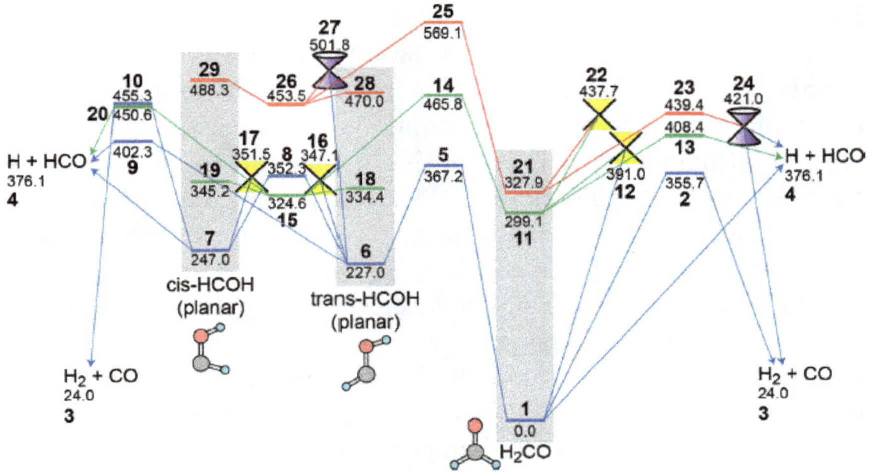

Figure 18.15. Summary of the PESs (in kJ/mol) for the S_0 (blue), S_1 (red), and T_1 (green) of formaldehyde at the CASPT2 level. Yellow/black crosses represent MESX points and cones mark MECIX points that were found.

Source: From Ref. [44].

this method and that it can handle both triplet and singlet electronic states.

18.10 CIS with FSSH dynamics

A promising and potentially computationally modest method that is common to many chemists studying excited-state systems is to use the so-called time-dependent density functional theory (TD-DFT). Subotnik *et al.* have been working with both TD-DFT and time-dependent Hartree–Fock (TD-HF) to develop these methods for use in understanding photochemistry, from ultrafast spectroscopy to energy transfer processes on excited states.[48–53] While they have made many contributions to the advancement of using TD-DFT and TD-HF to study photochemistry, which they recently reviewed,[51] we will focus on their work to model triplet energy transfer in organic chromophores using configuration interaction singles (CIS).[51,53] Utilizing the previous work of Closs *et al.*, who studied electron transfer in donor–bridge–acceptor organic chromophores,

Landry *et al.* focused on 4-(2-naphthylmethyl)benzaldehyde, called compound M, using CIS and a fewest switches surface hopping (FSSH) dynamics simulation to gain insight into the coupled nuclear-electronic dynamics of this electronic energy transfer (EET) process.[53–55] Compound M is interesting in that the electronic energy transfer cannot be described by Marcus theory. This break from Marcus theory arises in part because the flexibility of the methylene bridge allows the diabatic couplings to vary with the orientation of the donor and acceptor π-systems, shown in Fig. 18.16. The experimental evidence, using pump-probe spectroscopy, suggests that the electronic states involved are the initial ground state singlet; photoexcitation of the molecule generates the first excited singlet state (S_1), which then rapidly crosses intersystem to the first excited triplet state (T1).[55] This triplet state can exist in two energy wells: one with the excitation of the donor (red line in Fig. 18.17), and a lower energy

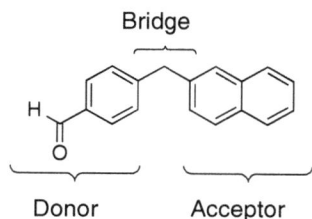

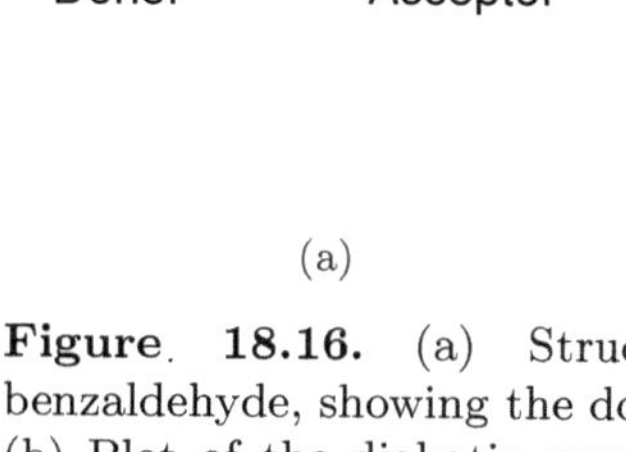

(a)

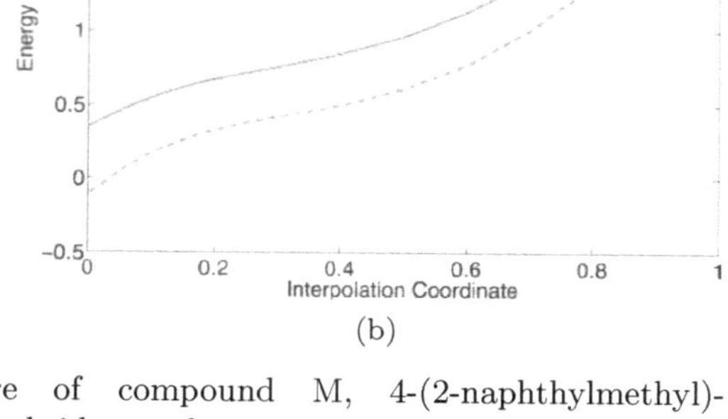

(b)

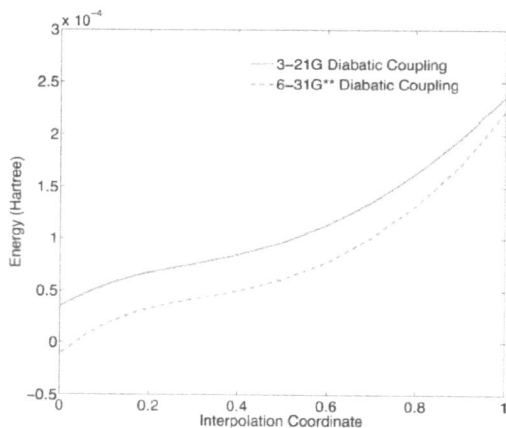

Figure. 18.16. (a) Structure of compound M, 4-(2-naphthylmethyl)-benzaldehyde, showing the donor, bridge, and acceptor portions of the molecule. (b) Plot of the diabatic coupling for compound M interpolated from the optimized geometry for excitation of the donor on the left to the optimized geometry for the excitation of the acceptor on the right, using CIS/3-21G, solid line, and CIS/6-31G**, dashed line.

Source: From Ref. [53].

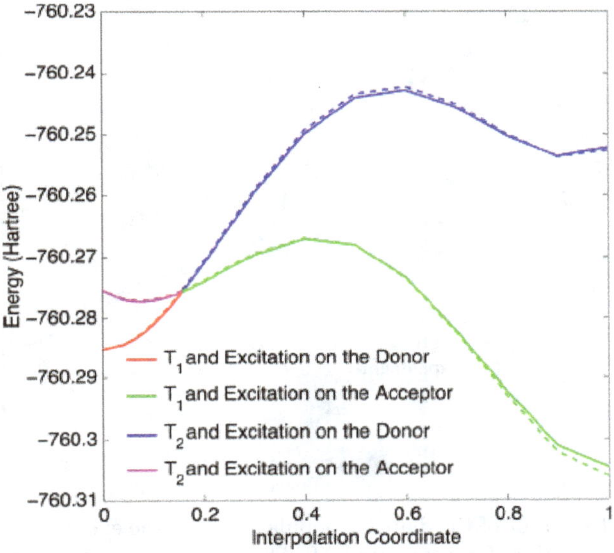

Figure 18.17. Plot of the energies of the T_1 and T_2 states as the optimized geometry is interpolated between the optimized geometry for the excitation on the donor to the optimized geometry for the excitation of the acceptor. The calculations were performed using CIS with two different bases: the 3-21G, solid lines, and 6-31G**, dotted lines. The energies of the larger basis set, 6-31G**, were scaled by a constant energy to have both bases agree at the zero interpolation coordinate.

Source: From Ref. [53].

well that is the excitation of the acceptor (green line in Fig. 18.17). For this system, there is a conical intersection between the T1 and second excited triplet state (T2) very close to the transition between the two energy wells of the T1 energy surface, shown in Fig. 18.17. Landry and Subotnik[56] then use both FSSH and a modified version of FSSH that they developed which incorporates stochastic decoherence events, called augmented FSSH (A-FSSH), to carry out dynamic simulations starting from the energy well of the optimized donor on the T1 surface. Using A-FSSH yields a much closer prediction of the experimental data shown in Fig. 18.18; however, for compound M, both A-FSSH and FSSH are within a factor of 2 of the experimental value. An interesting rule of thumb is that including decoherence in

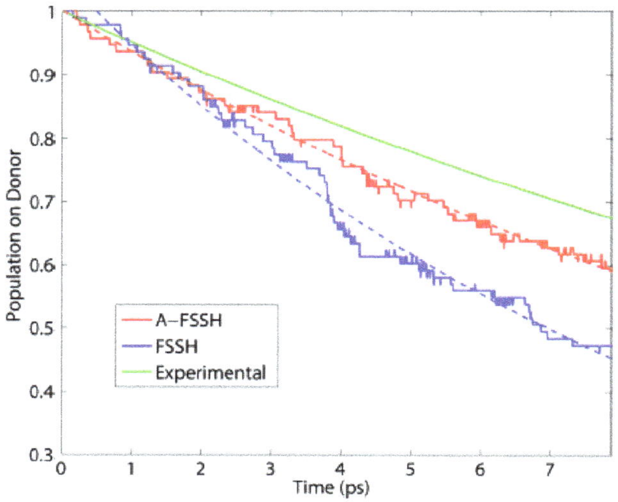

Figure 18.18. Plot of the diabatic population with the excitation on the donor as a function of time for compound M. The rate of the experimental decay of compound M, given by the green line, is 20 ps.[55] Fitting the computed decays, FSSH given by blue line and A-FSSH given by red line, with exponential fits, given by the dashed lines, yields decay times of 15 ps for A-FSSH and 9.3 ps for FSSH.

Source: From Ref. [53].

the FSSH slows down the nonadiabatic transfer rates and therefore possibly yields more accurate predictions. A more complete discussion of the importance of including decoherence is presented by these authors and interested readers are directed to Refs. [56, 57].

To understand the topology of the EET, on-the-fly electronic structure calculations during the surface hopping dynamic simulations, were carried out. The results of the trajectories are plotted in Fig. 18.19(a), showing that the T_1 and T_2 adiabatic surfaces appear to touch forming a conical intersection. Similar to the benchmarking method of Olivucci and coworkers presented above, to verify that the surface touching point is a CIX the relative magnitude of the derivative couplings around a circle about the CIX were calculated and found to have a nonvanishing curl indicating a CIX, shown in Fig. 18.19(b). In this final example, Landry *et al.* are able to model

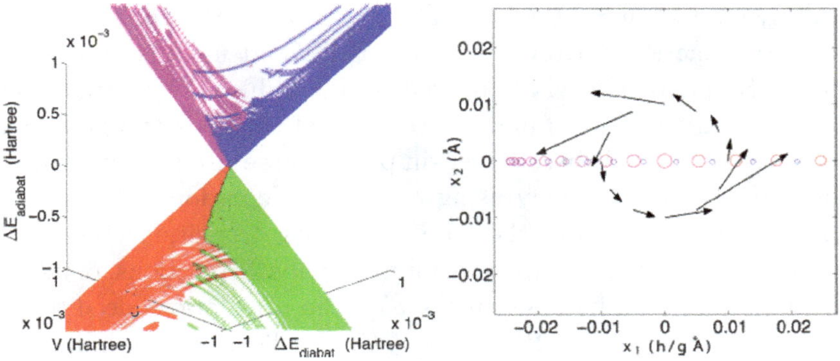

Figure 18.19. (a) Plot of the A-FSSH trajectories close to the crossing point. The three-dimensional scatter plot is generated by plotting the difference between adiabatic energies ($\Delta E_{\text{adiabat}}$) as a function of the difference between diabatic energies (ΔE_{diabat}) and the diabatic coupling (V). The color scheme is the same as in Fig. 18.17. (b) Plot of the magnitude of the derivative couplings (black arrows) around the point of closest approach to the CIX (central largest red circle). The circles, representing the projection of the nuclear coordinates into the branching plane, show the path of the closest approach trajectory, entering from the right and passing by the CoIn before turning around and passing again to exit on the right. The color of the circle indicates time, starting with red and turning to blue. The size of the circle is inversely proportional to the distance from the branching plane at that time step. The labels x_1 and x_2 represent the rescaled vectors spanning the plane of the adiabatic force difference and the derivative coupling.

Source: From Ref. [53].

the dynamics of energy transfer in compound M and demonstrate that this transfer occurs through a CIX.

18.11 Practical considerations for choosing a method for a specific problem

As described earlier, photochemistry presents a challenging problem to those wanting to model the mechanisms in detail. The problems involved are both human (lack of imagination and intuition on thinking about where CIXs might occur), and computational (difficulty in modeling excited states accurately and surface crossing topologies and dynamics, in particular). The nonadiabatic nature of photochemical reactions involves multiple PESs and understanding how the PESs interact is vital to predicting product distributions and future chemistry. Hopefully, the theoretical methods to understand photochemistry will become as simple as those used to understand thermal reactions, namely using *ab initio* calculations to search for the stationery points of the PES, reactants, transition states, and products. The added complexity for nonadiabatic reactions, of involving multiple PESs, means that the *ab initio* calculations need to also find crossing points of PESs, the CIX. While there are numerous reports and great strides have come from developing very high level *ab initio* calculations to map out the PESs, it will be the development and use of dynamic simulations with the PESs that will allow for chemical accuracy and comparison to experimental values.

For smaller systems of a few atoms, much of the pioneering work to study CIX using extremely high levels of theory has yielded results comparable to ultrafast spectroscopy.[5,58–60] However, this level of theory remains difficult for systems that are typically studied for organic photochemistry and is often not coded in the widely used commercial quantum mechanical software packages. However, given that studying photochemistry theoretically is still in its infancy and yet to grow into a widely utilized tool as in thermal chemistry, this means much of the software and methods used to carry out the dynamic simulations for surface hopping is still being developed and may not be widely distributed in commercial packages. However, an attempt is provided here to at least give a reasonable reference for what levels of theory and which software packages are commonly used to study organic photochemistry, using the above examples as a starting point and assuming the reader has access to a sufficiently large computer cluster.

Robb and Bearpark *et al.*[22,61–64] have been instrumental in the development and use of the Gaussian software to study CIX using CASSCF and RASSCF. In the example mentioned earlier, the relatively small molecules of *cis*-butadiene and cyclopentadiene are used with modest basis sets of 6-31G* and an expanded 6-31G* basis set. For structures where the active space is around 12 electrons in 12 orbitals, CASSCF with these basis sets is a reasonable choice for determining geometries and energies using many different codes, including Gaussian, GAMESS, MOLPRO, or MOLCAS, to name a few of the common computational chemistry packages. However, as stated previously, dynamic simulations on the PESs are required to fully understand and predict a photochemical outcome. Unfortunately, the combination of the dynamic simulations using the calculated PESs is not yet widely available in some of those computational chemistry packages, and much work has been done to develop methods that require the user to understand some computer programming to implement. Richter *et al.*[31,65] have developed SHARC, a semiclassical *ab initio* method which includes dynamic simulations using a surface-hopping algorithm and studied the relaxation of uracil. This method, using MOLPRO 2012, used CASSCF(14,10) with a 6-31G* basis set to optimize the geometries; however, the more accurate CASPT2 was used to calculate single-points energies from the CASSCF geometries. This work highlights a common theme of using the more accurate CASPT2 method on geometries obtained from optimizations with lower levels of theory. Depending on the theoretical method that is used, it is possible to obtain spurious CIX geometries that, when tested at higher levels of theory, are no longer a CIX. Therefore it is advised, similar to Olivucci, Martinez, and Landry, that proving the PESs do cross at a CIX of the correct dimensionality is important, and this can be easily done by computing a property, such as the PES of the h and g vectors in a radius about the CIX geometry yields only a single point of energy degeneracy.[32,37,51,53,66] Levine *et al.*[67] have developed AIMS, a method that has been implemented into MOLPRO, and interested readers should review the publication by the research group discussing the use of AIMS in MOLPRO. JADE, developed by Du and Lan, has been interfaced with several popular quantum mechanical modeling software packages such as Turbomole, Gaussian, and GAMESS. This method is a dynamics simulation that utilizes Tully's trajectory surface-hopping strategy with TD-DFT methods for determining molecular geometries and energies.[68] While the initial publication of JADE is primarily concerned with TD-DFT, it has been coded to also use CIS and TD-HF as well.[68] GRRM, which is licensed from the developers, has been interfaced with MOLPRO, GAMESS, and Gaussian, and could be interfaced with any computational chemistry program by the user, is an automated way of searching multiple PESs to find critical points which are then further reoptimized with a higher level of theory.[44–46,69,70] Finally, the effects of solvent or the environment around the chromophore may influence the photochemistry. While it is impossible to accurately model the large number of atoms that make up the solvent, enzyme pocket, or other local environment that the model photochemistry occurs in, it is possible to use a layered approach or hybrid approach using quantum mechanical calculations to model the essential atoms and either lower level quantum mechanics or molecular mechanics to model the surrounding atoms. The models, typically referred to as ONIOM models, can be very useful for partitioning the calculation into different layers. Recently, Manathunga *et al.*[71] used this approach for the PSB3 model system of Rhodopsin with the PSB3 portion treated quantum mechanically, the surrounding 4 Å enzyme pocket treated molecular mechanically and the rest of the enzyme fixed. The quantum mechanical calculations were performed using CASSCF followed by CASPT2 for more accurate energies, while the AMBER force field is used for the molecular mechanics. These calculations were carried out using the TINKER software package coupled with MOLCAS.[71] We anticipate many new developments in the modeling of photochemical reactions in the coming years to accommodate the growing interest in studying these important reactions.

References

1. Schapiro, I., Melaccio, F., Laricheva, E. N. and Olivucci, M. (2011). Using the computer to understand the chemistry of conical intersections, *Photochem. Photobiol. Sci.*, 10, pp. 867–886.
2. Yarkony, D. (2012). Nonadiabatic quantum chemistry — past, present, and future, *Chem. Rev.*, 112, pp. 481–498.

3. Yarkony, D. (2001). Conical intersections: The new conventional wisdom, *J. Phys. Chem. A*, 105, pp. 6277–6293.
4. Yarkony, D. (1998). Conical intersections: Diabolical and often misunderstood, *Acc. Chem. Res.*, 31, pp. 511–518.
5. Domcke, W., Yarkony, D., Johnson, M. and Martinez, T. (2012). Role of conical intersections in molecular spectroscopy and photoinduced chemical dynamics, *Annu. Rev. Phys. Chem.*, 63, pp. 325–352.
6. Domcke, W., Yarkony, D. R. and Koppel, H. (2004). *Conical Intersections Electronic Structure, Dynamics and Spectroscopy*, Vol. 15 (World Scientific Publishing Co. Pte. Ltd., Singapore).
7. Paterson, M., Bearpark, M., Robb, M., Blancafort, L. and Worth, G. (2005). Conical intersections: A perspective on the computation of spectroscopic Jahn-Teller parameters and the degenerate 'intersection space,' *Phys. Chem. Chem. Phys.*, 7, pp. 2100–2115.
8. Lasorne, B., Worth, G. and Robb, M. (2011). Excited-state dynamics, *Wiley Interdiscip. Rev. Comput. Mol. Sci.*, 1, pp. 460–475.
9. Blancafort, L., Lasorne, B., Bearpark, M. J., Worth, G. A. and Robb, M. A. (2009). In *The Jahn-Teller Effect Fundamentals and Implications of Physics and Chemistry*, eds. Koppel, H., Yarkony, D. R. and Barentzen, H., (Springer), pp. 169–200.
10. Klessinger, M. (2001). Introduction to the workshop on theoretical organic (photo)chemistry, *J. Photochem. Photobiol. A Chem.*, 144, pp. 217–219.
11. Haas, T., Klessinger, M. and Zilberg, S. (2000). Conical intersections in photochemistry, spectroscopy, and chemical dynamics — Preface, *Chem. Phys.*, 259, pp. 121–122.
12. Serrano-Andrés, L., Merchán, M. and Lindh, R. (2005). Computation of conical intersections by using perturbation techniques, *J. Chem. Phys.*, 122, p. 104107.
13. González, L., Escudero, D. and Serrano-Andrés, L. (2012). Progress and challenges in the calculation of electronic excited states, *Chemphyschem*, 13, pp. 28–51.
14. Levine, B. and Martinez, T. (2007). Isomerization through conical intersections, *Annu. Rev. Phys. Chem.*, 58, pp. 613–634.
15. Turro, N. J. (1991). *Modern Molecular Photochemistry* (University Science Books), www.uscibooks.com.
16. Turro, N. J. & Schuster, G. (1975). Photochemical reactions as a tool in organic syntheses, *Science*, 187, pp. 303–312.
17. Hammond, G. S. and Turro, N. J. (1963). Organic photochemistry, *Science*, 142, pp. 1541–1553.
18. Klessinger, M. and Michl, J. (1995). *Excited States and Photo-Chemistry of Organic Molecules* (Wiley).
19. van der Lugt, W. T. A. M. and Oosterhoff, L. J. (1969). Symmetry control and photoinduced reactions, *J. Am. Chem. Soc.*, 91, pp. 6042–6049.
20. Devaquet, A., Sevin, A. and Bigot, B. (1978). Avoided crossigns in excited-states potential-energy surfaces, *J. Am. Chem. Soc.*, 100, pp. 2009–2011.

21. Devaquet, A. (1975). Avoided crossigns in photochemistry, *Pure Appl. Chem.*, 41, pp. 455–473.
22. Santolini, V., Malhado, J., Robb, M., Garavelli, M. and Bearpark, M. (2015). Photochemical reaction paths of cis-dienes studied with RASSCF: The changing balance between ionic and covalent excited states, *Mol. Phys.*, 113, pp. 1978–1990.
23. Olsen, J., Roos, B., Jorgensen, P. and Jensen, H. (1988). Determinant based configuration-interaction algorithms for complete and restricted configuration-interaction spaces, *J. Chem. Phys.*, 89, pp. 2185–2192.
24. Malmqvist, P., Rendell, A. and Roos, B. (1990). The restricted active space self-consistent-field method, implemented with a split graph unitary-group approach, *J. Phys. Chem.*, 94, pp. 5477–5482.
25. Shahi, A., Cramer, C. and Gagliardi, L. (2009). Second-order perturbation theory with complete and restricted active space reference functions applied to oligomeric unsaturated hydrocarbons, *Phys. Chem. Chem. Phys.*, 11, pp. 10964–10972.
26. Sinanoglu, O. (1964). Many-electron theory of atoms, molecules and their interactions, *Adv. Chem. Phys.*, 6, pp. 315–412.
27. Oksuz, I. and Sinanoglu, O. (1969). Theory of atomic structure including electron correlation 3. kinds of correlation in ground and excited configurations, *Phys. Rev.*, 181, pp. 42–53.
28. Foster, J. P. and Weinhold, F. (1980). Natural hybrid orbitals, *J. Am. Chem. Soc.*, 102, pp. 7211–7218.
29. Ben-Nun, M., Quenneville, J. and Martinez, T. (2000). Ab initio multiple spawning: Photochemistry from first principles quantum molecular dynamics, *J. Phys. Chem. A*, 104, pp. 5161–5175.
30. Tavernelli, I. (2015). Nonadiabatic molecular dynamics simulations: synergies between theory and experiments, *Acc. Chem. Res.*, 48, pp. 792–800.
31. Richter, M., Mai, S., Marquetand, P. and González, L. (2014). Ultrafast intersystem crossing dynamics in uracil unravelled by ab initio molecular dynamics, *Phys. Chem. Chem. Phys.*, 16, pp. 24423–24436.
32. Gozem, S. *et al.* (2014). Shape of multireference, equation-of-motion coupled-cluster, and density functional theory potential energy surfaces at a conical intersection, *J. Chem. Theory Comput.*, 10, pp. 3074–3084.
33. Gozem, S. *et al.* (2013). Mapping the excited state potential energy surface of a retinal chromophore model with multireference and equation-of-motion coupled-cluster methods, *J. Chem. Theory Comput.*, 9, pp. 4495–4506.
34. Gozem, S., Krylov, A. and Olivucci, M. (2013) Conical intersection and potential energy surface features of a model retinal chromophore: Comparison of EOM-CC and multireference methods, *J. Chem. Theory Comput.*, 9, pp. 284–292.
35. Ben-Nun, M. and Martinez, T. (2002), Ab initio quantum molecular dynamics, *Adv. Chem. Phys.*, 121, pp. 439–512.

36. Ben-Nun, M. and Martinez, T. (1998). Nonadiabatic molecular dynamics: Validation of the multiple spawning method for a multidimensional problem, *J. Chem. Phys.*, 108, pp. 7244–7257.
37. Liu, L., Liu, J. and Martinez, T. J. (2016). Dynamical correlation effects on photoisomerization: Ab initio multiple spawning dynamics with MS-CASPT2 for a model trans-protonated Schiff base, *J. Phys. Chem. B*, 120, pp. 1940–1949.
38. Tapavicza, E., Tavernelli, I., Rothlisberger, U., Filippi, C. and Casida, M. E. (2008). Mixed time-dependent density-functional theory/classical trajectory surface hopping study of oxirane photochemistry, *J. Chem. Phys.*, 129, p. 124108.
39. Mori, T. and Martinez, T. (2013). Exploring the conical intersection seam: The seam space nudged elastic band method, *J. Chem. Theory Comput.*, 9, pp. 1155–1163.
40. Gozem, S. *et al.* (2012). Dynamic electron correlation effects on the ground state potential energy surface of a retinal chromophore model, *J. Chem. Theory Comput.*, 8, pp. 4069–4080.
41. Liu, R. (2002). Photoisomerization by Hula-twist. Photoactive biopigments, *Pure Appl. Chem.*, 74, pp. 1391–1396.
42. Warshel, A. (1976). Bicycle-pedal model for first step in vision process, *Nature*, 260, pp. 679–683.
43. Du, L. and Lan, Z. (2015). An on-the-fly surface-hopping program JADE for nonadiabatic molecular dynamics of polyatomic systems: Implementation and applications, *J. Chem. Theory Comput.*, 11, pp. 1360–1374.
44. Maeda, S., Taketsugu, T., Ohno, K. and Morokuma, K. (2015). From roaming atoms to hopping surfaces: mapping out global reaction routes in photochemistry, *J. Am. Chem. Soc.*, 137, pp. 3433–3445.
45. Maeda, S., Ohno, K. and Morokuma, K. (2009). Automated global mapping of minimal energy points on seams of crossing by the anharmonic downward distortion following method: A case study of H2CO, *J. Phys. Chem. A*, 113, pp. 1704–1710.
46. Suzuki, S., Maeda, S. and Morokuma, K. (2015). Exploration of quenching pathways of multiluminescent acenes using the GRRM method with the SF-TDDFT method, *J. Phys. Chem. A*, 119, pp. 11479–11487.
47. Townsend, D. *et al.* (2004). The roaming atom: Straying from the reaction path in formaldehyde decomposition, *Science*, 306, pp. 1158–1161.
48. Subotnik, J. E. and Rhee, Y. M. (2015). On surface hopping and time-reversal, *J. Phys. Chem. A*, 119, pp. 990–995.
49. Petit, A. and Subotnik, J. (2015). Appraisal of surface hopping as a tool for modeling condensed phase linear absorption spectra, *J. Chem. Theory Comput.*, 11, pp. 4328–4341.
50. Jain, A. and Subotnik, J. (2015). Does nonadiabatic transition state theory make sense without decoherence? *J. Phys. Chem. Lett.*, 6, pp. 4809–4814.

51. Subotnik, J. E., Alguire, E. C., Ou, Q., Landry, B. R. and Fatehi, S. (2015). The requisite electronic structure theory to describe photoexcited nonadiabatic dynamics: Nonadiabatic derivative couplings and diabatic electronic couplings, *Acc. Chem. Res.*, 48, pp. 1340–1350.
52. Petit, A. S. and Subotnik, J. E. (2014). Calculating time-resolved differential absorbance spectra for ultrafast pump-probe experiments with surface hopping trajectories, *J. Chem. Phys.*, 141, p. 154108.
53. Landry, B. R. and Subotnik, J. E. (2014). Quantifying the lifetime of triplet energy transfer processes in organic chromophores: A case study of 4-(2-Naphthylmethyl)benzaldehyde, *J. Chem. Theory Comput.*, 10, pp. 4253–4263.
54. Closs, G. L., Johnson, M. D., Miller, J. R. and Piotrowiak, P. (1989). A connection between intramolecular long-rangre electron, hole, and triplet energy transfers, *J. Am. Chem. Soc.*, 111, pp. 3751–3753.
55. Closs, G. L., Piotrowiak, P., Macinnis, J. M. and Fleming, G. R. (1988). Determination of long-distance intramolecular triplet energy-transfer rates — A quantitative comparison with electron-transfer, *J. Am. Chem. Soc.*, 110, pp. 2652–2653.
56. Landry, B. R. and Subotnik, J. E. (2012). How to recover Marcus theory with fewest switches surface hopping: add just a touch of decoherence, *J. Chem. Phys.*, 137, p. 22A513.
57. Landry, B. R. and Subotnik, J. E. (2011). Communication: Standard surface hopping predicts incorrect scaling for Marcus' golden-rule rate: The decoherence problem cannot be ignored, *J. Chem. Phys.*, 135, p. 191101.
58. Zhu, X. and Yarkony, D. R. (2016). On the elimination of the electronic structure bottleneck in on the fly nonadiabatic dynamics for small to moderate sized (10-15 atom) molecules using fit diabatic representations based solely on ab initio electronic structure data: The photodissociation of phenol, *J. Chem. Phys.*, 144, p. 024105.
59. Zhu, X. and Yarkony, D. R. (2016). Constructing diabatic representations using adiabatic and approximate diabatic data–Coping with diabolical singularities, *J. Chem. Phys.*, 144, p. 044104.
60. Malbon, C. L. and Yarkony, D. R. (2015). Nonadiabatic photodissociation of the hydroxymethyl radical from the 2(2)A state. Surface hopping simulations based on a full nine-dimensional representation of the 1,2,3(2)A potential energy surfaces coupled by conical intersections, *J. Phys. Chem. A*, 119, pp. 7498–7509.
61. Klene, M., Robb, M., Blancafort, L. and Frisch, M. (2003). A new efficient approach to the direct restricted active space self-consistent field method, *J. Chem. Phys.*, 119, pp. 713–728.
62. Bearpark, M. *et al.* (1996). The azulene S-1 state decays via a conical intersection: A CASSCF study with MMVB dynamics, *J. Am. Chem. Soc.*, 118, pp. 169–175.
63. Vacher, M., Meisner, J., Mendive-Tapia, D., Bearpark, M. J. and Robb, M. A. (2015). Electronic control of initial nuclear dynamics adjacent to a conical intersection, *J. Phys. Chem. A*, 119, pp. 5165–5172.

64. Vacher, M., Mendive-Tapia, D., Bearpark, M. J. and Robb, M. A. (2015). Electron dynamics upon ionization: Control of the timescale through chemical substitution and effect of nuclear motion, *J. Chem. Phys.*, 142, p. 094105.
65. Richter, M., Marquetand, P., González-Vázquez, J., Sola, I. and González, L. (2011). SHARC: Ab initio molecular dynamics with surface hopping in the adiabatic representation including arbitrary couplings, *J. Chem. Theory Comput.*, 7, pp. 1253–1258.
66. Ou, Q., Alguire, E. C. and Subotnik, J. E. (2015). Derivative couplings between time-dependent density functional theory excited states in the random-phase approximation based on pseudo-wavefunctions: behavior around conical intersections, *J. Phys. Chem. B*, 119, pp. 7150–7161.
67. Levine, B., Coe, J., Virshup, A. and Martinez, T. (2008). Implementation of ab initio multiple spawning in the MOLPRO quantum chemistry package, *Chem. Phys.*, 347, pp. 3–16.
68. Du, L. and Lan, Z. (2015). An on-the-fly surface-hopping program JADE for nonadiabatic molecular dynamics of polyatomic systems: Implementation and applications (11, p. 1360, 2015), *J. Chem. Theory Comput.*, 11, pp. 4522–4523.
69. Xiao, H., Maeda, S. and Morokuma, K. (2013). CASPT2 study of photodissociation pathways of ketene, *J. Phys. Chem. A*, 117, pp. 7001–7008.
70. Maeda, S., Ohno, K. and Morokuma, K. (2009). An automated and systematic transition structure explorer in large flexible molecular systems based on combined global reaction route mapping and microiteration methods, *J. Chem. Theory Comput.*, 5, pp. 2734–2743.
71. Manathunga, M. *et al.* (2016). Probing the Photodynamics of rhodopsins with reduced retinal chromophores. *J. Chem. Theory Comput.*, 12, pp. 839–850.

Chapter 19

Challenges in Predicting Stereoselectivity

Elizabeth H. Krenske

School of Chemistry and Molecular Biosciences
The University of Queensland, Australia

19.1 Introduction

Quantum chemistry has been responsible for many important discoveries in the fields of stereochemistry and asymmetric synthesis. Computational modeling of stereoselective reactions taps into the ability of quantum mechanical (QM) methods to characterize species that are not directly observable by experiment, such as transition states (TSs) and short-lived intermediates. Such calculations allow one to understand why a reaction favors a certain stereoisomer over another — and to what degree.

This chapter will describe some of the conceptual and technical challenges encountered in the quantitative modeling of stereoselectivity. The examples discussed will follow the development of the field over its 40-year history, highlighting several major milestones. Emphasis will be given to studies where computations have been used not only to rationalize the stereochemical outcomes of reactions but also to discover new chemistry ahead of experiment.

There are several types of computational approaches for predicting stereoselectivity. Most of this chapter will focus on the approach that utilizes quantum chemical exploration of reaction mechanisms, in particular transition-state modeling. Several alternative approaches, which utilize statistical analyses of structure–selectivity relationships or molecular mechanics modeling of TS structures, will also be briefly discussed.

19.2 Principles governing quantum chemical predictions of stereoselectivity

The major stereoisomeric product of a stereoselective reaction that is under kinetic control, and the level of stereoselectivity, can be predicted by comparing the computed activation barriers and energetics of the pathways leading to different stereoisomers. The simplest type of stereoselective reaction to model is one that gives two stereoisomeric products, (e.g., **A** → **X** + **Y**, where **X** and **Y** are diastereomers) and occurs in a single step under ordinary kinetic control (Fig. 19.1(a)). According to TS theory, the ratio of stereoisomeric products **X**:**Y** is determined by the difference between the Gibbs free energies of the TS structures **TSX** and **TSY** ($\Delta\Delta G^{\ddagger}$). In terms of the activation barriers ($\Delta G^{\ddagger}$), the stereoisomer ratio is given by Eq. (19.1).

$$\mathbf{X}:\mathbf{Y} = e^{\frac{-\Delta\Delta G^{\#}}{RT}} \quad \text{where} \quad \Delta\Delta G^{\#} = \Delta G^{\#}_{\mathbf{X}} - \Delta G^{\#}_{\mathbf{Y}} \qquad (19.1)$$

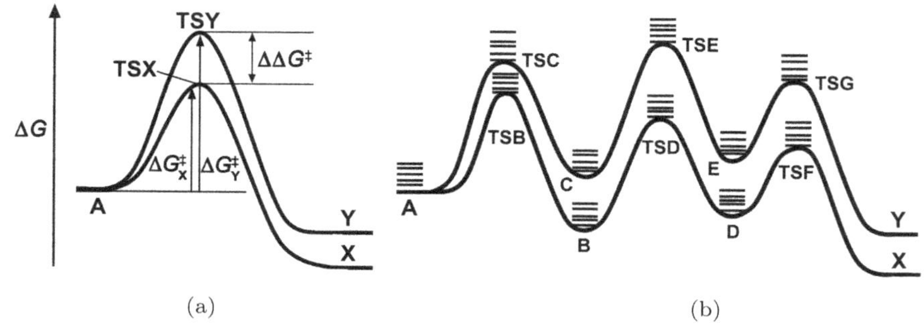

Figure 19.1. Free energy diagrams for (a) a one-step mechanism and (b) a multistep mechanism for a stereoselective reaction that converts **A** into stereoisomeric products **X** and **Y**.

Most reactions of interest to synthetic chemists are more complicated than the simple scenario depicted in Fig. 19.1(a). Figure 19.1(b) depicts a multistep mechanism typical of that which might occur, for example, in a transition-metal–catalyzed asymmetric transformation. Stereochemical predictions for such reactions are more challenging. Before TS theory can be applied, the rate- and stereoselectivity-determining steps must be identified. Ideally, computations of the mechanism would be validated against experimental data. Some other considerations include: (i) whether the reactant(s) that lead to different stereoisomers are in rapid equilibrium (Curtin–Hammett principle) or not; (ii) how to properly sample the conformational space of the relevant energy minima and TSs (see Chapter 5); (iii) how interactions with the solvent influence barrier heights (see Chapters 1 and 4); and (iv) whether the topology of the energy surface is simple enough that TS theory is applicable, or a more complex dynamical treatment is required (see Chapter 12).

Apart from the need to choose an appropriate QM method (see Chapter 2) for geometry optimizations and potential energy calculations, another important consideration is the calculation of entropy. The gas-phase free energies typically reported in the output of harmonic vibrational frequency calculations represent an attempt to approximate the effects of vibrational motion and temperature on enthalpy and entropy.[1]

Furthermore, most reactions are carried out in solution, and for synthetic reactions, computational chemists typically simulate the effects of solvent by means of a continuum solvent model (see Chapter 4). It is well known that continuum solvent models overestimate the entropies of species in solution. There are no definitive guidelines about how to correct for this problem, although numerous strategies have been suggested. Some of these have been summarized recently by Singleton[2] in his detailed study of the Morita Baylis–Hillman reaction. Singleton's study showed that the errors in computed barriers and energies of polar reactions in solution can be enormous. Fortunately, calculations of stereoselectivities depend on relative, rather than absolute, barriers, and therefore tend to benefit from error cancellation.

For many stereoselective reactions, the values of $\Delta\Delta G^\ddagger$ that discriminate between alternative stereoisomeric pathways are small — on the order of a few kcal/mol. For example, at 25°C, the difference between an unselective reaction (stereoisomer ratio = 1:1) and a highly selective one (>99:1) hinges on increasing $\Delta\Delta G^\ddagger$ by only 2.7 kcal/mol. In order to achieve the best possible estimate of such small energy differences, all important factors, such as conformational sampling, vibrational entropy, solvent effects, dispersion, and basis set superposition error, must be carefully considered.

19.3 Some milestones in the prediction of stereoselectivity

19.3.1 *Nucleophilic additions to chiral carbonyl compounds*

It is only within the last 40 years or so that the structures of the TSs of most reactions have been understood in any detail. For example, the typical lengths of partially formed bonds, the way in which bond angles change as reactants combine, and the preferred conformations of nearby groups only became open to investigation following the advent of *ab initio* molecular orbital theory. These properties are fundamentally linked to stereoselectivity.

One of the earliest demonstrations that QM calculations could be used to model stereoselectivity at all was Anh and Eisenstein's classic paper on nucleophilic additions to chiral carbonyl compounds.[3] Anh and Eisenstein used STO-3G calculations (cutting-edge at the time) to evaluate different stereochemical models for these reactions (Fig. 19.2). By calculating the energies of 12 different conformers of each diastereomeric "TS" for addition of H^- to 2-chloropropanal and 2-methylbutanal, they showed that the lowest energy structures bore closest resemblance to Felkin's model, while structures corresponding to TSs from the Cram, Cornforth, and Karabatsos models were higher in energy.

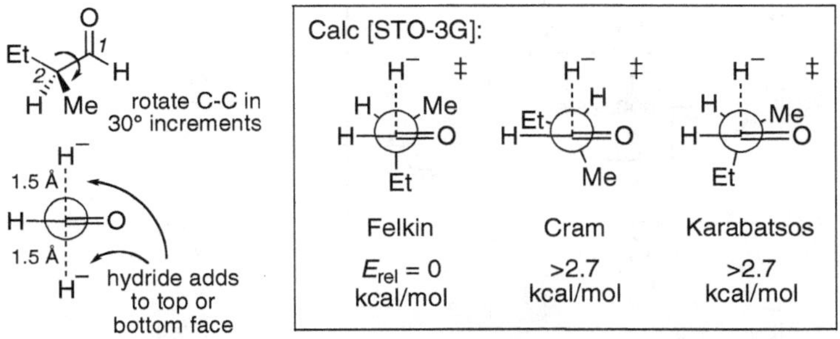

Figure 19.2. Approach used by Anh and Eisenstein to evaluate different stereochemical models for nucleophilic additions to chiral carbonyl compounds.

These calculations employed rigid models of the TSs. The carbonyl carbon was held planar, the hydride ion was held at a fixed distance from carbon, and only rotation about the C1–C2 bond and a few possible attack angles of the hydride ion were explored. Nonetheless, the general conformational principle revealed by Anh and Eisenstein's TS models was later confirmed by full optimizations at higher levels of theory,[4] and the Felkin–Anh model is now a central tenet of organic chemistry.

19.3.2 *Torquoselectivity*

In the decade that followed Anh and Eisenstein's paper, increases in computer power allowed researchers to perform complete geometry optimizations of TS structures for the first time. The first quantitative predictions of stereoselectivity were reported during this period. One of the pioneering studies was Houk's work on the electrocyclic ring opening of substituted cyclobutenes (Fig. 19.3).[5] On the basis of *ab initio* calculations (HF/3-21G or 6-31G(d)), Houk developed a general theory that accounted for the preferred direction of rotation by different C3 substituents ("torquoselectivity"). The rule is: donors rotate outward, while acceptors, if strong enough, rotate inwards. Molecular orbital calculations were the key to this discovery, since torquoselectivity is determined by orbital interactions between the HOMO of the breaking C–C σ-bond and the p (or π) type orbital on

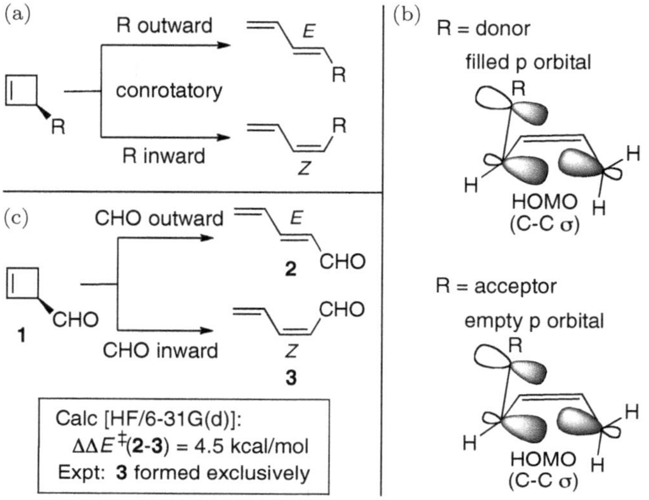

Figure 19.3. (a) Stereochemically distinct modes of conrotatory electrocyclic ring opening of substituted cyclobutenes. (b) Destabilizing interaction between the transition-state HOMO (C–C σ) and the filled p-orbital on a donor substituent when the donor rotates inward; and stabilizing interaction between the transition-state HOMO and the vacant p-orbital on an acceptor substituent when the acceptor rotates inward. (c) Stereoselective ring opening of 3-formylcyclobutene, as predicted theoretically and subsequently measured experimentally.

R in the TS structure (Fig. 19.3(b)). Inward rotation by a donor R group is disfavored because the TS contains an interaction between these two orbitals that is destabilizing since they are both filled. Conversely, inward rotation by an acceptor is favored because the TS contains a stabilizing donor–acceptor interaction between the HOMO and the empty p orbital on R.

Out of several dozen R groups computed, only a few were predicted to be strong enough acceptors to favor inward rotation. One of these was the formyl group, which had not yet been experimentally studied at the time. Transition-state modeling predicted a 4.5 kcal/mol preference for inward rotation by R = CHO (Fig. 19.3(c)). This prediction was confirmed by experiment: ring opening of 3-formylcyclobutene **1** gave exclusively the (Z)-dienal **3**.[6] This result could not have been predicted without TS modeling, since **3** is the less stable of the two dienal isomers and the TS leading to **3** is more sterically crowded than the TS leading to **2**.

19.3.3 *Proline-catalyzed aldol reactions*

Until the 1990s, QM computations were normally restricted to relatively small molecules, often truncated versions of experimentally studied systems. This situation began to change as density functional theory (DFT) established itself as a cost-effective alternative to correlated *ab initio* methods for calculations on larger molecules. The popularity of DFT, and the development of new functionals, continue today. An early proving-ground for DFT in the study of stereoselectivity was Houk's work on the (*S*)-proline-catalyzed intramolecular aldol reaction of Hajos, Parrish, Eder, Sauer, and Wiechert (Fig. 19.4).[7] At the time, the reason why this reaction favors the (*S*,*S*) enantiomer of the intermediate **5** (and eventually the dehydrated product *S*-**6**) was not understood.

Houk's group answered this question by performing B3LYP calculations. One major obstacle (common to many other stereochemical problems) was that the mechanism of the reaction was unknown. In the first place, it was not known whether the intermediate that undergoes C–C bond formation is a carbinolamine (as in **TS4a**) or an enamine (as in **TS4b**–**TS4d**). Computations definitively supported the involvement of an enamine, as the carbinolamine was >12 kcal/mol higher in energy than the lowest energy enamine TS. Furthermore, three enamine-based mechanisms had previously been proposed, differing with respect to whether proton transfer to the developing alkoxide involves an enaminium NH (**TS4b**), the proline carboxylic acid (**TS4c**), or a second molecule of proline (**TS4d**). The computed barriers pointed definitively to the carboxylic acid as the proton donor.

The stereoselectivity of ring closure was traced to two features.[8] First, the favored TS (*S*,*S*)-**TS4c** benefits from a more stabilizing electrostatic interaction between the developing alkoxide and one of the CH_2 protons adjacent to nitrogen (see Chapter 11). Second, in the disfavored TS (*R*,*R*)-**TS4c**, the carboxylic acid and the developing

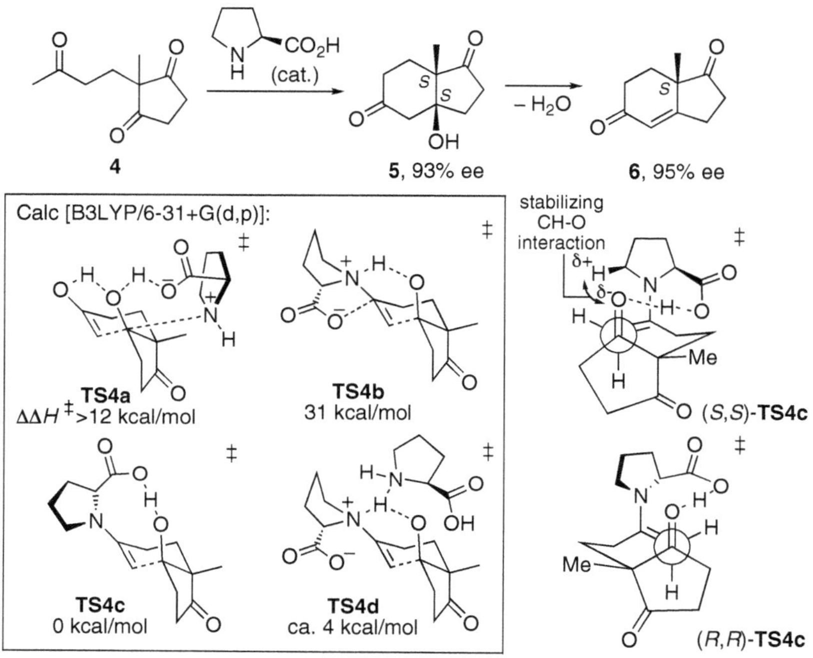

Figure 19.4. TSs for various possible mechanisms of the (*S*)-proline-catalyzed Hajos–Parrish–Eder–Sauer–Wiechert aldol reaction.

alkoxide are not aligned favorably. In order to allow the proton transfer to occur, the developing C=N double bond has to twist into a high-energy (nonplanar) conformation.

From these calculations and related studies of intermolecular reactions emerged a general model for rationalizing the enantio- and diastereoselectivities of proline-catalyzed aldol reactions. The key features of the model are: (i) the reaction involves addition to the *re* face of the s-*trans* (*E*)-enamine, (ii) the forming C–C bond adopts the staggered arrangement that best allows proton transfer from CO_2H to the developing alkoxide, and (iii) the facial selectivity with respect to the carbonyl group depends on how the two substituents of the carbonyl compound are best positioned in the chair-like six-membered TS.

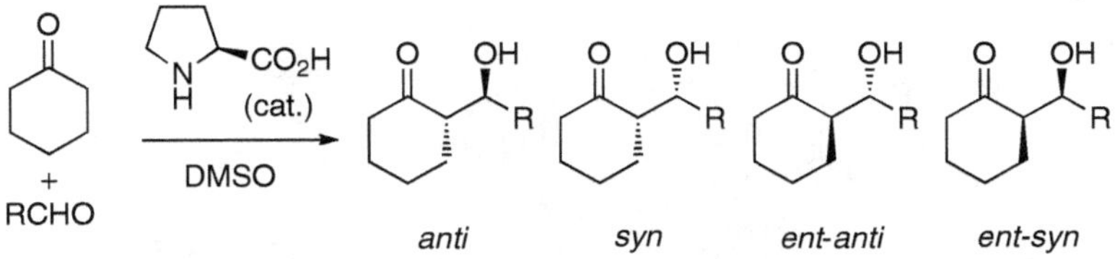

	R = Ph	*anti*	*syn*	*ent-anti*	*ent-syn*
B3LYP/ 6-31G(d)	Calc $\Delta H^{\ddagger}_{rel}$	0	0.4 ± 0.4	2.3 + 0.4	3.6 ± 0.4
	Calc %	58-80	20-50	1-4	<1
	Expt %	45-47	43-45	3-5	5-7
	R = ^{i}Pr				
	Calc $\Delta H^{\ddagger}_{rel}$	0	6.7 ± 0.4	4.6 + 0.4	7.8 ± 0.4
	Calc %	>99	<1	<1	<1
	Expt %	97-100	<1	0-3	<1

Figure 19.5. Theoretically predicted and experimentally measured product ratios in the proline-catalyzed aldol reactions of cyclohexanone with benzaldehyde and isobutyraldehyde ($\Delta H^{\ddagger}$ in kcal/mol).

Houk and List used these principles to perform *a priori* predictions of the stereoselectivities of two novel proline-catalyzed aldol reactions (Fig. 19.5).[9] B3LYP calculations gave values of $\Delta H^{\ddagger}_{gas}$, $\Delta G^{\ddagger}_{gas}$, and $\Delta G^{\ddagger}_{soln}$ for the four stereoisomeric pathways of the aldol reactions of cyclohexanone with benzaldehyde or isobutyraldehyde. The $\Delta H^{\ddagger}_{gas}$ values, which had the lowest estimated error (±0.4 kcal/mol), led to the predicted product ratios shown. Subsequent experiments showed that these calculations correctly predicted the sense and level of selectivity of both reactions.

19.3.4 *Organocatalytic Mannich reactions*

The predictive value of the Houk–List stereochemical model was further tested in 2006 when Houk, Tanaka, and Barbas reported the first computationally designed stereoselective organocatalyst. The target was to develop an *anti*-selective Mannich reaction (Fig. 19.6).[10] Prior to that time, most of the known chiral catalysts for Mannich reactions (including proline) gave primarily *syn* products. The challenge was: could theory be used to rationally design a

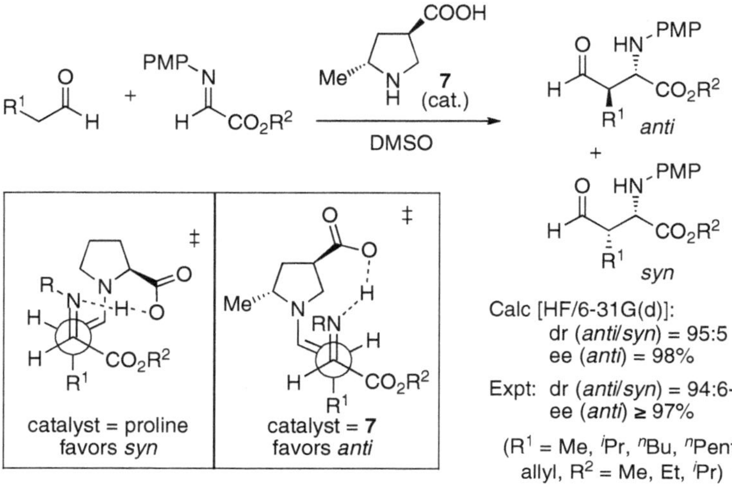

Figure 19.6. Theoretically predicted and experimentally measured diastereo- and enantioselectivities of Mannich-type reactions catalyzed by the designed organocatalyst **7**. Abbreviation: PMP = *para*-methoxyphenyl.

catalyst that would favor *anti* products? Extension of the Houk–List model to imines revealed that a key to switching the selectivity from *syn* to *anti* might lie in encouraging the reaction to occur via the s-*cis* rather than the s-*trans* enamine. To achieve this, a blocking group (Me) was introduced at the five-position of the pyrrolidine ring, while the carboxylic acid was moved to the three-position, where it could still effect proton transfer to the imine. HF/6-31G(d) calculations predicted that the designed catalyst **7** would favor the *anti* product shown with a dr of 95:5 and an ee of 98%. These predictions were shown to be excellent: experimentally, **7** catalyzed the formation of *anti* products from a range of imines, giving values of dr $\geq 94:6$ and ee $\geq 97\%$.

19.3.5 *Enantioselective Rh-catalyzed hydrogenation of enamides*

Computational modeling of organotransition metal chemistry is a challenging field, characterized by complex reaction mechanisms, changes in metal spin states, and various subtleties of substrate control (see Chapter 17).[11–13] In 2000, Landis and Feldgus[14,15] achieved the ambitious goal of modeling the enantioselective hydrogenation of prochiral enamides, catalyzed by a chiral rhodium diphosphine complex (Fig. 19.7). The study was designed to explain an intriguing stereochemical observation. Thus, it was known from experiments that complexation of the enamide to rhodium affords two diastereomeric complexes; unexpectedly, the major (*R*) enantiomer of product arises from the less stable complex while the minor (*S*) enantiomer originates from the more stable complex.

Building upon earlier calculations of small model systems, which had explored the mechanisms of Rh-catalyzed hydrogenations, Landis and Feldgus employed a three-layer ONIOM scheme (B3LYP:HF:UFF). Eight diastereomeric reaction pathways were modeled, four for each diastereomer of the Rh–enamide complex. The computed difference in energies between the diastereomeric complexes **8** and **9** ($\Delta\Delta G = 3.6\,\text{kcal/mol}$, 500:1 dr) agreed closely with experiments, which had shown that the concentration of the minor diastereomer was too small to be detectable by NMR.

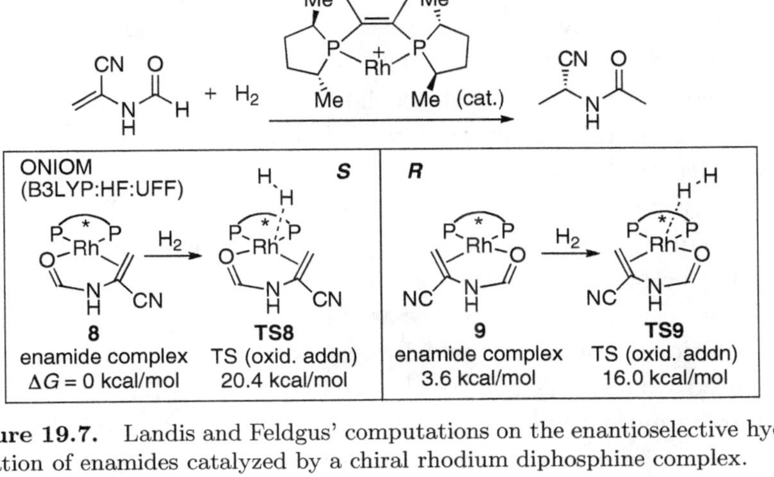

Figure 19.7. Landis and Feldgus' computations on the enantioselective hydrogenation of enamides catalyzed by a chiral rhodium diphosphine complex.

The calculations indicated that the rate- and stereoselectivity-determining step was the oxidative addition of H_2 to rhodium in **8** and **9**. The free energy barrier for formation of the R product via complex **9** (**TS9**) was 4.4 kcal/mol lower than that for formation of the S enantiomer via complex **8** (**TS8**). This $\Delta\Delta G^{\ddagger}$ value corresponded to an ee of 99.8%, in excellent agreement with the experimentally measured ee's, which were in excess of 99% for this type of substrate. The more facile oxidative addition of H_2 to the less stable Rh–enamide complex **9** was traced to the fact that the addition of H_2 could occur with little geometrical reorganization; in contrast, addition of H_2 to **8** induced a structural change where the enamide moved into a sterically congested region of space near one of the catalyst methyl groups.

Other notable successes in the prediction of stereoselective transition-metal–catalyzed reactions include studies of Sharpless' dihydroxylation of alkenes (Os),[16] Grubbs' Z-selective olefin metathesis (Ru),[17] and Noyori's transfer hydrogenation of ketones.[18]

19.3.6 *Enantioselective guanidinium-catalyzed Claisen rearrangement*

The studies described above demonstrated the potential for quantum chemistry to predict the stereoselectivities of both known and new reactions. Many more papers have since been published that have used theory to understand stereochemical problems. However, there have been notably fewer theory-guided predictions of new chemistry. Part of the reason why *a priori* prediction is challenging is that many synthetically interesting reactions involve relatively large, flexible substrates, which can adopt numerous conformations. Predictions for such reactions are more difficult than calculations on small models. Conformational sampling (see Chapter 5) is a major computational bottleneck, which must be carefully addressed before computed values of $\Delta\Delta G^{\ddagger}$ can be treated with confidence.

Furthermore, the larger the molecule, the greater the influence of noncovalent interactions (in particular dispersion) on geometry and energy (see Chapter 10). While some popular DFT methods such as B3LYP often perform quite well for reactions involving relatively simple molecules (closed shell, no low-lying electronic states), they do not properly model nonbonded interactions that involve dispersion, such as $\pi - \pi$ and CH–π interactions. Newer functionals have been developed that account for dispersion in various ways, such as through parameterization (e.g., Truhlar's M05 and M06 suites of functionals) or by empirical corrections (e.g., Grimme's DFT-D methods). Achieving a balance of all the covalent and noncovalent interactions that influence the energies of the transition-state structures of stereoselective reactions is a significant challenge.

A 2011 study by Uyeda and Jacobsen[19] utilized computations of noncovalent interactions to guide the development of chiral catalysts. Experimental studies on the enantioselective Claisen rearrangements catalyzed by chiral guanidinium salts **10** (Fig. 19.8) had shown that, compared to catalyst **10a** (R = Me), the catalyst **10b** (R = Ph) gave a higher level of asymmetric induction. DFT calculations traced this selectivity to an attractive noncovalent interaction between one of

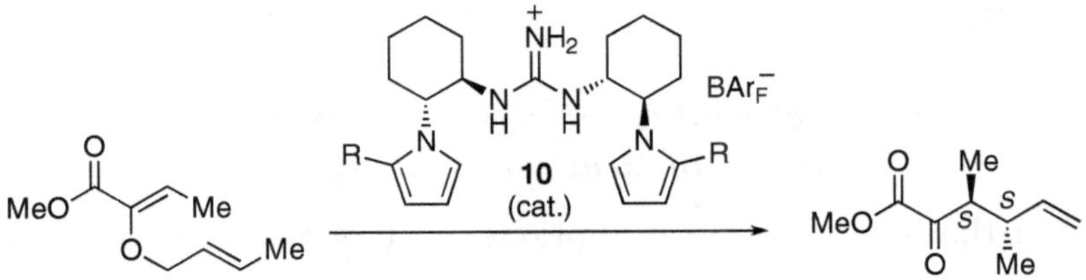

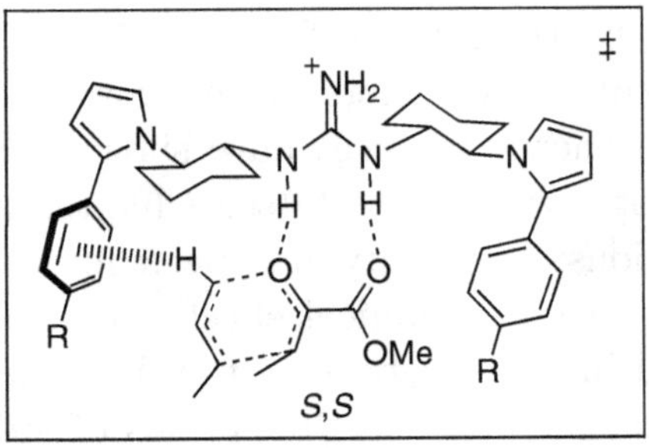

M05-2X/6-31G(d)

	R	Calc $\Delta\Delta H^\ddagger$ (kcal/mol)	Expt ee (%)
10a	Me	-	41
10b	Ph	2.5	73
10c	p-NH_2-C_6H_4	3.4	78
10d	p-F-C_6H_4	1.6	67

Figure 19.8. Theoretically predicted and experimentally measured enantioselectivities of Claisen rearrangements catalyzed by chiral guanidinium salts.

the Ph groups of **10b** and the allyl fragment of the substrate (which bears a partial positive charge) in the TS leading to the favored enantiomer.

M05-2X (but not B3LYP) calculations revealed that the enantioselectivity could be further improved by strengthening this noncovalent interaction. Thus, adding a *p*-NH_2 group onto the Ph ring (**10c**) was predicted to increase $\Delta\Delta H^{\ddagger}$ by 0.9 kcal/mol, while a *p*-F substituent (**10d**) lowered $\Delta\Delta H^{\ddagger}$. These theoretical predictions were confirmed by experiments, which showed the amino-substituted guanidinium salt to be a superior enantioselective catalyst, improving the ee's of Claisen rearrangements of a diverse group of substrates by 3–8%. The overestimation of the enantioselectivity by the calculations was suggested to be due to the neglect of entropy and solvent effects.

Modeling of noncovalent interactions is an active area of research (see Chapter 10).[20] Continued developments in this area should provide valuable opportunities for the design of new chiral auxiliaries and catalysts, taking advantage of the ability to engineer specific noncovalent interactions (such as hydrogen bonding, π–π, CH–π, cation–π, CH–O, and halogen bonding) within TS structures.

19.3.7 *Dynamical control of stereoselectivity in nucleophilic substitution at an sp² carbon*

One of the fundamental assumptions in all of the above studies — and indeed most theoretical studies of stereoselectivity — is that the kinetic selectivity is determined by the relative free energies of the TS structures that lead to different stereoisomers. Recent discoveries about the topologies of reaction energy surfaces have shown, however, that this assumption is not always correct (see Chapter 12). For a growing number of reactions, selectivity has been found to depend not on the TS energies but on dynamical effects, related to the shape of the energy surface in regions beyond the TS and/or motions within certain vibrational modes.[21,22] Predictions of stereoselectivity for such reactions require sophisticated computational treatments.

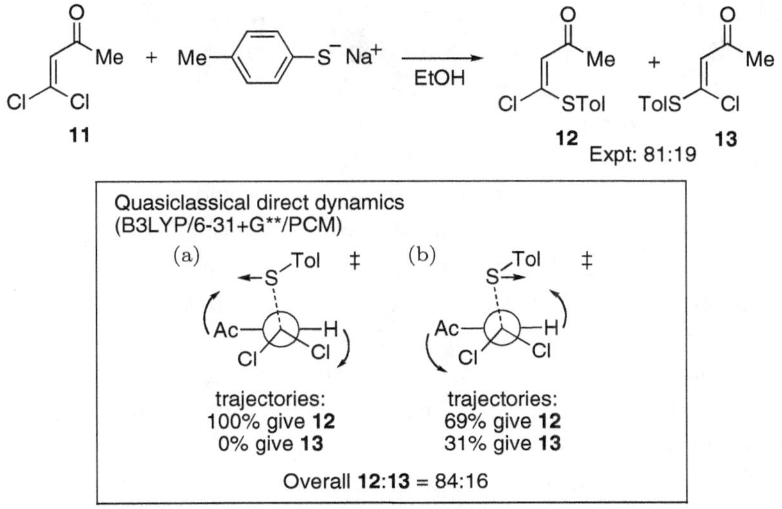

Figure 19.9. Dynamically controlled stereoselectivity in a nucleophilic substitution at an sp^2 carbon. The two directions of motion in the stereoselectivity controlling vibrational mode are labeled as (a) and (b).

One example is the nucleophilic substitution of the dichloride **11** by sodium *p*-tolylthiolate (Fig. 19.9), studied by Bogle and Singleton.[23] Two stereoisomeric products, **12** and **13**, are possible, depending on which of the chlorides acts as the leaving group. The experimental ratio of **12** to **13** was 81:19. DFT calculations (B3LYP and M06-2X) revealed that the reaction has an energy surface with a post-TS bifurcation. Both products are accessed via the same TS, and there is no intermediate. At the TS, there is no obvious distinction between the two chlorides of any sort that would suggest which one will ultimately depart. The steepest downhill pathway leading away from the TS structure (determined by IRC calculations) gives **12**. Thus, conventional modeling is qualitatively inaccurate for this reaction, being unable to predict that any **13** is formed.

Prediction of the stereoselectivity was achieved by means of trajectory calculations. Quasiclassical direct-dynamics simulations were initiated from the TS region. Out of 185 trajectories that led to product formation, 156 gave **12** and 29 gave **13**. The resulting ratio of **12** to **13** (84:16) agrees closely with experiment. The origin of the selectivity was traced to the motion occurring within a particular vibrational mode, as shown in Fig. 19.9. None of the trajectories in which this mode was initially assigned the direction labeled (a) led to formation of the minor product **13**. This product was only formed when the mode was assigned the direction labeled (b), and then only in 31% of the trajectories — a "dynamic matching" effect.

Trajectory calculations provide sophisticated insights into the factors that control stereoselective reactions, beyond the capabilities of conventional TS modeling. The discoveries of recent dynamics studies likely represent just the tip of the iceberg. New examples can be expected to be found in many other reactions. Tantillo's research group has recently shown, for example, that dynamical effects contribute to product selectivities in carbocationic cascade reactions involved in terpene biosynthesis.[24]

19.4 Alternative approaches to predicting stereoselectivity

Detailed QM computations of reaction mechanisms are not the only viable way to predict stereoselectivity. A number of research groups have devised alternative approaches, which attempt to relate structure to selectivity in a less computationally demanding way.[25] These techniques have the advantage of being suited to high-throughput *in silico* screening applications. Some examples include:

- Methods that use QM data to guide the parameterization of molecular mechanics forcefields for TS structures. This category includes the Q2MM method of Norrby and Wiest, which has, for example, provided reliable predictions of enantioselectivity in Rh-catalyzed hydrogenation reactions.[26]
- The ACE method of Moitessier, which attempts to approximate the TS structure as a weighted average of the reactant and product. The "TS" energy is computed using molecular mechanics.[27]
- Approaches based on statistical modeling of structure–selectivity data. For example, Harper and Sigman[28] used polynomial fitting of

substituent steric and electronic parameters to develop predictive models for Cr-catalyzed propargylations of carbonyl compounds. Another example of a quantitative structure–selectivity relationship based approach is Kozlowski's work on asymmetric alkylation reactions, described below.

19.4.1 *Enantioselective alkylation of aldehydes*

Kozlowski *et al.*[29] pioneered the use of quantitative structure–selectivity relationships (QSSRs) to predict stereoselectivities. Their work examined the enantioselective additions of Et_2Zn to aldehydes, catalyzed by chiral β-amino alcohols **14** (Fig. 19.10), a reaction whose mechanism had previously been explored by a number of groups. Kozlowski's group computed TS geometries for a variety of catalysts using PM3 semiempirical theory and then calculated probe interaction energies at a series of grid points around the catalyst in each TS. Least-squares regression analyses revealed that as few as two grid points could be used to predict the experimental enantioselectivities ($\Delta\Delta G^{\ddagger}$). The QSSR method gave good predictions of the level of selectivity (low, medium, or high) for a range of chiral β-amino alcohols.

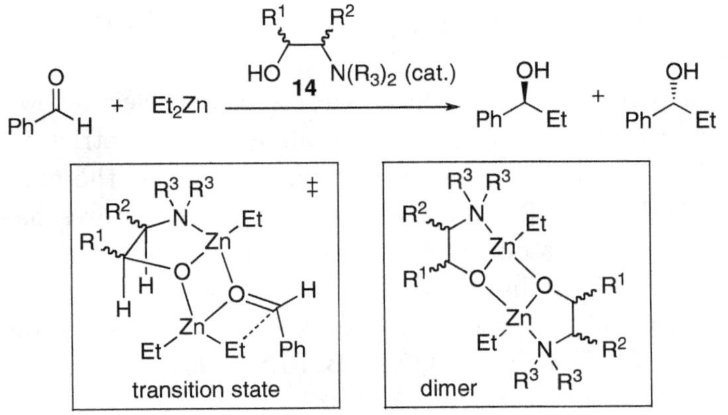

Figure 19.10. Enantioselective addition of Et_2Zn to aldehydes, catalyzed by chiral β-amino alcohols.

Subsequently, a more economical method was developed that used the ground-state geometries of the catalyst dimers, rather than TS geometries.[30] This simplification was possible because key structural features of the catalyst dimer resembled those of the TS structure (particularly the tetrahedral zinc centres). One of the attractive features of these QSSR approaches is that the statistical fitting produces useful chemical insights. The grid points identified as most important by the least-squares analysis correspond to the sites that are the most crucial to stereochemical induction. This offers the opportunity for improving stereoselectivity by rationally modifying the steric or electronic properties of substituents at those positions.

19.5 Practical considerations

Given sufficient computer resources, it is now possible to address challenging stereochemical problems in a way that leads to a reasonable expectation of a reliable match between theoretical predictions and experiment. Achieving the best possible confidence when calculating small values of $\Delta\Delta G^{\ddagger}$ demands that the computational method employed provides chemical accuracy (<1 kcal/mol) or better. Such a high benchmark requires careful attention to QM method selection, conformational sampling (see Chapter 5), vibrational entropy (see Sec. 19.2), basis set superposition error, and solvent effects.

It is now routinely possible, and advisable, to select a QM method that includes dispersion. Dispersion influences geometries and energies and becomes increasingly important the larger the molecule is. Dispersion-dominated interactions within TS structures have been shown to act as stereocontrol elements in numerous reactions. Some of the best QM methods currently available for such applications include Grimme's empirically corrected DFT (of which some examples are B3LYP-D3, B97-D3, and B2PLYP-D3),[31] Zhao and Truhlar's Minnesota suite of functionals (e.g., M06 and M06-2X),[32] Chai and Head-Gordon's long-range corrected ωB97X-D functional,[33] and

Antony and Grimme's spin-component scaled SCS-MP2 *ab initio* method.[34] Validation of the chosen QM method against either experimental results, or high-level (e.g., CCSD(T)) calculations, is good practice.

Ideally, a triple-zeta quality (or higher) basis set would be used for both geometry optimizations and energy calculations. Sometimes, however, a double-zeta basis set can provide good performance at a lower cost for geometry optimizations, vibrational frequency calculations (thermochemistry), and solvation energies, and these terms can be added to a single-point potential energy calculated with a larger basis set. A promising economical alternative reported by Grimme is B3LYP-D3-gCP/6-31G(d).[35] This method includes an empirical dispersion term and a geometrical counterpoise correction to correct for basis set superposition error. It has been shown to be superior to the long-popular B3LYP/6-31G(d) for calculations of activation barriers and thermodynamics. Another cost-effective alternative to a full QM treatment is a hybrid QM/MM approach such as ONIOM.[36]

Whether or not the solvent influences stereoselectivity, it is generally considered desirable to include a treatment of solvent effects. Continuum solvation models such as CPCM,[37] SMD,[38] and COSMO,[39] are popular. Sometimes, if specific interactions (e.g., hydrogen bonding) with the solvent are important, these can be modeled by incorporation of a few explicit solvent molecules.

With the above technical considerations in mind, it is possible to attempt predictions for challenging stereochemical problems. Computational modeling of stereoselectivity has evolved dramatically over its 40-year history, which began with STO-3G calculations on simple model "TSs" and has progressed to full QM calculations on multistep catalytic asymmetric transformations. Continued developments will enable attention to be devoted to newly emerging challenges, such as multifunctional catalysis, multicatalyst systems, redox processes, enzyme-mediated reactions, and photochemical reactions.

References

1. Zhao, Y. and Truhlar, D. G. (2008). Computational characterization and modeling of buckyball tweezers: Density functional study of concave–convex $\pi \cdots \pi$ interactions, *Phys. Chem. Chem. Phys.*, 10, pp. 2813–2818.
2. Plata, R. E. and Singleton, D. A. (2015). A case study of the mechanism of alcohol-mediated Morita Baylis–Hillman reactions. The importance of experimental observations, *J. Am. Chem. Soc.*, 137, pp. 3811–3826.
3. Anh, N. T. and Eisenstein, O. (1977). Theoretical interpretation of 1–2 asymmetric induction. The importance of antiperiplanarity, *Nouv. J. Chim.*, 1, pp. 61–70.
4. Houk, K. N., Paddon-Row, M. N., Rondan, N. G., Wu, Y.-D., Brown, F. K., Spellmeyer, D. C., Metz, J. T., Li, Y. and Loncharich, R. J. (1986). Theory and modeling of stereoselective organic reactions, *Science*, 231, pp. 1108–1117.
5. Dolbier, W. R., Jr., Koroniak, H., Houk, K. N. and Sheu, C. (1996). Electronic control of stereoselectivities of electrocyclic reactions of cyclobutenes: A triumph of theory in the prediction of organic reactions, *Acc. Chem. Res.*, 29, pp. 471–477.
6. Rudolf, K., Spellmeyer, D. C. and Houk, K. N. (1987). Prediction and experimental verification of the stereoselective electrocyclization of 3-formylcyclobutene, *J. Org. Chem.*, 52, pp. 3708–3710.
7. Clemente, F. R. and Houk, K. N. (2004). Computational evidence for the enamine mechanism of intramolecular aldol reactions catalyzed by proline, *Angew. Chem. Int. Ed.*, 43, pp. 5766–5768.
8. Bahmanyar, S. and Houk, K. N. (2001). The origin of stereoselectivity in proline-catalyzed intramolecular aldol reactions, *J. Am. Chem. Soc.*, 123, pp. 12911–12912.
9. Bahmanyar, S., Houk, K. N., Martin, H. J. and List, B. (2003). Quantum mechanical predictions of the stereoselectivities of proline-catalyzed asymmetric intermolecular aldol reactions, *J. Am. Chem. Soc.*, 125, pp. 2475–2479.
10. Mitsumori, S., Zhang, H., Cheong, P. H.-Y., Houk, K. N., Tanaka, F. and Barbas, C. F., III (2006). Direct asymmetric *anti*-Mannich-type reactions catalyzed by a designed amino acid, *J. Am. Chem. Soc.*, 128, pp. 1040–1041.
11. Balcells, D. and Maseras, F. (2007). Computational approaches to asymmetric synthesis, *New J. Chem.*, 31, pp. 333–343.
12. Hopmann, K. H. (2015). Quantum chemical studies of asymmetric reactions: historical aspects and recent examples, *Int. J. Quantum Chem.*, 115, pp. 1232–1249.
13. Sperger, T., Sanhueza, I. A., Kalvet, I. and Schoenebeck, F. (2015). Computational studies of synthetically relevant homogeneous organometallic catalysis involving Ni, Pd, Ir, and Rh: An overview of commonly employed DFT methods and mechanistic insights, *Chem. Rev.*, 115, pp. 9532–9586.

14. Landis, C. R. and Feldgus, S. (2000). A simple model for the origin of enantioselection and the anti "lock-and-key" motif in asymmetric hydrogenation of enamides as catalyzed by chiral diphosphine complexes of Rh(I), *Angew. Chem. Int. Ed.*, 39, pp. 2863–2866.
15. Feldgus, S. and Landis, C. R. (2000). Large-scale computational modeling of [Rh(DuPHOS)]+-catalyzed hydrogenation of prochiral enamides: Reaction pathways and the origin of enantioselection, *J. Am. Chem. Soc.*, 122, pp. 12714–12727.
16. Ujaque, G., Maseras, F. and Lledós, A. (1999). Theoretical study on the origin of enantioselectivity in the bis(dihydroquinidine)-3,6-pyridazine· osmium tetroxide-catalyzed dihydroxylation of styrene, *J. Am. Chem. Soc.*, 121, pp. 1317–1323.
17. Liu, P., Xu, X., Dong, X., Keitz, B. K., Herbert, M. B., Grubbs, R. H. and Houk, K. N. (2012). *Z*-Selectivity in olefin metathesis with chelated Ru catalysts: Computational studies of mechanism and selectivity, *J. Am. Chem. Soc.*, 134, pp. 1464–1467.
18. Yamakawa, M., Yamada, I. and Noyori, R. (2001). CH/π attraction: The origin of enantioselectivity in transfer hydrogenation of aromatic carbonyl compounds catalyzed by chiral η^6-arene-ruthenium(II) complexes, *Angew. Chem. Int. Ed.*, 40, pp. 2818–2821.
19. Uyeda, C. and Jacobsen, E. N. (2011). Transition-state charge stabilization through multiple non-covalent interactions in the guanidinium-catalyzed enantioselective Claisen rearrangement, *J. Am. Chem. Soc.*, 133, pp. 5062–5075.
20. Krenske, E. H. and Houk, K. N. (2013). Aromatic interactions as control elements in stereoselective organic reactions, *Acc. Chem. Res.*, 46, 979–989.
21. Doubleday, C., Nendel, M., Houk, K. N., Thweatt, D. and Page, M. (1999). Direct dynamics quasiclassical trajectory study of the stereochemistry of the vinylcyclopropane–cyclopentene rearrangement, *J. Am. Chem. Soc.*, 121, pp. 4720–4721.
22. A recent example: Biswas, B. and Singleton, D. A. (2015). Controlling selectivity by controlling the path of trajectories, *J. Am. Chem. Soc.*, 137, pp. 14244–14247.
23. Bogle, X. S. and Singleton, D. A. (2012). Dynamic origin of the stereoselectivity of a nucleophilic substitution reaction, *Org. Lett.*, 14, pp. 2528–2531.
24. Pemberton, R. P., Ho, K. C. and Tantillo, D. J. (2015). Modulation of inherent dynamical tendencies of the bisabolyl cation *via* preorganization in *epi*-isozizaene synthase, *Chem. Sci.*, 6, pp. 2347–2353.
25. Jover, J. and Fey, N. (2014). The computational road to better catalysts, *Chem. Asian J.*, 9, pp. 1714–1723.
26. Donoghue, P. J., Helquist, P., Norrby, P.-O. and Wiest, O. (2009). Prediction of enantioselectivity in rhodium catalyzed hydrogenations, *J. Am. Chem. Soc.*, 131, pp. 410–411.

27. Corbeil, C. R., Thielges, S., Schwartzentruber, J. A. and Moitessier, N. (2008). Toward a computational tool predicting the stereochemical outcome of asymmetric reactions: Development and application of a rapid and accurate program based on organic principles, *Angew. Chem. Int. Ed.*, 47, pp. 2635–2638.
28. Harper, K. C. and Sigman, M. S. (2011). Three-dimensional correlation of steric and electronic free energy relationships guides asymmetric propargylation, *Science*, 333, pp. 1875–1878.
29. Kozlowski, M. C., Dixon, S. L., Panda, M. and Lauri, G. (2003). Quantum mechanical models correlating structure with selectivity: Predicting the enantioselectivity of β-amino alcohol catalysts in aldehyde alkylation, *J. Am. Chem. Soc.*, 125, pp. 6614–6615.
30. Ianni, J. C., Annamalai, V., Phuan, P.-W., Panda, M. and Kozlowski, M. C. (2006). A priori theoretical prediction of selectivity in asymmetric catalysis: Design of chiral catalysts by using quantum molecular interaction fields, *Angew. Chem. Int. Ed.*, 45, pp. 5502–5505.
31. Grimme, S. (2011). Density functional theory with London dispersion corrections, *WIREs Comput. Mol. Sci.*, 1, pp. 211–228.
32. Zhao, Y. and Truhlar, D. G. (2008). Density functionals with broad applicability in chemistry, *Acc. Chem. Res.*, 41, pp. 157–167.
33. Chai, J.-D. and Head-Gordon, M. (2008). Long-range corrected hybrid density functionals with damped atom–atom dispersion corrections, *Phys. Chem. Chem. Phys.*, 10, pp. 6615–6620.
34. Antony, J. and Grimme, S. (2007). Is spin-component scaled second-order Møller–Plesset perturbation theory an appropriate method for the study of noncovalent interactions in molecules? *J. Phys. Chem. A*, 111, pp. 4862–4868.
35. Kruse, H., Goerigk, L. and Grimme, S. (2012). Why the standard B3LYP/6-31G* model chemistry should not be used in DFT calculations of molecular thermochemistry: understanding and correcting the problem, *J. Org. Chem.*, 77, pp. 10824–10834.
36. Dapprich, S., Komáromi, I., Byun, K. S., Morokuma, K. and Frisch, M. J. (1999). A new ONIOM implementation in Gaussian98. Part I. The calculation of energies, gradients, vibrational frequencies and electric field derivatives, *J. Mol. Struct. (THEOCHEM)*, 461–462, pp. 1–21.
37. Cossi, M., Rega, N., Scalmani, G. and Barone, V. (2003). Energies, structures, and electronic properties of molecules in solution with the C-PCM solvation model, *J. Comput. Chem.*, 24, pp. 669–681.
38. Marenich, A. V., Cramer, C. J. and Truhlar, D. G. (2009). Universal solvation model based on solute electron density and on a continuum model of the solvent defined by the bulk dielectric constant and atomic surface tensions, *J. Phys. Chem. B*, 113, pp. 6378–6396.
39. Klamt, A. and Schüürmann, G. (1993). COSMO: A new approach to dielectric screening in solvents with explicit expressions for the screening energy and its gradient, *J. Chem. Soc., Perkin Trans.*, 2, pp. 799–805.

Index

A

ab initio multiple spawning (AIMS), 562
ab initio, 31
absolutely localized molecular orbitals, 191
activation strain, 213
activation strain model, 374
adaptive natural density partitioning (AdNDP), 467
adiabatic, 463
anharmonic, 10
anion–π, 290
anisotropy of the induced current density (ACID), 284
anomeric, 453
antiperiplanar, 454
aromatic ring current shielding (ARCS), 284
aromatic stabilization energies (ASE), 279
Arrhenius plots, 13
artificial force induced reaction (AFIR), 333
Austin model 1 (AM1), 44
avoided crossing, 543
asynchronous, 20

B

B3LYP, 37
basis set, 2, 50
basis set superposition error (BSSE), 6
Bell, 413
benchmarked, 14
Berny algorithm, 322
bifurcation, 85
Bigeleisen and Mayer, 403
block localized wavefunction, 191
Boltzmann distribution, 10
Born, 109
Born–Oppenheimer approximation, 3, 545

C

caldera, 84
catalysis, 520
cation–π, 53
CH–π, 595
CH–O, 290, 596
charge transfer (CT), 227
ChElPG, 121
chemical shift, 166
cluster-continuum, 506
complete active space self-consistent field (CASSCF), 34
conceptual DFT, 373
concerted, 20
configurations, 9
conformations, 8
conical intersection, 544
conjugation, 201

Connolly surface, 123
conservation of momentum, 360
constrained DFT (CDFT), 231
continuum, 15, 97
correlation energy, 32
correlation-consistent basis set, 31
COSMO, 116
Coulomb's law, 108
coupling constant, 176
crystal packing, 229
Curtin–Hammett principle, 585

D

deformation density, 196
degenerate, 545
degrees of freedom, 10
density functional theory (DFT), 31
density of states, 255
diabatic, 463
dielectric, 104
diffuse functions, 51
dispersion, 16
distortion, 371
distortion–interaction, 213
DP4, 180
dynamic effects, 84
dynamic matching, 84

E

effective fragment potential (EFP), 132
electron affinity (EA), 229
electron localization function (ELF), 469
electron sharing indexes (ESI), 468
electronegativity, 274, 452
electronic energy, 10
electrostatic, 197
electrostatic potential, 19
emission, 228
energy decomposition, 191
ensembles, 12
enthalpy, 10
entropy, 12
error, 19
excited-state, 543
excitons, 228
extended HMO theory (EHT), 43
extended transition-state (ETS), 192
Eyring equation, 11, 527

F

Fermi Golden Rule, 544
force field, 9
free energy, 10
frequency calculation, 10
frontier molecular orbital (FMO), 371
functional, 4

G

gauche effect, 453
gauge-including atomic orbitals (GIAOs), 166
gauge problem, 166
Gauss's law, 106
Gaussian functions, 2
Gaussian-type, 32
geometry optimization, 3
global minimum, 22
global reaction route mapping (GRRM), 336
Green function, 113

H

Hückel molecular orbital method (HMO), 42
half-life, 357
halogen bonding, 596
Hamiltonian operator, 2
Hammett plot, 507
Hartree–Fock (HF), 32
Heisenberg uncertainty principle, 11
Hessian, 322
HOMO–LUMO, 193
homodesmotic, 456
homolytic, 208
hybridization, 455

hydrogen bond, 46
hyperconjugation, 197

I

imaginary frequency, 12
individual gauges for localized orbitals (IGLO), 167
interaction energy, 192
intramolecular vibrational energy redistribution (IVR), 351
ion–dipole complex, 75
ionic character, 197
ionic liquids, 509
ionization potential (IP), 229
IR spectra, 73
IRC, 7
irreducible representation, 197
iso-chemical shielding surfaces (ICSS), 285
isodesmic, 201, 456
isogyric, 456
isolobal, 210
isomerization stabilization energy (ISE), 280
isotope exchange, 403
isotropic, 174

K

Kohn–Sham, 193

L

Lennard–Jones, 76, 134
lifetimes, 367
linear free energy relationships (LFER), 24
linear-scaling, 175
lone-pair/π, 290
low-mode sampling (LMOD), 149

M

Möbius, 70
magnetic, 273
Marcus theory, 570
master equation (ME), 355
Merz–Kollman–Singh, 121
metadynamics, 333, 509
microsolvation, 129
minima, 8
minimum energy pathway (MEP), 321
Minnesota functionals, 37
MM-GBSA, 137
MM-PBSA, 137
model chemistry, 2
molecular dynamics (MD), 8
molecular mechanics (MM), 9
molecular orbitals, 2
Monte Carlo, 9
multi-configurational self-consistent field (MCSCF), 34
multi-standard (MTSD), 174
multipole expansion, 291
multireference, 75
Møller and Plesset (MP), 33

N

nanotube, 253
natural atomic orbitals (NAOs), 465
natural bond orbital (NBO), 300
natural energy decomposition analysis (NEDA), 191
natural orbital for chemical valence (NOCV), 192
natural population analysis (NPA), 466
NCI, 529
neglect of differential overlap, 43
nonadiabatic, 541
noncrossing rule, 543
nonstatistical dynamics, 84, 351
normal mode, 367
nucleus-independent chemical shifts (NICS), 78
nudged elastic band, 323

O

ONIOM, 31
Onsager, 109

optimized effective potential method (OEP), 172
orbital interaction, 196
orbital symmetry, 371
orthogonal, 464

P

π-stacking, 291
$\pi - \pi$, 595
pK_a, 503
parametric method 3 (PM3), 46
Pariser–Parr–Pople method (PPP), 42
partition function, 11
parts, 166–167, 171, 173–174
Pauli repulsion, 193
PBE0, 41
PCM, 16
permittivity, 107
perturbation theory, 232
phase space theory (PST), 355
photoisomerization, 562
Poisson, 107
Poisson–Boltzmann equation, 108
polarizability, 15
polarization functions, 51
Pople basis set, 31
post-transition state bifurcations (PTSBs), 7
potential energy, 5
potential energy surface (PES), 5
prediction, 23

Q

QSAR, 508
quantum harmonic oscillator (QHO), 10
quantum theory of atoms-in-molecules (QTAIM), 300
quasi-equilibrium, 8

R

rate-determining, 525
Recife model 1 (RM1), 47
reorganization energy, 397
repulsion, 282
resonance, 201
resonance assisted hydrogen bonding (RAHB), 213
resonance energy, 277
restricted active space (RAS), 547
Rice–Ramsperger–Kassel–Marcus (RRKM) theory, 7, 351
ring current, 282

S

saddle point, 5, 292
scan, 323
Schrödinger equation, 1
SCS-MP2, 299
seam, 567
self-consistent field (SCF), 3
self-consistent reaction field (SCRF), 17
semi-empirical, 31
semiconductor, 247
shielding, 166
single point, 3
singlet–triplet gap, 72
SMD, 16
SMx, 121
solvation shell, 99
solvent-accessible surface, 123
SOMO, 193
spin, 19
split-valence, 51
stability, 21
stationary point, 5
stepwise, 20
stereoelectronic, 451
steric, 451
strain, 282
string method, 323
super-atomic-molecular-orbitals (SAMOs), 230
symmetry adapted perturbation theory (SAPT), 191
symmetry numbers, 406

synchronous, 20
synchronous transit, 323

T

Taft, 371
thermodynamic cycle, 504
through-bond, 487
through-space, 293
time-dependent DFT (TD-DFT), 38
TIP, 135
torquoselectivity, 587
torsion, 149
trajectories, 8
transition state, 5
transition state theory (TST), 7
tunneling, 85

U

unrestricted, 32

V

valence bond, 13
valley-ridge inflection (VRI) point, 7
van der Waals, 202
vibrational modes, 10

W

ωB97X-D, 39
wave function, 2
Wigner's expression, 412
Woodward–Hoffmann rules, 43

Z

zero-point energy, 10

www.ingramcontent.com/pod-product-compliance
Lightning Source LLC
Chambersburg PA
CBHW070711220326
41598CB00026B/3692
* 9 7 8 1 8 0 0 6 1 0 6 4 4 *